Graduate Texts in Mathematics 258

For other titles published in this series, go to
www.springer.com/series/136

Osman Güler

Foundations of Optimization

Osman Güler
Department of Mathematics & Statistics
University of Maryland, Baltimore County
Hilltop Circle 1000
Baltimore, MD 21250
USA
guler@umbc.edu

ISSN 0072-5285
ISBN 978-1-4614-2647-9 ISBN 978-0-387-68407-9 (eBook)
DOI 10.1007/978-0-387-68407-9
Springer New York Dordrecht Heidelberg London

Mathematics Subject Classification (2010): 90-01, 90C30, 90C46, 90C25, 90C05, 90C20, 90C34, 90C47, 49M37, 49N15, 49J53, 49J50, 49M15, 49K35, 65K05, 65K10, 52A05, 52A07, 52A35, 52A41, 52A40, 52A37

Printed on acid-free paper

Springer is part of Springer Science+Business Media (www.springer.com)

I dedicate this book to my mother, Zülfiye Güler, and to the memory of my father, Eşref Güler.

Preface

Optimization is everywhere. It is human nature to seek the best option among all that are available. Nature, too, seems to be guided by optimization—many laws of nature have a variational character. Among geometric figures in the plane with a fixed perimeter, the circle has the greatest area. Such isoperimetric problems involving geometric figures date back to ancient Greece. Fermat's principle, discovered in 1629, stating that the tangent line is horizontal at a minimum point, seems to have influenced the development of calculus. The proofs of Rolle's theorem and the mean value theorem in calculus use the Weierstrass theorem on the existence of maximizers and minimizers. The introduction of the brachistochrone problem in 1696 by Johann Bernoulli had a tremendous impact on the development of the calculus of variations and influenced the development of functional analysis. The variational character of laws of mechanics and optics were discovered in the seventeenth and eighteenth centuries. Euler and Lagrange forged the foundations of the calculus of variations in the eighteenth century. In the nineteenth century, Riemann used Dirichlet's principle, which has a variational character, in his investigations in complex analysis. The simplex method for linear programming was discovered shortly after the advent of computers in the 1940s, and influenced the subsequent development of mathematical programming. The emergence of the theory of optimal control in the 1950s was in response to the need for controlling space vehicles and various industrial processes. Today, optimization is a vast subject with many subfields, and it is growing at a rapid pace. Research is proceeding in various directions—advancement of theory, development of new applications and computer codes, and establishment or renewal of ties with many fields in science and industry.

The main focus of this book is optimization in finite-dimensional spaces. In broad terms, this is the problem of optimizing a function f in n variables over a subset of $\mathbb{R}^n$. Thus, the decision variable $x = (x_1, \ldots, x_n)$ is finite-dimensional. A typical problem in this area is a mathematical program (P), which concerns the minimization (or maximization) of a function $f(x)$ subject to finitely many functional constraints of the form $g_i(x) \leq 0$, $h_j(x) = 0$ and a set constraint of the form $x \in C$, where f, g_i, h_j are real-valued functions defined on some subsets of $\mathbb{R}^n$ and C is a subset of $\mathbb{R}^n$. Any finite-dimensional vector space may be substituted for $\mathbb{R}^n$ without any loss of generality. If the domain of f is an open set and there are no constraints (or more generally if the domains of g_i, h_j, and C are open sets), then we have an unconstrained optimization problem. If all the functions are affine and C is defined by linear equations and inequalities, then (P) is called a linear program. If f, g_i are convex functions, h_j is an affine function, and C is a convex set, then (P) is a convex program. If the number of functional constraints is infinite, then (P) is called a semi-infinite program. Mathematical programs have many real-life applications. In particular, linear programming, and more recently semidefinite programming, are enormously popular and have many industrial and scientific applications. The latter problem optimizes a linear function subject to linear equality constraints over the cone of symmetric positive semidefinite matrices.

The main goal of the theory of mathematical programming is to obtain optimality conditions (necessary and sufficient) for a local or global minimizer of (P). This is an impossible task unless some kind of regularity is assumed about the data of (P)—the functions f, g_i, h_j and the set C. This can be differentiability (in some form) of the functions, or the convexity of the functions as well as of the set C. In this book, we will assume that the functions f, g_i, h_j are differentiable as many times as needed (except in cases where there is no advantage to do so), and do not develop nonsmooth analysis in any systematic way. Optimization from the viewpoint of nonsmooth analysis is competently covered in several recent books; see for example *Variational Analysis* by Rockafellar and Wets, and *Variational Analysis and Generalized Differentiation* by Mordukhovich. Another goal of the theory, important especially in convex programming, is the duality theory, whereby a second convex program (D) is associated with (P) such that the pair (P)-(D) have remarkable properties which can be exploited in several useful ways. If the problem (P) has a lot of structure, it may be possible to use the optimality conditions to solve analytically for the solutions to (P). This desirable situation is very valuable when it is successful, but it is rare, so it becomes necessary to devise numerical optimization techniques or algorithms to search for the optimal solutions. The process of designing efficient algorithms requires a great deal of ingenuity, and the optimality conditions contribute to the process in several ways, for example by suggesting the algorithm itself, or by verifying the correctness of the numerical solutions returned by the algorithm. The role of the duality theory in designing algorithms is similar, and often more decisive.

All chapters except Chapter 14 are concerned with the theory of optimization. We have tried to present all the major results in the theory of finite-dimensional optimization, and strived to provide the best available proofs whenever possible. Moreover, we include several independent proofs of some of the most important results in order to give the reader and the instructor of a course using this book flexibility in learning or teaching the key subjects.

On several occasions we give proofs that may be new. Not all chapters deal exclusively with finite-dimensional spaces, however. Chapters 3, 5, 6, 14, and Appendices A and C contain, in part or fully, important results in nonlinear analysis and in the theory of convexity in infinite-dimensional settings.

Chapter 14 may be viewed as a short course on three basic optimization algorithms: the steepest descent method, Newton's method, and the conjugate-gradient method. In particular, the conjugate-gradient method is presented in great detail. The three algorithms are chosen to be included because many computational schemes in mathematical programming have their origins in these algorithms.

Audience and background

The book is suitable as a textbook for a first course in the theory of optimization in finite-dimensional spaces at the graduate level. The book is also suitable for self-study or as a reference book for more advanced readers. It evolved out of my experience in teaching a graduate-level course twelve times since 1993, eleven times at the University of Maryland, Baltimore County (UMBC), and once in 2001 at Bilkent University, Ankara, Turkey. An important feature of the book is the inclusion of over two hundred carefully selected exercises as well as a fair number of completely solved examples within the text.

The prerequisites for the course are analysis and linear algebra. The reader is assumed to be familiar with the basic concepts and results of analysis in finite-dimensional vector spaces—limits, continuity, completeness, compactness, connectedness, and so on. In some of the more advanced chapters and sections, it is necessary to be familiar with the same concepts in metric and Banach spaces. The reader is also assumed to be familiar with the fundamental concepts and results of linear algebra—vector space, matrix, linear combination, span, linear independence, linear map (transformation), and so on.

Suggestions for using this book in a course

Ideally, a first course in finite-dimensional optimization should cover the first-order and second-order optimality conditions in unconstrained optimization, the fundamental concepts of convexity, the separation theorems involving convex sets (at least in finite-dimensional spaces), the theory of linear inequalities and convex polyhedra, the optimality conditions in nonlinear programming, and the duality theory of convex programming. These are treated in Chapters 2, 4, 6, 7, 9, and 11, respectively, and can be covered in a one-semester course. Chapter 1 on differential calculus can be covered in such a course, or referred to as needed, depending on the background of the students. In any case, it is important to be familiar with the multivariate Taylor formulas, because they are used in deriving optimality conditions and in differentiating functions.

In my courses, I cover Chapter 1 (Sections 1.1–1.5), Chapter 2 (Sections 2.1–2.5), Chapter 4 (Sections 4.1–4.5), Chapter 6 (Sections 6.1–6.5, and assuming the results from Chapter 5 that are used in some proofs), Chapter 7 (Sections 7.1–7.4), Chapter 9 (Sections 9.1–9.2, 9.4–9.9), and Chapter 11 (Sections 11.1–11.6). This course emphasizes the use of separation theorems for convex sets for deriving the optimality conditions for nonlinear programming. This approach is both natural and widely applicable—it is possible to use the same idea to derive optimality conditions for many types of problems, from nonlinear programming to optimal control problems, as was shown by Dubovitskii and Milyutin.

Several other possibilities exist for covering most of this core material. If the goal is to cover quickly the basics of nonlinear programming but not of convexity, then one can proceed as above but skip Chapter 6 and the first two sections of Chapter 7, substitute Appendix A for Sections 7.3 and 7.4, and skip Section 9.1. In this approach, one needs to accept the truth of Theorem 11.15 without proof.

A third possibility is to follow Chapter 3 to cover the theory of linear inequalities and the basic theorems of nonlinear analysis, and then cover Chapter 9 (Sections 9.3–9.9). Still other possibilities exist for covering the core material.

If more time is available, an instructor may choose to cover Chapter 14 on algorithms, Chapter 8 on linear programming, Chapter 10 on nonlinear programming, or Chapter 12 on semi-infinite programming. In a course oriented more toward convexity, the instructor may cover Chapter 5, 6, or 13 for a more in-depth study of convexity. In particular, Chapters 5 and 6 contain very detailed, advanced results on convexity.

Chapters 4–8, 11, and 13 can be used for a stand-alone one-semester course on the theory of convexity. If desired, one may supplement the course by presenting the theory of Fenchel duality using, for example, Chapters 1–3 and 6 of the book *Convex Analysis and Variational Problems* by Ekeland and Temam. The theory of convexity has an important place in optimization. We already mentioned the role of the separation theorems for convex sets in deriving optimality conditions in mathematical programming. The theory of duality is a powerful tool with many uses, both in theory of optimization and in the design of numerical optimization algorithms. The role of convexity in the complexity theory of optimization is even more central; since the work of Nemirovskii and Yudin in the 1970s on the ellipsoid method, we know that convex programming (and some close relatives) is the only known class of problems that are computationally tractable, that is, for which polynomial-time methods can be developed.

The major pathways through the book are indicated in the following diagram.

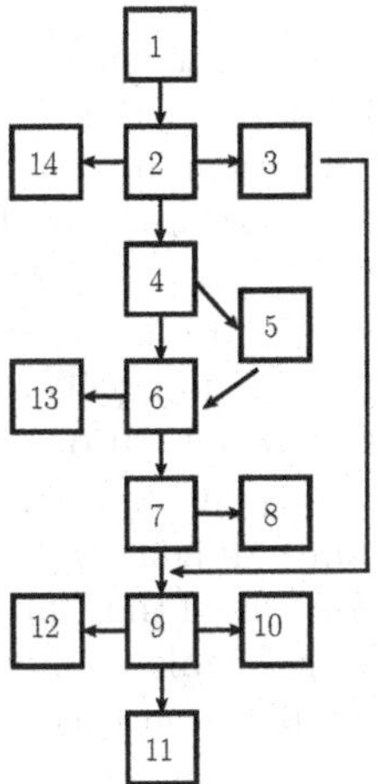

Comments on the contents of individual chapters

Chapter 1 includes background material on differential calculus. Two novel features of the chapter are the converse of Taylor's formula and Danskin's theorem. The first result validates the role of Taylor's formula for computing derivatives, and Danskin's formula is a useful tool in optimization.

Chapter 2 develops the first-order and second-order optimality conditions in unconstrained optimization. Section 2.4 deals with quadratic forms and symmetric matrices. We recall the spectral decomposition of a symmetric matrix, give the eigenvalue characterizations of definite and semidefinite matrices, state Descartes's exact rule of sign (whose proof is given in Appendix B), and use it as tool for recognizing definite and semidefinite matrices. We also include a proof of Sylvester's theorem on the positive definiteness of a symmetric matrix. (An elegant optimization-based proof is given in an exercise at the end of the chapter.) In Section 2.5, we give the proofs of the inverse and implicit function theorems and Lyusternik's theorem using an optimization-based approach going back at least to Carathéodory. A proof of Morse's lemma is given in Section 2.6 because of the light it throws on the second-order optimality conditions.

Chapter 3 is devoted to Ekeland's ϵ-variational principle (and its relatives) and its applications. We use it to prove the central result on linear inequalities (Motzkin's transposition theorem), and the basic theorems of nonlinear analysis in a general setting. Variational principles are fascinating, and their importance in optimization is likely to grow even more in the future.

The next three chapters are devoted to convexity. Chapter 4 treats the fundamentals of convex analysis. We include Section 4.1 on affine geometry because of its intrinsic importance, and because it helps make certain results in convexity more transparent.

Chapter 5 delves into the structure of convex sets. A proper understanding of concepts such as the relative interior, closure, and the faces of convex sets is essential for proving separation theorems involving convex sets and much else. The concept of the relative interior is developed in both algebraic and topological settings.

Chapter 6 is devoted to the separation of convex sets, the essential source of duality, at least in convex programming. The chapter is divided into two parts. Sections 6.1–6.5 deal with the separation theorems in finite dimensions and do not depend heavily on Chapter 5. They are sufficient for somebody who is interested in only the finite-dimensional situation. Section 6.5 is devoted to the finite-dimensional version of the Dubovitskii–Milyutin theorem, a convenient separation theorem, applicable to the separation of several convex sets. Sections 6.6–6.8 treat the separation theorems involving two or several convex sets in a very general setting. Chapter 5 is a prerequisite for these sections, which are intended for more advanced readers.

Chapters 7 and 8 treat the theories of convex polyhedra and linear programming, respectively. Two sections, Section 7.5 on Tucker's complementarity theorem and Section 8.5 on the existence of strictly complementary solutions in linear programming, are important in interior-point methods.

Chapters 9 and 10 treat nonlinear programming. The standard, basic theory consisting of first-order (Fritz John and KKT) and second-order conditions for optimality is given in Chapter 9. A novel feature of the chapter is the inclusion of a first-order sufficient optimality condition that goes back to Fritz John, and several completely solved examples of nonlinear programs. Chapter 10 gives complete solutions for seven structured optimization problems. These problems are chosen for their intrinsic importance and to demonstrate that optimization techniques can resolve important problems.

Chapter 11 deals with duality theory. We have chosen to treat duality using the Lagrangian function. This approach is completely general for convex programming, because it is equivalent to the approach by Fenchel duality in that context, and more general because it is sometimes applicable beyond convex programming. We establish the general correspondence between saddle point and duality in Section 11.2 and apply it to nonlinear programming in Section 11.3. The most important result of the chapter is the strong duality theorem for convex programming given in Section 11.4, under very weak conditions. It is necessary to use sophisticated separation theorems to achieve this result. After treating several examples of duality in Section 11.5, we turn to the duality theory of conic programming in Section 11.6. As a novel application, we give a proof of Hoffman's lemma using duality in Section 11.8.

Semi-infinite programming is the topic of Chapter 12. This subject is not commonly included in most optimization textbooks, but many important problems in finite dimensions require it, such as the problem of finding the extremal-volume ellipsoids associated with a convex body in $\mathbb{R}^n$. We derive the Fritz John optimality conditions for these problems using Danskin's theorem when the set indexing the constraints is compact. In the rest of the chapter we solve several specific, important semi-infinite programming problem rather than giving a systematic theory. Another method to treat convex semi-infinite programs, using Helly's theorem, is given in Section 13.2.

Chapter 13 is devoted to several special topics in convexity that we deem interesting: the combinatorial theory of convex sets, homogeneous convex functions, decomposition of convex cones, and norms of polynomials. The last topic finds an interesting application to self-concordant functions in interior-point methods.

The focus of Chapter 14 is on algorithms. The development of numerical algorithms for optimization problems is a highly intricate art and science, and anything close to a proper treatment would require several volumes. This chapter is included in our book out of the conviction that there should be a place in a book on theory for a chapter such as this, which treats in some depth a few select algorithms. This should help the reader put the theory in perspective, and accomplish at least three goals: the reader should see how theory and algorithms fit together, how they are different, and whether there are differences in the thought processes that go into developing each part. It should also give the reader additional incentive to learn more about algorithms.

We choose to treat three fundamental optimization algorithms: the steepest-descent (and gradient projection), Newton's, and conjugate-gradient methods. We develop each in some depth and provide convergence rate estimates where possible. For example, we provide the convergence rate for the steepest-descent method for the minimization of a convex quadratic function, and for the minimization of a convex function with Lipschitz gradient. The convergence theory of Newton's method is treated, including the convergence theory due to Kantorovich. Finally, we give a very extensive treatment of the conjugate-gradient method. We prove its remarkable convergence properties and show its connection with orthogonal polynomials.

In Appendix A, we give the theory for the consistency of a system of finitely many linear (both strict and weak) inequalities in arbitrary vector spaces. The algebraic proof has considerable merits: it is very general, does not need any prerequisites, and does not use the completeness of the field over which the vector space is defined. Consequently, it is applicable to linear inequalities with rational coefficients.

In Appendix B, we give a short proof of Descartes's exact rule of sign, and in Appendix C, the classical proofs of the open mapping theorem and Graves's theorem.

Acknowledgements

I am indebted to many people for their help at various stages of writing this book. First of all, I would like to thank my graduate students, who during the last twelve years have given me valuable feedback by alerting me to numerous inaccuracies, typos, ambiguities, and difficulties in the exposition in the different versions of the manuscript. I would like to thank Gürkan Üstünkar and Ömer E. Kundakçıoğlu for typing my initial set of lecture notes. My colleague Jinglai Shen taught twice using my manuscript, and gave me a number of excellent suggestions. My brother-in-law Wilson Renne kept me on track with his inquiries about the status of my book. Achi Dosanjh at Springer Verlag has been a constant help ever since our initial correspondence. Most of all, I would like to thank my wife, Colleen, my daughter, Aylin, and my son, Timur, for their encouragement, and for the many sacrifices they have made during the long time it took to write this book. Finally, I am grateful to the National Science Foundation for summer support during the writing of this book.

Ellicott City, Maryland, February 2010 — Osman Güler

Contents

1

Differential Calculus

This chapter is devoted to differential calculus. Tools from differential calculus are widely used in many branches of analysis, including in optimization. In optimization, our main concern in this book, they are used, among other things, to derive optimality conditions in extremal problems which are described by differentiable functions.

We treat the calculus of scalar-valued functions $f : U \to \mathbb{R}$ or vector-valued functions (maps) $f : U \to \mathbb{R}^m$, where U is an open set in $\mathbb{R}^n$. The vector spaces $\mathbb{R}^n$ and $\mathbb{R}^m$ can replaced with any finite-dimensional vector spaces over $\mathbb{R}$ without changing any of our results or methods. In fact, most of our results remain true (with some minor changes) if $\mathbb{R}^n$ and $\mathbb{R}^m$ are replaced by Banach spaces. Although this extension can be done in a straightforward manner, we desire to keep our presentation fairly concrete, and the finite-dimensional vector space setting is sufficient for our needs. We deviate from this rule only in Chapter 3, where we consider differentiable functions in Banach spaces.

The interested reader is referred to the books by Edwards [84] and Spivak [245] for more detailed treatments of calculus in finite-dimensional vector spaces, and Dieudonné [77] and Hörmander [140] in Banach spaces. Surveys of differential calculus in even more general vector spaces may be found in the references [11, 12].

1.1 Taylor's Formula

Taylor's formula in one or several variables is needed to obtain necessary and sufficient conditions for local optimal solutions to unconstrained and constrained optimization problems. In this section, we treat Taylor's formula for functions of a single variable. The several-variable version of the formula is treated in later sections of this chapter.

We start with Taylor's formula in *Lagrange's form*.

O. Güler, *Foundations of Optimization*, Graduate Texts in Mathematics 258,
DOI 10.1007/978-0-387-68407-9_1,

Theorem 1.1. *Let $f : I = (c, d) \to \mathbb{R}$ be a n-times differentiable function. If a, b are distinct points in I, then there exists a point $\overline{x}$ strictly between a and b such that*

$$\begin{aligned} f(b) &= f(a) + f'(a)(b-a) + \frac{f''(a)}{2}(b-a)^2 + \cdots + \frac{f^{(n-1)}(a)}{(n-1)!}(b-a)^{n-1} \\ &\quad + \frac{f^{(n)}(\overline{x})}{n!}(b-a)^n \\ &= \sum_{i=0}^{n-1} \frac{f^{(i)}(a)}{i!}(b-a)^i + \frac{f^{(n)}(\overline{x})}{n!}(b-a)^n. \end{aligned}$$

Note that the case $n = 1$ is precisely the mean value theorem.

Proof. The idea of the proof is similar to that in the case $n = 1$: create a function $g(t)$ such that $g^{(k)}(a) = 0$, $k = 0, \ldots, n-1$, $g(b) = 0$, and apply Rolle's theorem repeatedly.

The $(n-1)$th-degree Taylor approximation (polynomial) at a,

$$P_{n-1}(t) = f(a) + f'(a)(t-a) + \frac{f''(a)}{2}(t-a)^2 + \cdots + \frac{f^{(n-1)}(a)}{(n-1)!}(t-a)^{n-1},$$

satisfies the conditions $P_{n-1}^{(k)}(a) = f^{(k)}(a)$, $k = 0, \ldots, n-1$. Thus, the function $h(t) := f(t) - P_{n-1}(t)$ satisfies the condition $h^{(k)}(a) = 0$, $k = 0, \ldots, n-1$. However, $h(b)$ may not vanish. We rectify the situation by defining

$$g(t) = f(t) - P_{n-1}(t) - \frac{K}{n!}(t-a)^n,$$

with the constant K chosen such that $g(b) = 0$, that is,

$$f(b) = P_{n-1}(b) + \frac{K}{n!}(b-a)^n. \tag{1.1}$$

Then,

$$g^{(k)}(a) = f^{(k)}(a) - P_{n-1}^{(k)}(a) = 0, \quad k = 0, \ldots, n-1, \qquad g(b) = 0.$$

Rolle's theorem implies that there exists x_1 strictly between a and b such that $g'(x_1) = 0$. Since $g'(a) = g'(x_1) = 0$, Rolle's theorem applied to g' implies that there exists x_2 strictly between a and x_1 such that $g''(x_2) = 0$. We continue in this fashion and obtain $\{x_i\}_1^{n-1}$ such that $g^{(i-1)}(x_{i-1}) = 0$. Finally, $g^{(n-1)}(a) = g^{(n-1)}(x_{n-1}) = 0$, and applying Rolle's theorem once more, we obtain a point x_n strictly between a and x_{n-1} such that $g^{(n)}(x_n) = 0$. Since $g^{(n)}(t) = f^{(n)}(t) - K$, we have $g^{(n)}(x_n) = K$. Equation (1.1) implies the theorem. □

This proof is adapted from [268].

In practice, the most useful cases of Taylor's theorem correspond to $n = 1$ and $n = 2$.

Corollary 1.2. *Let $f : I = (c, d) \to \mathbb{R}$ be a function and a, b be distinct points in I.*

(i) **(*Mean value theorem*)** *If f is differentiable on I, then there exists a point ξ strictly between a and b such that*

$$f(b) = f(a) + f'(\xi)(b - a).$$

(ii) If f is twice differentiable on I, then there exists a point ζ strictly between a and b such that

$$f(b) = f(a) + f'(a)(b - a) + \frac{f''(\zeta)}{2}(b - a)^2.$$

Another form of Taylor's theorem expresses the remainder term as an iterated integral. Its proof is perhaps conceptually the simplest, and it can be used to easily deduce other forms of Taylor's theorem from it.

Theorem 1.3. *Let f satisfy the conditions of Theorem 1.1. We have*

$$\begin{aligned} f(b) = f(a) + f'(a)(b - a) + \frac{f''(a)}{2!}(b - a)^2 + \cdots \\ + \frac{f^{(n-1)}(a)}{(n-1)!}(b - a)^{n-1} + \int_a^b \int_a^{s_1} \cdots \int_a^{s_{n-1}} f^{(n)}(s_n) ds_n \cdots ds_1. \end{aligned} \tag{1.2}$$

Proof. The idea of the proof is to use the fundamental theorem of calculus repeatedly. First, we have $f(b) - f(a) = \int_a^b f'(x)dx$, or

$$f(b) = f(a) + \int_a^b f'(s_1)ds_1.$$

Similarly, $f'(s_1) = f'(a) + \int_a^{s_1} f''(s_2)ds_2$, and this gives

$$\begin{aligned} f(b) &= f(a) + \int_a^b f'(s_1)ds_1 = f(a) + \int_a^b \left(f'(a) + \int_a^{s_1} f''(s_2)ds_2 \right) ds_1 \\ &= f(a) + f'(a)(b - a) + \int_a^b \int_a^{s_1} f''(s_2)ds_2ds_1. \end{aligned}$$

Continuing in this fashion, we obtain

$$\begin{aligned} f(b) = f(a) + f'(a)(b - a) + f''(a) \int_a^b \int_a^{s_1} \int_a^{s_2} ds_3ds_2ds_1 + \cdots \\ + f^{(n-1)}(a) \int_a^b \int_a^{s_1} \cdots \int_a^{s_{n-1}} ds_n \cdots ds_1 \\ + \int_a^b \int_a^{s_1} \cdots \int_a^{s_{n-1}} f^{(n)}(s_n)ds_n \cdots ds_1. \end{aligned}$$

Now it is easy to prove by induction that

$$\int_a^b \int_a^{s_1} \cdots \int_a^{s_{k-1}} ds_k \cdots ds_1 = \frac{(b-a)^k}{k!}. \tag{1.3}$$

Indeed, this is trivially true for $k = 1$, and assuming that it holds for k, we have

$$\int_a^b \int_a^{s_1} \cdots \int_a^{s_k} ds_{k+1} ds_k \cdots ds_1 = \int_a^b \frac{(s_1-a)^k}{k!} ds_1 = \frac{(b-a)^{k+1}}{(k+1)!},$$

where the first equation follows from the induction hypothesis. □

As an easy corollary, we obtain Taylor's formula in Lagrange's form.

Corollary 1.4. *Theorem 1.1 follows from Theorem 1.3.*

Proof. By the mean value theorem there exists $\overline{x} \in (a, s_{n-1})$ such that

$$\int_a^{s_{n-1}} f^{(n)}(s_n) ds_n = f^{(n)}(\overline{x})(s_{n-1} - a) = \int_a^{s_{n-1}} f^{(n)}(\overline{x}) ds_n.$$

The proof is completed by substituting this in the iterated integral in the statement of Theorem 1.3 and using (1.3). □

Next, we give Taylor's formula in *Cauchy's form.*

Theorem 1.5. *Let f satisfy the conditions of Theorem 1.1. We have*

$$f(b) = f(a) + f'(a)(b-a) + \frac{f''(a)}{2}(b-a)^2 + \cdots + \frac{f^{(n-1)}(a)}{(n-1)!}(b-a)^{n-1}$$
$$+ \frac{1}{(n-1)!} \int_a^b f^{(n)}(x)(b-x)^{n-1}\, dx. \tag{1.4}$$

Proof. The domain of the iterated integral in the statement of Theorem 1.3 is $\{(s_1, \ldots, s_n) : a \le s_n \le s_{n-1} \le \cdots \le s_1 \le b\}$. By Fubini's theorem, this integral can be written as

$$\int_a^b f^{(n)}(s_n) \int_{s_n}^b \cdots \int_{s_2}^b ds_1 \cdots ds_{n-1} ds_n. \tag{1.5}$$

We claim that

$$\int_{s_k}^b \cdots \int_{s_2}^b ds_1 \cdots ds_{k-1} = \frac{(b-s_k)^{k-1}}{(k-1)!}.$$

Indeed, this is trivial to check for $k = 1$. If it holds for k, then

$$\int_{s_{k+1}}^b \cdots \int_{s_2}^b ds_1 \cdots ds_k ds_{k+1} = \int_{s_{k+1}}^b \frac{(b-s_k)^{k-1}}{(k-1)!} ds_k = \frac{(b-s_{k+1})^k}{k!},$$

where the first equality follows from the induction hypothesis. This proves the claim for $k+1$. Then the integral (1.5) becomes $\int_a^b f^{(n)}(s_n)(b-s_n)^{n-1}/(n-1)! ds_n$, and coincides with the one in (1.4). The theorem is proved. □

We now give a simple demonstration of the use of Taylor's formula in optimization. Suppose that x^* is a *critical point*, that is, $f'(x^*) = 0$. The quadratic Taylor's formula gives

$$f(x) = f(x^*) + f'(x^*)(x - x^*) + \frac{f''(\overline{x})}{2}(x - x^*)^2,$$

for some $\overline{x}$ strictly between x and x^*. Now, if $f''(\overline{x}) \geq 0$ for all $\overline{x} \in I$, then

$$f(x) \geq f(x^*) \quad \text{for all} \quad x \in I.$$

This shows that x^* is a *global minimizer* of f. A function f with $f''(x)$ nonnegative at all points is a convex function. If $f''(\overline{x}) \geq 0$ only in a neighborhood of x^*, then x^* is a *local minimizer* of f. If $f'(x^*) = 0$ and $f''(x) \leq 0$ for all x, then $f(x) \leq f(x^*)$, for all x, that is, x^* is a global maximizer of f. Such a function f is a concave function. Chapter 4 treats convex and concave (not necessarily differentiable) functions in detail.

1.2 Differentiation of Functions of Several Variables

Definition 1.6. *Let $f : U \to \mathbb{R}$ be a function on an open set $U \subseteq \mathbb{R}^n$. If $x \in U$, the limit*

$$\frac{\partial f}{\partial x_i}(x) := \lim_{t \to 0} \frac{f(x_1, \ldots, x_{i-1}, x_i + t, x_{i+1}, \ldots, x_n) - f(x)}{t},$$

if it exists, is called the partial derivative *of f at x with respect to x_i. If all the partial derivatives exist, then the vector*

$$\nabla f(x) := (\partial f/\partial x_1, \ldots, \partial f/\partial x_n)^T$$

is called the gradient *of f.*

Let $d \in \mathbb{R}^n$ be a vector $d = (d_1, \ldots, d_n)^T$. Denoting by e_i the ith coordinate vector

$$e_i := (0, \ldots, 1, 0, \ldots, 0)^T,$$

where the only nonzero entry 1 is in the ith position, we have

$$d = d_1 e_1 + \cdots + d_n e_n.$$

Definition 1.7. *The* directional derivative *of f at $x \in U$ along the direction $d \in \mathbb{R}^n$ is*

$$f'(x; d) := \lim_{t \searrow 0} \frac{f(x + td) - f(x)}{t},$$

provided the limit on the right-hand side exists as $t \geq 0$ approaches zero.

Clearly, $f'(x;\alpha d) = \alpha f'(x;d)$ for $\alpha \geq 0$, and we note that if $f'(x;-d) = -f'(x;d)$, then we have

$$f'(x;d) = \lim_{t\to 0} \frac{f(x+td)-f(x)}{t},$$

because

$$f'(x;d) = -f'(x,-d) = -\lim_{t\searrow 0} \frac{f(x-td)-f(x)}{t} = \lim_{s\nearrow 0} \frac{f(x+sd)-f(x)}{s}.$$

Definition 1.8. *A function $f : U \to \mathbb{R}$ is* Gâteaux differentiable *at $x \in U$ if the directional derivative $f'(x;d)$ exists for all directions $d \in \mathbb{R}^n$ and is a* linear *function of d.*

Let f be Gâteaux differentiable at x. Since $d = d_1e_1 + \cdots + d_ne_n$, and $f'(x;d)$ is linear in d, we have

$$\begin{aligned} f'(x;d) &= f'(x;d_1e_1 + \cdots + d_ne_n) = d_1f'(x;e_1) + \cdots + d_nf'(x;e_n) \\ &= \langle d, \nabla f(x)\rangle = d^T\nabla f(x). \end{aligned}$$

Definition 1.9. *The function $f : U \to \mathbb{R}$ is* Fréchet differentiable *at the point $x \in U$ if there exists a linear function $\ell : \mathbb{R}^n \to \mathbb{R}$, $\ell(x) = \langle l, x\rangle$, such that*

$$\lim_{\|h\|\to 0} \frac{f(x+h)-f(x)-\langle l,h\rangle}{\|h\|} = 0. \tag{1.6}$$

Intuitively speaking, this means that the function f can be "well approximated" around x by an affine function $h \to f(x) + \langle l,h\rangle$, that is,

$$f(x+h) \approx f(x) + \langle l,h\rangle.$$

This approximate equation can be made precise using Landau's little "oh" notation, where we call a vector $o(h) \in \mathbb{R}^n$ if

$$\lim_{h\to 0} \frac{\|o(h)\|}{\|h\|} = 0.$$

With this notation, the Fréchet differentiability of f at x is equivalent to stating $f(x+h) - f(x) - \langle l,h\rangle = o(h)$, or

$$f(x+h) = f(x) + \langle l,h\rangle + o(h). \tag{1.7}$$

The $o(h)$ notation is very intuitive and convenient to use in proofs involving limits.

Clearly, if f is Fréchet differentiable at x, then it is continuous at x, because (1.7) implies that $\lim_{h\to 0} f(x+h) = f(x)$.

The vector l in the definition of Fréchet differentiability can be calculated explicitly. Choosing $h = te_i$ $(i = 1,\ldots,n)$ in (1.6) gives

$$\lim_{t\to 0} \frac{f(x+te_i) - f(x) - tl_i}{t} = 0.$$

We have $l_i = \partial f(x)/\partial x_i$, and thus

$$l = \nabla f(x). \tag{1.8}$$

Then (1.7) becomes

$$f(x+h) = f(x) + \langle \nabla f(x), h \rangle + o(h).$$

This also gives us the following Theorem.

Theorem 1.10. *If $U \subseteq \mathbb{R}^n$ is open and $f : U \to \mathbb{R}$ is Fréchet differentiable at x, then f is Gâteaux differentiable at x.*

Thus, Fréchet differentiability implies Gâteaux differentiability, but the converse is not true; see the exercises at the end of the chapter. Consequently, Fréchet differentiability is a *stronger* concept than Gâteaux differentiability. In fact, the former concept is a *uniform* version of the latter: it is not hard to see that f is Fréchet differentiable at x if and only if f is Gâteaux differentiable and the limit

$$\lim_{t\to 0} \frac{f(x+td) - f(x) - \langle \nabla f(x), d \rangle}{t}$$

converges *uniformly* to zero for all $\|d\| \le 1$, that is, given $\varepsilon > 0$, there exists $\delta > 0$ such that

$$\left\| \frac{f(x+td) - f(x) - \langle \nabla f(x), d \rangle}{t} \right\| < \varepsilon,$$

for all $|t| < \delta$ *and* for all $\|d\| \le 1$.

Definition 1.11. *If x, y are two points in $\mathbb{R}^n$, we denote by $[x, y]$ the closed* line segment *between x and y, that is,*

$$[x, y] = \{x + t(y - x) : 0 \le t \le 1\}.$$

Similarly, we use (x, y) to denote the "open" line segment $\{x + t(y - x) : 0 < t < 1\}$. Note that if $x = y$, we have $[x, y] = (x, y) = \{x\}$.

Lemma 1.12. (*Mean value theorem*) *Let $f : U \to \mathbb{R}$ be a Gâteaux differentiable function on an open set U in $\mathbb{R}^n$. If x, y are distinct points in U such that the line segment $[x, y]$ lies in U, then there exists a point z on (x, y) strictly between x and y, such that*

$$f(y) = f(x) + \langle \nabla f(z), y - x \rangle.$$

Proof. Define the function $h(t) = f(x + t(y - x))$. Since f is Gâteaux differentiable, $h(t)$ is differentiable and

$$h'(t) = \langle \nabla f(x + t(y - x)), y - x \rangle.$$

It follows from the mean value theorem in one variable (Corollary 1.2) that there exists $0 < \bar{t} < 1$ such that $h(1) - h(0) = h'(\bar{t})$. Define $z = x + \bar{t}(y - x)$. Since $h(0) = f(x)$ and $h(1) = f(y)$, we see that

$$f(y) - f(x) = \langle \nabla f(z), y - x \rangle.$$

□

Theorem 1.13. *Let $f : U \to \mathbb{R}$ be a function on an open set $U \subseteq \mathbb{R}^n$. If $f(x)$ is Gâteaux differentiable at $x_0 \in U$ and the partial derivatives $\partial f / \partial x_j$ $(j = 1, \ldots, n)$ are continuous at x_0, then f is Fréchet differentiable at x_0.*

Proof. The mean value theorem (Lemma 1.12) implies that there exists a point $\overline{x}$ strictly between x_0 and $x_0 + h$ such that

$$f(x_0 + h) - f(x_0) - \langle \nabla f(x_0), h \rangle = \langle \nabla f(\overline{x}) - \nabla f(x_0), h \rangle,$$

and the continuity of ∇f at x_0 implies that $\nabla f(\overline{x}) - \nabla f(x_0) \to 0$ as $h \to 0$. Therefore,

$$f(x_0 + h) - f(x_0) - \langle \nabla f(x_0), h \rangle = o(h),$$

proving that f is Fréchet differentiable at x_0. □

1.3 Differentiability of Vector-Valued Functions

Let $F : U \to \mathbb{R}^m$ be a function, where U is an open subset of $\mathbb{R}^n$. We write

$$F(x) = F(x_1, \ldots, x_n) = \begin{pmatrix} f_1(x) \\ f_2(x) \\ \vdots \\ f_m(x) \end{pmatrix},$$

where f_i is called the ith *coordinate function* of F.

Definition 1.14. *The function F is called* Gâteaux differentiable *at $x \in U$ if*

$$\lim_{t \to 0} \frac{F(x + td) - F(x)}{t}$$

is linear in d, that is, there exists a linear map $T : \mathbb{R}^n \to \mathbb{R}^m$, say $T(x) = Ax$, where A is an $m \times n$ matrix, such that

$$\lim_{t \to 0} \frac{F(x + td) - F(x)}{t} = Ad.$$

This equation means that

$$\lim_{t \to 0} \frac{f_i(x + td) - f_i(x)}{t} = \langle a^i, d \rangle,$$

where a^i is the ith row of A. Thus, F is Gâteaux differentiable if and only if each coordinate function f_i is. It follows from (1.8) that

$$a^i = \nabla f_i(x)^T = (\partial f_i/\partial x_1, \ldots, \partial f_i/\partial x_n).$$

Therefore,

$$A = \begin{bmatrix} \partial f_1/\partial x_1 & \ldots & \partial f_1/\partial x_j & \ldots & \partial f_1/\partial x_n \\ \vdots & & \vdots & & \vdots \\ \partial f_i/\partial x_1 & \ldots & \partial f_i/\partial x_j & \ldots & \partial f_i/\partial x_n \\ \vdots & & \vdots & & \vdots \\ \partial f_m/\partial x_1 & \ldots & \partial f_m/\partial x_j & \ldots & \partial f_m/\partial x_n \end{bmatrix} = \begin{bmatrix} \nabla f_1(x)^T \\ \vdots \\ \nabla f_m(x)^T \end{bmatrix} = [\partial f_i/\partial x_j].$$

We denote A by $DF(x)$ and call it the *Jacobian* of F at x. Note that if $m = 1$, that is, F is a scalar-valued function, then

$$DF(x) = \nabla F(x)^T.$$

Definition 1.15. *The function F is called* Fréchet differentiable *at $x \in U$ if there exists a linear map $B : \mathbb{R}^n \to \mathbb{R}^m$ satisfying*

$$\lim_{\|h\| \to 0} \frac{\|F(x+h) - F(x) - Bh\|}{\|h\|} = 0. \tag{1.9}$$

The map B, denoted by $DF(x)$, is called the Fréchet derivative of F.

This equation means, as above,

$$\lim_{\|h\| \to 0} \frac{|f_i(x+h) - f_i(x) - \langle b^i, h \rangle|}{\|h\|} = 0,$$

for each $i = 1, \ldots, m$. In other words, F is Fréchet differentiable at x if and only if each coordinate function f_i is. We have $b^i = \nabla f_i(x) = (\partial f_i/\partial x_1, \ldots, \partial f_i/\partial x_n)^T$, and

$$B = DF(x) = [\partial f_i(x)/\partial x_j].$$

If F is Fréchet differentiable at x, then F is Gâteaux differentiable at x. Since F is Fréchet (Gâteaux) differentiable at x if and only if each coordinate function f_i is Fréchet (Gâteaux) differentiable at x, this follows from previously proved results about scalar functions.

1.4 The Chain Rule

Theorem 1.16. *Let $F : U \to V$, $G : V \to \mathbb{R}^k$, where $U \subseteq \mathbb{R}^n$ and $V \subseteq \mathbb{R}^m$ are open sets, and let $H = G \circ F : U \to \mathbb{R}^k$, $H(x) = G(F(x)) = (G \circ F)(x)$ be their composition. If F is Gâteaux differentiable at x and G is Fréchet differentiable at $y = F(x)$, then H is Gâteaux differentiable at x and*

$$DH(x) = DG(y) \circ DF(x).$$

Moreover, if F and G are Fréchet differentiable at x and $y = F(x)$, respectively, then H is Fréchet differentiable at x.

Proof. Set $A = DF(x)$ and $B = DG(y)$. We have

$$F(x + td) = F(x) + tDF(x)d + o(t) = F(x) + tAd + o(t)$$

and

$$\begin{aligned} H(x + td) &= G(F(x + td)) = G\big(F(x) + (tAd + o(t))\big) \\ &= G(F(x)) + DG(F(x))\big(tAd + o(t)\big) + o\big(tAd + o(t)\big) \\ &= H(x) + tBAd + Bo(t) + o\big(tAd + o(t)\big) \\ &= H(x) + tBAd + o(t). \end{aligned}$$

Comparing the above equation with

$$H(x + td) = H(x) + tDH(x)d + o(t),$$

we conclude that

$$DH(x) = BA = DG(y) \circ DF(x).$$

This proves the first part of the theorem.

If F is Fréchet differentiable at x, similar calculations show that

$$F(x + h) = F(x) + DF(x)h + o(h)$$

and

$$\begin{aligned} &H(x + h) = G(F(x + h)) = G\left(F(x) + DF(x)h + o(h)\right) \\ &= G(F(x)) + DG(F(x))\left[DF(x)h + o(h)\right] + o\left(DF(x)h + o(h)\right) \\ &= H(x) + DG(F(x)) \circ DF(x)h + DG(F(x))o(h) \\ &\quad + o(DF(x)h + o(h)) \\ &= H(x) + DG(y) \circ DF(x)h + o(h). \end{aligned}$$

This shows that H is Fréchet differentiable. □

Exercise 16 on page 26 shows that the chain rule may fail under Gâteaux differentiability.

The mean value theorem fails for vector-valued functions (see Exercise 13), but we have the following two very important substitutes.

Lemma 1.17. *Let $f : I \to \mathbb{R}^m$ be a map on an interval $I = (a, b)$. If f is differentiable at every point in I, then*

$$\|f(y) - f(x)\| \leq |y - x| \cdot \sup_{0\leq t\leq 1} \|Df(x + t(y - x))\|, \quad x, y \in I.$$

Proof. Let $M > \sup_{0\leq t\leq 1} \|Df(x + t(y - x))\|$ and set

$$K := \{t \,:\, 0 \leq t \leq 1,\ \|f(x + t(y - x)) - f(x)\| \leq Mt|y - x|\}.$$

The set K is closed, since f is continuous, and $0 \in K$. Let s be the largest element of K. We claim that $s = 1$. Otherwise, $s < 1$, and choosing $t \in (s, 1)$ such that $t - s$ is small enough, we have

$$\begin{aligned}
&\|f(x + t(y - x)) - f(x)\| \\
&\leq \|f(x + t(y - x)) - f(x + s(y - x))\| + \|f(x + s(y - x)) - f(x)\| \\
&\leq \Big\|Df(x + s(y - x))(t - s)(y - x) + o((t - s)|y - x|)\Big\| + Ms|y - x| \\
&\leq M(t - s)|y - x| + Ms|y - x| \\
&= Mt|y - x|,
\end{aligned}$$

that is, $t \in K$, a contradiction. This proves the claim and the lemma. □

Theorem 1.18. *Let $f : U \to \mathbb{R}^m$ be Gâteaux differentiable on an open set U in $\mathbb{R}^n$. If x, y are points in U such that the line segment $[x, y]$ lies in U, and $T : \mathbb{R}^n \to \mathbb{R}^m$ is a linear map, then*

$$\|f(y) - f(x) - T(y - x)\| \leq \|y - x\| \cdot \sup_{0\leq t\leq 1} \|Df(x + t(y - x)) - T\|.$$

Proof. The map $g(t) = f(x + t(y - x)) - tT(y - x)$ is differentiable with the derivative

$$Dg(t) = [Df(x + t(y - x)) - T](y - x).$$

Lemma 1.17 gives $\|g(1) - g(0)\| \leq \sup_{0\leq t\leq 1} \|Dg(t)\|$, so that

$$\begin{aligned}
\|f(y) - f(x) - T(y - x)\| &\leq \sup_{0\leq t\leq 1} \|Df(x + t(y - x))(y - x) - T(y - x)\| \\
&\leq \|y - x\| \cdot \sup_{0\leq t\leq 1} \|Df(x + t(y - x)) - T\|.
\end{aligned}$$

□

Lemma 1.17 and Theorem 1.18 hold in a Banach space setting, with the same proof.

Theorem 1.19. *Let $F : U \to \mathbb{R}^m$ be a map on an open set $U \subseteq \mathbb{R}^n$. If $F(x) = (f_1(x), \ldots, f_m(x))^T$ is Gâteaux differentiable at $x_0 \in U$ and the partial derivatives $\partial f_i/\partial x_j$ ($i = 1, \ldots, m$, $j = 1, \ldots, n$) are continuous at x_0, then F is Fréchet differentiable at x_0.*

Proof. By Theorem 1.18, we have

$$\|f(x+h) - f(x) - Df(x)h\| \leq \sup_{0 \leq t \leq 1} \|Df(x+th) - Df(x)\| \cdot \|h\| = o(h)$$

as $h \to 0$, because Df is continuous. □

Definition 1.20. *Let $f : U \to \mathbb{R}^m$ be a map on an open set $U \subseteq \mathbb{R}^n$. We call f* twice *Fréchet differentiable on U if both f and Df are Fréchet differentiable on U, and denote by $D^2 f := D(Df)$ the second derivative of f. By induction, we say that f is k-times Fréchet differentiable on U if f is $(k-1)$-times Fréchet differentiable and $D^{k-1}f$ is Fréchet differentiable. We denote by $D^k f = D(D^{k-1}f)$ the kth derivative of f.*

If $a \in U$ and $u, v \in \mathbb{R}^n$, then we denote by $D^2 f(a)[u,v]$ the directional derivative of the function $h(x) := f'(x;u)$ along direction v, that is,

$$D^2 f(a)[u,v] := h'(x;v).$$

The operation of taking successive directional derivatives is commutative under suitable differentiability assumptions.

Theorem 1.21. *If $f : U \to \mathbb{R}$ is twice Fréchet differentiable on an open set U in $\mathbb{R}^n$, then $D^2 f(a)$ is a symmetric bilinear form for all $a \in U$, that is,*

$$D^2 f(a)[u,v] = D^2 f(a)[v,u] \quad \text{for all} \quad u, v \in \mathbb{R}^n.$$

Proof. Define $g(t) := f(a+u+tv) - f(a+tv)$. We have

$$g'(t) = Df(a+u+tv)(v) - Df(a+tv)(v),$$

and Lemma 1.17 applied to $g(t) - tg'(0)$ gives

$$\|g(1) - g(0) - g'(0)\| \leq \sup_{0 \leq t \leq 1} \|g'(t) - g'(0)\|. \tag{1.10}$$

Since Df is Fréchet differentiable, we have

$$\begin{aligned} Df(a+u+tv)(v) - Df(a)(v) - D^2 f(a)[v, u+tv] &= o(\|v\| \cdot \|u+tv\|) \\ &\leq o((\|u\| + \|v\|)^2), \\ Df(a+tv)(v) - Df(a)(v) - D^2 f(a)[v, tv] &= o(\|v\| \cdot \|tv\|) \\ &\leq o((\|u\| + \|v\|)^2). \end{aligned}$$

Subtracting the second equation from the first one gives

$$Df(a+u+tv)(v) - Df(a+tv)(v) - D^2 f(a)[v,u] = o((\|u\| + \|v\|)^2),$$

that is,

$$g'(t) - D^2 f(a)[v,u] = o\big((\|u\| + \|v\|)^2\big).$$

Using this in (1.10) gives the equations $g'(t) - g'(0) = o\big((\|u\| + \|v\|)^2\big)$ and

$$g(1) - g(0) - D^2 f(a)[v,u] = o\big((\|u\| + \|v\|)^2\big).$$

Since $g(1) - g(0) = f(a+u+v) - f(a+v) - f(a+u) + f(a)$ is symmetric in u and v, we similarly have

$$g(1) - g(0) - D^2 f(a)[u,v] = o\big((\|u\| + \|v\|)^2\big).$$

Consequently,

$$B[u,v] := D^2 f(a)[u,v] - D^2 f(a)[v,u] = o\big((\|u\| + \|v\|)^2\big).$$

Let $\|u\| = \|v\| = 1$, and let $t \to 0$. We have $B[tu,tv] = o(t^2)$. Thus, $B[u,v] = o(t^2)/t^2 \to 0$, that is, $B[u,v] = 0$. This proves the symmetry of $D^2 f(a)$. $\square$

Exercise 12 shows that $D^2 f(x)[u,v] = D^2 f(x)[v,u]$ may fail in the absence of sufficient differentiability assumptions.

Corollary 1.22. *If $f : U \to \mathbb{R}$ is k-times Fréchet differentiable on an open set U in $\mathbb{R}^n$ and $a \in U$, then $D^k f(a)$ is a symmetric k-linear form, that is,*

$$D^k f(a)[u_{\sigma(1)}, \ldots, u_{\sigma(k)}] = D^k f(a)[u_1, \ldots, u_k] \quad \textit{for all} \quad u_1, \ldots, u_k \in E,$$

where σ is a permutation of the set $\{1, 2, \ldots, k\}$.

The proof is obtained from Theorem 1.21 by induction.

1.5 Taylor's Formula for Functions of Several Variables

Taylor's formula for a function of a single variable extends to a function $f : U \to \mathbb{R}$ defined on an open set $U \subseteq \mathbb{R}^n$. For this purpose, we restrict f to line segments in U. Let $\{x + td : t \in \mathbb{R}\}$ be a line passing through x and having the direction $d \neq 0 \in \mathbb{R}^n$. Define the function $h(t) = f(g(t)) = f(x+td) = (f \circ g)(t)$, where $g(t) = x + td$. It follows from the chain rule that

$$h'(t) = \langle \nabla f(x+td), d \rangle = \frac{\partial f(x+td)}{\partial x_1} d_1 + \cdots + \frac{\partial f(x+td)}{\partial x_n} d_n.$$

Differentiating h' using the chain rule again, we obtain

$$\begin{aligned}
h''(t) &= \left\langle \nabla(\frac{\partial f(x+td)}{\partial x_1}), d \right\rangle d_1 + \cdots + \left\langle \nabla(\frac{\partial f(x+td)}{\partial x_n}), d \right\rangle d_n \\
&= \left(\frac{\partial}{\partial x_1}(\frac{\partial f(x+td)}{\partial x_1})d_1 + \cdots + \frac{\partial}{\partial x_n}(\frac{\partial f(x+td)}{\partial x_1})d_n \right) d_1 + \cdots \\
&\quad + \left(\frac{\partial}{\partial x_1}(\frac{\partial f(x+td)}{\partial x_n})d_1 + \cdots + \frac{\partial}{\partial x_n}(\frac{\partial f(x+td)}{\partial x_n})d_n \right) d_n \\
&= \left(\frac{\partial^2 f(x+td)}{\partial x_1^2} d_1^2 + \frac{\partial^2 f(x+td)}{\partial x_2 \partial x_1} d_2 d_1 + \cdots + \frac{\partial^2 f(x+td)}{\partial x_n \partial x_1} d_n d_1 \right) \\
&\quad + \cdots + \left(\frac{\partial^2 f(x+td)}{\partial x_1 \partial x_n} d_1 d_n + \cdots + \frac{\partial^2 f(x+td)}{\partial x_n^2} d_n^2 \right) \\
&= \sum_{j=1}^{n} \sum_{i=1}^{n} \frac{\partial^2 f(x+td)}{\partial x_i \partial x_j} d_i d_j.
\end{aligned}$$

Therefore,

$$\begin{aligned}
h''(t) &= (d_1, \ldots, d_n) \begin{bmatrix} \frac{\partial^2 f(x+td)}{\partial x_1^2} & \frac{\partial^2 f(x+td)}{\partial x_j \partial x_1} & \frac{\partial^2 f(x+td)}{\partial x_n \partial x_1} \\ \vdots & \vdots & \vdots \\ \frac{\partial^2 f(x+td)}{\partial x_1 \partial x_i} & \frac{\partial^2 f(x+td)}{\partial x_j \partial x_i} & \frac{\partial^2 f(x+td)}{\partial x_n \partial x_i} \\ \vdots & \vdots & \vdots \\ \frac{\partial^2 f(x+td)}{\partial x_1 \partial x_n} & \frac{\partial^2 f(x+td)}{\partial x_j \partial x_n} & \frac{\partial^2 f(x+td)}{\partial x_n^2} \end{bmatrix} \begin{pmatrix} d_1 \\ \vdots \\ d_n \end{pmatrix} \\
&= d^T \left[\frac{\partial^2 f(x+td)}{\partial x_j \partial x_i} \right] d = d^T D^2 f(x+td) d.
\end{aligned}$$

The matrix

$$Hf(x) := D^2 F(x) = D(\nabla f(x)) = [\partial^2 f(x)/\partial x_i \partial x_j]$$

is called the Hessian matrix of f at x. If the second-order partial derivatives $\partial^2 f(x)/\partial x_i \partial x_j$ are continuous, then the mixed derivatives are equal, that is,

$$\frac{\partial^2 f(x)}{\partial x_i \partial x_j} = \frac{\partial^2 f(x)}{\partial x_j \partial x_i},$$

and the Hessian matrix $Hf(x)$ is symmetric, that is, $Hf(x)^T = Hf(x)$.

One can keep differentiating $h(t)$ to obtain

$$h^{(3)}(t) = \sum_{i=1}^{n} \sum_{j=1}^{n} \sum_{k=1}^{n} \frac{\partial^3 f(x+td)}{\partial x_i \partial x_j \partial x_k} d_i d_j d_k = D^3 f(x+td)[d,d,d],$$

and in general

$$h^{(k)}(t) = \sum_{i_1=1}^{n} \cdots \sum_{i_k=1}^{n} \frac{\partial^k f(x+td)}{\partial x_{i_1} \cdots \partial x_{i_k}} d_{i_1} d_{i_2} \cdots d_{i_k} = D^k f(x+td) \underbrace{[d, \ldots, d]}_{k \text{ times}}.$$

We note that $A[d,\ldots,d] = D^k f(x)[d,\ldots,d]$ is a *k-linear form*, which gives rise to a *k-linear functional* $A[d_1,\ldots,d_k]$. If the kth-order partial derivatives are continuous, then $A[d_1,\ldots,d_k]$ is symmetric, that is,

$$A[d_{\sigma(1)},\ldots,d_{\sigma(k)}] = A[d_1,\ldots,d_k],$$

for any permutation σ of the set $\{1,\ldots,k\}$.

After these preparations, we can display the multivariate Taylor's formula.

Theorem 1.23. (***Multivariate Taylor's formula***) *Let U be an open subset of $\mathbb{R}^n$, and let x, y be distinct points in U such that the line segment $[x, y]$ lies in U. If $f : U \to \mathbb{R}$ has continuous kth-order partial derivatives on U, then there exists a point $z \in (x, y)$ such that*

$$\begin{aligned} f(y) &= \sum_{i=0}^{k-1} \frac{1}{i!} D^i f(x) \underbrace{[y-x,\ldots,y-x]}_{i \text{ times}} + \frac{1}{k!} D^k f(z) \underbrace{[y-x,\ldots,y-x]}_{k+1 \text{ times}} \\ &= f(x) + Df(x)[y-x] + \frac{1}{2} D^2 f(x)[y-x, y-x] + \cdots \\ &\quad + \frac{1}{(k-1)!} D^{k-1} f(x)[y-x,\ldots,y-x] + \frac{1}{k!} D^k f(z)[y-x,\ldots,y-x]. \end{aligned}$$

Here

$$\begin{aligned} Df(x)[y-x] &= \langle \nabla f(x), y-x \rangle, \\ D^2 f(x)[y-x, y-x] &= (y-x)^T H f(x)(y-x) = d^T H(x) d. \end{aligned}$$

Proof. It follows from Taylor's formula for h that there exists $0 < \bar{t} < 1$ such that

$$h(1) = h(0) + h'(0) + \frac{h''(0)}{2!} + \cdots + \frac{h^{(k-1)}(0)}{(k-1)!} + \frac{h^{(k)}(\bar{t})}{k!}.$$

We note that $h(1) = f(y)$, $h(0) = f(x)$, $h'(0) = \langle \nabla f(x), y-x \rangle$, and $h''(0) = D^2 f(x)[d, d] = d^T H(x) d$. In general,

$$h^{(i)}(t) = D^i f(x + t(y-x))[y-x,\ldots,y-x].$$

Setting $z = x + \bar{t}(y-x)$, we see that the Taylor's formula in the statement of the theorem holds. □

Corollary 1.24. *Let U be an open subset of $\mathbb{R}^n$, and let $f : U \to \mathbb{R}$ have continuous kth-order partial derivatives on U. Then, as y approaches x,*

$$f(y) = \sum_{i=0}^{k} \frac{1}{i!} D^i f(x) \underbrace{[y-x,\ldots,y-x]}_{i \text{ times}} + o(\|y-x\|^k).$$

We remark that in light of Exercise 1, the corollary holds under the weaker assumption that f is k-times Fréchet differentiable.

Usually, Taylor's formula is considered for $k = 1, 2, 3$,

$$\begin{aligned} f(y) &= f(x) + \langle \nabla f(z_1), y - x \rangle, \\ f(y) &= f(x) + \langle \nabla f(x), y - x \rangle + \frac{1}{2}(y-x)^T Hf(z_2)(y-x), \\ f(y) &= f(x) + \langle \nabla f(x), y - x \rangle + \frac{1}{2}(y-x)^T Hf(x)(y-x) \\ &\quad + \frac{1}{6} D^3 f(z_3)[y-x, y-x, y-x], \end{aligned}$$

where $z_i \in (x, y)$, $i = 1, 2, 3$. As $y \to x$, we have

$$\begin{aligned} f(y) &= f(x) + \langle \nabla f(x), y - x \rangle + o(\|y - x\|), \\ f(y) &= f(x) + \langle \nabla f(x), y - x \rangle + \frac{1}{2}(y-x)^T Hf(x)(y-x) + o(\|y-x\|^2), \\ f(y) &= f(x) + \langle \nabla f(x), y - x \rangle + \frac{1}{2}(y-x)^T Hf(x)(y-x) \\ &\quad + \frac{1}{6} D^3 f(x)[y-x, y-x, y-x] + o(\|y-x\|^3). \end{aligned}$$

1.6 The Converse of Taylor's Theorem

Taylor's theorem has the following converse.

Theorem 1.25. *Let U be an open subset of $\mathbb{R}^n$. If a continuous function $f : U \to \mathbb{R}$ satisfies*

$$f(x+y) = a_0(x) + a_1(x)[y] + \frac{1}{2}a_2(x)[y^2] + \cdots + \frac{1}{k!}a_k(x)[y^k] + o(\|y\|^k), \quad (1.11)$$

where $a_i(x)$ is a symmetric i-linear form on $\mathbb{R}^n$ and where we have written $a_k[y^k]$ for $a_k(x)[y, \ldots, y]$ to simplify notation, then f is k-times Fréchet differentiable and $a_i(x) = D^i f(x)$, $i = 1, \ldots, k$.

Proof. We use induction on k. For $k = 0, 1$, the theorem is true by the continuity of f and the definition of Fréchet differentiability, respectively. Suppose the theorem is true for $k - 1$. Then, $a_k(x)[y^k] = o(\|y\|^{k-1})$ in (1.11), and the induction hypothesis implies that $a_j(x) = D^j f(x)$ for $j = 0, \ldots, k-1$. We now expand $f(x + y + z)$ in two ways,

$$\begin{aligned} f(x+y+z) &= f(x+y) + Df(x+y)[z] + \cdots + \frac{D^{k-1}f(x+y)[z^{k-1}]}{(k-1)!} \\ &\quad + \frac{1}{k!}a_k(x+y)[z^k] + o(\|y\|^k) \end{aligned}$$

and

$$f(x+y+z) = f(x) + Df(x)[y+z] + \cdots + \frac{D^{k-1}f(x)[(y+z)^{k-1}]}{(k-1)!} + \frac{1}{k!}a_k(x)[(y+z)^k] + o(\|y+z\|^k).$$

Fix x, and restrict z such that the ratio $\|y\|/2 \le \|z\| \le \|y\|$. Then $o(\|z\|^i)$, $o(\|y\|^i)$, and $o(\|y+z\|^i)$ are all equivalent. Subtracting the second equation above from the first one, collecting coefficients of $[z^i]$, and denoting by $g_i(y)$ the coefficient of $[z^i]$, we have

$$g_0(y) + g_1(y)[z] + \cdots + g_k(y)[z^k] = o(\|y\|^k). \tag{1.12}$$

Note that

$$g_k(y)[z^k] = \frac{1}{k!}\left(a_k(x+y) - a_k(x)\right)[z^k] = o(\|y\|^k),$$

since $a_k(x+y) - a_k(x) = o(1)$ by the continuity of a_k. Thus, this term may be dropped from equation (1.12). We claim that each remaining term in (1.12) is also $o(\|y\|^k)$. To prove this, we replace z with $t_i z$, where $\{t_i\}_1^k$ are all distinct. Then the resulting equations may be written as

$$\begin{bmatrix} 1 & t_1 & \cdots & t_1^k \\ 1 & t_2 & \cdots & t_2^k \\ \vdots & \vdots & & \vdots \\ 1 & t_k & \cdots & t_k^k \end{bmatrix} \begin{pmatrix} g_0(y) \\ g_1(y)[z] \\ \vdots \\ g_{k-1}(y)[z^{k-1}] \end{pmatrix} = \begin{pmatrix} o(\|y\|^k) \\ o(\|y\|^k) \\ \vdots \\ o(\|y\|^k) \end{pmatrix}.$$

Since t_i are distinct, the *Vandermonde matrix* above has determinant $\prod_{i<j}(t_i - t_j) \neq 0$, so that it is nonsingular. It follows by Cramer's rule that $g_i(y)[z^i] = o(\|y\|^k)$ for $i = 1, \ldots, k-1$. In particular,

$$g_{k-1}(y)[z^{k-1}] = \left[\frac{D^{k-1}f(x+y)}{(k-1)!} - \frac{D^{k-1}f(x)}{(k-1)!} - \frac{ka_k(x)[y]}{k!}\right][z^{k-1}] = o(\|y\|^k),$$

where the expression involving $a_k(x)$ follows since the symmetry of $a_k(y)$ gives $a_k(y)[(y+z)^k] = ka_k(y)[y, z, z, \ldots, z] + \cdots$. This gives

$$D^{k-1}f(x+y) - D^{k-1}f(x) - a_k(x)[y] = o(\|y\|),$$

which implies $a_k(x) = D^k f(x)$. □

The simple proof above is in [203]. See also [3] for a more general form of this theorem.

The significance of Theorem 1.25 is that Taylor's formula is often a very efficient tool for computing the derivatives $D^k f(x)$ of a multivariate function.

If we can develop the function $t \mapsto f(x+td)$ as a series in t in some way, then the coefficient of t^k is $D^k f(x)[d,\dots,d]/k!$ by Theorem 1.25, and we can recover $D^k f$ easily.

We illustrate this technique with some important examples.

Example 1.26. **(Quadratic functions)** Let

$$f(x) = \frac{1}{2}\langle Ax, x\rangle + \langle c, x\rangle + \alpha = \frac{1}{2}\sum_{j=1}^{n}\sum_{i=1}^{n} a_{ij}x_i x_j + \sum_{j=1}^{n} c_j x_j + \alpha$$

be a quadratic function in $\mathbb{R}^n$, where A is a symmetric $n \times n$ matrix, $c \in \mathbb{R}^n$, and α is a scalar. It is easy but tedious to compute the gradient and Hessian of f by computing the first-order and second-order partial derivatives of f.

Alternatively,

$$\begin{aligned} f(x+td) &= \frac{1}{2}\langle A(x+td), x+td\rangle + \langle c, x+td\rangle + \alpha \\ &= \frac{1}{2}\langle Ax, x\rangle + t\langle Ax, d\rangle + \frac{t^2}{2}\langle Ad, d\rangle + \langle c, x\rangle + t\langle c, d\rangle + \alpha \\ &= f(x) + t\langle Ax + c, d\rangle + \frac{t^2}{2}\langle Ad, d\rangle. \end{aligned}$$

Theorem 1.25 implies that

$$\nabla f(x) = Ax + c, \quad Hf(x) = D^2 f(x) = A.$$

These formulas apply without change to any quadratic function in an inner product space E if A is a self-adjoint (symmetric) linear operator $A : E \to E$.

Example 1.27. **(Logarithm of the determinant function)**

Let us first consider the vector space $R^{n\times n}$ of $n \times n$ matrices. This is a vector space of dimension n^2. A natural inner product on $\mathbb{R}^{n\times n}$ is given by

$$\langle X, Y\rangle := \sum_{i,j=1}^{n} x_{ij}y_{ij} = \sum_{i,j=1}^{n} (X^T)_{ji}y_{ij} = \operatorname{tr}(X^T Y).$$

Next, we consider the linear subspace S^n of symmetric $n \times n$ matrices of $\mathbb{R}^{n\times n}$. This is a vector space of dimension $n(n+1)/2$. The trace inner product on $\mathbb{R}^{n\times n}$ induces an inner product on S^n given by

$$\langle X, Y\rangle := \operatorname{tr}(XY).$$

Let us now consider the function

$$f(X) = \ln\det X,$$

where $X \in S^n$ is positive definite, an important function in semidefinite programming.

We compute the Taylor series of f at a given positive definite matrix $X \in S^n$ and a given direction $D \in S^n$. We have

$$\begin{aligned} f(X+tD) &= \ln\det(X+tD) = \ln\det(X^{1/2}(I+tX^{-1/2}DX^{-1/2})X^{1/2}) \\ &= \ln\det X + \ln\det(I+tX^{-1/2}DX^{-1/2}). \end{aligned}$$

Writing $\overline{D} := X^{-1/2}DX^{-1/2} = Q^T \Lambda Q$, where Q is an orthogonal matrix and $\Lambda = \operatorname{diag}(\lambda_1, \ldots, \lambda_n)$, we obtain

$$\begin{aligned} f(X+tD) - f(X) &= \ln\det(I+t\overline{D}) = \ln\det(Q^T(I+t\Lambda)Q) \\ &= \ln\det(I+t\Lambda) = \ln\prod_i^n(1+t\lambda_i) = \sum_1^n \ln(1+t\lambda_i). \end{aligned}$$

Since

$$\ln(1+s) = \int \frac{ds}{1+s} = \int (1 - s + s^2 - s^3 + \cdots)ds = s - \frac{s^2}{2} + \frac{s^3}{3} + \cdots,$$

we have

$$f(X+tD) - f(X) = \sum_1^n t\lambda_i - \frac{t^2}{2}\lambda_i^2 + o(t^2) = t\operatorname{tr}(\Lambda) - \frac{t^2}{2}\operatorname{tr}(\Lambda^2) + o(t^2).$$

Noting that

$$\operatorname{tr}(\Lambda) = \operatorname{tr}(Q^T\Lambda Q) = \operatorname{tr}(\overline{D}) = \operatorname{tr}(X^{-1}D) = \langle X^{-1}, D\rangle,$$

and similarly

$$\operatorname{tr}(\Lambda^2) = \operatorname{tr}(X^{-1/2}DXD^{-1/2})^2 = \operatorname{tr}(X^{-1}DX^{-1}D) = \langle X^{-1}DX^{-1}, D\rangle,$$

we obtain

$$f(X+tD) = f(X) + t\langle X^{-1}, D\rangle - \frac{t^2}{2}\langle X^{-1}DX^{-1}, D\rangle + o(t^2).$$

Theorem 1.25 again implies that $\nabla f(X) = X^{-1}$ and $D^2 f(X)(D) = X^{-1}DX^{-1}$. Here $D^2 f(X)(D)$ can be written, using the tensor (Kronecker) product notation,

$$(X^{-1} \otimes X^{-1})(D) := X^{-1}DX^{-1}.$$

In summary, we have

$$\nabla f(X) = X^{-1}, \quad D^2 f(X) = -X^{-1} \otimes X^{-1}. \tag{1.13}$$

Higher-order derivatives of f can be obtained by computing more terms in the Taylor expansion above.

The derivative of the determinant of not necessarily symmetric matrices is considered in Exercise 22 at the end of the chapter. See also Exercise 23.

Example 1.28. **(Matrix inversion)**

Denote by $\mathrm{GL}(n,\mathbb{R})$ the set of nonsingular matrices in the vector space $V = R^{n\times n}$ of $n\times n$ real matrices. We will show that the matrix inversion map

$$\mathrm{Inv} : \mathrm{GL}(n,\mathbb{R}) \to \mathrm{GL}(n,\mathbb{R}), \quad \mathrm{Inv}(A) = A^{-1}$$

is infinitely differentiable, and compute its derivatives.

First, we have from $(I + A)(I - A + A^2 - A^3 + (-1)^k A^k) = I - A^{k+1}$, Neumann's formula

$$(I + A)^{-1} = I - A + A^2 - A^3 + \cdots \quad \text{for } \|A\| < 1.$$

It follows that for $|t|$ small,

$$(I + tA)^{-1} = I - tA + t^2A^2 - t^3A^3 + \cdots = \sum_{t=0}^{\infty}(-1)^k t^k A^k,$$

which immediately gives that

$$D^k \,\mathrm{Inv}(I)[A, \ldots, A] = (-1)^k k! A^k.$$

Now, if $A \in \mathrm{GL}(n,\mathbb{R})$ and $H \in V$ are arbitrary matrices, and $K := A^{-1}H$, we have

$$(A + tH)^{-1} = [A(I + A^{-1}H)]^{-1} = (I + tK)^{-1}A^{-1} = \sum_{t=0}^{\infty}(-1)^k t^k K^k A^{-1},$$

and we obtain

$$D^k \,\mathrm{Inv}(A)[H, \ldots, H] = (-1)^k k! A^{-1}HA^{-1}H \cdots A^{-1}HA^{-1},$$

where the right-hand side contains $k + 1$ A's and k H's.

These results extend verbatim to inversion of nonsingular continuous linear mappings $A : X \to X$ where X is a Banach space.

1.7 Danskin's Theorem

Danskin's theorem [65, 66] is one of the fundamental theorems in optimization theory. We will use it in Chapters 9 and 12 to derive optimality conditions for nonlinear programming and semi-infinite programming, respectively.

Theorem 1.29. **(*Danskin*)** *Suppose $f : X \times Y \to \mathbb{R}$ is a continuous function, where $X \subseteq \mathbb{R}^n$ is an open set, Y is a compact set of a topological space F, and $\nabla_x f(x, y)$ exists and is continuous. Then the marginal function*

$$\varphi(x) := \max_{y\in Y} f(x, y)$$

is continuous and has directional derivatives in every direction, which are given by the formula

$$\varphi'(x;h) = \max_{y \in Y(x)} \langle \nabla_x f(x,y), h \rangle, \tag{1.14}$$

where $Y(x) = \{y \in Y : \varphi(x) = f(x,y)\}$ *is the set of maximizers in the definition of* $\varphi(x)$.

Proof. We first prove that $\varphi(x)$ is continuous. Let $x_0 \in X$ and let $\{x_k\}_1^\infty$ be a sequence converging to x_0. Pick $y_k \in Y$ such that $\varphi(x_k) = f(x_k, y_k)$. Since Y is compact, we may assume without loss of generality that $y_k \to y_0 \in Y$. Noting that $f(x_k, y_k) \geq f(x_k, y)$ for any $y \in Y$, we have

$$\lim_{k\to\infty} \varphi(x_k) = \lim_{k\to\infty} f(x_k, y_k) = f(x_0, y_0) \geq \lim_{k\to\infty} f(x_k, y) = f(x_0, y),$$

for all $y \in Y$. The inequality $f(x_0, y_0) \geq f(x_0, y)$ implies $\varphi(x_0) = f(x_0, y_0)$, and we have $\varphi(x_0) = f(x_0, y_0) = \lim_{k\to\infty} \varphi(x_k)$, that is, $\varphi(x)$ is continuous.

Let $0 \neq h \in \mathbb{R}^n$ be a direction, and let $\{x_k\}_1^\infty$, $x_k = x_0 + t_k h$, $t_k \geq 0$, be a sequence converging to x_0. Let $y \in Y(x_0)$ be an arbitrary point. If $\varphi(x_k) = f(x_k, y_k)$ $(k \geq 1)$, we have

$$\begin{aligned} \frac{\varphi(x_k) - \varphi(x_0)}{t_k} &= \frac{f(x_k, y_k) - f(x_0, y)}{t_k} \\ &= \frac{f(x_k, y_k) - f(x_k, y)}{t_k} + \frac{f(x_k, y) - f(x_0, y)}{t_k} \\ &\geq \frac{f(x_k, y) - f(x_0, y)}{t_k} \\ &= \langle \nabla_x f(x_0 + t_k' h, y), h \rangle, \end{aligned}$$

where the inequality follows since $f(x_k, y_k) \geq f(x_k, y)$, and the last equality follows from the mean value theorem. Taking limits, we obtain

$$\varliminf_{k\to\infty} \frac{\varphi(x_k) - \varphi(x_0)}{t_k} \geq \langle \nabla_x f(x_0, y), h \rangle \quad \text{for all} \quad y \in Y(x_0).$$

This implies

$$\varliminf_{k\to\infty} \frac{\varphi(x_k) - \varphi(x_0)}{t_k} \geq \max_{y \in Y(x_0)} \langle \nabla_x f(x_0, y), h \rangle. \tag{1.15}$$

Similarly, if $\varphi(x_k) = f(x_k, y_k)$ $(k \geq 0)$, where $y_k \to y_0$, we also have

$$\begin{aligned} \frac{\varphi(x_k) - \varphi(x_0)}{t_k} &= \frac{f(x_k, y_k) - f(x_0, y_k)}{t_k} + \frac{f(x_0, y_k) - f(x_0, y_0)}{t_k} \\ &\leq \frac{f(x_k, y_k) - f(x_0, y_k)}{t_k} \\ &= \langle \nabla_x f(x_0 + t_k'' h, y_k), h \rangle. \end{aligned}$$

This implies that

$$\varlimsup_{k\to\infty}\frac{\varphi(x_k)-\varphi(x_0)}{t_k} \le \varlimsup_{i\to\infty}\langle\nabla_x f(x_0+t'_{k_i}h, y_{k_i}), h\rangle$$
$$= \langle\nabla_x f(x_0, y_0), h\rangle$$
$$\le \max_{y\in Y(x_0)}\langle\nabla_x f(x_0, y), h\rangle.$$

This inequality and (1.15) prove (1.14). □

We remark that Danskin's theorem and its proof above hold verbatim if X is an open set in a Banach space. It can also be generalized in other ways; for example, in the case where Y is a finite set, the directional differentiability of $x \mapsto f(x,y)$ for each y is enough to guarantee the directional differentiability of φ.

Corollary 1.30. *Let $\{f_i\}_1^k$ be functions defined on a set X in $\mathbb{R}^n$, and let $\varphi(x) := \max\{f_i(x) : i = 1,\dots,k\}$ be their pointwise maximum. If all f_i are directionally differentiable at x_0 in the direction $h \in \mathbb{R}^n$, then φ is directionally differentiable at x_0 in the direction h, and*

$$\varphi'(x;h) = \max_{y\in I} f'_i(x;h),$$

where $I = \{i : \varphi(x) = f_i(x)\}$.

This can be proved by mimicking the proof of Theorem 1.29. We leave it to the reader; see Exercise 24.

1.8 Exercises

1. Let $f : I = (c,d) \to \mathbb{R}$ be an n-times differentiable function. Show that

$$f(x) = f(a) + f'(a)(x-a) + \cdots + \frac{f^{(n)}(a)}{n!}(x-a)^n + o((x-a)^n).$$

The point of the exercise is to prove the above equality without assuming that f n-times continuously differentiable, because if $f^{(n)}$ is continuous, then the equality follows readily from Theorem 1.1.
Hint: Prove the equality

$$\lim_{x\to a}\frac{f(x)-f(a)-f'(a)(x-a)-\cdots-\frac{f^{(n)}(a)}{n!}(x-a)^n}{(x-a)^n} = 0,$$

using induction on n, passing from n to $n+1$ using L'Hospital's rule.

2. This exercise gives a fairly simple approach to Taylor's formula in Cauchy's form (Theorem 1.5) using integration by parts. The idea is to write

$$f(b) - f(a) = \int_a^b f'(x)dx = -\int_a^b f'(x)d(b-x),$$

and then use integration by parts on the last integral. This gives

$$\begin{aligned} f(b) &= f(a) - f'(x)(b-x)|_a^b + \int_a^b (b-x)f''(x)dx \\ &= f(a) + f'(a)(b-a) + \int_a^b (b-x)f''(x)dx, \end{aligned}$$

which is Theorem 1.5 for $n = 2$.

(a) Use integration by parts on the last integral above to prove the theorem for $n = 3$.

(b) Use induction on n to complete the proof of Theorem 1.5.

3. This exercise outlines an interesting approach to Taylor's formula in Cauchy's form.

Let $f : J \to \mathbb{R}$ be a function on an open interval J, differentiable enough times. Consider the operations

$$A : f(x) \mapsto \int_a^x f(t)dt, \quad B : f(x) \mapsto f'(x), \quad I : f(x) \mapsto f(x).$$

Show that $BA(f(x)) = f(x)$, but $AB(f(x)) = f(x) - f(a)$, so that $AB \neq BA$, that is, A and B do not *commute*, when $f(a) \neq 0$. Obviously, $B^k(f(x)) = f^{(k)}(x)$. The formula for A^k is more complicated. Show that

$$A^2(f(x)) = \int_a^x \int_a^s f(t)dt\,ds = \int_a^x \int_t^x f(t)ds\,dt = \int_a^x (x-t)f(t)dt,$$

where the second equality follows from Fubini's theorem for multiple integrals. More generally, show that

$$A^k(f(x)) = \int_a^x \frac{(x-t)^{k-1}}{(k-1)!} f(t)dt,$$

a formula due to Cauchy.

Observe that

$$\sum_{k=0}^{n-1} A^k(I - AB)B^k = \sum_{k=0}^{n-1} (A^kB^k - A^{k+1}B^{k+1}) = I - A^nB^n.$$

Noting that $(I - AB)(f(x)) = f(a)$, show that the above telescoping formula gives

$$\sum_{k=0}^{n-1} \frac{(x-a)^k}{k!} f^k(a) = f(x) - \int_a^x \frac{(x-t)^{n-1}}{(n-1)!} f^{(n)}(t)dt,$$

which is precisely Taylor's formula in Cauchy's form.
This problem is taken from [261], which contains simple derivations of certain other formulas in analysis.

4. Here is an interesting approach, using determinants, to Taylor's formula in Lagrange's form.
Let $f(x)$, $\{f_i(x)\}_1^{n+2}(x)\}$ be $(n+1)$-times continuously differentiable functions. Then

$$\begin{vmatrix} f(x) & f_1(x) & \cdots & f_{n+2}(x) \\ f(0) & f_1(0) & \cdots & f_{n+2}(0) \\ f'(0) & f_1'(0) & \cdots & f_{n+2}'(0) \\ \vdots & \vdots & \cdots & \vdots \\ f^{(n)}(0) & f_1^{(n)}(0) & \cdots & f_{n+2}^{(n)}(0) \\ f^{(n+1)}(h) & f_1^{(n+1)}(h) & \cdots & f_{n+2}^{(n+1)}(h) \end{vmatrix} = 0$$

for some h strictly between 0 and x. To prove this, consider x as a constant and let $D^{(i)}(h)$ denote the function of h by replacing the last row of the determinant with $f^{(i)}(h), f_1^{(i)}(h), \ldots, f_{n+2}^{(i)}(h)$.

(a) Show that the derivative of $D^{(i)}(h)$ with respect to h is $D^{(i+1)}(h)$ for $i = 0, 1, \ldots, n$, and the determinant above is $D^{(n+1)}(h)$.

(b) Show that $D^{(0)}(0) = 0$ and $D^{(0)}(x) = 0$.

(c) Use Rolle's theorem to prove the existence of h_1 strictly between 0 and x such that $D^{(1)}(h_1) = 0$. Also, show that $D^{(1)}(0) = 0$. Use Rolle's theorem again to prove the existence of h_2 strictly between 0 and h_1 such that $D^{(2)}(h_2) = 0$.

(d) Continue in this fashion to show that there exists a point h strictly between 0 and x such that $D^{(n+1)}(h) = 0$.

As an application, show that there exists a point h strictly between 0 and x such that

$$\begin{vmatrix} f(x) & 1 & \frac{x}{1!} & \frac{x^2}{2!} & \cdots & \frac{x^n}{n!} & \frac{x^{n+1}}{(n+1)!} \\ f(0) & 1 & 0 & 0 & \cdots & 0 & 0 \\ f'(0) & 0 & 1 & 0 & \cdots & 0 & 0 \\ \vdots & \vdots & & \vdots & \cdots & \vdots & \vdots \\ f^{(n)}(0) & 0 & 0 & 0 & \cdots & 1 & 0 \\ f^{(n+1)}(h) & 0 & 0 & 0 & \cdots & 0 & 1 \end{vmatrix} = 0,$$

and that the above determinant is

$$f(x) - f(0) - f'(0)x - \frac{f''(0)}{2}x^2 - \cdots - \frac{f^{(n)}(0)}{n!}x^n - \frac{f^{(n+1)}(h)}{(n+1)!}x^{n+1}.$$

5. Let $f : \mathbb{R}^n \to \mathbb{R}$ be a function satisfying the inequality $|f(x)| \le \|x\|^2$. Show that f is Fréchet differentiable at 0.

6. Define a function $f : \mathbb{R}^2 \to \mathbb{R}$ as follows:

$$f(x,y) = \begin{cases} x & \text{if } y = 0, \\ y & \text{if } x = 0, \\ 0 & \text{otherwise.} \end{cases}$$

Show that the partial derivatives

$$\frac{\partial f(0,0)}{\partial x} := \lim_{t \to 0} \frac{f(t,0) - f(0,0)}{t}, \quad \text{and} \quad \frac{\partial f(0,0)}{\partial y} := \lim_{t \to 0} \frac{f(0,t) - f(0,0)}{t}$$

exist, but that f is not Gâteaux differentiable at $(0,0)$.

7. ***(Genocchi-Peano)*** Define the function $f : \mathbb{R}^2 \to \mathbb{R}$

$$f(x,y) = \begin{cases} \frac{xy^2}{x^2+y^4} & \text{if } (x,y) \neq (0,0), \\ 0 & \text{if } (x,y) = (0,0). \end{cases}$$

   (a) Show that f is directionally differentiable at $(0,0)$, that is, f has directional derivatives at the origin along all directions.
   (b) Show that f is not Gâteaux differentiable at the origin.
   (c) Show that, even though f is continuous when restricted to lines passing thorough the origin, f is not continuous at the origin.

8. Define a function $f : \mathbb{R}^2 \to \mathbb{R}$ as follows:

$$f(x,y) = \begin{cases} \frac{2y\exp(-x^{-2})}{y^2+\exp(-2x^{-2})} & \text{if } x \neq 0, \\ 0 & \text{otherwise.} \end{cases}$$

Show that f is Gâteaux differentiable at $(0,0)$, but that f is not continuous there.

9. Define a function $f : \mathbb{R}^2 \to \mathbb{R}$ as follows:

$$f(x,y) = \begin{cases} \frac{x^3y}{x^4+y^2} & \text{if } (x,y) \neq (0,0), \\ 0 & \text{if } (x,y) = (0,0). \end{cases}$$

Show that f is Gâteaux differentiable but not Fréchet differentiable at $(0,0)$.

10. Define a function $f : \mathbb{R}^2 \to \mathbb{R}$ as follows:

$$f(x,y) = \begin{cases} \frac{y(x^2+y^2)^{3/2}}{(x^2+y^2)^2+y^2} & \text{if } (x,y) \neq (0,0), \\ 0 & \text{if } (x,y) = (0,0). \end{cases}$$

Show that f is Gâteaux differentiable but not Fréchet differentiable at $(0,0)$.

11. Define a function $f : \mathbb{R}^2 \to \mathbb{R}$ as follows:

$$f(x,y) = \begin{cases} \frac{xy}{r} \sin\left(\frac{1}{r}\right) & \text{if } (x,y) \neq (0,0), \\ 0 & \text{if } (x,y) = (0,0), \end{cases}$$

where $r = \|(x,y)\| = (x^2+y^2)^{1/2}$. Show that $\partial f/\partial x$ and $\partial f/\partial y$ exist at every point $(x,y) \in \mathbb{R}^2$, and the four functions $x \mapsto \partial f(x,b)/\partial x$, $y \mapsto \partial f(a,y)/\partial x$, $x \mapsto \partial f(x,b)/\partial y$, $y \mapsto \partial f(a,y)/\partial y$ are continuous for any $(a,b) \in \mathbb{R}^2$, but f is not Fréchet differentiable at $(0,0)$.

12. Let $f : \mathbb{R}^2 \to \mathbb{R}$ be a function defined by the formula

$$f(x,y) = \begin{cases} \frac{xy(x^2-y^2)}{x^2+y^2} & \text{if } (x,y) \neq (0,0), \\ 0 & \text{if } (x,y) = (0,0). \end{cases}$$

Show that all four second-order partial derivatives $\partial^2 f/\partial x^2$, $\partial^2 f/\partial x \partial y$, $\partial^2 f/\partial y \partial x$, and $\partial^2 f/\partial y^2$ exist everywhere on $\mathbb{R}^2$, but $\partial^2 f/\partial x \partial y \neq \partial^2 f/\partial y \partial x$ at the point $(0,0)$.

13. Define a function $F : \mathbb{R}^2 \to \mathbb{R}^2$ as follows:

$$F(x,y) = (x^3, y^2).$$

Let $x = (0,0)$ and $y = (1,1)$. Show that there is no vector z on the line segment between x and y such that

$$F(y) - F(x) = DF(z)(y-x).$$

This shows that the mean value theorem (Lemma 1.12) does not generalize, at least in the same form.

14. Let $f : \mathbb{R}^n \to \mathbb{R}^m$ be a Gâteaux differentiable map such that the Jacobian Df vanishes identically, that is, $Df(x) = 0$ for all $x \in \mathbb{R}^n$. Use Theorem 1.18 to give a short proof that f must be a constant function. More generally, use the same theorem to prove that if Df is a constant matrix, then f must be an affine transformation.

15. For a given scalar $p \in [1, \infty)$, let

$$f(x) \equiv \|x\|_p \equiv \left(\sum_{i=1}^{n} |x_i|^p \right)^{1/p}, \quad x \in \mathbb{R}^n,$$

denote the l_p-norm for vectors in $\mathbb{R}^n$. Compute the partial derivatives $\partial f/\partial x_i$, $i = 1, 2, \ldots, n$, for any vector x with no zero component. Does f have a Fréchet or Gâteaux derivative at such a point? At the point $x = 0$? What more can be said for the case $p = 2$?

16. This exercise shows that Gâteaux differentiability may not be enough for the chain rule to hold.

(a) (***Fréchet***) Define the functions $f : \mathbb{R} \to \mathbb{R}^2$, $f(t) = (t, t^2)$, and $g : \mathbb{R}^2 \to \mathbb{R}$,

$$g(x, y) = \begin{cases} x & \text{if } y = x^2, \\ 0 & \text{otherwise.} \end{cases}$$

Show that g is Gâteaux differentiable at $(0, 0)$ with gradient $\nabla g(0, 0) = (0, 0)$, but $g \circ f$ is the identity function on $\mathbb{R}$, to conclude that the chain rule for $g \circ f$ fails at $t = 0$.

(b) Define the functions $f : \mathbb{R} \to \mathbb{R}^2$, $f(t) = (t \cos t, t \sin t)$, and $g : \mathbb{R}^2 \to \mathbb{R}$ given (in polar coordinates) by

$$g(r, \theta) = \begin{cases} \frac{r^2}{\theta^3} & \text{if } 0 < \theta < 2\pi, \\ 0 & \text{if } \theta = 0. \end{cases}$$

Show that g is Gâteaux differentiable at $(0, 0)$ with gradient $\nabla g(0, 0) = (0, 0)$, but $(g \circ f)(t) = 1/t$, so that the chain rule for $g \circ f$ again fails at $t = 0$.

17. Let $f : V \to \mathbb{R}$ be an infinitely differentiable function on a vector space V. Let f be *n-homogeneous*, that is,

$$f(tx) = t^n f(x).$$

Show that

$$D^k f(x) \underbrace{[x, \ldots, x]}_{k \text{ times}} = \begin{cases} \frac{n!}{(n-k)!} f(x), & k = 0, \ldots, n, \\ 0, & k > n. \end{cases}$$

The formula for the case $k = 1$, $Df(x)[x] = nf(x)$, is known as *Euler's formula.*

Hint: Write the Taylor series for $f(x + tx)$, and note that $f(x + tx) = f((1 + t)x) = (1 + t)^n f(x)$.

18. Let $M : \mathbb{R}^{n_1} \times \mathbb{R}^{n_2} \times \cdots \times \mathbb{R}^{n_k} \to \mathbb{R}^m$ be a multilinear map, that is, $x_i \mapsto M(x_1, \ldots, x_{i-1}, x_i, x_{i+1}, \ldots, x_k)$ is linear when all variables x_j other than x_i are fixed. Show that

$$M'(x; h) = M(h_1, x_1, \ldots, x_k) + M(x_1, h_2, x_3, \ldots, x_k) + \cdots + M(x_1, x_2, \ldots, x_{k-1}, h_k),$$

where we have used the notation $x = (x_1, \ldots, x_k)$, $h = (h_1, \ldots, h_k)$, and $M'(x; h) = M'(x_1, \ldots, x_k; h_1, \ldots, h_k)$. Then, compute $D^2 M(x)[h, h]$ and $D^3 M(x)[h, h, h]$. How do the formulas simplify when M is a symmetric multilinear mapping?

Hint: Compute $M(x_1 + th_1, x_2 + th_2, \ldots, x_k + th_k)$ using multilinearity of M.

19. Let $F : \mathbb{R}^n \to \mathbb{R}^m$ be a map with Lipschitz derivative, that is, there exists $L \geq 0$ such that

$$\|DF(y) - DF(x)\| \leq L\|y - x\| \text{ for all } x, y \in \mathbb{R}^n.$$

Show that

$$\|F(y) - F(x) - DF(x)(y - x)\| \leq \frac{L}{2}\|y - x\|^2 \text{ for all } x, y \in \mathbb{R}^n.$$

Notice that the slightly weaker inequality, with the constant $L/2$ replaced by L, follows immediately from Theorem 1.18.
Hint: Define the function $\varphi(t) = F(x + t(y - x)) - tDF(x)(y - x)$. Show that $\varphi'(t) = (DF(x + t(y - x)) - DF(x))(y - x)$, and use the inequality $\|\int_0^1 \varphi'(t)\, dt\| \leq \int_0^1 \|\varphi'(t)\| dt$.

20. Let $f : I = (c, d) \to \mathbb{R}$ be such that $0 \in I$.
(a) If $f \in C^1$ (continuously differentiable) on I, then show that there exists a continuous function a on I such that

$$f(x) = f(0) + a(x)x.$$

Moreover, show that if $f \in C^2$, then $a \in C^1$.
(b) If $f \in C^2$ (twice continuously differentiable) on I, then show that there exists a continuous function b on I such that

$$f(x) = f(0) + f'(0)x + b(x)x^2.$$

Hint: If $x \neq 0$, the above equations define $a(x)$ and $b(x)$,

$$a(x) = \frac{f(x) - f(0)}{x}, \quad b(x) = \frac{f(x) - f(0) - f'(0)x}{x^2}.$$

Use L'Hospital's rule to show that $a(0)$ and $b(0)$ can be defined in such a way that the functions a, b are continuous at $x = 0$.
(c) Let $f : U \to \mathbb{R}$ be a C^1 (continuous partial derivatives) function in a neighborhood U of the origin in $\mathbb{R}^n$. Prove that there exist continuous functions $\{a_i(x)\}_1^n$ on U such that

$$f(x_1, \ldots, x_n) = f(0) + \sum_{i=1}^{n} x_i a_i(x_1, \ldots, x_n).$$

Moreover, show that if $f \in C^2$, then $a_i \in C^1$.
Hint: Show that (a) guarantees the existence of a continuous function $a(x)$ such that $f(x_1, \ldots, x_n) = f(0, x_2, \ldots, x_n) + x_1 a_1(x_1, \ldots, x_n)$. Then, use induction on n.
(d) Let $f : U \to \mathbb{R}$ be a C^2 (second partial derivatives continuous) function on U. Using (b), show that there exists a continuous function $b(x)$ on U such that

$$f(x_1, \ldots, x_n) = f(0, x_2, \ldots, x_n) + x_1 \frac{\partial f(0, x_2, \ldots, x_n)}{\partial x_1} + x_1^2 b(x_1, \ldots, x_n).$$

(e) Let f be a function as in (d) and assume that $\nabla f(0) = 0$. Prove that there exist continuous functions $\{b_{ij}(x)\}_{i,j=1}^n$ on U such that

$$f(x_1,\ldots,x_n) = f(0) + \sum_{i,j=1}^{n} x_i x_i b_{ij}(x_1,\ldots,x_n).$$

Hint: Use (c) on the C^1 function $\partial f(0,x_2,\ldots,x_n)/\partial x_1$ in (d) to obtain a representation

$$f(x_1,\ldots,x_n) = f(0,x_2,\ldots,x_n) + \sum_{j=1}^{n} x_1 x_j b_{1j}(x_1,\ldots,x_n).$$

Complete the proof by induction on n.

21. Compute the first two derivatives of the determinant function $f(x) = \det X$ on S^n, the space of $n \times n$ symmetric real matrices
    (a) From scratch, mimicking the derivation in Example 1.27.
    (b) Using the chain rule and the results of Example 1.27.
22. Let $\mathbb{R}^{n\times n}$ be the space of $n \times n$ real matrices. Show that if $A(t) \in \mathbb{R}^{n\times n}$ is a differentiable function of t then $d(\det A(t))/dt$ is the sum of the determinants of n matrices, in which the ith matrix is $A(t)$ except that the ith row is differentiated. Use this result to prove that the directional derivative of the determinant function at the matrix $A \in \mathbb{R}^{n\times n}$ along the direction $B \in \mathbb{R}^{n\times n}$ is given by

$$(\det)'(A;B) = \operatorname{tr}(\operatorname{Adj}(A)B) = \langle \operatorname{Adj}(A)^T, B\rangle,$$

where $\operatorname{Adj}(A)$ is the adjoint of A, and where the inner product on $\mathbb{R}^{n\times n}$ is the trace inner product given by $\langle X, Y\rangle = \operatorname{tr}(X^T Y)$. Conclude that

$$D(\det)(A) = \operatorname{Adj}(A)^T.$$

Hint: Use the determinant formula $\det X = \sum_\sigma \operatorname{sgn}(\sigma) x_{1\sigma(1)} \cdots x_{n\sigma(n)}$ to compute $d(\det A(t))/dt$, and Laplace's expansion formula for determinants to compute $(\det)'(A;B)$.

23. Let $A \in \mathbb{R}^{n\times n}$. Show that
    (a) $(\det)'(I;A) = \operatorname{tr}(A)$, where I is the identity matrix.
    Hint: Show that $\det(I+tA) = 1 + t(a_{11} + a_{22} + \cdots + a_{nn}) + \cdots$, using the formula $\det A = \sum_\sigma \operatorname{sgn}(\sigma) a_{1\sigma(1)} \cdots a_{n\sigma(n)}$.
    (b) Show that if A is a nonsingular matrix, then

$$(\det)'(A;B) = \det(A)\operatorname{tr}(A^{-1}B) = \langle \det(A)A^{-T}, B\rangle.$$

Consequently, show that

$$A^{-1} = \frac{\operatorname{Adj}(A)}{\det A}.$$

Hint: Use $\det(A + tB) = \det(A)\det(I + A^{-1}B)$ and the previous problem.

24. Prove Corollary 1.30.

2

Unconstrained Optimization

In optimization theory, the optimality conditions for interior points are usually much simpler than the optimality conditions for boundary points. In this chapter, we deal with the former, easier case. Boundary points appear more prominently in constrained optimization, when one tries to optimize a function, subject to several functional constraints. For this reason, the optimality conditions for boundary points are generally discussed in constrained optimization, whereas the optimality conditions for interior points are discussed in unconstrained optimization, regardless of whether the optimization problem at hand has constraints.

In this chapter, we first establish some basic results on the existence of global minimizer or maximizers of continuous functions on a metric space. These are the famous Weierstrass theorem and its variants, which are essentially the only general tools available for establishing the existence of optimizers.

The rest of the chapter is devoted to obtaining the fundamental first-order and second-order necessary and sufficient optimality conditions for minimizing or maximizing differentiable functions. Since the tools here are based on differentiation, and differentiation is a local theory, the optimality conditions generally apply to *local* optimizers. The necessary and sufficient conditions play different, usually complementary, roles. A typical necessary condition for a local minimizer, say, states that certain conditions, usually given as equalities or inequalities, must be satisfied at a local minimizer. A typical sufficient condition for a local minimizer, however, states that if certain conditions are satisfied at a given point, then that point must be a local minimizer.

The nature (local minimum, local maximum, or saddle point) of a critical point x of a twice differentiable function f is deduced from the definiteness properties of the quadratic form $q(d) = \langle D^2 f(x)d, d\rangle$ involving the Hessian matrix $D^2 f(x)$. Thus, there is a need for an efficient recognition of a symmetric matrix. Several tools are developed in Section 2.4 for this purpose. A novel feature of this section is that we give an exposition of a simple tool, *Descartes's*

O. Güler, *Foundations of Optimization*, Graduate Texts in Mathematics 258,
DOI 10.1007/978-0-387-68407-9_2, © Springer Science+Business Media, LLC 2010

rule of sign, that can be used to count *exactly* the number of positive and negative eigenvalues of a symmetric matrix, including $D^2 f(x)$.

The inverse function theorem and the closely related implicit function theorem are important tools in many branches of analysis. Another closely related result, Lyusternik's theorem [191], is an important tool in optimization, where it is used in the derivation of optimality conditions in constrained optimization. We give an elementary proof of the implicit function theorem in finite-dimensional vector spaces in Section 2.5, following Carathéodory [54], and use it to prove the inverse function theorem and Lyusternik's theorem in finite dimensions. The proof of the same theorems in Banach spaces is given in Chapter 3 using Ekeland's ϵ-variational principle. If one is interested only in finite-dimensional versions of these results, it suffices to read only Section 2.5.

The local behavior of a C^2 function f around a nondegenerate critical point x ($D^2 f(x)$ is nonsingular) is determined by the Hessian matrix $D^2 f(x)$. This is the content of Morse's lemma, which is treated in Section 2.6. Morse's lemma is a basic result in Morse theory, which investigates the relationships between various types of critical points of a function f; see, for example, Milnor [197] for an introduction to Morse theory.

2.1 Basic Results on the Existence of Optimizers

We start by defining various types of optimal points.

Definition 2.1. *Let $f : U \to \mathbb{R}$ be a function on a set $U \subseteq \mathbb{R}^n$. Let $x^* \in U$ be an arbitrary point, and let $B_r(x^*) := \{x \in U : \|x - x^*\| < r\}$ be the open ball of radius r around x^*. The point x^* is called*

(a) a local minimizer *of f if*

$$f(x^*) \le f(x) \quad \text{for all } x \text{ in some ball } B_r(x^*),$$

and a strict local minimizer *of f if*

$$f(x^*) < f(x) \quad \text{for all } x \in B_r(x^*),\ x \neq x^*;$$

(b) a global minimizer *of f on U if*

$$f(x^*) \le f(x) \quad \text{for all } x \in U,$$

and a strict global minimizer *of f on U if*

$$f(x^*) < f(x) \quad \text{for all } x \in U,\ x \neq x^*;$$

(c) a critical point *of f if f is Gâteaux differentiable at x^* and $\nabla f(x^*) = 0$;*

(d) a saddle point *of f if it is a critical point and there exist points y, z in any ball $B_r(x^*)$ such that $f(y) < f(x^*) < f(z)$.*

Parallel definitions apply for a *maximizer* of f. We call a point x^* an *optimizer* of f if x^* is a minimizer or a maximizer in any of the senses above.

The most basic result on the existence of optimizers is the following theorem, due to Weierstrass.

Theorem 2.2. (***Weierstrass***) *Let $f : K \to \mathbb{R}$ be a continuous function defined on a compact metric space K. Then there exists a global minimizer $x^* \in K$ of f on K, that is,*

$$f(x^*) \leq f(x) \quad \text{for all} \quad x \in K.$$

Proof. Let $\{x_k\}$ in K be a minimizing sequence for f, that is, $f(x_k) \to \inf\{f(x) : x \in K\} =: f^*$, where we may have $f^* = -\infty$. Since K is compact, there exists a subsequence $\{x_{k_i}\}$ converging to $x^* \in K$. Since f is continuous, we have $f(x^*) = \lim_{i\to\infty} f(x_{k_i}) = f^* \in \mathbb{R}$, and thus the point x^* is a global minimizer of f on K. □

An alternative proof runs as follows:

Proof. Define $K_n := \{x \in K : f(x) > n\}$. Then K_n is open, and $K = \cup_{n=-\infty}^{\infty} K_n$, that is, $\{K_n\}_{n=-\infty}^{\infty}$ is an open cover of K. Since K is compact, a finite subset $\{K_{n_i}\}_{i=1}^{k}$ also covers K, that is, $K = \cup_{i=1}^{k} K_{n_i}$. Then $K = K_n$, where $n := \min\{n_i : i = 1, \dots, k\}$, and $f^* := \inf\{f(x) : x \in K\} > -\infty$. Thus, f is bounded from below on K.

Suppose that f does not have a global minimizer on K. Define $F_n := \{x \in K : f(x) > f^* + 1/n\}$. Then F_n is an open subset of K and $K = \cup_{n=1}^{\infty} F_n$. As above, we have $K = F_n$ for some $n > 1$, that is, $f(x) > f^* + 1/n$ for all $x \in K$, a contradiction to the definition of f^*. □

We remark that the second proof is more general, since it is valid verbatim on all compact topological spaces, not only compact metric spaces.

The compactness assumption can be relaxed somewhat.

Theorem 2.3. *Let $f : E \to \mathbb{R}$ be a continuous function defined on a metric space E. If f has a nonempty, compact sublevel set $\{x \in E : f(x) \leq \alpha\}$, then f achieves a global minimizer on E.*

Proof. Let $\{x_n\}$ be a minimizing sequence for f, that is,

$$f(x_n) \to \inf\{f(x) : x \in E\} = \inf_E f =: f^*.$$

Denote by D the sublevel set above, that is, $D = \{x \in E : f(x) \leq \alpha\}$. Clearly, there exists N such that $x_n \in D$ for all $n \geq N$. Since D is compact, $\{x_n\}_N^{\infty}$ has a convergent subsequence $x_{n_k} \to x^* \in D$. Since f is continuous, we have

$$f(x^*) = \lim_{n\to\infty} f(x_n) = f^*.$$

This means that f achieves its minimum on E at the point x^*. □

Definition 2.4. *A function $f : D \to \mathbb{R}$ on a subset D of a normed vector space E is called* coercive *if*

$$f(x) \to \infty \quad \textit{as } \|x\| \to \infty.$$

Corollary 2.5. *If $f : D \to \mathbb{R}$ is a continuous coercive function defined on a closed set $D \subseteq \mathbb{R}^n$, then f achieves a global minimum on D.*

Proof. The sublevel sets $l_\alpha(f) = \{x \in D : f(x) \leq \alpha\}$ are closed, since f is continuous, and bounded since f is coercive. Thus, f achieves its minimum on L at a point x^*, which is also a global minimizer of f on D. □

Example 2.6. (***The fundamental theorem of algebra***)

This famous theorem states that every polynomial

$$p(z) := a_n z^n + a_{n-1} z^{-1} + \cdots + a_1 z + a_0,$$

with leading coefficient $a_n \neq 0$ and where the coefficients a_i are complex numbers, has a complex root, hence n complex roots counting multiplicities. The problem has a fascinating history, and it is generally agreed that the first rigorous proof of it was given by the great mathematician Gauss in 1797, when he was just 20 years old, and appeared in his doctoral thesis of 1799. Here, we give an elementary proof of this result. This very short proof from [253] uses optimization techniques, but the essential idea is already in Fefferman [92], and probably in earlier works.

Consider minimizing the function

$$f(z) = |p(z)|$$

over the complex numbers. We have

$$|p(z)| = |z|^n \cdot \left| a_n + \frac{a_{n-1}}{z} + \frac{a_{n-2}}{z^2} + \cdots + \frac{a_1}{z^{n-1}} + \frac{a_0}{z^n} \right|.$$

As $|z| \to \infty$, the norm of the sum above converges to $|a_n| > 0$. Thus, $f(z)$ is a coercive function, and so has a minimizer z^* in $\mathbb{C}$.

Without loss of any generality, we may assume that $z^* = 0$; otherwise, we can consider the polynomial $q(z) = p(z + z^*)$. We have

$$|a_0| = f(0) \leq f(z) = \left| \sum_{k=0}^{n} a_k z^k \right|, \quad z \in \mathbb{C}.$$

If $a_0 = 0$, $z = 0$ is a root of p, and we are done. We claim that in fact, $a_0 = 0$. Suppose $a_0 \neq 0$ and let

$$p(z) = a_0 + a_k z^k + z^{k+1} q(z),$$

where $a_k \neq 0$ is the first nonzero coefficient after a_0 and q is a polynomial. Choose a kth root $w \in \mathbb{C}$ of $-a_0/a_k$. Then

$$p(tw) = a_0 + a_k t^k w^k + t^{k+1} w^{k+1} q(tw) = (1 - t^k) a_0 + t^k [t w^{k+1} q(tw)].$$

If $0 < t < 1$ is small enough, then $t|w^{k+1} q(tw)| < |a_0|$, and

$$|p(tw)| < (1 - t^k)|a_0| + t^k |a_0| = |a_0|,$$

a contradiction.

2.2 First-Order Optimality Conditions

Theorem 2.7. (***First-order necessary condition for a local optimizer***) *Let $f : U \to \mathbb{R}$ be a Gâteaux differentiable function on an open set $U \subseteq \mathbb{R}^n$. A local optimizer is a critical point, that is,*

$$x \text{ a local optimizer} \quad \Longrightarrow \quad \nabla f(x) = 0.$$

Clearly, the theorem holds verbatim if $U \subseteq \mathbb{R}^n$ is an arbitrary set with a nonempty interior, f is Gâteaux differentiable on $\operatorname{int} U$, and $x \in \operatorname{int} U$. We will not always point out such obvious facts in the interest of not complicating the statements of our theorems.

Proof. We first assume that x is a local minimizer of f. If $d \in \mathbb{R}^n$, then

$$f'(x; d) = \lim_{t \to 0} \frac{f(x + td) - f(x)}{t} = \langle \nabla f(x), d \rangle.$$

If $|t|$ is small, then the numerator above is nonnegative, since x is a local minimizer. If $t > 0$, then the difference quotient is nonnegative, so in the limit as $t \searrow 0$, we have $f'(x; d) \geq 0$. However, if $t < 0$, the difference quotient is nonpositive, and we have $f'(x; d) \leq 0$. Thus, we conclude that $f'(x; d) = \langle \nabla f(x), d \rangle = 0$. If x is a local maximizer of f, then $\langle \nabla f(x), d \rangle = 0$, since x is a local minimizer of $-f$. Picking $d = \nabla f(x)$ gives $f'(x; d) = \|\nabla f(x)\|^2 = 0$, that is, $\nabla f(x) = 0$. □

We note that Theorem 2.7 proves the following more general result.

Corollary 2.8. *Let $f : U \to \mathbb{R}$ be a function on an open set $U \subseteq \mathbb{R}^n$. If $x \in U$ is a local minimizer of f and the directional derivative $f'(x; d)$ exists for a direction $d \in \mathbb{R}^n$, then $f'(x; d) \geq 0$.*

Remark 2.9. Functions that have directional derivatives but are not necessarily differentiable occur naturally in optimization, for example in minimizing a function that is the pointwise maximum of a set of differentiable functions. See Danskin's theorem, Theorem 1.29, on page 20.

In fact, it is possible use this approach to derive optimality conditions for constrained optimization problems. See Section 12.1 for the derivation of optimality conditions in semi-infinite programming.

Example 2.10. Here is an optimization problem from the theory of *orthogonal polynomials*; see [250], whose solution is obtained using a novel technique, a differential equation.

We determine the minimizers and the minimum value of the function

$$f(x_1, \ldots, x_n) = \frac{1}{2}\sum_1^n x_j^2 - \sum_{1 \le i < j \le n} \ln|x_i - x_j|.$$

Differentiate f with respect to each variable x_j and set to zero to obtain

$$\frac{\partial f}{\partial x_j} = x_j - \sum_{i \neq j} \frac{1}{x_j - x_i} = 0.$$

To solve for x, consider the polynomial

$$g(x) = \prod_1^n (x - x_j),$$

which has roots at the point $x = x_1, \ldots, x_n$. Differentiating this function gives

$$g'(x_j) = \prod_{i \neq j}(x_j - x_i), \quad \frac{g''(x_j)}{g'(x_j)} = 2\prod_{i \neq j}\frac{1}{x_j - x_i},$$

so that $\partial f / \partial x_j = 0$ can be written as

$$g''(x_j) - 2x_j g'(x_j) = 0,$$

meaning that the polynomial

$$g''(x) - 2xg'(x)$$

of order n has the same roots as the polynomial $g(x)$, so must be proportional to $g(x)$. Comparing the coefficients of x^n gives

$$g''(x) - 2xg'(x) + 2ng(x) = 0.$$

The solution to this differential equation is the *Hermite polynomial* of order n,

$$H_n(x) = n! \sum_0^{[n/2]} \frac{(-1)^k (2x)^{n-2k}}{k!(n-2k)!}.$$

Therefore, the solutions x_j are the roots of the Hermite polynomial $H_n(x)$. The discriminant of H_n is given by

$$\prod_{i<j}(x_i - x_j)^2 = 2^{-(n(n-1)/2} \prod_1^n j^j,$$

and the above formula for H_n gives

$$\sum_1^n x_j^2 = n(n-1)/2.$$

Thus, the minimum value of f is

$$\frac{1}{4}n(n-1)(1+\ln 2) - \frac{1}{2}\sum_1^n j\ln j.$$

2.3 Second-Order Optimality Conditions

Definition 2.11. *An $n \times n$ matrix A is called* positive semidefinite *if*

$$\langle Ad, d\rangle \geq 0 \quad \textit{for all} \quad d \in \mathbb{R}^n.$$

It is called positive definite *if*

$$\langle Ad, d\rangle > 0 \quad \textit{for all} \quad d \in \mathbb{R}^n,\ d \neq 0.$$

Note that if A is positive semidefinite, then $a_{ii} = \langle Ae_i, e_i\rangle \geq 0$, and if A is positive definite, then $a_{ii} > 0$. Similarly, choosing $d = te_i + e_j$ gives $q(t) := a_{ii}t^2 + 2a_{ij}t + a_{jj} \geq 0$ for all $t \in \mathbb{R}$. Recall that the quadratic function $q(t)$ is nonnegative (positive) if and only if its discriminant $\Delta = 4(a_{ij}^2 - a_{ii}a_{jj})$ is nonpositive (negative). Thus, $a_{ii}a_{jj} - a_{ij}^2 \geq 0$ if A is positive semidefinite, and $a_{ii}a_{jj} - a_{ij}^2 > 0$ if A is positive definite.

Theorem 2.12. (***Second-order necessary condition for a local minimizer***) *Let $f : U \to \mathbb{R}$ be twice Gâteaux differentiable on an open set $U \subseteq \mathbb{R}^n$ in the sense that there exist a vector $\nabla f(x)$ and a symmetric matrix $Hf(x)$ such that for all $h \in \mathbb{R}^n$,*

$$f(x+th) = f(x) + t\langle \nabla f(x), h\rangle + \frac{t^2}{2}\langle Hf(x)h, h\rangle + o(t^2). \tag{2.1}$$

(This condition is satisfied if f has continuous second-order partial derivatives, that is, if $f \in C^2$.)

If $x \in U$ is a local minimizer of f, then the matrix $Hf(x)$ is positive semidefinite.

Proof. The first-order necessary condition implies $\nabla f(x) = 0$. Since x is a local minimizer, we have $f(x+th) \geq f(x)$ if $|t|$ is small enough. Then, (2.1) gives

$$\frac{t^2}{2}\langle Hf(x)h, h\rangle + o(t^2) \geq 0.$$

Dividing by t^2 and letting $t \to 0$ gives

$$h^T Hf(x)h \geq 0 \quad \text{for all} \quad h \in \mathbb{R}^n,$$

proving that $Hf(x)$ is positive semidefinite. □

We remark that the converse does not hold; see Exercise 9 on page 56. However, we have the following theorem.

Theorem 2.13. (***Second-order sufficient condition for a local minimizer***) *Let $f : U \to \mathbb{R}$ be C^2 on an open set $U \subseteq \mathbb{R}^n$. If $x \in U$ is a critical point and $Hf(x)$ is positive definite, then x is a strict local minimizer of f on U.*

Proof. Define $A := Hf(x)$. Since $g(d) := \langle Ad, d\rangle > 0$ for all d on the unit sphere $S := \{d \in \mathbb{R}^n : \|d\| = 1\}$ and S is compact, it follows that there exists $\alpha > 0$ such that $g(d) \geq \alpha > 0$ for all $d \in S$. Since g is homogeneous, we have $g(d) \geq \alpha \|d\|^2$ for all $d \in \mathbb{R}^n$.

Let $\|d\|$ be sufficiently small. It follows from the multivariate Taylor's formula (Corollary 1.24) and the fact $\nabla f(x) = 0$ that

$$\begin{aligned} f(x+d) &= f(x) + \langle \nabla f(x), d\rangle + \frac{1}{2}\langle Ad, d\rangle + o(\|d\|^2) \\ &\geq f(x) + \|d\|^2 \left(\frac{\alpha}{2} + \frac{o(\|d\|^2)}{\|d\|^2}\right) \\ &> f(x). \end{aligned}$$

This proves that x is a strict local minimizer of f. □

The positive definiteness condition on A is really needed. Exercise 9 describes a problem in which a critical point x has $Hf(x)$ positive semidefinite, but x is actually a saddle point.

However, a global positive semidefiniteness condition on $Hf(x)$ has strong implications.

Theorem 2.14. (***Second-order sufficient condition for a global minimizer***) *Let $f : U \to \mathbb{R}$ be a function with positive semidefinite Hessian on an open convex set $U \subseteq \mathbb{R}^n$. If $x \in U$ is a critical point, then x is a global minimizer of f on U.*

Proof. Let $y \in U$. It follows from the multivariate Taylor's formula (Theorem 1.23) that there exists a point $z \in (x, y)$ such that

$$f(y) = f(x) + \langle \nabla f(x), y - x\rangle + \frac{1}{2}(y-x)^T Hf(z)(y-x).$$

Since $\nabla f(x) = 0$ and $Hf(z)$ is positive semidefinite, we have $f(y) \geq f(x)$ for all $y \in D$. Thus, x is a global minimizer of f on U. □

Remark 2.15. We remark that a function with a positive semidefinite Hessian is a convex function. If the Hessian is positive definite at every point, then the function is strictly convex. In this case, the function f has at most one critical point, which is the unique global minimizer. Chapter 4 treats convex (not necessarily differentiable) functions in detail.

Theorem 2.16. (***Second-order sufficient condition for a saddle point***) *Let $f : U \to \mathbb{R}$ be twice Gâteaux differentiable on an open set $U \subseteq \mathbb{R}^n$ in the sense of* (2.1)*. If $x \in U$ is a critical point and $Hf(x)$ is indefinite, that is, it has at least one positive and one negative eigenvalue, then x is a saddle point of f on U.*

Proof. Define $A := Hf(x)$. If $\lambda > 0$ is an eigenvalue of A with a corresponding eigenvector $d \in \mathbb{R}^n$, $\|d\| = 1$, then $\langle Ad, d\rangle = \langle \lambda d, d\rangle = \lambda$, and it follows from Corollary 1.24 that for sufficiently small $t > 0$,

$$\begin{aligned} f(x+td) &= f(x) + t\langle \nabla f(x), d\rangle + \frac{t^2}{2}\langle Ad, d\rangle + o(t^2) \\ &= f(x) + \frac{t^2}{2}\lambda + o(t^2) > f(x). \end{aligned}$$

Similarly, if $\lambda < 0$ is an eigenvalue of A with a corresponding eigenvector d, $\|d\| = 1$, then $f(x+td) < f(x)$ for small enough $t > 0$. This proves that x is a saddle point. □

Definition 2.17. *Let $f : U \to \mathbb{R}$ be a C^2 function on an open set $U \subseteq \mathbb{R}^n$. A critical point $x \in U$ is called* nondegenerate *if the Hessian matrix $D^2 f(x)$ is nonsingular.*

A well-known result, Morse's lemma [202], states that if x is a nondegenerate critical point, then the Hessian $Df(x_0)$ *determines* the behavior of f around x_0. More precisely, it states that if $f : U \to \mathbb{R}$ is at least C^{2+k} $(k \geq 1)$ on an open set $U \subseteq \mathbb{R}^n$, and if $x_0 \in U$ is a nondegenerate critical point of f, then there exist open neighborhoods $V \ni x_0$ and $W \ni 0$ in $\mathbb{R}^n$ and a one-to-one and onto C^k map $\varphi : V \to W$ such that

$$f(x) = f(x_0) + \frac{1}{2}\langle D^2 f(x_0)\varphi(x), \varphi(x)\rangle.$$

This is the content of Theorem 2.32 on page 49. See also Corollary 2.33.

We end this section by noting that the second-order tests considered above, and especially Morse's lemma, give conclusive information about a critical point except when the Hessian matrix is degenerate. In these degenerate cases, nothing can be deduced about the critical point in general: it could be a local minimizer, local maximizer, or a saddle point. For example, the origin $(x, y) = (0, 0)$ is a critical point of the function $f(x, y) = x^3 - 3xy^2$ (the real part of the complex function $(x + iy)^3$), with $D^2 f(0, 0) = 0$. It is a saddle point, and the graph of this function is called a *monkey saddle*. A computer plot of the graph of f will reveal that this saddle is different from the familiar horse saddle in that there is also a third depression for the tail of the monkey.

Example 2.18. Consider the family of problems

$$\min f(x, y) := x^2 + y^2 + \beta xy + x + 2y.$$

We have

$$\nabla f(x,y) = \begin{pmatrix} 2x + \beta y + 1 \\ 2y + \beta x + 2 \end{pmatrix}, \quad Hf(x,y) = \begin{bmatrix} 2 & \beta \\ \beta & 2 \end{bmatrix}.$$

We have $\nabla f(x,y) = 0$ if and only if

$$\begin{aligned} 2x + \beta y &= -1, \\ \beta x + 2y &= -2. \end{aligned}$$

If $\beta \neq \mp 2$, then the unique solution to the above equations is $(x^*, y^*) = (2\beta - 2, \beta - 4)/(4 - \beta^2)$. If $\beta = 2$, the above equations become $2x + 2y = -1$ and $2x + 2y = -2$, thus inconsistent. Similarly, if $\beta = -2$, we also have an inconsistent system of equations. Therefore, no critical points exist for $\beta = \mp 2$.

The eigenvalues of $A := Hf(x,y)$ can be calculated explicitly: the characteristic polynomial of A is

$$\det(A - \lambda I) = (2 - \lambda)^2 - \beta^2 = 0,$$

which has solutions $\lambda = 2 \mp \beta$. These are the eigenvalues of A. Thus, the eigenvalues of A are positive for $-2 < \beta < 2$. In this case, the optimal solution (x^*, y^*) calculated above is a global minimizer of f by Theorem 2.13 and Corollary 2.20 below. In the case $|\beta| > 2$, one eigenvalue of A is positive and the other negative, so that the corresponding optimal solution $z^* := (x^*, y^*)$ is a saddle point by Theorem 2.16.

Finally, let us consider the behavior of f when $\beta = \mp 2$, when it has no critical point. If $\beta = 2$, then $f(x,y) = (x+y)^2 + x + 2y$; thus $f(x,-x) = -x$ and $f(x,-x) \to \mp\infty$ as $x \to \pm\infty$. When $\beta = -2$, f has a similar behavior.

2.4 Quadratic Forms

We have seen that symmetric positive semidefinite and positive definite matrices are important in the second-order optimality conditions for a local minimizer. In this section, we give characterizations of such matrices.

We recall the *spectral decomposition* or *orthogonal diagonalization* of symmetric matrices.

Theorem 2.19. (Spectral decomposition of a symmetric matrix) *Let A be an $n \times n$ real symmetric matrix. There exist a real diagonal matrix $\Lambda = \mathrm{diag}(\lambda_1, \ldots, \lambda_n)$ and a real orthogonal matrix $U = [u_1, \ldots, u_n]$ such that*

$$A = U \Lambda U^T.$$

The scalar λ_i is an eigenvalue of A, and u_i is an eigenvector of A corresponding to λ_i.

Proof. It is well known from linear algebra that A has n real eigenvalues $\{\lambda_i\}_1^n$ with corresponding eigenvectors $\{u_i\}_1^n$, $\|u_i\| = 1$, which are mutually orthogonal, that is, $\langle u_i, u_j \rangle = 0$ for $i \neq j$. From $Au_i = \lambda_i u_i$, we obtain

$$AU = A[u_1, \ldots, u_n] = [Au_1, \ldots, Au_n] = [\lambda_1 u_1, \ldots, \lambda_n u_n] = U\Lambda,$$

where $U = [u_1, \ldots, u_n]$ and $\Lambda = \mathrm{diag}(\lambda_1, \ldots, \lambda_n)$. Since the eigenvalues are orthogonal, we have

$$U^T U = \begin{bmatrix} u_1^T \\ \vdots \\ u_n^T \end{bmatrix} [u_1, \ldots, u_n] = [\langle u_i, u_j \rangle] = I,$$

that is, U is an orthogonal matrix with inverse U^T. It follows that $A = A(UU^T) = (AU)U^T = U\Lambda U^T$. □

In Section 10.1 (page 251), we will give an optimization proof of this theorem. This approach provides a variational characterization of the eigenvalues, which has many applications.

Corollary 2.20. *Let A be an $n \times n$ symmetric matrix. Then A is positive semidefinite if and only if all eigenvalues of A are nonnegative. Moreover, A is positive definite if and only if all eigenvalues of A are positive.*

Proof. We have

$$d^T Ad = d^T U\Lambda U^T d = (U^T d)^T \Lambda (U^T d).$$

Since U is nonsingular, we see that $d^T Ad \geq 0$ for all $d \in \mathbb{R}^n$ if and only if $d^T \Lambda d \geq 0$ for all $d \in \mathbb{R}^n$. In other words, A is positive semidefinite if and only if Λ is. Since

$$d^T \Lambda d = \sum_{i=1}^{n} \lambda_i d_i^2,$$

Λ is positive semidefinite if and only $\lambda_i \geq 0$ for each $i = 1, \ldots, n$. This proves the first part of the theorem. The proof of the second part is similar. □

Although this result characterizes the symmetric positive semidefinite and positive definite matrices, the determination of the eigenvalues of A is not an easy computational task unless n is small. However, here we are interested only in the *signs* of the eigenvalues and not their exact numerical values.

It is also possible to simultaneously "diagonalize" two symmetric matrices, provided one of them is positive definite. This result is frequently useful in optimization. For example, it may be used to give a quick proof of the fact that the function $F(X) = -\ln \det X$ is convex on the cone of positive definite matrices.

Theorem 2.21. *Let A and B be symmetric $n \times n$ matrices such that at least one of the matrices is positive definite. The matrices can be* simultaneously diagonalized *in the sense that there exists a nonsingular matrix $X \in \mathbb{R}^{n\times n}$ such that*

$$X^T AX = \mathrm{diag}\{\lambda_1, \ldots, \lambda_n\}, \quad X^T BX = \mathrm{diag}\{\delta_1, \ldots, \delta_n\}.$$

We remark that the notion of diagonalization in this theorem is different from its usual definition linear algebra where diagonalizing a square matrix A means finding an invertible matrix X such that $X^{-1}AX$ is a diagonal matrix. The diagonalization above is more appropriate quadratic forms, because substituting $x = Xy$ in the quadratic from $q_1(x) = \langle Ax, x\rangle$ gives the quadratic form $q_2(y) := q_1(Xx) = \langle Cy, y\rangle$ where $C = X^T AX$.

Proof. Suppose that B is positive definite. Then B has the spectral decomposition $U^T BU = D$, where $U \in \mathbb{R}^{n\times n}$ is orthogonal and $D = \mathrm{diag}\{d_1, \ldots, d_n\}$ is a diagonal matrix with all $d_i > 0$. Define the square root of B,

$$C := U\, \mathrm{diag}\{\sqrt{d_1}, \ldots, \sqrt{d_n}\}\, U^T.$$

Note that $C^{-1}BC^{-1} = I$. Now $\overline{A} := C^{-1}AC^{-1}$ has the spectral decomposition $V^T\overline{A}V = \Lambda = \mathrm{diag}\{\lambda_1, \ldots, \lambda_n\}$. Setting $X = C^{-1}V$, we see that

$$X^T AX = \Lambda, \quad X^T BX = V^T C^{-1}BC^{-1}V = V^T V = I,$$

completing the proof. □

2.4.1 Counting Roots of Polynomials in Intervals

The number of positive (and negative) eigenvalues can be counted by a simple rule dating back to Descartes in seventeenth century.

Definition 2.22. *Let $a_0, a_1, \ldots, a_n$ be a sequence of real numbers. If all the numbers in the sequence are nonzero, the total number of variations of sign in the sequence, denoted by $V(a_0, a_1, \ldots, a_n)$, is the number of times consecutive numbers a_{k-1} and a_k differ in sign, that is,*

$$V(a_0, a_1, \ldots, a_n) := |\{k : a_{k-1}a_k < 0, k = 1, \ldots, n\}|.$$

If the sequence $a_0, a_1, \ldots, a_n$ contains zeros, then $V(a_0, a_1, \ldots, a_n)$ is defined to be the variations of the reduced sequence by ignoring all zero elements in the sequence. Also, we define $V(a_0) = 0$ for any $a_0 \in \mathbb{R}$.

For example, $V(1, 0, 0, -3, 2, 0, 1, -7, 3) = V(1, -3, 2, 1, -7, 3) = 4$.

Theorem 2.23. **(*Descartes's rule of sign*)** *Let $p(x) = a_0 + a_1x + a_2x^2 + \cdots + a_nx^n$ be a polynomial of degree n with real coefficients. Then the number of positive roots $N_p(0, \infty)$ of p is given by*

$$N_p(0,\infty) = V(a_0, a_1, \ldots, a_n) - 2\kappa$$

for some nonnegative integer κ.

Moreover, if the roots of p *are all real, then* $\kappa = 0$, *that is,*

$$N_p(0,\infty) = V(a_0, a_1, \ldots, a_n).$$

A simple proof of the theorem is given in Appendix B.

Corollary 2.24. *Let* $A_{n\times n}$ *be a symmetric matrix and let* $p(\lambda) = \det(\lambda I - A) = a_0 + a_1\lambda + \cdots + a_n\lambda^n$ *be the characteristic polynomial of* A. *The number of positive eigenvalues of* A *is given by*

$$N_p(0,\infty) = V(a_0, a_1, \ldots, a_n),$$

and the number of negative eigenvalues by

$$N_p(-\infty,0) = V(a_0, -a_1, a_2, \ldots, (-1)^n a_n).$$

Proof. The characteristic polynomial has only real roots, these being the eigenvalues of A. This proves the first equality. The second equality follows by considering the polynomial $q(\lambda) = -p(\lambda)$ and noting that the k coefficient of q is $(-1)^k a_k$. □

Alternatively, $N_p(-\infty,0)$ can be computed by noting that the positive, negative, and zero eigenvalues (counted according to its multiplicity) of A add up to n.

2.4.2 Sylvester's Theorem

There is also a remarkable determinant test due to Sylvester to recognize a symmetric positive definite matrix. We first need to introduce some concepts.

Let A be an $n \times n$ symmetric matrix. The submatrix

$$A_k := \begin{bmatrix} a_{11} & \ldots & a_{1k} \\ \vdots & & \vdots \\ a_{k1} & \ldots & a_{kk} \end{bmatrix}$$

consisting of the first k rows and columns of A is called the kth *leading principal submatrix* of A, and its determinant $\det A_k$ is called the kth *leading principal minor* of A.

Theorem 2.25. (*Sylvester*) *Let* A *be an* $n \times n$ *symmetric matrix. Then* A *is positive definite if and only if all the leading principal minors of* A *are positive, that is,* A *is positive definite if and only if* $\det A_i > 0$, $i = 1, \ldots, n$.

Proof. We first prove that if A is positive definite, then all leading principal minors of A are positive. We use induction on n, the dimension of A. The proof is trivial for $n = 1$. Assuming that the result is true for n, we will prove it for $n+1$. Let A be an $(n+1) \times (n+1)$ symmetric, positive definite matrix. We write

$$A = \begin{bmatrix} B & b \\ b^T & c \end{bmatrix},$$

where B is a symmetric $n \times n$ matrix, $b \in \mathbb{R}^n$, and $c \in \mathbb{R}$. Choosing $0 \neq d \in \mathbb{R}^n$, we have

$$0 < (d^T, 0) A \begin{pmatrix} d \\ 0 \end{pmatrix} = (d^T, 0) \begin{bmatrix} B & b \\ b^T & c \end{bmatrix} \begin{pmatrix} d \\ 0 \end{pmatrix} = d^T B d,$$

that is, B is positive definite. By the induction hypothesis, we have $\det A_i > 0$, $i = 1, \ldots, n$. Since A is positive definite, its eigenvalues $\{\lambda_i\}_{i=1}^{n+1}$ are all positive. Thus, we also have $\det A_{n+1} = \det A = \lambda_1 \cdots \lambda_{n+1} > 0$.

Conversely, let us prove that if all $\det A_i > 0$, $i = 1, \ldots, n+1$, then A is positive definite. The proof is again by induction on n. The proof is trivial for $n = 1$. Suppose the theorem is true for n; we will prove it for $n+1$.

Since $\det A_i > 0$ for $i = 1, \ldots, n$ we see by the induction hypothesis that B is positive definite. Suppose A is *not* positive definite. Then $\lambda_{n+1} < 0$, and since $\det A = \lambda_1 \cdots \lambda_{n+1} > 0$, we must also have $\lambda_n < 0$. Let u_n and u_{n+1} be the eigenvectors of A corresponding to λ_n and λ_{n+1}, respectively. We have $\langle u_n, u_{n+1} \rangle = 0$, so that we can choose scalars α_n and α_{n+1} such that $u = \alpha_n u_n + \alpha_{n+1} u_{n+1}$ is not zero but has the last $((n+1)\text{th})$ component equal to zero, say $u = (v, 0)^T$ where $v \neq 0$. Then $u^T A u = v^T B v > 0$, since B is positive definite. However, we also have

$$\begin{aligned} 0 < u^T A u &= \langle \alpha_n u_n + \alpha_{n+1} u_{n+1}, A(\alpha_n u_n + \alpha_{n+1} u_{n+1}) \rangle \\ &= \langle \alpha_n u_n + \alpha_{n+1} u_{n+1}, \lambda_n \alpha_n u_n + \lambda_{n+1} \alpha_{n+1} u_{n+1} \rangle \\ &= \lambda_n \alpha_n^2 \langle u_n, u_n \rangle + \lambda_{n+1} \alpha_{n+1}^2 \langle u_{n+1}, u_{n+1} \rangle < 0, \end{aligned}$$

where the last inequality follows from the facts $\lambda_i < 0$ and $\|u_i\| = 1$, $i = n, n+1$. This contradiction shows that all eigenvalues of A are positive. Corollary 2.20 implies that A is positive definite. □

This simple proof is taken from Carathéodory [54], p. 187.

Another elegant proof of Sylvester's theorem, more in the spirit of optimization techniques, is outlined in Exercise 12 at the end of the chapter.

2.5 The Inverse Function, Implicit Function, and Lyusternik Theorems in Finite Dimensions

In this section, we first give an elementary proof of the implicit function theorem in finite-dimensional vector spaces. This proof has a variational flavor, and is used to prove the inverse function theorem and Lyusternik's theorem. The implicit function theorem will also be utilized to prove Morse's lemma in Section 2.6.

Theorem 2.26. (***Implicit function theorem***) *Let $f : U \times V \to \mathbb{R}^m$ be a C^1 mapping, where $U \subseteq \mathbb{R}^n$ and $V \subseteq \mathbb{R}^m$ are open sets. Let $(x_0, y_0) \in U \times V$ be a point such that $f(x_0, y_0) = 0$ and $D_y f(x_0, y_0) : \mathbb{R}^m \to \mathbb{R}^m$, the derivative of f with respect to y, is nonsingular.*

Then there exist neighborhoods $U_1 \ni x_0$ and $V_1 \ni y_0$ and a C^1 mapping $y : U_1 \to V_1$ such that a point $(x, y) \in U_1 \times V_1$ satisfies $f(x, y) = 0$ if and only if $y = y(x)$. The derivative of y at x_0 is given by

$$Dy(x_0) = -D_y f(x_0, y_0)^{-1} D_x f(x_0, y_0).$$

Moreover, if f is k-times continuously differentiable, that is, $f \in C^k$, then $y(x) \in C^k$.

The linear case should help one to remember the form of the implicit function theorem: if $f(x, y) = Ax + By$ and $D_y f = B$ is an invertible matrix, then the equation $f(x, y) = \alpha$ gives $Ax + By = \alpha$. This may be solved for y by premultiplying it by B^{-1}, giving $y(x) = B^{-1}(\alpha - Ax)$.

Proof. Assume without loss of generality that $x_0 = 0$ and $y_0 = 0$, by considering the function $(x, y) \mapsto f(x + x_0, y + y_0) - f(x_0, y_0)$ if necessary. Let $f(x) = (f_1(x, y), \ldots, f_m(x, y))$, where f_i is the ith coordinate function of f. Since Df is continuous, there exist neighborhoods U_0 and V_0 of the origin in $\mathbb{R}^n$ and $\mathbb{R}^m$, respectively, such that the matrix

$$\begin{bmatrix} \nabla_y f_1(x, y_1)^T \\ \nabla_y f_2(x, y_2)^T \\ \vdots \\ \nabla_y f_m(x, y_m)^T \end{bmatrix} \tag{2.2}$$

is invertible for all $(x, y_i) \in U_0 \times V_0$.

We claim that for every $x \in U_0$, there exists at most one $y \in V_0$ such that $f(x, y) = 0$. Otherwise, there would exist $y, z \in V_0$, $y \neq z$, such that $f(x, y) = f(x, z) = 0$. The mean value theorem (Lemma 1.12) implies that there exists $y_i \in (y, z)$ such that

$$f_i(x, z) - f_i(x, y) = \langle \nabla_y f_i(x, y_i), z - y \rangle = 0, \quad i = 1, \ldots, m.$$

Since the matrix in (2.2) is nonsingular, we obtain $y = z$, a contradiction that proves our claim.

Let $\overline{B}_r(0) \subseteq V_0$. Since $f(0, 0) = 0$, we have $f(0, y) \neq 0$ for $y \in S_r(0) := \{y \in \mathbb{R}^l : \|y\| = r\}$, and since f is continuous on $U_0 \times V_0$, there exists $\alpha > 0$ such that $\|f(0, y)\| \geq \alpha$ for all $y \in S_r(0)$. It follows that the function

$$F(x,y) := \|f(x,y)\|^2 = \sum_{i=1}^{m} f_i(x,y)^2$$

satisfies the properties

$$F(0,y) \geq \alpha > 0 \ \text{ for } \ y \in S_r(0) \ \text{ and } \ F(0,0) = 0.$$

Since F is continuous, there exists an open neighborhood $U_1 \subseteq U_0$ of $0 \in \mathbb{R}^n$ such that

$$F(x,y) \geq \frac{\alpha}{2}, \ \ F(x,0) \leq \frac{\alpha}{2} \ \text{ for all } \ x \in U_1, \ y \in S_r(0).$$

Thus, for a fixed $x \in U_1$, the function $y \mapsto F(x,y)$ achieves its minimum on $\overline{B}_r(0)$ at a point $y(x)$ in the interior of $\overline{B}_r(0)$, and we have

$$D_y F(x,y(x)) = 2D_y f(x,y(x)) f(x,y(x)) = 0,$$

and since the matrix $D_y f(x,y(x))$ is nonsingular, we conclude that

$$f(x,y(x)) = 0.$$

Writing $\Delta y := y(x+\Delta x) - y(x)$, we have by the mean value theorem

$$0 = D_x f(\tilde{x},\tilde{y})\Delta x + D_y f(\tilde{x},\tilde{y})\Delta y$$

for some point $(\tilde{x},\tilde{y})$ on the line segment between $(x,y(x))$ and $(x+\Delta x, y(x+\Delta x))$. This implies that as $\|\Delta x\|$ goes to zero, so does $\|\Delta y\|$, proving that $y(x)$ is a continuous function.

The function $y(x)$ is actually C^1, since by Taylor's formula

$$\begin{aligned} 0 &= f(x+\Delta x, y(x+\Delta x)) - f(x,y(x)) \\ &= D_x f(x,y(x))\Delta x + D_y f(x,y(x))\Delta y + o((\Delta x, \Delta y)), \end{aligned}$$

and since $o((\Delta x, \Delta y)) = o(\Delta x)$ by the continuity of $y(x)$, we have

$$\Delta y = -D_y^{-1} f(x,y(x)) D_x f(x,y(x))\Delta x + o(\Delta x).$$

This proves that $y(x)$ is Fréchet differentiable at x with

$$Dy(x) = -D_y^{-1} f(x,y(x)) D_x f(x,y(x)).$$

If $f \in C^2$, then $D_y^{-1} f(x,y(x)) = \operatorname{Adj} D_y f(x,y(x)) / \det D_y f(x,y(x))$ and $D_x f(x,y(x))$ are C^1, and the above formula shows that the function $y(x)$ is C^2. In general, if C^k, we prove by induction on k that $y(x)$ is C^k. □

This elementary proof is taken from Carathéodory [54], pp. 10–13. A similar kind of proof, using penalty functions, will used in Chapter 9 to obtain optimality conditions for constrained optimization problems.

Corollary 2.27. (***Inverse function theorem***) *Let f be a C^1 map from a neighborhood of $x_0 \in \mathbb{R}^n$ into $\mathbb{R}^n$.*

If $Df(x_0)$ is nonsingular, then there exist neighborhoods $U \ni x_0$ and $V \ni y_0 = f(x_0)$ such that $f : U \to V$ is a C^1 diffeomorphism, and

$$Df^{-1}(y) = Df(x)^{-1} \text{ for all } (x, y) \in U \times V, \; y = f(x).$$

Moreover, if f is C^k, then f is a C^k diffeomorphism on U.

Proof. Define the function $F(x, y) = f(x) - y$, and note that $D_x F(x_0, y) = Df(x_0)$ is nonsingular. Apply Theorem 2.26 to F. □

The map $f : \mathbb{R}^2 \to \mathbb{R}^2$ given by $f(x, y) = (e^x \cos y, e^x \sin y)$ has the Jacobian $\det Df(x, y)) = e^x \neq 0$, hence locally one-to-one around every point $(x, y) \in \mathbb{R}^2$. However, f is clearly not one-to-one globally.

Definition 2.28. *Let M be a nonempty subset of $\mathbb{R}^n$ and $x \in M$. A vector $d \in \mathbb{R}^n$ is called a* tangent direction *of M at x if there exist a sequence $x_n \in M$ converging to x and a nonnegative sequence α_n such that*

$$\lim_{n \to \infty} \alpha_n (x_n - x) = d.$$

The tangent cone *of M at x, denoted by $T_M(x)$, is the set of all tangent directions of M at x.*

This definition is sufficient for our purposes. We remark that the same definition is valid in a topological vector space. A detailed study of this and several related concepts is needed in nonsmooth analysis; see [230] and [199, 200].

Theorem 2.29. (***Lyusternik***) *Let $f : U \to \mathbb{R}^m$ be a C^1 map, where $U \subset \mathbb{R}^n$ is an open set. Let $M = f^{-1}(f(x_0))$ be the level set of a point $x_0 \in U$.*

If the derivative $Df(x_0)$ is a linear map onto *$\mathbb{R}^m$, then the tangent cone of M at x_0 is the null space of the linear map $Df(x_0)$, that is,*

$$T_M(x_0) = \{d \in \mathbb{R}^n : Df(x_0)d = 0\}.$$

Remark 2.30. Let $f = (f_1, \ldots, f_m)$, where $\{f_i\}$ are the components functions of f. It is easy to verify that

$$\operatorname{Ker} Df(x_0) = \{d \in \mathbb{R}^n : \langle \nabla f_i(x_0), d \rangle = 0, i = 1, \ldots, m\},$$

and that the surjectivity of $Df(x_0)$ is equivalent to the linear independence of the gradient vectors $\{\nabla f_i(x_0)\}_1^m$.

Proof. We may assume that $x_0 = 0$ and $f(x_0) = 0$, by considering the function $x \mapsto f(x + x_0) - f(x_0)$ if necessary. Define $A := Df(0)$. The proof of the inclusion $T_M(0) \subseteq \operatorname{Ker} A$ is easy: if $d \in T_M(0)$, then there exist points $x(t) = td + o(t) \in M$, and we have

$$0 = f(0 + td + o(t)) = f(0) + tDf(0)(d) + o(t) = tDf(0)(d) + o(t).$$

Dividing both sides by t and letting $t \to 0$, we obtain $Df(0)(d) = 0$.

The proof of the reverse inclusion $\operatorname{Ker} A \subseteq T_M(0)$ is based on the idea that the equation $f(x) = 0$ can be written as $f(y, z) = 0$ in a form that is suitable for applying the implicit function theorem.

Define $K := \operatorname{Ker} A$ and $L := K^{\perp}$. Since A is onto $\mathbb{R}^m$, we can identify K and L with $\mathbb{R}^{n-m}$ and $\mathbb{R}^m$, respectively, by introducing a suitable basis in $\mathbb{R}^n$. We write a point $x \in \mathbb{R}^n$ in the form $x = (y, z) \in K \times L$. We have $A = [D_y f(0), D_z f(0)]$, and

$$0 = A(K) = \{A(d_1, 0) : d_1 \in \mathbb{R}^{n-m}\} = D_y f(0)(\mathbb{R}^{n-m}),$$

so that $D_y f(0) = 0$. Since A has rank m, it follows that $D_z f(0)$ is nonsingular.

Theorem 2.26 implies that there exist neighborhoods $U_1 \subseteq \mathbb{R}^m$ and $U_2 \subseteq \mathbb{R}^{n-m}$ around the origin and a C^1 map $\alpha : U_1 \to U_2$, $\alpha(0) = 0$, such that $x = (y, z) \in U_1 \times U_2$ satisfies $f(x) = 0$ if and only if $z = \alpha(y)$. The equation $f(x) = 0$ can then be written as $f(y, \alpha(y)) = 0$. Differentiating this equation and using the chain rule, we obtain

$$0 = D_y f(y, \alpha(y)) + D_z f(y, \alpha(y)) D\alpha(y).$$

At the origin $x = 0$, $D_y f(0) = 0$, and $D_z f(0)$ nonsingular, so that $D\alpha(0) = 0$. If $|y|$ is small, we have

$$\alpha(y) = \alpha(0) + D\alpha(0)y + o(y) = o(y).$$

Let $d = (d_1, 0) \in K$. As $t \to 0$, the point $x(t) := (td_1, \alpha(td_1)) = (td_1, o(t))$ lies in M, that is, $f(x(t)) = 0$, and satisfies $(x(t) - td)/t = (0, o(t))/t \to 0$. This implies that $K \subseteq T_M(0)$, and the theorem is proved. □

2.6 Morse's Lemma

Let $f : U \to \mathbb{R}$ be a C^{2+k} $(k \geq 0)$ function on an open set $U \subseteq \mathbb{R}^n$. Recall that a critical point $x \in U$ is called *nondegenerate* if the Hessian matrix $D^2 f(x)$ is nonsingular. Morse's lemma, due originally to Morse [202], states that after a local, possibly nonlinear, change of coordinates, the function f is *identical* to its quadratic form $q(x) := f(x_0) + \langle D^2 f(x_0)(x - x_0), x - x_0 \rangle$. Thus, the quadratic function $q(x)$ determines the behavior of the function f around x_0.

Morse's original proof uses the Gram–Schmidt process. A modern version of the proof can be found in Milnor [197]. The simple proof below is from [6]. It has the virtue that the same proof, with obvious modifications, works in Banach spaces.

The following technical result is needed in the proof of Morse's lemma.

Lemma 2.31. *Let S^n be the space of $n \times n$ symmetric matrices, $A \in S^n$ nonsingular, and let S^n_A be the vector space of $n \times n$ matrices X such that AX is symmetric. The quadratic map*

$$q_A : S^n_A \to S^n \qquad \text{defined by } q_A(X) = X^T AX$$

is locally one-to-one around $I \in S^n_A$. Consequently, there exist open neighborhoods $U \ni I$ and $V \ni A$ such that $q_A^{-1} : V \to U$ is a well-defined, infinitely differentiable map.

Proof. We have

$$\begin{aligned} q(I + tH) &:= q_A(I + tH) = (I + tH^T)A(I + tH) \\ &= A + t(H^T A + AH) + t^2 H^T AH = A + 2tAH + t^2 AH^2, \end{aligned}$$

so that $Dq(I)(H) = 2AH$. The mapping $Dq(I)$ is one-to-one, since $Dq(I)(H) = AH = 0$ implies $H = 0$, due to the fact that A is nonsingular.

The map $Dq(I)$ is also onto, since given $Y \in S^n$, the matrix $X := A^{-1}Y/2$ is in S^n_A and satisfies $Dq(I)(X) = Y$. The rest of the lemma follows from the inverse function theorem (Corollary 2.27). □

Theorem 2.32. (***Morse's lemma***) *Let $k \geq 1$ and $f : U \to \mathbb{R}$ be a C^{2+k} function on an open set $U \subseteq \mathbb{R}^n$. If $x_0 \in U$ is a nondegenerate critical point of f, then there exist open neighborhoods $V \ni x_0$ and $W \ni 0$ in $\mathbb{R}^n$ and a C^k diffeomorphism $\varphi : V \to W$ such that*

$$f(x) = f(x_0) + \frac{1}{2}\langle D^2 f(x_0)\varphi(x), \varphi(x)\rangle.$$

Proof. We may assume without any loss of generality that U is a convex set, $x_0 = 0$, and $f(0) = 0$. Let $0 \neq x \in U$, and define $\alpha(t) := f(tx)$. We have

$$\alpha(1) = \alpha(0) + \alpha'(0) + \int_0^1 (1 - t)\alpha''(t)\, dt$$

by Theorem 1.5, and since $\alpha'(t) = \langle \nabla f(tx), x\rangle$, $\nabla f(0) = 0$ and $\alpha''(t) = \langle D^2 f(tx)x, x\rangle$, we obtain

$$f(x) = \frac{1}{2}\langle A(x)x, x\rangle, \qquad \text{where} \qquad A(x) := 2\int_0^1 (1 - t)D^2 f(tx)\, dt.$$

Note that $A : U \to S^n$ is a C^k map, and $A(0) = 2(\int_0^1 (1 - t)\, dt)D^2 f(0) = D^2 f(0)$. Consequently, the map

$$H : V_0 \to Z \qquad \text{defined by } H = q^{-1}_{A(0)} \circ A,$$

where V_0 is a neighborhood of $0 \in \mathbb{R}^n$ and Z is a neighborhood of $I \in S^n_{A(0)}$ as in Lemma 2.31, is also C^k.

We have $A = q_{A(0)} \circ H$, that is, $A(x) = q_{A(0)}(H(x)) = H(x)^T A(0) H(x)$ for $x \in V_0$, and

$$\begin{aligned} f(x) &= \langle A(x)x, x\rangle = \langle H(x)^T A(0) H(x) x, x\rangle \\ &= \langle A(0)H(x)x, H(x)x\rangle = \frac{1}{2}\langle A(0)\varphi(x), \varphi(x)\rangle, \end{aligned}$$

where

$$\varphi(x) := H(x)x, \quad \varphi : V_0 \to \mathbb{R}^n,$$

is a C^k map. Since $H(0) = I$, we have

$$\begin{aligned} x + o(\|x\|) &= H(0)x + o(\|x\|) = (H(0) + DH(0)x + o(\|x\|))x \\ &= H(x)x = \varphi(x) = \varphi(0) + D\varphi(0)x + o(\|x\|), \end{aligned}$$

where the third and fifth equalities follow from Taylor's formula. This proves that $D\varphi(0) = I$, and hence is nonsingular. Thus, the inverse function theorem implies that there exist neighborhoods V, W of $0 \in \mathbb{R}^n$, $V \subseteq V_0$, such that $\varphi : V \to W$ is a C^k diffeomorphism. □

Corollary 2.33. *Let $f : U \to \mathbb{R}$ be a C^{2+k} function as in Theorem 2.32, and let $x_0 \in U$ be a nondegenerate critical point of f such that the Hessian matrix $A = Df(x_0)$ has k $(0 \le k \le n)$ positive and $n - k$ negative eigenvalues.*

Then there exists a local, nonlinear coordinate transformation $y = \psi(x)$ ($\psi : W \to V$ is a C^k diffeomorphism between some neighborhoods $W \ni 0$ and $V \ni x_0$) such that

$$f(\psi(y)) = f(x_0) + y_1^2 + \cdots + y_k^2 - y_{k+1}^2 - \cdots - y_n^2. \tag{2.3}$$

Proof. Let $A := Df(x_0)$ have the spectral decomposition $A = U^T \Lambda U$, where $\Lambda = \operatorname{diag}(\lambda_1, \ldots, \lambda_k, \ldots, \lambda_n)$ with $\lambda_i > 0$ for $i \le k$ and $\lambda_i < 0$ for $i > k$. Let $\varphi : V \to W$ be the C^k mapping in Theorem 2.32, where V and W are open neighborhoods of x_0 and 0, respectively. Define

$$y = \psi^{-1}(x) := \frac{1}{\sqrt{2}}|\Lambda|^{1/2} U\varphi(x), \quad x \in V,$$

where $|\Lambda|^{1/2}$ is the diagonal matrix with diagonal entries $\sqrt{|\lambda_i|}$, $i = 1, \ldots, n$. Theorem 2.32 and a straightforward computation give the representation (2.3). □

The proofs in this section work for functions f that are at least C^3. However, appropriate versions of Morse's lemma exist for C^2 functions; see, for example, [254]. There also exist higher-order versions of Morse's lemma for critical points x_0 such that there exists $k \ge 2$ such that $D^i f(x_0) = 0$ for $i = 1, \ldots, k-1$ and $D^k f(x_0)$ is nondegenerate in a certain sense; see [51].

2.7 Semicontinuous Functions

Semicontinuous functions are of independent interest in analysis. They also play an important role in optimization, since they appear in Ekeland's variational principle and in the theory of convex functions.

The concept of semicontinuous functions can be defined on a topological space. For our purposes, it will be sufficient to consider metric spaces. In this section, E will denote a metric space with a distance function d.

We start with some notions of limits. In optimization theory, various operations converge to $\pm\infty$, thus making it convenient to consider extended real numbers by adding ∞ and/or $-\infty$ to real numbers.

Definition 2.34. *Let $\{x_n\}$ be a sequence of extended real numbers, that is, $x_n \in \mathbb{R} \cup \{\pm\infty\}$. The* limit inferior *of $\{x_n\}$ is*

$$\varliminf_{n\to\infty} x_n := \lim_{n\to\infty} \inf\{x_n, x_{n+1}, \ldots\} = \sup_n \inf_{k\geq n} x_k,$$

where the second equality follows since $\{\inf_{k\geq n} x_k\}$ is an increasing sequence in n. Similarly, the limit superior *of $\{x_n\}$ is*

$$\varlimsup_{n\to\infty} x_n := \lim_{n\to\infty} \sup_{k\geq n} x_k = \inf_n \sup_{k\geq n} x_k.$$

Let $f : E \to \mathbb{R} \cup \{\pm\infty\}$ be an extended real-valued function. The limit inferior of f as $x \in E$ converges to $x_0 \in E$ is defined by

$$\varliminf_{x\to x_0} f(x) := \lim_{\delta\to 0} \inf_{d(x,x_0)<\delta} f(x) = \sup_{\delta\to 0} \inf_{d(x,x_0)<\delta} f(x),$$

and its limit superior by

$$\varlimsup_{x\to x_0} f(x) := \lim_{\delta\to 0} \sup_{d(x,x_0)<\delta} f(x) = \inf_{\delta\to 0} \sup_{d(x,x_0)<\delta} f(x).$$

Lemma 2.35. *Let $f : E \to \mathbb{R} \cup \{\pm\infty\}$. We have*

$$\varliminf_{x\to x_0} f(x) = \inf_{\{x_n\}} \varliminf_{n\to\infty} f(x_n),$$

where the infimum on the right-hand side is taken over all sequences $x_n \to x_0$. Similarly,

$$\varlimsup_{x\to x_0} f(x) = \sup_{\{x_n\}} \varlimsup_{n\to\infty} f(x_n).$$

Proof. We prove only the first equality, since the second one follows immediately from it. Define

$$M := \varliminf_{x\to x_0} f(x), \quad L := \inf_{\{x_n\}} \varliminf_{n\to\infty} f(x_n), \quad N_\delta := \{x \in E : d(x, x_0) < \delta\}.$$

First, we consider the case $M = -\infty$. Note that it is enough to show the existence of a sequence $x_n \to x_0$ such that $f(x_n) \to -\infty$. Since $M = -\infty$, it follows from the definition of $\underline{\lim}_{x\to x_0} f(x)$ above that $\inf x \in N_{1/n} f(x) = -\infty$ for all $n > 0$. Thus, we can find $x_n \in N_{1/n}$ satisfying $f(x_n) < -n$, proving the lemma.

Next, consider the case $M = \infty$. Let $\{x_n\}$ be an arbitrary sequence converging to x_0. We claim that $f(x_n) \to \infty$, from which the lemma follows immediately. Since $M = \infty$, for a given $\alpha > 0$, there exists $\delta > 0$ such that $\alpha < \inf_{N_\delta} f(x)$. Since $x_n \to x_0$, there exists N such that $x_n \in N_\delta$ for all $n \geq N$. Thus, $f(x_n) > \alpha$ for all $n \geq N$, and the claim is proved.

Finally, consider the case $-\infty < M < \infty$. On the one hand, given $\varepsilon > 0$, there exists $\delta > 0$ such that $\inf_{N_\delta} f(x) > M - \varepsilon$. Thus, $f(x) > M - \varepsilon$ for all $x \in N_\delta$. Let $\{x_n\}$ be an arbitrary sequence converging to x_0. Since $x_n \in N_\delta$ for all large enough n, we have $\underline{\lim}\, f(x_n) \geq M - \varepsilon$, for any $\varepsilon > 0$. Thus, $\underline{\lim}_{n\to\infty} f(x_n) \geq M$ for *any* sequence converging to x_0, proving the inequality $L \geq M$. On the other hand, since $\inf_{x\in N_\delta} f(x) \nearrow M$ as $\delta \searrow 0$, we have $\inf_{x\in N_{1/n}} f(x) \nearrow M$ as $n \to \infty$. If we pick $x_n \in N_{1/n}$ such that $f(x_n) \leq (\inf_{x\in N_{1/n}} f(x)) + 1/n$, then

$$L \leq \underline{\lim}_{n\to\infty} f(x_n) \leq \underline{\lim}_{n\to\infty} \left((\inf_{x\in N_{1/n}} f(x)) + 1/n \right) = \lim_{n\to\infty} \inf_{x\in N_{1/n}} f(x) = M.$$

This proves the reverse inequality $L \leq M$. □

We are now ready to define semicontinuous functions.

Definition 2.36. *Let $f : E \to \mathbb{R} \cup \{\pm\infty\}$. The function f is called* lower semicontinuous *at a point $x_0 \in E$ if*

$$f(x_0) \leq \underline{\lim}_{x\to x_0} f(x).$$

Equivalently, by virtue of Lemma 2.35, f is lower semicontinuous at x_0 if

$$f(x_0) \leq \underline{\lim}_{n\to\infty} f(x_n),$$

for every sequence $x_n \to x_0$.

The function f is called upper semicontinuous *at x_0 if $-f$ is lower semicontinuous at x_0, that is,*

$$f(x_0) \geq \overline{\lim}_{x\to x_0} f(x),$$

or

$$f(x_0) \geq \overline{\lim}_{n\to\infty} f(x_n),$$

for every sequence $x_n \to x_0$.

The function f is called lower semicontinuous or closed *on E if it is lower semicontinuous at every point in E. Similarly, f is called upper semicontinuous on E if it is upper semicontinuous at every point in E.*

Remark 2.37. We always have $\underline{\lim}_{x\to x_0} f(x) \le f(x_0)$, since x_0 lies in every neighborhood N_δ, so that f is lower semicontinuous at x_0 if and only if

$$f(x_0) = \underline{\lim}_{x\to x_0} f(x).$$

Similarly, f is upper semicontinuous at x_0 if and only if

$$f(x_0) = \overline{\lim}_{x\to x_0} f(x).$$

Also, note that any function is lower semicontinuous at a point x with $f(x) = -\infty$, and similarly any function is upper semicontinuous at a point x_0 with $f(x) = \infty$.

It will be seen shortly that the semicontinuity properties of f are tied up with the closedness of its epigraph.

Definition 2.38. *If $f : E \to \mathbb{R} \cup \{\pm\mathbb{R}\}$ is a function, the set*

$$\text{epi}(f) := \{(x,t) \in E \times \mathbb{R} : f(x) \le t\}$$

is called the epigraph *of f. Similarly, the set*

$$\text{hypo}(f) := \{(x,t) \in E \times \mathbb{R} : f(x) \ge t\}$$

is called the hypograph *of f.*

Theorem 2.39. *Let $f : E \to \mathbb{R} \cup \{\pm\infty\}$. The following are equivalent:*

(a) f is lower semicontinuous (upper semicontinuous) on E,
(b) epi(f) (hypo(f)) is a closed subset of $E \times \mathbb{R}$,
(c) The sublevel set $\{x \in E : f(x) \le \alpha\}$ ($\{x \in E : f(x) \ge \alpha\}$) is closed for all $\alpha \in \mathbb{R}$.

Proof. We prove the theorem only for a lower semicontinuous function, since the upper semicontinuous case follows immediately.

(a) implies (b): Let (x_n, y_n) be a sequence in epi(f) converging to a point (x, y). Since f is lower semicontinuous at x, $f(x) \le \underline{\lim}\, f(x_n) \le \lim y_n = y$, proving that $(x, y) \in \text{epi}(f)$.

(b) implies (c): Let x_n be a sequence in $L := \{z : f(z) \le \alpha\}$ converging to a point $x \in E$. We have $(x_n, \alpha) \in \text{epi}(f)$ converging to $(x, \alpha) \in \text{epi}(f)$, meaning that $x \in L$. Thus, L is closed.

(c) implies (a): Let $f(x) \in \mathbb{R}$. We claim that f is lower semicontinuous at x. Otherwise, there exists $\varepsilon > 0$ such that $\sup_{\delta>0} \inf_{N_\delta} f(x) = f(x) - 2\varepsilon$. Thus, for any $\delta > 0$, we have $\inf_{N_\delta} f(x) \le f(x) - 2\varepsilon$, meaning that we can find a sequence $x_n \to x$ such that $f(x_n) \le f(x) - \varepsilon$. Since the set $S = \{z : f(z) \le f(x) - \varepsilon\}$ is closed and $x_n \in S$, we have $x \in S$. This implies that $f(x) \le f(x) - \varepsilon$, a contradiction that proves our claim.

Since f is automatically lower semicontinuous at a point where $f(x) = -\infty$, it remains to consider the case $f(x) = \infty$. If f is not lower semicontinuous at such a point x, we have $\sup_{\delta>0} \inf_{N_\delta} f(x) = \alpha \in \mathbb{R}$. Then $\inf_{N_\delta} f(x) \le \alpha$ for any $\delta > 0$. Let $\beta \in \mathbb{R}$, $\beta > \alpha$. We can find a sequence $x_n \to x$ such that $f(x_n) \le \beta$. Since $S = \{z : f(z) \le \beta\}$ is closed and $x_n \in S$, we have $x \in S$, that is, $f(x) \le \beta < \infty$, a contradiction. □

Figure 2.1 illustrates the epigraph of a function whose function value jumps up at the point x, making the function not lower semicontinuous there. If we had $f(x) = \lim_{y \nearrow x} f(y)$ instead, the function f would be lower semicontinuous at x, although it would still be discontinuous at x.

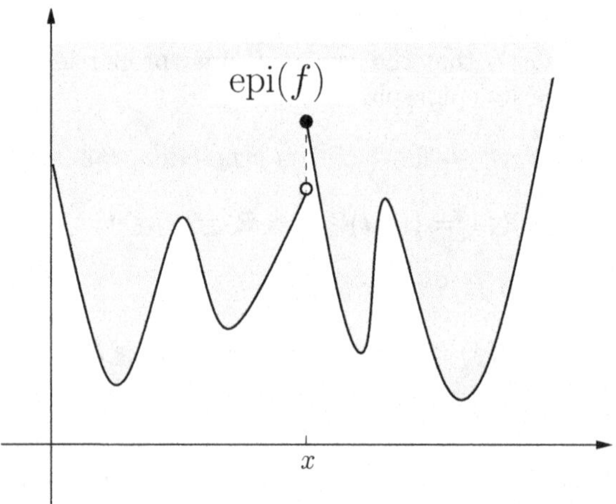

Fig. 2.1. Epigraph of a function.

Corollary 2.40. *If the functions $f, g : E \to \mathbb{R} \cup \{+\infty\}$ are lower semicontinuous, then so is $f + g$.*

Proof. We claim that

$$\{x : f(x) + g(x) > t\} = \cup_{\alpha \in \mathbb{R}} \left(\{x : f(x) > t - \alpha\} \cap \{x : g(x) > \alpha\}\right).$$

If $f(x) + g(x) = t + 2\varepsilon > t$ and $g(x) = \alpha + \varepsilon > \alpha$, then $f(x) = t - \alpha + \varepsilon > t - \alpha$. This proves that the set on the left-hand side is a subset of the one on the right-hand side. The reverse inclusion is trivial, and the claim is proved. Hence the set $\{x : f(x) + g(x) > t\}$ is open, since it is a union of open sets. □

Theorem 2.41. *Let $f : E \to \mathbb{R} \cup \{\infty\}$ be a lower semicontinuous function defined on a metric space E. If f has a nonempty compact sublevel set,*

$$l_\alpha(f) := \{x \in E : f(x) \le \alpha\},$$

then f achieves its global minimum on E.

Proof. Let $\{x_n\}$ be a minimizing sequence for f, that is,

$$f(x_n) \searrow \inf\{f(x) : x \in E\} =: \inf_E f.$$

Clearly, there exists an integer N such that $x_n \in l_\alpha(f)$ for all $n \geq N$. Since $l_\alpha(f)$ is compact, $\{x_n\}_N^\infty$ has a convergent subsequence $x_{n_k} \to x^* \in l_\alpha(f)$. Since f is lower semicontinuous at x^*, we have

$$f(x^*) \leq \underline{\lim}\, f(x_n) = \inf_E f(x).$$

This means that $f(x^*) = \inf_E f$, that is, f achieves its minimum on E at the point x^*. □

We remark that the second proof of Theorem 2.2 can be extended without any changes to give an alternative proof of this theorem.

The following extension of Theorem 2.2 follows immediately.

Corollary 2.42. *A lower semicontinuous function $f : K \to \mathbb{R}$ on a compact metric space K achieves its global minimum on K.*

Corollary 2.43. *Let $f : D \to \mathbb{R}$ be a lower semicontinuous function defined on a topological space D.*

(a) If D is compact, or
(b) D is a subset of a finite-dimensional normed vector space E and f is coercive,

then f achieves a global minimum on D.

Proof. In either case, all sublevel sets of f are compact. In (ii), this follows from the fact that the sublevel sets of f are closed and bounded, hence compact. □

We note that Theorem 2.2 follows immediately from part (i) of this corollary.

2.8 Exercises

1. (a) Show that for all values of a, the function $f(x, y) = x^3 - 3axy + y^3$ has no global minimizers or global maximizers.
   (b) For each value of a, find all the critical point(s) of f and determine their nature, that is, determine whether each critical point is a local minimum, local maximum, or saddle point.
2. Consider the function $f(x, y) = e^{x^2+y^2} - x^2 - 2y^2$.
   (a) Find the critical points of f.
   (b) Find the local (and global) minima and maxima of f as well as its saddle points.

3. Find the critical points of the function $f(x, y, z) = xyze^{-x-y-z}$, and determine their nature.
4. Solve the geometric programming problem

$$\min_{t_1>0, t_2>0} \quad \frac{1}{t_1 t_2} + t_1 + t_2.$$

5. Consider the function

$$f(x, y, z) = 2x^2 + xy + y^2 + yz + z^2 - 6x - 7y - 8z + 9.$$

   (a) Using the first-order necessary conditions, find a critical point of f.
   (b) Verify that the point found in (a) is a local minimum of f by verifying the second-order sufficient conditions.
   (c) Prove that the point is a global minimum of f.
6. Let $f : \mathbb{R}^n \to \mathbb{R}$ be a Gâteaux differentiable function. If $\lim_{\|x\|\to\infty} \frac{f(x)}{\|x\|} = \infty$, then show that the gradient function $\nabla f(x)$ is onto, that is, given $u \in \mathbb{R}^n$, there exists a point x such that $\nabla f(x) = u$.
   Hint: Consider functions $f(x) - \langle c, x \rangle$ that are bounded from below by choosing suitable $c \in \mathbb{R}^n$.
7. A *strong minimizer* of a function f is a point x_0 satisfying the condition $f(x_0) = \inf f > -\infty$ and $x_n \to x_0$ whenever $f(x_n) \to \inf f$.
   (a) Show that a strong minimizer is a strict minimizer.
   (b) Show that a strict minimizer is not necessarily a strong minimizer. (Consider the function $f(x) = x^2 e^x$.)
   (c) Show that x_0 is a strong minimizer if and only if $\operatorname{diam}(S(f, \epsilon)) \searrow 0$ as $\epsilon \to 0$, where $S(f, \epsilon) = \{x : f(x) \le \inf f + \epsilon\}$.
8. Consider the following problems.
   (a) Let $f : \mathbb{R} \to \mathbb{R}$ be a function with a continuous derivative. Show that if f has a local minimizer that is not a global minimizer, then it must have another critical point.
   (b) Contrast this with the function $f : \mathbb{R}^2 \to \mathbb{R}$ given by $f(x, y) = e^{3y} - 3xe^y + x^3$. Show that f has a unique critical point that is a local minimizer but not a global one. Show that the polynomial $g(x, y) = x^2(1 + y^3) + y^2$ also has a unique local minimizer that is not a global one.
   (c) Show that the polynomial $g(x, y) = (xy - x - 1)^2 + (x^2 - 1)^2$ has two local minimizers.

   Parts (b) and (c) seem counterintuitive; plotting their graphs using computer software should be helpful in revealing their unusual properties.
9. Consider the function $f(x, y) = x^2 - \alpha xy^2 + 2y^4$. Show that for all parameter values except two, the origin $(0, 0)$ is the only critical point of f.
   (a) Find the exceptional α's and show that f has infinitely many critical points for these α values. Determine the nature of these critical points.
   (b) Consider the values of α's for which the origin is the only critical point. For each α, determine the nature of the critical point. Show

that in some cases, the origin is a local minimum, but in other cases, it is a saddle point.

(c) Show that even when the origin is a saddle point, $(0,0)$ is a local strict minimizer of f on every line passing through the origin. In fact, show that, except for one line, the function $g(t) := f(td)$ satisfies $g'(0) = 0$ and $g''(0) > 0$.

10. Consider the quadratic function $f(x) = \frac{1}{2}\langle Ax, x\rangle + \langle c, x\rangle + a$, where A is a symmetric $n \times n$ matrix. If f is bounded from below on $\mathbb{R}^n$, show that A is positive semidefinite, and that f achieves its minimum on $\mathbb{R}^n$.
Hint: Diagonalize A.

11. Let $f : C \to \mathbb{R}$ be a twice Fréchet differentiable function on an open set in $\mathbb{R}^n$. Suppose that $\Delta f(x) := \sum_{i=1}^n \frac{\partial^2 f(x)}{\partial x_i^2} = 0$ for all $x \in C$, that is, f is a *harmonic function*. If p is a critical point of f, that is, $\nabla f(p) = 0$, and the Hessian $Hf(p)$ is not identically zero, then p is must be a saddle point of f.

12. Sylvester's theorem, Theorem 2.25 on page 43, states that a symmetric matrix A is positive definite if and only if all the leading principal minors of A are positive. The purpose of this problem is to give an elegant proof of this result using optimization techniques.
Let A be an $(n+1) \times (n+1)$ symmetric matrix in the form $A = \left[\begin{smallmatrix} B & b \\ b^T & c \end{smallmatrix}\right]$, where B is a positive definite $n \times n$ matrix, $b \in \mathbb{R}^n$, and $c \in \mathbb{R}$.

(a) Consider the quadratic function

$$p(x) := (x^T, 1)\begin{bmatrix} B & b \\ b^T & c \end{bmatrix}\begin{pmatrix} x \\ 1 \end{pmatrix} = \langle Bx, x\rangle + 2\langle b, x\rangle + c$$

on $\mathbb{R}^n$. Show that the point $x^* = -B^{-1}b$ is the unique global minimizer of p on $\mathbb{R}^n$, and $p(x^*) = c - \langle B^{-1}b, b\rangle$. Thus, p is positive on $\mathbb{R}^n$ if and only if $c - \langle B^{-1}b, b\rangle > 0$.

(b) Show that $\det A = \det B \cdot (c - \langle B^{-1}b, b\rangle)$.
Hint: Find a suitable vector $d \in \mathbb{R}^n$ such that

$$\begin{bmatrix} I & 0 \\ d^T & 1 \end{bmatrix}\begin{bmatrix} B & b \\ b^T & c \end{bmatrix}\begin{bmatrix} I & d \\ 0 & 1 \end{bmatrix} = \begin{bmatrix} B & 0 \\ 0 & c - b^T B^{-1} b \end{bmatrix}.$$

This is related to the notion of Schur complement in linear algebra.

(c) Prove Sylvester's theorem by induction on the dimension of A using parts (a) and (b).

13. The purpose of this problem is to demonstrate that the behavior of a smooth function around a regular point x is determined by its derivative $Df(x) \neq 0$. Thus, it is a first-order version of Morse's lemma.
Let f be C^k in a neighborhood of the origin in $\mathbb{R}^n$, and $f(0) = 0$. Suppose that 0 is a regular point, that is, $l := \nabla f(0) \neq 0$. Let $x_0 \in \mathbb{R}^n$ such $l(x_0) = 1$.

(a) Show that the linear map $T : \mathbb{R}^n \to (\operatorname{Ker} l) \times \mathbb{R}$ defined by

$$T(x) := (x - l(x)x_0, l(x))$$

is one-to-one and onto, thus an isomorphism between $\mathbb{R}^n$ and $(\operatorname{Ker} l) \times \mathbb{R}$.

(b) Show that the C^k map

$$\psi(x) := (x - l(x)x_0, f(x))$$

has derivative $D\psi(0) = T$. Conclude using the inverse function theorem that ψ is a C^k diffeomorphism between a neighborhood U of 0 and its image $\psi(U)$.

(c) Define the C^k diffeomorphism

$$\varphi := T^{-1} \circ \psi,$$

so that $\psi = T \circ \varphi$. Show that the equation $\psi(x) = T(\varphi(x))$ gives the sought-after formula

$$f(x) = l(\varphi(x)) \quad \text{for all} \quad x \in U.$$

We remark that this result is valid in Banach spaces, since the inverse function theorem holds in that setting.

14. Let $f : U \to \mathbb{R}$ be a C^2 function on an open set $U \subseteq \mathbb{R}^n$. Show that the nondegenerate critical points of f are isolated: if $x_0 \in U$ is a nondegenerate critical point of f, then there exists an open neighborhood $V \ni x_0$ such that x_0 is the only critical point of f on V.
Hint: The inverse function theorem may be helpful.
15. ***(Jordan and von Neumann [149])*** If f is a quadratic form, that is, $f(x) = \langle Ax, x \rangle$, where A is a symmetric $n \times n$ matrix, then f satisfies the properties
(i) $f(x + y) + f(x - y) = 2f(x) + 2f(y)$ for all $x, y \in \mathbb{R}^n$,
(ii) f is continuous.
The property (i) is called the *parallelogram law*. These two properties characterize quadratic forms, even in more general spaces than $\mathbb{R}^n$ [149, 104] and [240], pp. 275–276.
For a function $f : \mathbb{R}^n \to \mathbb{R}$, define

$$B(x, y) := \frac{1}{4} \left(f(x + y) - f(x - y) \right).$$

(Note that if f is a quadratic form, then $B(x, y) = \langle Ax, y \rangle$.)
Parts (a)–(c) below prove that if f satisfies only (i), then B is symmetric, $B(x, y) = B(y, x)$, and additive in each variable, $B(x + y, z) = B(x, z) + B(y, z)$.
(a) Show that $f(0) = 0$, $f(-x) = x$, and $f(2x) = 4f(x)$. Use these to show that B is symmetric and $B(x, x) = f(x)$.

(b) Show that

$$\begin{aligned} 8B(x,z) + 8B(y,z) &= 2f(x+z) + 2f(y+z) - 2f(x-z) - 2f(y-z) \\ &= f(x+y+2z) + f(x-y) - f(x+y-2z) \\ &\quad - f(x-y) \\ &= 4B(x+y, 2z). \end{aligned}$$

Consequently,

$$B(x,z) + B(y,z) = \frac{1}{2}B(x+y, 2z). \tag{2.4}$$

(c) Show directly from the definition of B that $B(0,z) = B(z,0) = 0$, and use it and (2.4) to prove that $B(x,z) = B(x,2z)/2$. Then show that this and (2.4) give

$$B(x+y,z) = B(x,z) + B(y,z),$$

that is, the function $B(x,y)$ is additive in the first variable, and similarly in the second variable.

Now, use property (ii) to prove that

(d) B is homogeneous in each variable, that is,

$$B(tx,y) = tB(x,y), \; B(x,ty) = tB(x,y),$$

for all $x, y \in \mathbb{R}^n$, and for all $t \in \mathbb{R}$.
Hint: Use (c) to show that $B(nx,y) = nB(x,y)$ for all integers n. Next, if $t = m/n$ is a rational number, define $z := tx = mx/n$. Then $nz = mx$, and $nB(z,y) = mB(x,y)$ or $B(tx,y) = tB(x,y)$. Finally, use continuity of f to show that $B(tx,y) = tB(x,y)$ for all $t \in \mathbb{R}$.

(e) For each fixed y, the function $x \mapsto B(x,y)$ is linear. Show that there exists $l(y) \in \mathbb{R}^n$ such that $B(x,y) = \langle x, l(y)\rangle$. Show that l is a linear function of y, and that $l(y) = Ay$ for some $n \times n$ matrix A. Show that, without losing any generality, A may be assumed to be a symmetric matrix.

(f) Prove that a norm $\|\cdot\|$ is Euclidean, that is, it comes from an inner product, if and only if the function $f(x) = \|x\|^2$ satisfies the parallelogram law. This is the motivation of the paper of Jordan and von Neumann [149].

16. Let $f : U \to \mathbb{R}^m$ be a C^1 mapping on an open set $U \subseteq \mathbb{R}^n$. Suppose that at a point $x_0 \in U$, $Df(x_0) : \mathbb{R}^n \to \mathbb{R}^m$ is one-to-one, so that it is an isomorphism between $\mathbb{R}^n$ and $L := Df(x_0)(\mathbb{R}^n)$. Assume without loss of generality that $x_0 = 0$ and $f(0) = 0$. The purpose of this problem is to prove that $f(U)$ is C^1 diffeomorphic to a neighborhood of the origin in L, that is, there exists a local C^1 diffeomorphism of $\mathbb{R}^m$ around the origin such that $g \circ f$ is a C^1 diffeomorphism between a neighborhood of $0 \in \mathbb{R}^n$ and a neighborhood of the origin in L. (Thus, g "straightens out" the image of f around 0.)

(a) Let M be a subspace of $\mathbb{R}^m$ complementary to L, that is, $\mathbb{R}^m = L+M$ and $L \cap M = \{0\}$. Show that the linear map $T : \mathbb{R}^m \to L \times M$, given by $T(u) = Df(0)^{-1}v + w$, where $u = v + w$, $v \in L$, $w \in M$, is a linear isomorphism, and that the C^1 map $\overline{f} := T \circ f : U \to L \times M$ satisfies $D\overline{f}(0)(h) = (h, 0)$. Conclude that it is enough to prove our claim for $\overline{f}$, or equivalently, to assume that $f : \mathbb{R}^n \to \mathbb{R}^n \times \mathbb{R}^k = \mathbb{R}^{n+k}$, $f(0) = 0$, and $Df(0)(h) = (h, 0)$.

(b) Consider the map $\varphi : \mathbb{R}^n \times \mathbb{R}^k \to \mathbb{R}^n \times \mathbb{R}^k$ given by $\varphi(x, y) = f(x) + (0, y)$. Show that $D\varphi(0, 0)$ is the identity mapping on $\mathbb{R}^{n+k}$, so that φ must be a local diffeomorphism between two neighborhoods of the origin in $\mathbb{R}^{n+k}$. Call its inverse g.

(c) Show that

$$(g \circ f)(x) = g(f(x)) = g(\varphi(x, 0)) = (x, 0).$$

Show that this implies that g satisfies the required properties.

3

Variational Principles

Consider a C^1 function $f : \mathbb{R}^n \to \mathbb{R}$. If f has a (global) minimizer $\overline{x}$, then we know that $\nabla f(\overline{x}) = 0$. Suppose, however, that f is bounded below, but does not have a minimizer. Is it possible to find a minimizing sequence $\{x_n\}$ satisfying $f(x_n) \to \inf f$ that also satisfies $\nabla f(x_n) \to 0$? If this is true, then it should be possible to obtain new optimality conditions even when minimizers do not exist. A celebrated result of Ekeland [86] known as *Ekeland's ϵ-variational principle* ensures that the answer to the above question is yes. Moreover, this principle is valid in a much more general context and has turned out to be one of the most important of the recent contributions to analysis. In this chapter, we give a fairly detailed exposition of this important principle and some of its applications. These include a short proof of Banach's fixed point theorem, a characterization of the consistency of a system of linear inequalities, and the proofs of some of the most basic theorems of analysis (the open mapping theorem, Graves's theorem, Lyusternik's theorem, the inverse function theorem, and the implicit function theorem). Another significant application to the derivation of the Fritz John conditions in nonlinear programming is postponed to Section 9.3.

We will also consider lower semicontinuous functions $f : M \to \mathbb{R} \cup \{+\infty\}$ on a metric space M. The function is allowed to take on the value $+\infty$. This is a convenient device when a real-valued function f is defined on only a proper subset D of M. We then extend f by declaring $f(x) = \infty$ for $x \notin D$. Of course, the function f that is identically equal to $+\infty$ on all of M is not interesting, and is called an *improper function*. In the contrary case, that is, when $f(x)$ is finite for at least one point x in M, we call f a *proper function*.

3.1 Ekeland's ϵ-Variational Principle

Let (M, d) be a metric space, and let $f : M \to \mathbb{R}$ be any function. Define a relation $y \preceq x$ on M by the condition

O. Güler, *Foundations of Optimization*, Graduate Texts in Mathematics 258,
DOI 10.1007/978-0-387-68407-9_3, © Springer Science+Business Media, LLC 2010

$$y \preceq x \quad \iff \quad f(y) + d(x, y) \le f(x).$$

This is a partial ordering on M, that is, for all x, y, z in M, we have

$$\begin{aligned}
&(i) \quad x \preceq x, \\
&(ii) \quad x \preceq y \text{ and } y \preceq x \text{ implies } x = y, \\
&(iii) \quad x \preceq y \text{ and } y \preceq z \text{ implies } x \preceq z.
\end{aligned}$$

Now, property (i) is trivially true, (ii) is equivalent to the property of the distance function that $d(x, y) = 0$ implies $x = y$, and (iii) follows from the triangle inequality $d(x, z) \le d(x, y) + d(y, z)$. We call a minimal point in the partial order $\preceq$ a *d-point*. Thus, a point $x \in M$ is a d-point if $y \preceq x$ implies $y = x$, or equivalently,

$$f(x) < f(y) + d(x, y) \text{ for all } y \in M, \ y \neq x.$$

Define the set

$$S(x) := \{y \in M : y \preceq x\} = \{y \in M : f(y) + d(x, y) \le f(x)\}.$$

We have $x \in S(x)$, so that $S(x) \neq \emptyset$. Since $\preceq$ is a partial order, it follows that $y \preceq x$ if and only if $S(y) \subseteq S(x)$. We note that if f is a lower semicontinuous function, then $S(x)$ is a closed subset of M. Also, a d-point x is characterized by the condition that $S(x)$ is a singleton, that is, $S(x) = \{x\}$.

Ekeland's ϵ-variational principle is an easy corollary of the following important theorem.

Theorem 3.1. *Let (M, d) be a metric space. The following conditions are equivalent:*

(a) (M, d) is a complete metric space,
(b) For any proper lower semicontinuous function $f : M \to \mathbb{R} \cup \{\infty\}$ bounded below, and any point $x_0 \in M$, there exists a d-point x satisfying $x \preceq x_0$.

Proof. (a) $\Longrightarrow$ (b): We may assume that $f(x_0) \in \mathbb{R}$, since otherwise we can replace x_0 with any $z \in M$ such that $f(z) < \infty$. Generate a sequence $\{x_n\}_0^\infty$ recursively such that given x_n, choose x_{n+1} to be any point in $S(x_n)$. We claim that $\{x_n\}$ is a Cauchy sequence. If $n > m$, then $x_n \preceq x_{n-1} \preceq \cdots \preceq x_m$, so that $x_n \preceq x_m$ and $f(x_n) + d(x_n, x_m) \le f(x_m)$. Thus, $\{f(x_n)\}$ is a decreasing sequence of real numbers bounded from below, say $f(x_n) \searrow \alpha$ for some $\alpha \in \mathbb{R}$. Since $d(x_n, x_m) \le f(x_m) - f(x_n) \to 0$ as $m, n \to \infty$, the claim is proved. Since M is a complete metric space, x_n converges to a point $x \in M$. Since $x_n \in S(x_n)$ for all $k \ge n$ and $S(x_n)$ is closed, we see that $x \in S(x_n)$. Thus, $x \preceq x_n \preceq x_{n-1}$ for all $n \ge 1$, or equivalently $x \in \cap_{n=1}^\infty S(x_n)$. Consequently, $S(x) \subseteq \cap_{n=1}^\infty S(x_n)$, but this is not strong enough to conclude that $S(x)$ is a singleton. For that, we need to pick $x_{n+1} \in S(x_n)$ more carefully.

Choose $x_{n+1} \in S(x_n)$ such that

$$f(x_{n+1}) \leq \inf_{S(x_n)} f + \frac{1}{n}.$$

Now, if $z \in S(x)$, we have $z \preceq x \preceq x_{n-1} \preceq x_n$, and

$$f(z) + d(z, x_n) \leq f(x_n) \leq \inf_{S(x_{n-1})} f + \frac{1}{n} \leq f(z) + \frac{1}{n},$$

so that

$$d(z, x_n) \leq \frac{1}{n} \to 0 \text{ as } n \to \infty.$$

This gives $x_n \to z = x$, proving that $S(x) = \{x\}$, that is, x is a d-point.

(b) $\implies$ *(a):* Let $\{x_n\}_1^\infty$ be a Cauchy sequence in M. Consider the function

$$f(x) := 2 \lim_{n\to\infty} d(x, x_n).$$

Here the numerical sequence $\{d(x, x_n)\}_1^\infty$ is a Cauchy sequence, since $|d(x, x_m) - d(x, x_n)| \leq d(x_m, x_n) \to 0$ as $m, n \to \infty$, hence converges, so that the function $f(x)$ is well-defined. The function f is continuous, since $|d(x, x_n) - d(y, x_n)| \leq d(x, y)$ implies that $|f(x) - f(y)| \leq d(x, y)$.

Note that $f(x_n) \to 0$. Let $x \in M$ be a d-point of f. Then

$$f(x) \leq f(x_n) + d(x, x_n) \quad \text{for all} \quad n \geq 1.$$

Letting $n \to \infty$ gives $f(x) \leq f(x)/2$, that is, $f(x) = 0$. This implies $d(x, x_n) \to 0$ or $x_n \to x$, proving that (M, d) is a complete metric space. □

The proof of (a) implies (b) is due to Ekeland [86]. See also Ekeland [87] for the origins of the improved proof above. The proof of (b) implies (a) is given in Weston [265] and in Sullivan [249].

Theorem 3.2. (***Ekeland's ϵ-variational principle***) *Let (M, d) be a complete metric space, and let $f : M \to \mathbb{R} \cup \{+\infty\}$ be a proper lower semicontinuous function that is bounded from below.*

Then for every $\epsilon > 0$, $\lambda > 0$, and $x \in M$ such that

$$f(x) \leq \inf_M f + \epsilon,$$

there exists an element $x_\epsilon \in M$ satisfying the following three properties:

$$\begin{aligned} f(x_\epsilon) &\leq f(x), \\ d(x_\epsilon, x) &\leq \lambda, \\ f(x_\epsilon) &< f(z) + \frac{\epsilon}{\lambda} d(z, x_\epsilon) \quad \text{for all} \quad z \in M, \ z \neq x_\epsilon. \end{aligned} \tag{3.1}$$

Ekeland's ϵ-variational principle is illustrated in Figure 3.1. At a d-point (such as x_ϵ), the cone lies completely below the graph of f, since $f(x_\epsilon) - \frac{\epsilon}{\lambda} d(z, x_\epsilon) < f(z)$ for all $z \neq x_\epsilon$.

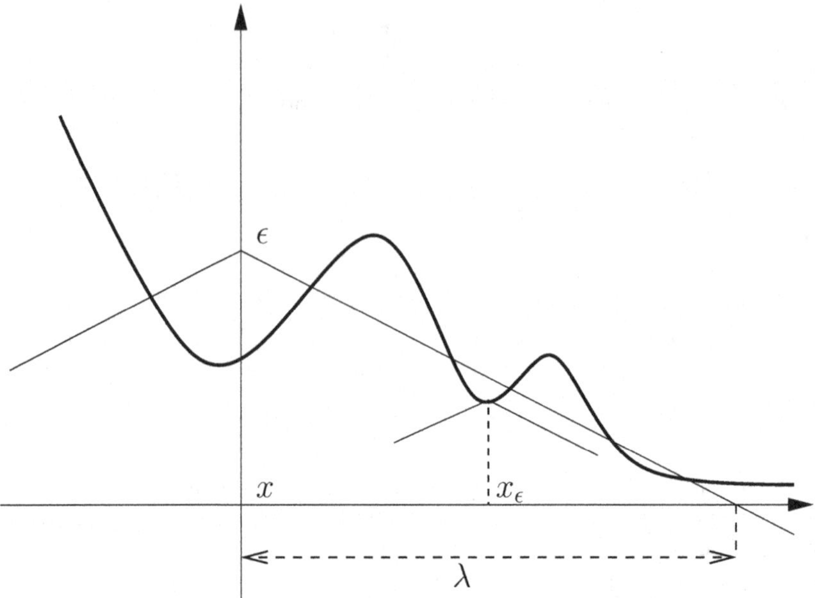

Fig. 3.1. Ekeland's ϵ-variational principle.

Proof. It suffices to prove the theorem for $\lambda = 1$ and $\epsilon = 1$; the general case follows by replacing the distance function d by the equivalent distance function d/λ and the function f by the function f/ϵ.

It follows from Theorem 3.1 that there exists a d-point $\overline{x} \preceq x$. We claim that the point $\overline{x}$ satisfies the properties in (3.1). The third condition states that $\overline{x}$ is a d-point, while the first two conditions follow from the inequalities

$$f(\overline{x}) + d(x, \overline{x}) \leq f(x) \leq \inf_M f + 1 \leq f(\overline{x}) + 1,$$

where first inequality is a consequence of $\overline{x} \preceq x$. □

There is a shorter proof of Ekeland's ϵ-variational principle due to Hiriart-Urruty [133] in finite dimensions, that is, when $M = \mathbb{R}^n$ equipped with the distance function $d(x, y) = \|x - y\|$ given by any norm, say the Euclidean norm. The proof in [133] gives a slightly weaker statement than the third inequality in (3.1). The proof below is from [41].

Proof. We assume that $\lambda = 1$ without loss of generality. Define the function

$$g(z) := f(z) + \epsilon\|z - x\|.$$

The function $g(z)$ is clearly lower semicontinuous, and is coercive, that is, $\lim_{\|z\|\to\infty} f(z) = \infty$. Thus, the set K of minimizers of g is a compact subset of $\mathbb{R}^n$. Let $x_\epsilon \in K$ be a point that minimizes f on K.

We have $g(x_\epsilon) \le g(z)$, that is,

$$f(x_\epsilon) + \epsilon\|x_\epsilon - x\| \le f(z) + \epsilon\|z - x\| \quad \text{for all} \quad z \in \mathbb{R}^n.$$

Putting $z = x$ in the above inequality gives

$$f(x_\epsilon) + \epsilon\|x_\epsilon - x\| \le f(x) \le \inf_{\mathbb{R}^n} f + \epsilon \le f(x_\epsilon) + \epsilon.$$

These immediately yield the first and second conditions in (3.1).

To prove the third inequality in (3.1), we note that for points in $z \in K$, $z \neq x_\epsilon$,

$$f(x_\epsilon) \le f(z) < f(z) + \epsilon\|z - x_\epsilon\|,$$

while if $z \notin K$, then

$$f(x_\epsilon) + \epsilon\|x_\epsilon - x\| < f(z) + \epsilon\|z - x\| \le f(z) + \epsilon(\|z - x_\epsilon\| + \|x_\epsilon - x\|),$$

again yielding the inequality $f(x_\epsilon) < f(z) + \epsilon\|z - x_\epsilon\|$. □

Taking $\lambda = \sqrt{\epsilon}$ in Theorem 3.2, we immediately obtain the following corollary.

Corollary 3.3. *Let the function f and the point x satisfy the conditions in Theorem 3.2. Then there exists a point x_ϵ satisfying the following conditions:*

$$\begin{aligned} f(x_\epsilon) &\le f(x), \\ d(x_\epsilon, x) &\le \sqrt{\epsilon}, \\ f(z) &> f(x_\epsilon) - \sqrt{\epsilon} d(z, x_\epsilon) \quad \text{for all} \quad z \in M, z \neq x_\epsilon. \end{aligned}$$

Ekeland's ϵ-variational principle has turned out to be one of the most important and versatile of the recent tools in analysis. We will discuss several of its applications below, but these barely scratch the surface on its possible uses. The interested reader may consult Ekeland [87, 88] for a fairly comprehensive review of results on this topic up to about 1990, and the growing mathematical literature for its more recent applications.

Definition 3.4. *Let X, Y be Banach spaces, and $f : U \to Y$ a map on an open subset $U \subseteq X$. The map f is called* Gâteaux differentiable *at $x \in U$ if*

$$\lim_{t \to 0} \frac{f(x + td) - f(x)}{t} = Ad \quad \text{for all} \quad d \in X,$$

where $A : X \to Y$ is a continuous linear map. If f is Gâteaux differentiable at every point $x \in U$, we say f is Gâteaux differentiable on U. The map A, denoted by $Df(x)$, is called the Gâteaux derivative of f.

The map f is called Fréchet differentiable *at $x \in U$ if there exists a continuous linear map $B : X \to Y$ such that*

$$\lim_{\|h\|\to 0} \frac{\|f(x+h) - f(x) - Bh\|}{\|h\|} = 0. \tag{3.2}$$

The map B, denoted by $Df(x)$, is called the Fréchet derivative of f. If f is Fréchet differentiable at every point $x \in U$, we say f is Fréchet differentiable on U. Then we have $Df : U \to L(X,Y)$, where $L(X,Y)$ is the Banach space of continuous linear maps from X into Y endowed with the operator norm $\|L\| := \sup_{\|x\|=1} \|Lx\|$.

If Df is continuous on U, then we say f is continuously differentiable on U and write $f \in C^1$. If $Df \in C^1$, then we call f twice continuously differentiable and write $f \in C^2$. A kth-order continuously differentiable map f, $f \in C^k$, is defined similarly by induction.

It is easy to show that if f Fréchet differentiable at $x \in U$, then it is Gâteaux differentiable at x and the two derivatives of f agree.

Corollary 3.5. *Let $f : X \to \mathbb{R}$ be a function on a Banach space X that is Gâteaux differentiable, lower semicontinuous, and bounded from below. Let $\epsilon > 0$, and let $x \in X$ be a point such that*

$$f(x) \le \inf_X f + \epsilon.$$

Then there exists a point $x_\epsilon \in X$ such that

$$\begin{aligned} f(x_\epsilon) &\le f(x), \\ \|x - x_\epsilon\| &\le 1, \\ \|\nabla f(x_\epsilon)\| &\le \epsilon. \end{aligned}$$

Consequently, there exists a minimizing sequence $\{x_n\}$ in X satisfying the conditions

$$f(x_n) \to \inf_X f \quad \text{and} \quad \nabla f(x_n) \to 0.$$

Proof. Theorem 3.2 gives a point x_ϵ satisfying the first two conditions above. To prove the third condition, note that for an arbitrary direction $d \in X$, $\|d\| = 1$, we have

$$t\langle \nabla f(x_\epsilon), d\rangle + o(t) = f(x_\epsilon + td) - f(x_\epsilon) \ge -t\epsilon$$

as $t \to 0$, where the equality follows since f is Gâteaux differentiable and the inequality follows from the third inequality in Theorem 3.2. This gives $\langle \nabla f(x_\epsilon), d\rangle \ge -\epsilon$, or equivalently $\langle \nabla f(x_\epsilon), d\rangle \le \epsilon$ for all $d \in X$, $\|d\| = 1$. Therefore, $\|\nabla f(x_\epsilon)\| \le \epsilon$.

Let y_n be a point in X satisfying $f(y_n) \le \inf_X f + 1/n$. We have already shown that there exists a point x_n satisfying the conditions $f(x_n) \le f(y_n) \le \inf_X f + 1/n$ and $\|\nabla f(x_n)\| \le 1/n$. The sequence $\{x_n\}_1^\infty$ satisfies the required properties. □

It is also possible to investigate the asymptotic properties of functions using Ekeland's ϵ-variational principle.

Theorem 3.6. *Let $f : X \to \mathbb{R}$ be a function on a Banach space X that is Gâteaux differentiable, lower semicontinuous, bounded from below, but that is* not coercive. *Define*

$$\alpha := \underline{\lim}_{\|x\|\to\infty} f(x) = \sup_{n\geq 1} \inf_{\|x\|\geq n} f(x).$$

Then there exists a sequence $\{x_n\}$ in X satisfying the conditions

$$\|x_n\| \to \infty, \quad f(x_n) \to \alpha, \quad \text{and} \quad \nabla f(x_n) \to 0.$$

Proof. Let (M_n, d) be the complete metric space such that $M_n := \{x \in X : \|x\| \geq n\}$ and the distance function d is given by the norm on X. Let $\|y_n\| \geq n+2$ be a point satisfying

$$f(y_n) \leq \inf_{M_{n+2}} f + \frac{1}{n} = \inf_{M_n} f + \delta_n + \frac{1}{n},$$

where $\delta_n := \inf_{M_{n+2}} f - \inf_{M_n} f$. We have $\delta_n \searrow 0$, since $\inf_{M_n} f \nearrow \alpha$. Applying Theorem 3.2 to the function f on M_n with $\epsilon = \epsilon_n := \delta_n + 1/n$ and $\lambda = 1$, we obtain a point $x_n \in M_n$ satisfying the conditions

$$\begin{aligned} f(x_n) &\leq f(y_n), \\ \|x_n - y_n\| &\leq 1, \\ f(x_n) &< f(x) + \epsilon_n \|x - x_n\| \quad \text{for all } x \in M_n,\ x \neq x_n. \end{aligned}$$

It follows that $\|x_n\| \geq \|y_n\| - \|x_n - y_n\| \geq n+1$, so that x_n is in the interior of the region $\{x \in X : \|x\| \geq n\}$. Let $h \in X$, $\|h\| = 1$ be a unit direction vector. We have

$$t\langle -\nabla f(x_n), h\rangle + o(t) = f(x_n) - f(x_n + th) \leq t\epsilon_n$$

as $t \to 0$. Dividing the end terms above by t, letting $t \searrow 0$, and then maximizing over all unit directions h, we obtain $\|\nabla f(x_n)\| \leq \epsilon_n \to 0$ as $n \to \infty$. Since

$$f(x_n) \leq f(y_n) \leq \inf_{M_{n+2}} f + \frac{1}{n} \leq \alpha + \frac{1}{n},$$

the sequence $\{x_n\}$ satisfies all the required properties. □

See Brézis and Nirenberg [49] for the original proof, and related interesting results.

The well-known *Banach fixed point theorem* also follows easily from Ekeland's ϵ-variational principle.

Definition 3.7. *Let $\varphi : M \to M$ be a mapping on a metric space (M, d). A point $\overline{x} \in M$ is called a* fixed point *of φ if $\varphi(\overline{x}) = \overline{x}$.*

The mapping φ is called a contractive mapping *if there exist a constant $0 \leq \alpha < 1$ such that*

$$d(\varphi(x), \varphi(y)) \leq \alpha d(x, y) \quad \textit{for all} \quad x, y \in M.$$

Theorem 3.8. (***Banach fixed point theorem***) *A contractive mapping $\varphi : M \to M$ on a complete metric space (M, d) has a unique fixed point.*

Proof. Define the function $f(x) := d(x, \varphi(x))$, and choose $\epsilon \in (0, 1 - \alpha)$. Theorem 3.2 implies that there exists a point $\bar{x} \in M$ such that

$$f(\bar{x}) \leq f(x) + \epsilon d(x, \bar{x}) \quad \text{for all} \quad x \in M.$$

We claim that $\varphi(\bar{x}) = \bar{x}$. Otherwise, choosing $x = \varphi(\bar{x})$ above, we have

$$d(\varphi(\bar{x}), \bar{x}) \leq d(\varphi(\bar{x}), \varphi^2(\bar{x})) + \epsilon d(\varphi(\bar{x}), \bar{x}) \leq (\alpha + \epsilon) d(\varphi(\bar{x}), \bar{x}).$$

This gives the contradiction $1 \leq \alpha + \epsilon < 1$. The uniqueness of the fixed point follows from the usual argument: if x_1 and x_2 are two distinct fixed points, then we get a contradiction,

$$d(x_1, x_2) = d(\varphi(x_1), \varphi(x_2)) \leq \alpha d(x_1, x_2) < d(x_1, x_2).$$

□

3.2 Borwein–Preiss Variational Principle

The perturbation in Ekeland's ϵ-variational principle is nonsmooth. There are smooth versions of variational principles pioneered by Borwein and Preiss [42]. We state a generalization of their results due to Li and Shi [189].

Definition 3.9. *Let (M, d) be a metric space. We call a function $\rho : M \times M \to [0, \infty]$ a* gauge-type *function if it is continuous, $\rho(x, x) = 0$ for all $x \in M$, and given $\epsilon > 0$, there exists a $\delta > 0$ such that for all $y, z \in M$, we have $\rho(y, z) \leq \delta$ implies $d(y, z) < \epsilon$.*

Theorem 3.10. (***Borwein–Preiss Variational Principle***) *Let (M, d) be a complete metric space and let $f : M \to \mathbb{R} \cup \{+\infty\}$ be a proper lower semicontinuous function, bounded from below. Let ρ be a gauge-type function and $\{\delta_k\}_{k=0}^{\infty}$ a sequence of positive numbers.*

If $\epsilon > 0$ and $x \in M$ satisfies

$$f(x) \leq \inf_M f \, + \, \epsilon,$$

then there exist an element $x_\epsilon \in M$ and a sequence $\{x_n\} \subset M$ such that

$$\rho(x, x_\epsilon) \leq \frac{\epsilon}{\delta_0}, \quad \rho(x_n, x_\epsilon) \leq \frac{\epsilon}{2^k \delta_0},$$

$$f(x_\epsilon) + \sum_{k=0}^{\infty} \delta_k \rho(x_\epsilon, x_n) \leq f(x),$$

$$f(z) + \sum_{k=0}^{\infty} \delta_k \rho(z, x_n) > f(x_\epsilon) + \sum_{k=0}^{\infty} \delta_k \rho(x_\epsilon, x_n) \quad \textit{for all} \quad z \in M, \ z \neq x_\epsilon.$$

The proof of this theorem resembles that of Theorem 3.2, which the interested reader can find in the book [43]. We will instead give a version of this theorem in finite-dimensional Euclidean spaces that is much simpler to prove, but which is sufficient for our purposes.

Theorem 3.11. (***Smooth variational principle in finite-dimensional spaces***) *Let $f : \mathbb{R}^n \to \mathbb{R} \cup \{+\infty\}$ be a proper lower semicontinuous function, bounded from below. Let λ and $p \geq 1$.*

If $\epsilon > 0$ and $x \in \mathbb{R}^n$ satisfies

$$f(x) \leq \inf_{\mathbb{R}^n} f + \epsilon,$$

then there exists an $x_\epsilon \in \mathbb{R}^n$ such that

$$\|x - x_\epsilon\| \leq \lambda,$$

$$f(x_\epsilon) + \frac{\epsilon}{\lambda^p}\|x_\epsilon - x\|^p \leq f(z) + \frac{\epsilon}{\lambda^p}\|z - x\|^p \quad \textit{for all} \quad z \in \mathbb{R}^n.$$

Proof. Note that the function $g(z) := f(z) + \frac{\epsilon}{\lambda^p}\|z - x\|^p$ is coercive, and so has a minimizer $x_\epsilon \in \mathbb{R}^n$. This proves the second assertion of the theorem. If we set $z = x$ in this inequality, we obtain

$$f(x_\epsilon) + \frac{\epsilon}{\lambda^p}\|x - x_\epsilon\|^p \leq f(x) \leq \inf_{\mathbb{R}^n} f + \epsilon \leq f(x_\epsilon) + \epsilon.$$

This gives $\|x - x_\epsilon\| \leq \lambda$. □

Corollary 3.12. *If a C^2 function $f : \mathbb{R}^k \to \mathbb{R}$ is bounded from below, then there exists a sequence $\{x_n\}_1^\infty$ in $\mathbb{R}^k$ satisfying the properties*

$$f(x_n) \to \inf_{\mathbb{R}^k} f,$$

$$\nabla f(x_n) \to 0,$$

$$\underline{\lim}_{n \to \infty} \langle D^2 f(x_n) d, d \rangle \geq 0 \quad \textit{for each} \quad d \in \mathbb{R}^k.$$

Proof. Let $p = 2$ and $\lambda = 1$ in Theorem 3.11. Let x, $\epsilon > 0$, and $x_\epsilon > 0$ be as in that theorem. Note that for $d \in \mathbb{R}^k$, the point x_ϵ satisfies $\|x_\epsilon - x\| \leq 1$ and the conditions

$$f(x_\epsilon) \le \inf_{\mathbb{R}^k} f + \epsilon,$$
$$\nabla g(x_\epsilon) := \nabla f(x_\epsilon) + 2\epsilon(x_\epsilon - x) = 0,$$
$$f(x_\epsilon + td) - f(x_\epsilon) \ge \epsilon \left(\|x_\epsilon - x\|^2 - \|(x_\epsilon - x) + td\|^2 \right).$$

From the last inequality above, we obtain for $\|d\| = 1$,

$$\begin{aligned} -2\epsilon &= \frac{\epsilon \left(2\|x_\epsilon - x\|^2 - \|(x_\epsilon - x) + td\|^2 - \|(x_\epsilon - x) - td\|^2 \right)}{t^2} \\ &\le \frac{f(x_\epsilon + td) + f(x_\epsilon - td) - 2f(x_\epsilon)}{t^2}. \end{aligned}$$

It is easily verified, using Taylor's formula, that the last term above tends to $\langle D^2 f(x_\epsilon)d, d\rangle$ as $t \to 0$.

Altogether, x_ϵ has the properties

$$f(x_\epsilon) \le \inf_{\mathbb{R}^k} f + \epsilon,$$
$$\|\nabla f(x_\epsilon)\| \le 2\epsilon \|x_\epsilon - x\| \le 2\epsilon,$$
$$\langle D^2 f(x_\epsilon)d, d\rangle \ge -2\epsilon \quad \text{for each} \quad d \in \mathbb{R}^k, \ \|d\| = 1,$$

where we used the fact $\|x - x_\epsilon\| \le 1$ in the middle inequality.

The sequence $\{x_n\}$, where x_n is x_ϵ corresponding to $\epsilon = 1/n$, satisfies the required properties. □

It is also possible to improve Theorem 3.6 for twice differentiable functions.

Corollary 3.13. *If a C^2 function $f : \mathbb{R}^k \to \mathbb{R}$ is bounded from below and is not coercive, then there exists a sequence $\{x_n\}_1^\infty$ in $\mathbb{R}^n$, $\|x_n\| \to \infty$, such that*

$$f(x_n) \to \varliminf_{\|x\| \to \infty} f(x),$$
$$\nabla f(x_n) \to 0,$$
$$\varliminf_{n \to \infty} \langle D^2 f(x_n)d, d\rangle \ge 0 \quad \textit{for all} \quad d \in \mathbb{R}^k.$$

Proof. Define

$$\alpha := \varliminf_{\|x\| \to \infty} f(x) = \sup_{n \ge 1} \inf_{\|x\| \ge n} f(x).$$

Let (M_n, d) be the complete metric space $M_n := \{x \in X : \|x\| \ge n\}$ with the distance function d given by the Euclidean norm on $\mathbb{R}^k$. Let $\|y_n\| \ge n + 2$ be a point satisfying

$$f(y_n) \le \inf_{M_{n+2}} f + \frac{1}{n} = \inf_{M_n} f + \delta_n + \frac{1}{n},$$

where $\delta_n := \inf_{M_{n+2}} f - \inf_{M_n} f$. We have $\delta_n \searrow 0$, since $\inf_{M_n} f \nearrow \alpha$. Applying Theorem 3.11 to the function f in M_n, with the point y_n, the parameters $\epsilon = \epsilon_n := \delta_n + 1/n$, $\lambda = 1$, and $p = 2$, we obtain a point x_n such that

$$\begin{aligned} f(x_n) \le f(y_n) &\le \alpha + \frac{1}{n}, \\ \|x_n - y_n\| &\le 1, \\ f(x_n) + \epsilon_n \|x_n - y_n\|^2 &\le f(x) + \epsilon_n \|x - y_n\|^2 \text{ for all } \|x\| \ge n. \end{aligned} \tag{3.3}$$

The middle inequality in (3.3) gives

$$\|x_n\| \ge \|y_n\| - \|x_n - y_n\| \ge \|y_n\| - 1 \ge n + 1,$$

while the last inequality in (3.3) gives $\nabla f(x_n) + 2\epsilon_n(x_n - y_n) = 0$, implying

$$\|\nabla f(x_n)\| = 2\epsilon_n \|x_n - y_n\| \le 2\epsilon_n.$$

Also, the last inequality in (3.3), applied to the points $x = x_n \pm td$ for fixed but arbitrary $d \in \mathbb{R}^k$, $\|d\| = 1$, and sufficiently small $t > 0$, gives

$$f(x_n) + \epsilon_n \|x_n - y_n\|^2 \le f(x_n \pm td) + \epsilon_n \|x_n - y_n \pm td\|^2.$$

Upon summing and simplifying these two inequalities, we arrive at,

$$\begin{aligned} -2\epsilon_n &= \frac{\epsilon_n (2\|x_n - y_n\|^2 - \|(x_n - y_n) + td\|^2 - \|(x_n - y_n) - td\|^2)}{t^2} \\ &\le \frac{f(x_n + td) + f(x_n - td) - 2f(x_n)}{t^2} \\ &\xrightarrow{t \to 0} \langle D^2 f(x_n) d, d \rangle. \end{aligned}$$

In summary, the point x_n has the properties $\|x_n\| \ge n + 1$, $f(x_n) \le \alpha + \frac{1}{n}$, $\|\nabla f(x_n)\| \le \epsilon_n$, and $\langle D^2 f(x_n) d, d \rangle \ge -2\epsilon_n$ for each $d \in \mathbb{R}^k$, $\|d\| = 1$. Since $\epsilon_n \to 0$, the sequence $\{x_n\}$ satisfies the required properties. □

3.3 Consistency of Linear Equalities and Inequalities

Systems of finitely many linear equations and inequalities (finite linear systems) play prominent roles in many parts of optimization, with linear programming, and the theory of convex polyhedra being the primary examples. They also have a role to play in the derivation of optimality conditions for general mathematical programming problems. Because of their importance, and because it gives an instructor flexibility in covering them, we treat the solvability of finite linear systems from several independent points of view.

The oldest method for treating finite linear systems is the *Fourier–Motzkin elimination method*, which goes back to Fourier in the nineteenth century and Motzkin in the 1930s; see Stoer and Witzgall [247]. A special case of this method is given in Chapter 7. This approach has the merit that it is completely elementary, and moreover, the field over which the vector space is defined is more general, so that one can characterize solvability of linear systems over, say, the field of rational numbers using this approach.

Another method for treating finite linear systems in vector spaces over more general fields combinatorial. This goes back to Carver [55] in the 1920s, and is treated in Appendix A. This appendix does not have any prerequisites and can be read at any time.

The traditional approach to characterizing the solvability of linear equations and inequalities is to obtain them as consequences of separation theorems for convex sets. We treat this method in detail in Chapter 7.

Finally, it is possible to approach finite linear systems via variational principles; this is described below.

Gordan's lemma is the equivalence of parts (b) and (c) in the following theorem.

Theorem 3.14. *Define the function*

$$f(x) = \ln \sum_{i=1}^{m} e^{\langle a_i, x\rangle},$$

where $\{a_i\}_1^m$ *are vectors in* $\mathbb{R}^n$*. The following statements are equivalent:*

$$\begin{array}{ll} (a) & f(x) \text{ is bounded from below,} \\ (b) & \text{there exists } \lambda \in \mathbb{R}^m, \lambda \geq 0, \lambda \neq 0, \sum_{i=1}^{m} \lambda_i a_i = 0, \\ (c) & \text{there exists no } x \text{ satisfying } \langle a_i, x\rangle < 0,\ i = 1, \ldots, m. \end{array}$$

Proof. The proofs of $(b) \Rightarrow (c)$ and $(c) \Rightarrow (a)$ are trivial.

To prove $(a) \Rightarrow (b)$, note that Corollary 3.5 gives a sequence $\{x_n\}$ satisfying

$$\nabla f(x_n) = \sum_{i=1} \lambda_i^{(k)} a_i \to 0,$$

where $\lambda_i^{(k)} = \frac{e^{\langle a_i, x_n\rangle}}{\sum_{j=1}^{m} e^{\langle a_j, x_n\rangle}}$. Since $0 \leq \lambda_i^{(k)} \leq 1$, it has a convergent subsequence, and we may without loss of generality assume that $\lambda^{(k)} \to \lambda$, $\lambda \geq 0$, and $\lambda \neq 0$. This gives $\sum_{i=1}^{m} \lambda_i a_i = 0$ and proves Gordan's lemma. □

The idea of using variational principles to prove Gordan's lemma appears to be due to Hiriart-Urruty; see [41].

Theorem 3.15. (***Motzkin's transposition theorem, homogeneous version***) *Let* $\{a_i\}_1^l$, $\{b_j\}_1^m$, *and* $\{c_k\}_1^p$ *be vectors in* $\mathbb{R}^n$*. Then the linear system*

$$\begin{aligned} \langle a_i, x\rangle &< 0, \quad i = 1, \ldots, l, \\ \langle b_j, x\rangle &\leq 0, \quad j = 1, \ldots, m, \\ \langle c_k, x\rangle &= 0, \quad k = 1, \ldots, p, \end{aligned} \tag{3.4}$$

is inconsistent if and only if there exist vectors (multipliers)

$$\begin{aligned}\lambda &:= (\lambda_1, \ldots, \lambda_l) \geq 0, \ \lambda \neq 0, \\ \mu &:= (\mu_1, \ldots, \mu_m) \geq 0, \\ \delta &:= (\delta_1, \ldots, \delta_p),\end{aligned} \tag{3.5}$$

such that

$$\sum_1^l \lambda_i a_i + \sum_1^m \mu_j b_j + \sum_1^p \delta_k c_k = 0. \tag{3.6}$$

Proof. We may assume, without any loss of generality, that $p = 0$, that is, linear equalities are not present in (3.4), since each equality $\langle c_k, x\rangle = 0$ may be replaced by two inequalities $\langle c_k, x\rangle \leq 0$ and $\langle -c_k, x\rangle \leq 0$, and the multipliers for these inequalities can be then combined in (3.6).

It is easy to see that if (3.6) holds with the multipliers satisfying (3.5), then (3.4) must be inconsistent, since if a vector x satisfies it, then we obtain the contradiction

$$\begin{aligned}0 &= \Big\langle \sum_1^l \lambda_i a_i + \sum_1^m \mu_j b_j + \sum_1^p \delta_k c_k, x\Big\rangle \\ &= \sum_1^l \lambda_i \langle a_i, x\rangle + \sum_1^m \mu_j \langle b_j, x\rangle + \sum_1^p \delta_k \langle c_k, x\rangle < 0,\end{aligned}$$

where the last inequality follows since at least one λ_i is positive.

Conversely, assume that (3.4) is inconsistent. We use induction on m to establish the claim that there exist multipliers satisfying (3.5) such that (3.6) holds. If $m = 0$, then the claim follows directly from Theorem 3.14.

Assuming that it is true for $m-1$, we now prove the claim for m. Call the inequality system (3.4) I_m, and the system obtained from it by removing the last inequality $\langle b_m, x\rangle \leq 0$, I_{m-1}. Thus, I_m is inconsistent. We may assume that I_{m-1} is consistent, since otherwise we obtain, using the induction hypothesis, nonnegative multipliers $\lambda \in \mathbb{R}^l$, $\mu \in \mathbb{R}^{m-1}$, $\lambda \neq 0$ such that $\sum_1^l \lambda_i a_i + \sum_1^{m-1} \mu_j b_j = 0$. Then (3.6) holds with $\mu_m = 0$, and the claim is proved.

Thus assume that I_{m-1} is consistent. Adding to this system the equality $\langle b_m, x\rangle = 0$ renders it inconsistent. Defining

$$L := \{b_m\}^\perp = \{x : \langle b_m, x\rangle = 0\},$$

we can write this inconsistent system in the form

$$\langle \Pi_L a_i, x\rangle < 0, \ i = 1, \ldots, l, \ \langle \Pi_L b_j, x\rangle \leq 0, \ j = 1, \ldots, m-1 \quad (x \in L),$$

where Π_L denotes the orthogonal projection operator onto L. By the induction hypothesis, there exist nonnegative multipliers $\lambda \in \mathbb{R}^l$, $\mu \in \mathbb{R}^{m-1}$, $\lambda \neq 0$ such that $\sum_1^l \lambda_i \Pi_L a_i + \sum_1^{m-1} \mu_j \Pi_L b_j = 0$. Hence the vector $\sum_1^l \lambda_i a_i + \sum_1^{m-1} \mu_j b_j$ lies in $L^\perp = \operatorname{span}\{b_m\}$, and there exists $\mu_m \in \mathbb{R}$ such that

$$\sum_1^l \lambda_i a_i + \sum_1^m \mu_j b_j = 0.$$

It remains to show that $\mu_m \geq 0$. Let x satisfy I_{m-1}, so that $\langle a_i, x \rangle < 0$ for $i = 1, \ldots, l$, and $\langle b_j, x \rangle \leq 0$ for $j = 1, \ldots, m-1$. Since I_m is inconsistent, we also have $\langle b_m, x \rangle > 0$, and the above equality gives

$$0 = \sum_1^l \lambda_i \langle a_i, x \rangle + \sum_1^m \mu_j \langle b_j, x \rangle < \mu_m \langle b_m, x \rangle,$$

yielding $\mu_m > 0$. Here the inequality follows since at least one λ_i is positive. □

Corollary 3.16. (***Farkas's lemma, homogeneous version***) *Let $\{a_i\}_1^m$ and c be vectors in $\mathbb{R}^n$. The following statements are equivalent:*

(a) *if x satisfies $\langle a_i, x \rangle \leq 0$, $i = 1, \ldots, m$, then it also satisfies $\langle c, x \rangle \leq 0$,*

(b) *there exists $\lambda \geq 0$ such that $c = \sum_{i=1}^m \lambda_i a_i$.*

Proof. The corollary follows immediately from Theorem 3.15 by noting that the validity of (a) is equivalent to the inconsistency of the linear inequality system

$$\langle -c, x \rangle < 0, \ \langle a_i, x \rangle \leq 0, \ i = 1, \ldots, m.$$

□

The next theorem characterizes the solvability of the most general system of finitely many linear equations and inequalities, and thus must be considered one of the central result in this area.

Theorem 3.17. (***Motzkin's transposition theorem, affine version***) *Let $\{a_i\}_1^l$, $\{b_j\}_1^m$, $\{c_k\}_1^p$ be vectors in $\mathbb{R}^n$, and let $\{\alpha_i\}_1^l$, $\{\beta_j\}_1^m$, $\{\gamma_k\}_1^p$ be scalars. Then the linear system*

$$\begin{aligned} \langle a_i, x \rangle &< \alpha_i, \quad i = 1, \ldots, l, \\ \langle b_j, x \rangle &\leq \beta_j, \quad j = 1, \ldots, m, \\ \langle c_k, x \rangle &= \gamma_k, \quad k = 1, \ldots, p, \end{aligned} \tag{3.7}$$

is inconsistent *if and only if there exist vectors (multipliers) $\lambda_0 \in \mathbb{R}$, $\lambda \in \mathbb{R}^l$, $\mu \in \mathbb{R}^m$, $\delta \in \mathbb{R}^p$, satisfying*

$$\begin{gathered} \sum_1^l \lambda_i a_i + \sum_1^m \mu_j b_j + \sum_1^p \delta_k c_k = 0, \\ \sum_1^l \lambda_i \alpha_i + \sum_1^m \mu_j \beta_j + \sum_1^p \delta_k \gamma_k + \lambda_0 = 0, \\ (\lambda_0, \lambda, \mu) \geq 0, \quad (\lambda_0, \lambda) \neq 0. \end{gathered} \tag{3.8}$$

Proof. Note that the inconsistency of the system (3.7) is equivalent to that of the homogenized system $\langle a_i, x\rangle < t\alpha_i$, $\langle b_j, x\rangle \le t\beta_j$, $\langle c_k, x\rangle = t\gamma_k$, $t > 0$, that is, of the system

$$\begin{aligned} &\langle (0,-1),(x,t)\rangle < 0, \\ &\langle (a_i,-\alpha_i),(x,t)\rangle < 0, \quad i = 1,\ldots,l, \\ &\langle (b_j,-\beta_j),(x,t)\rangle \le 0, \quad j = 1,\ldots,m, \\ &\langle (c_k,-\gamma_k),(x,t)\rangle = 0, \quad k = 1,\ldots,p. \end{aligned} \tag{3.9}$$

This follows from the fact that if x solves (3.7), then $(x,1)$ solves the homogeneous system, and conversely, if (x,t) solves the homogeneous system, then x/t solves (3.7).

Theorem 3.15 implies that there exist multipliers $\lambda_0 \in \mathbb{R}$, $\lambda \in \mathbb{R}^l$, $\mu \in \mathbb{R}^m$, and $\delta \in \mathbb{R}^p$ satisfying the sign restrictions $0 \le (\lambda_0,\lambda) \ne 0$, $0 \le \mu$, and the equality

$$\lambda_0(0,-1) + \sum_1^l \lambda_i(a_i,-\alpha_i) + \sum_1^m \lambda_i(b_j,-\beta_j) + \sum_1^p \delta_k(c_k,-\gamma_k) = 0,$$

which is exactly (3.8). □

Corollary 3.18. (*Farkas's lemma, affine version*) *Let*

$$\langle a_i, x\rangle \le \alpha_i, \; i = 1,\ldots,m, \tag{3.10}$$

be a consistent *system of linear inequalities, where* $\{a_i\}_1^m \subset \mathbb{R}^n$. *The following statements are equivalent:*

(*a*) *if x satisfies $\langle a_i, x\rangle \le \alpha_i$, $i = 1,\ldots,m$, then it also satisfies $\langle c, x\rangle \le \gamma$,*

(*b*) *there exists $0 \le \lambda \in \mathbb{R}^m$ such that $\sum_{i=1}^m \lambda_i a_i = c$, and $\sum_{i=1}^m \lambda_i \alpha_i \le \gamma$.*

Proof. Note that the validity of (a) is equivalent to the inconsistency of the linear inequality system

$$\langle -c, x\rangle < -\gamma, \; \langle a_i, x\rangle \le \alpha_i, \; i = 1,\ldots,m.$$

It follows from Theorem 3.17 that there exist nonnegative multipliers $0 \ne (\mu_0,\mu_1) \in \mathbb{R}^2$ and $\lambda \in \mathbb{R}^m$ such that

$$-\mu_1 c + \sum_{i=1}^m \lambda_i a_i = 0, \quad -\mu_1\gamma + \sum_{i=1}^m \lambda_i \alpha_i + \mu_0 = 0.$$

If $\mu_1 > 0$, then we may assume that $\mu_1 = 1$ due to homogeneity, and the corollary follows. If $\mu_1 = 0$, then we have $\mu_0 > 0$, $\sum_{i=1}^m \lambda_i a_i = 0$, and $\sum_{i=1}^m \lambda_i \alpha_i < 0$. This, however, contradicts the consistency of (3.10), since if x satisfies it, then

$$0 \ge \sum_1^m \lambda_i(\langle a_i, x\rangle - \alpha_i) = -\sum_1^m \lambda_i \alpha_i > 0.$$

□

3.4 Variational Proofs of Some Basic Theorems of Nonlinear Analysis

In this section, we demonstrate that some of the most important results in nonlinear analysis can be proved using Ekeland's ϵ-variational principle. These include the open mapping theorem, Graves's theorem [110], the inverse function theorem, implicit function theorem, and Lyusternik's theorem. (The traditional proofs of the open mapping theorem and Graves's theorem are given in Appendix C for comparison and completeness.)

The inverse function theorem and the closely related implicit function theorem are important tools in many branches of analysis. Lyusternik's theorem [191] is an important tool in optimization, where it is used in the derivation of optimality conditions in constrained optimization. The open mapping theorem is a fundamental result in functional analysis.

All of these results are valid in Banach spaces. However, if one is interested only in using the inverse function, implicit function, and Lyusternik's theorems in a finite-dimensional setting, then there is a short cut. All these results are proved in an elementary manner in Section 2.5.

3.4.1 The Open Mapping and Graves's Theorems

We first define a useful concept due to De Giorgi, Marino, and Tosques [70].

Definition 3.19. *Let $f : X \to \mathbb{R}$ be a function on a metric space X. The* strong slope *of $|\nabla f|(x)$ of f at a point $x \in X$ is given by the formula*

$$|\nabla f|(x) = \begin{cases} \overline{\lim}_{z \to x} \frac{f(x)-f(z)}{d(x,z)} & \text{if } x \text{ is not a local minimizer of } f, \\ 0 & \text{otherwise.} \end{cases}$$

Thus, at a point x that is not a local minimizer of f, $|\nabla f|(x)$ measures the *largest local rate of decrease of f at x.*

The concept of strong slope interacts nicely with Ekeland's ϵ-variational principle: let $f : X \to \mathbb{R}$ be a lower semicontinuous function on a complete metric space X that is bounded from below. If $\sigma > 0$, then Theorem 3.1 implies that there exists a σd-point $\overline{x}$, and we notice from its definition that $\overline{x}$ satisfies the property

$$|\nabla f|(\overline{x}) \leq \sigma.$$

This proves the existence of points with small positive strong slope, that is,

$$\inf\{|\nabla f|(x) : \inf_M f < f(x) < \infty\} = 0.$$

We also need the following, partial, statement of the open mapping theorem.

Lemma 3.20. *Let $T : X \to Y$ be a continuous linear mapping from a Banach space X onto a Banach space Y. Then there exists a constant $\tau > 0$ such that $\tau \overline{B}_Y \subseteq \overline{A(\overline{B}_X)}$, where $\overline{B}_X = \{x \in X : \|x\| \leq 1\}$ and $\overline{B}_Y = \{y \in Y : \|y\| \leq 1\}$ are the closed unit balls in X and Y, respectively.*

Proof. Since A is onto,

$$Y = A(X) = A(\cup_{n=1}^{\infty} n\overline{B}_X) = \cup_{n=1}^{\infty} A(n\overline{B}_X) = \cup_{n=1}^{\infty} nA(\overline{B}_X).$$

It follows from the Baire category theorem that at least one set $\overline{nA(\overline{B}_X)}$ contains an open set. This implies that $\overline{A(\overline{B}_X)}$ contains an open set, say $y + \tau B_Y \subseteq \overline{A(\overline{B}_X)}$. Since $\overline{B}_X = -\overline{B}_X$, we also have $-y + \tau B_Y \subseteq \overline{A(\overline{B}_X)}$. If $z \in Y$ such that $\|z\| < \tau$, then there exist $\{u_k\}_1^{\infty}$ and $\{v_k\}_1^{\infty}$ in $\overline{B}_X$ such that $y + z = \lim Au_k$ and $-y + z = \lim Av_k$. But then $z = \lim A(u_k + v_k)/2 \in \overline{A(\overline{B}_X)}$, proving $\tau B_Y \subseteq \overline{A(\overline{B}_X)}$. □

We are now ready to state and prove Graves's theorem. It should be noticed that the proof below needs only Lemma 3.20 above and not the full statement of the open mapping theorem, that it is shorter, and that it proves a somewhat stronger result than Theorem C.2. Moreover, we deduce the full proof of the open mapping theorem from it.

Theorem 3.21. (***Graves's theorem***) *Let X and Y be Banach spaces, $r > 0$, and let $f : r\overline{B}_X \to Y$ be a mapping such that $f(0) = 0$. Let $A : X \to Y$ be a continuous linear mapping onto Y satisfying $\tau \overline{B}_Y \subseteq \overline{A(\overline{B}_X)}$. Let $f - A$ be Lipschitz continuous on D with a constant δ, $0 \leq \delta < \tau$, that is,*

$$\|f(x_1) - f(x_2) - A(x_1 - x_2)\| \leq \delta \|x_1 - x_2\| \text{ for all } x_1, x_2 \in r\overline{B}_X.$$

Then

$$(\tau - \delta) r B_Y \subseteq f(r\overline{B}_X), \tag{3.11}$$

that is, the equation $y = f(x)$ has a solution $\|x\| \leq r$ whenever $\|y\| < (\tau - \delta) r$.

Moreover,

$$c\, d(x, f^{-1}(y)) \leq \|f(x) - y\|, \quad \text{for all } \|x\| \leq r,\ \|y\| < cr, \tag{3.12}$$

where $c : \tau - \delta > 0$.

Proof. Define $D = r\overline{B}_X$, and for each point $y \in Y$, define the function f_y on D,

$$f_y(x) := \|f(x) - y\|.$$

We first claim that

$$|\nabla f_y|(x) \geq c > 0 \text{ for all } x \in D \text{ such that } f_y(x) \neq 0.$$

If $x \in D$ is such a point, then Lemma 3.20 implies that there exists a sequence $\{d_n\} \subset X$ with $\|d_n\| \leq 1$ such that

$$-\tau \frac{f(x) - y}{\|f(x) - y\|} = -\tau \frac{f(x) - y}{f_y(x)} = \lim_{n\to\infty} Ad_n.$$

We have

$$f(x + td_n) - y = \Big[(f - A)(x + td_n) - (f - A)(x)\Big] + t\Big[Ad_n + \tau \frac{f(x) - y}{f_y(x)}\Big] + \Big(1 - \frac{t\tau}{f_y(x)}\Big)(f(x) - y),$$

implying that for sufficiently small $t > 0$ and sufficiently large n,

$$\begin{aligned} f_y(x + td_n) &\le t\delta \|d_n\| + o(t) + \left(1 - \frac{t\tau}{f_y(x)}\right) f_y(x) \\ &\le f_y(x) + t(\delta - \tau) + o(t) \\ &< f_y(x), \end{aligned}$$

where the last equality follows from $\delta < \tau$. This shows that x is not a local minimizer of f_y. We have therefore

$$\begin{aligned} |\nabla f_y|(x) &= \overline{\lim_{z\to x}} \frac{f_y(x) - f_y(z)}{\|x - z\|} \ge \overline{\lim_{t\searrow 0, n\to\infty}} \frac{f_y(x) - f_y(x + td_n)}{t\|d_n\|} \\ &\ge \overline{\lim_{t\searrow 0, n\to\infty}} \frac{f_y(x) - f_y(x + td_n)}{t} \ge \tau - \delta = c, \end{aligned}$$

proving our claim.

If (3.11) is false, then there exists $y \in Y$, $\|y\| < cr$, such that f_y is positive on D. Choosing $\epsilon := \|y\| < cr$ and noting that $\|y\| = f_y(0) \le \inf_D f_y + \epsilon$, it follows from Theorem 3.2 (with $\lambda = r$) that there exists a point $x_\epsilon \in D$ satisfying

$$f_y(x_\epsilon) \le f_y(x) + \frac{\epsilon}{r}\|x_\epsilon - x\| \quad \text{for all } x \in D.$$

Therefore, $c \le |\nabla f_y|(x_\epsilon) \le \epsilon/r < c$, where the first inequality follows from our first claim, and the second one follows since x_ϵ is an $(\epsilon/r)d$-point. This is a contradiction that settles (3.11).

Finally, let $\|y\| < cr$ and $x \in D$ such that $f(x) \ne y$. Define $\delta := d(x, f^{-1}(y)) > 0$, $\epsilon := f_y(x) = \|f(x) - y\| > 0$, and pick $0 < \lambda < \delta$. Since $\inf_D f_y = 0$, we have $f_y(x) = \inf_D f_y + \epsilon$, and Theorem 3.2 implies that there exists a point $\overline{x} \in D$ satisfying $\|\overline{x} - x\| \le \lambda < \delta$, so that $f_y(\overline{x}) > 0$, and that

$$f_y(\overline{x}) \le f_y(z) + \frac{\epsilon}{\lambda}\|\overline{x} - z\| \quad \text{for all } z \in D,$$

so that $|\nabla f_y|(\overline{x}) \le \epsilon/\lambda$. Therefore,

$$c \le |\nabla f_y|(\overline{x}) \le \frac{\epsilon}{\lambda} = \frac{\|f(x) - y\|}{\lambda},$$

for all $0 < \lambda < \delta = d(x, f^{-1}(y))$. This proves (3.12). □

Corollary 3.22. (***Open mapping theorem***) *Let X and Y be Banach spaces and let $A : X \to Y$ be a continuous linear mapping onto Y. Then A is an open mapping, that is, if $O \subseteq X$ is open, then $A(O)$ is open in Y.*

Proof. Applying Theorem 3.21 with $f = A$, $y = 0$, and $\delta = 0$, we obtain

$$\tau\, d(x, A^{-1}0) \leq \|Ax\|,$$

for all x with small enough norm, and hence for all x, by the homogeneity of the above inequality. If $y = Ax$ satisfies $\|y\| < \tau$, then $d(x, A^{-1}0) < 1$. Thus there exists a point $u \in X$ such that $Au = 0$ and $\|x - u\| < 1$. Since $A(x-u) = y$, this shows that $\tau B_Y \subseteq A(B_X)$. Since A is linear, it follows that A is an open mapping. □

The references [16] and [145] contain more applications of Ekeland's ϵ-variational principle along these lines.

3.4.2 Lyusternik's Theorem

Theorem 3.23. (***Lyusternik's theorem***) *Let X and Y be Banach spaces, $U \subseteq X$ an open set, and $f : U \to Y$. Let $T_M(x_0)$ be the tangent cone of the level set $M = f^{-1}(f(x_0))$ at the point $x_0 \in U$.*

If f is C^1 in a neighborhood of a point x_0 such that $Df(x_0)$ is a linear mapping onto Y, then $T_M(x_0)$ is the null space of the linear map $Df(x_0)$, that is,

$$T_M(x_0) = \operatorname{Ker} Df(x_0) := \{d \in X : Df(x_0)(d) = 0\}. \tag{3.13}$$

Proof. As in the proof of Theorem 2.29, we may assume that $x_0 = 0$ and $f(x_0) = 0$. Define $A = Df(0)$. The proof of the inclusion $T_M(0) \subseteq \operatorname{Ker} A$ is the standard one given there, namely if $d \in T_M(0)$, then there exist points $x(t) = td + o(t)$ in M, so that

$$0 = f(td + o(t)) = f(0) + tDf(0)(d) + o(t) = tAd + o(t).$$

Dividing both sides by t and letting $t \to 0$, we obtain $Ad = 0$.

To prove the reverse inclusion $\operatorname{Ker} A \subseteq T_M(0)$, note that Theorem 1.18 implies

$$\|f(x_2) - f(x_1) - A(x_2 - x_1)\| \leq \|x_2 - x_1)\| \cdot \sup_{t\in[0,1]} \|Df(x_1 + t(x_2 - x_1)) - A\|.$$

Therefore, given $\epsilon > 0$, there exists a neighborhood $U \ni 0$ such that $f - A$ is Lipschitz continuous with a constant ϵ on U. It follows from Theorem 3.21 that there exists a constant $c > 0$ such that $cd(x, M) \leq \|f(x)\|$ for all x in a neighborhood $V \ni 0$.

If $d \in \operatorname{Ker} A$, then $f(td) = f(0) + tAd + o(t) = o(t)$, so that $d(td, M) = o(t)$. Thus, there exists $x(t) \in M$ satisfying $x(t) - td = o(t)$. We obtain

$$\frac{x(t) - 0}{t} = d + \frac{o(t)}{t} \to d \text{ as } t \to 0,$$

proving that $d \in T_M(0)$. □

The references [78, 146, 145] provide much more information on Lyusternik's theorem and its uses in optimization and related fields. The reference Ioffe [145] is a recent survey on Graves's theorem and the associated concept of *metric regularity*.

3.4.3 The Inverse and Implicit Function Theorems

The inverse function theorem and the closely related implicit function theorems are among the most important results in all of nonlinear analysis. We turn to these results next.

Theorem 3.24. (***Inverse function theorem***) *Let X and Y be Banach spaces, $x_0 \in X$, and f is a C^1 mapping from a neighborhood of x_0 into Y.*

If $Df(x_0) : X \to Y$ is invertible, then f is a C^1 diffeomorphism on a neighborhood of x_0.

Moreover, if f is C^k on Ω, then f is a C^k diffeomorphism on a neighborhood of x_0.

Proof. Define $A = Df(x_0)$. Corollary 3.22 implies that A^{-1} is a continuous linear map; thus $\|A^{-1}\| = \sup_{\|y\|=1} \|A^{-1}y\| < \infty$. As noted in the proof Theorem 3.23, given $\epsilon > 0$, there exists a neighborhood U of x_0 such that

$$\|f(x_2) - f(x_1) - A(x_2 - x_1)\| \le \epsilon\|x_2 - x_1\| \quad \text{for all} \quad x_1, x_2 \in U. \tag{3.14}$$

This implies that f is one-to-one in a neighborhood of x_0, because if $x_1 \neq x_2$ and $f(x_1) = f(x_2)$ in the above inequality, then

$$\begin{aligned} \|A(x_1 - x_2)\| &= \|f(x_1) - f(x_2) - A(x_1 - x_2)\| \le \epsilon\|x_1 - x_2\| \\ &\le \epsilon\|A^{-1}\| \cdot \|A(x_1 - x_2)\|, \end{aligned}$$

which cannot hold if $\epsilon > 0$ is small enough.

Moreover, the inclusion (3.11) in Theorem 3.21 implies that there exist open neighborhoods $U \ni x_0$ and $V \ni y_0$ such that $f : U \to V$ is one-to-one and onto, and the inequality (3.12) implies that $f^{-1} : V \to U$ is continuous. Setting $y_i := f(x_i)$, $i = 1, 2$, we have

$$\begin{aligned} &\|f^{-1}(y_2) - f^{-1}(y_1) - A^{-1}(y_2 - y_1)\| \\ &= \|x_2 - x_1 - A^{-1}(f(x_2) - f(x_1))\| \\ &\le \|A^{-1}\| \cdot \|f(x_2) - f(x_1) - A(x_2 - x_1)\| \\ &\le \epsilon\|A^{-1}\| \cdot \|x_2 - x_1\| \\ &\le \frac{\epsilon\|A^{-1}\|}{c}\|y_2 - y_1\|, \end{aligned}$$

for all $y_1, y_2 \in V$, where the second inequality follows from (3.14), and the last inequality from (3.12). This proves that f^{-1} is differentiable at y_0 with $Df^{-1}(y_0) = A^{-1} = Df(x_0)^{-1}$.

If the neighborhood U containing x_0 is chosen small enough such that $Df(x)$ is invertible for every $x \in U$, then f is a diffeomorphism between U and $V = f(U)$. It follows that if $y = f(x)$, then

$$Df^{-1}(y) = Df(x)^{-1},$$

that is,

$$D(f^{-1}) = \mathrm{Inv} \circ Df \circ f^{-1} \tag{3.15}$$

on $f(U)$, where Inv is the map sending a nonsingular matrix to its inverse. Since Inv is infinitely differentiable (see Example 1.28) and Df and f^{-1} are continuous, it follows that $f^{-1} \in C^1$. If, moreover, $f \in C^2$, that is, $Df \in C^1$, it follows from 3.15 that $D(f^{-1}) \in C^1$, that is, $f^{-1} \in C^2$. Induction on k settles the general case. □

Remark 3.25. The above proof of the inverse function theorem differs from the usual proofs in that Graves's theorem (for which we gave a variational as well as a classical proof) is used instead of Banach's fixed point theorem to establish the fact that f is an open mapping near x_0. The common proof based on the latter theorem can be found in many books; see for example Lang [183] for a clean presentation.

The inverse function theorem may fail if the continuity assumption on Df is removed; see an example in [77], page 273.

If X and Y are Banach spaces, $x_0 \in X$, $y_0 \in Y$, and f is a C^1 map in a neighborhood of (x_0, y_0), we denote by $D_x f(x_0, y_0)$ the derivative of f at (x_0, y_0) with respect to x, that is, $D_x f(x_0, y_0)$ is the derivative of the map $x \mapsto f(x, y_0)$ at x_0.

Theorem 3.26. (***Implicit function theorem***) *Let X, Y be Banach spaces, $x_0 \in X$, $y_0 \in Y$, and f a C^1 map in a neighborhood of (x_0, y_0). Define $w_0 = f(x_0, y_0)$.*

If $D_y f(x_0, y_0) : Y \to Y$ is a linear isomorphism, then there exist neighborhoods $U \ni x_0$ and $V \ni y_0$, and a C^1 mapping $y : U \to V$ such that a point $(x, y) \in U \times V$ satisfies $f(x, y) = w_0$ if and only if $y = y(x)$. The derivative of y at x_0 is given by

$$Dy(x_0) = -D_y f(x_0, y_0)^{-1} D_x f(x_0, y_0).$$

Moreover, if f is C^k, then so is $y(x)$.

Proof. Define the map $F(x, y) = (x, f(x, y))$. At a point $z_0 = (x_0, y_0)$, it follows from Taylor's formula

$$F(x_0+th, y_0+tk) = (x_0+th, f(x_0+th, y_0+tk))$$
$$= \Big(x_0+th, f(x_0,y_0)+t[D_xf(x_0,y_0)h+D_yf(x_0,y_0)k]+o(t)\Big)$$
$$= F(x_0,y_0)+t\Big(h, D_xf(x_0,y_0)h+D_yf(x_0,y_0)k\Big)+o(t),$$

that the derivative $DF(x_0, y_0) : X \times Y \to X \times Y$ given by

$$DF(x_0,y_0)(h,k) = \big(h, D_xf(x_0,y_0)h + D_yf(x_0,y_0)k\big)$$

is a linear isomorphism. It follows from Theorem 3.24 that there exist neighborhoods $U \ni x_0$, $V \ni y_0$, and $W \ni (x_0, w_0)$ such that $F : U \times V \to W$ is a bijection and F^{-1} is C^1 (C^k if f is C^k). Note that $F^{-1}(x,y) = (x, g(x,y))$ for some function $g \in C^1$.

Let $(x, y) \in U \times V$. We have $f(x,y) = w_0$ if and only $F(x,y) = (x, w_0)$, or $(x,y) = F^{-1}(x,w_0) = (x, g(x,w_0))$, that is, if and only if $y = g(x,w_0) =: y(x)$. Since $g \in C^1$, we have $y(x) \in C^1$ as well.

The chain rule applied to the function $f(x, y(x)) = w_0$ gives $D_xf(x,y(x))+ D_yf(x,y(x))Dy(x) = 0$, leading to the formula for $Dy(x_0)$. □

3.5 Exercises

1. Let $f : \mathbb{R}^n \to \mathbb{R}$ be a lower semicontinuous function that is bounded from below. Suppose that there exist $a > 0$ and b and $R > 0$ such that $f(x) \geq a\|x\| + b$ for $\|x\| \geq R$. Prove that the image $\nabla f(\mathbb{R}^n)$ is dense in the ball $\{x \in \mathbb{R}^n : \|x\| \leq a\}$.
   Hint: Apply Ekeland's ϵ-variational principle to the function $g(x) = f(x) - \langle x, u\rangle$, where $\|u\| \leq a$.
2. Let $f : X \to \mathbb{R}$ be a lower semicontinuous Gâteaux differentiable function on a Banach space X. If f is bounded from below and coercive, that is, $\lim_{\|x\|\to\infty} \frac{f(x)}{\|x\|} = \infty$, then the range $\{\nabla f(x) : x \in X\}$ is dense in X^*, the dual space of X.
   Remark: Compare this problem to Exercise 6 on page 56.
3. ***(Caristi)*** Let (X, d) be a complete metric space, and $f : X \to \mathbb{R} \cup \{\infty\}$ a lower semicontinuous function that is bounded from below. Let $T : X \to X$ be a multivalued mapping, that is, for each $x \in X$, $T(x)$ is a subset of X. Prove that if

$$f(y) \leq f(x) - d(x,y) \quad \text{for all} \quad y \in T(x),$$

   then there exists a fixed point of T, that is, there exists a point $\overline{x} \in X$ such that

$$\overline{x} \in T(\overline{x}).$$

   Hint: Set $\epsilon = \lambda = 1$ in Ekeland's ϵ-variational principle, and show that the point obtained is a fixed point of T.
   Caristi's theorem is in fact equivalent to Ekeland's ϵ-variational principle, because it is possible to obtain the latter from the former.

4. Prove the following refinement of Corollary 3.5: let f be a function as in that corollary, and $\{x_n\}$ a minimizing sequence for f. Prove that there exists another minimizing sequence $\{y_n\}$ for f such that

$$f(y_n) \le f(x_n), \quad \|x_n - y_y\| \to 0, \quad \text{and} \quad \nabla f(y_n) \to 0.$$

5. Let $f : \mathbb{R}^n \to \mathbb{R}$ be a C^1 function. The function f is said to satisfy the *Palais–Smale condition* if whenever a sequence $\{x_n\}_1^\infty$ satisfies the conditions

$$f(x_n) \to \alpha, \quad \nabla f(x_n) \to 0,$$

then $\{x_n\}$ contains a convergent subsequence.
Prove that if f is bounded from below and satisfies the Palais–Smale condition, then f must be a coercive function.
6. Let $f : E \to E$ be a contractive mapping on a Banach space E. Prove that the mapping $T = I - f$ is a homeomorphism of E, that is, $T : E \to E$ is one-to-one and onto, and T^{-1} is continuous.
Hint: Given $y \in E$, apply Banach's fixed point theorem to the mapping $x \mapsto y - f(x)$.
7. Let $f : U \to \mathbb{R}^m$ be a C^1 map, where $U \subset \mathbb{R}^n$ is an open set. If the gradient $Df(x_0)$ at a point $x_0 \in U$ is a linear map onto $\mathbb{R}^m$, prove that there exists a local right inverse of f around x_0, that is, there exist neighborhoods $V \ni x_0$ and $W \ni f(x_0)$ and a C^1 mapping $g : W \to V$ such that $f(g(y)) = y$ for all $y \in W$.
Hint: Define $A = Df(x_0)$, and find a linear map $B : \mathbb{R}^m \to \mathbb{R}^n$ such that $AB = I_m$. Then show that the inverse function theorem can be applied to $h = f \circ B$, and use h to define a suitable function g satisfying the required properties.
State and prove an analogous theorem when $Df(x_0)$ is a one-to-one linear map.

4

Convex Analysis

Convexity is an important part of optimization, and we devote several chapters to various aspects of it in this book. This chapter treats the most basic properties of convex sets and functions, while Chapter 5 is devoted to their deeper properties, such as the relative interior (both in the algebraic and topological senses) and boundary structure of convex sets, the continuity properties of convex functions, and homogenization of convex sets. Chapter 6 treats the separation properties of convex sets, a very important topic in optimization. The calculus of the relative interior of convex sets developed in Chapter 5 plays an important role here. Chapters 7 and 8 deal with two, related, special topics, the theory of convex polyhedra and the theory of linear programming, respectively. Both are important topics within optimization, and each has wide applicability within science, engineering, and technology. Many developments in optimization have in fact been inspired by linear programming. Finally, Chapter 13 investigates several special topics in convexity.

4.1 Affine Geometry

Convex sets contain the line segment between any two of its points, while affine sets contain the whole line between any two of its points. Thus, the natural setting for convex sets is an affine subset of a vector space. Besides, concepts such as the dimension and the relative interior of a convex set make essential use of ideas from affine geometry; as an example; see the statement of Theorem 4.13 below due to Carathéodory. For this reason, we review here the fundamental properties of affine sets before embarking on the study of convex sets.

Definition 4.1. *A nonempty subset A of a vector space E is called an* affine set *if for points x and y in A, the line passing through x and y is contained in A; in other words,*

$$x, y \in A \implies \ell := \{x + t(y - x) = (1 - t)x + ty : t \in \mathbb{R}\} \subseteq A.$$

O. Güler, *Foundations of Optimization*, Graduate Texts in Mathematics 258,
DOI 10.1007/978-0-387-68407-9_4,

Let A and B be affine sets in vector spaces E and F, respectively. A map $F : A \to B$ is called an affine map *if*

$$F((1-t)x + ty) = (1-t)F(x) + tF(y) \quad \text{for all } t \in \mathbb{R}.$$

Let $\{x_i\}_1^k$ be a finite set of points in E. An affine combination *of $\{x_i\}_1^k$ is any point of the form*

$$\sum_{i=1}^{k} \lambda_i x_i, \quad \sum_{i=1}^{k} \lambda_i = 1.$$

Let B be a nonempty set in E. The affine hull *(or span) of B is the set of all affine combinations of points from B, that is,*

$$\operatorname{aff}(B) := \left\{ \sum_{i=1}^{k} \lambda_i b_i \; : \; b_i \in B, \; \sum_{i=1}^{k} \lambda_i = 1, \; k = 1, 2, \ldots \right\}.$$

Lemma 4.2. *Let A be a nonempty set in E. Then $\operatorname{aff}(A)$ is an affine set, in fact, the smallest affine set containing A.*

Proof. It is clear that $A \subseteq \operatorname{aff}(A)$. Let us show that $\operatorname{aff}(A)$ is an affine set. If $t \in \mathbb{R}$ and $u, v \in \operatorname{aff}(A)$ have the forms

$$u = \sum_{i=1}^{k} \lambda_i x_i, \; v = \sum_{j=1}^{l} \mu_j y_j, \text{ where } x_i, y_j \in A, \sum_{i=1}^{k} \lambda_i = 1, \; \sum_{j=1}^{l} \mu_j = 1,$$

then

$$(1-t)u + tv = \sum_{i=1}^{k} (1-t)\lambda_i x_i + \sum_{j=1}^{l} t\mu_j y_j.$$

Since

$$\sum_{i=1}^{k} (1-t)\lambda_i + \sum_{j=1}^{l} t\mu_j = (1-t)\sum_{i=1}^{k} \lambda_i + t\sum_{j=1}^{l} \mu_j = (1-t) + t = 1,$$

we see that $(1-t)u + tv$ is an affine combination of points in A; this proves that $\operatorname{aff}(A)$ is an affine set.

Next, we show that if D is an affine set containing A, then $\operatorname{aff}(A) \subseteq D$, that is, every affine combination $\sum_{i=1}^{k} \lambda_i x_i$, $x_i \in A$, lies in D. We use induction on k, the number of elements in the affine combination. The result is trivial for $k = 1$, and easy for $k = 2$: if x_1 and x_2 are points in A, and $0 \neq t \in \mathbb{R}$, then $x_1, x_2 \in D$, and since D is affine, we have $(1-t)x_1 + tx_2 \in D$. Supposing we have shown the result for every $i < k$, we prove it for k. Consider an affine combination $x := \sum_{i=1}^{k} \lambda_i x_i$ with $\{x_i\}_1^k \subseteq A$, $\lambda_i \neq 0$, $i = 1, \ldots, k$. Write

$$x = \alpha\Big(\sum_{i=1}^{k-1} \frac{\lambda_i}{\alpha} x_i\Big) + (1-\alpha)x_k, \quad \text{where } \alpha := \sum_{i=1}^{k-1} \lambda_i = 1 - \lambda_k.$$

The point $\overline{x} := \sum_{i=1}^{k-1}(\lambda_i/\alpha)x_i$ is an affine combination of $\{x_i\}_{i=1}^{k-1}$, since $\sum_{i=1}^{k-1}\lambda_i/\alpha = 1$, and it lies in D by the induction hypothesis. Since $x = t\overline{x} + (1-t)x_k$, and $\overline{x}, x_k \in D$, we have $x \in D$. The lemma is proved. □

An immediate consequence of Lemma 4.2 is the following result.

Lemma 4.3. *If $A \subseteq E$ is an affine set, then* $\operatorname{aff}(A) = A$*, that is, all affine combinations of elements from A lie in A:*

$$\{\lambda_i\}_1^k \subset \mathbb{R}, \ \sum_{i=1}^k \lambda_i = 1, \ \{x_i\}_1^k \subset A \qquad \Longrightarrow \qquad \sum_{i=1}^k \lambda_i x_i \in A.$$

Affine sets are precisely the translations of linear subspaces. The following result makes this connection explicit.

Theorem 4.4. *If $A \subseteq E$ is an affine subset of E, and $a \in A$ is an arbitrary point, then*

$$L := A - a = \{y - a : y \in A\}$$

is a linear subspace of E, which is independent of $a \in A$; consequently,

$$A = a + L, \qquad \textit{and} \qquad L = A - A = \{y - z : y, z \in A\}. \tag{4.1}$$

Conversely, if $a \in E$ and L is a linear subspace of E, then $A := a + L$ is an affine subspace of E.

Proof. Let us first prove that L is a linear subspace of E. Let $\{x_1, x_2\} \subseteq L$ and $\{\alpha_1, \alpha_2\} \subset \mathbb{R}$. Writing $x_i = y_i - a$, $y_i \in A$, we have

$$\begin{aligned} y &:= a + \alpha_1 x_1 + \alpha_2 x_2 = a + \alpha_1(y_1 - a) + \alpha_2(y_2 - a) \\ &= (1 - \alpha_1 - \alpha_2)a + \alpha_1 y_1 + \alpha_2 y_2 \in A, \end{aligned}$$

by Lemma 4.3, since y is an affine combination of $a, y_1, y_2 \in A$. Thus, $\alpha_1 x_1 + \alpha_2 x_2 = y - a \in L$, and L is a linear subspace of E.

Next, we claim that if $a_1, a_2 \in A$, then $A - a_1 = A - a_2$. This is equivalent to proving that if $a \in A$, the b defined by the equation $a - a_1 = b - a_2$, that is, $b := a - a_1 + a_2$, lies in A. This follows again from Lemma 4.3, since b is an affine combination of $a, a_1, a_2 \in A$.

It remains to prove the converse statement. Let $A := a + L$, where $a \in E$ and $L \subseteq E$ is a linear subspace of E. If $y_i = a + x_i$, $x_i \in L$, $i = 1, 2$, then

$$(1 - \lambda)y_1 + \lambda y_2 = a + [(1 - \lambda)x_1 + \lambda x_2] \in a + L = A,$$

proving that A is an affine subset of E. □

Definition 4.5. *Let $A \subseteq E$ be an affine set. The* (affine) dimension *A is the dimension of the linear subspace $L = A - A$,*

$$\dim(A) = \dim(A - A).$$

A set of vectors $\{x_i\}_1^l$ *in* A *is called* affinely independent *if every affine combination of* $\{x_i\}_1^l$ *is uniquely written as an affine combination; in other words, if* $x = \sum_1^l \alpha_i x_i$, $\sum_1^l \alpha_i = 1$, *then the coefficients* $\{\alpha_i\}_1^l$ *are unique.*

A set of vectors $\{x_i\}_1^k$ *in* A *is called an* affine basis *of* A *if it is affinely independent and spans* A.

Lemma 4.6. *Let* $A \subseteq E$ *be an affine subset of* E *with the corresponding linear subspace* $L = A - A$.

A subset $\{x_i\}_1^l$ *of* A *is affinely independent if and only if*

$$\sum_1^l \alpha_i x_i = 0, \quad \sum_1^l \alpha_i = 0 \qquad \Longrightarrow \qquad \alpha_i = 0, \ i = 1, \ldots, l. \tag{4.2}$$

Moreover, $\{x_i\}_1^l$ *in* A *is affinely independent if and only if for any fixed* j, *the vectors* $\{x_i - x_j\}_{i \neq j}$ *in* L *are linearly independent.*

A set of vectors $\{x_i\}_1^m$ *affinely span* A *if and only if for any fixed* j, *the vectors* $\{x_i - x_j\}_{i \neq j}$ *linearly span* L. *Consequently, a set of vectors* $\{x_i\}_1^k$ *is a affine basis of* A *if and only if for any fixed* j, *the set of vectors* $\{x_i - x_j\}_{i \neq j}$ *is a linear basis of* L.

Proof. Notice that $x = \sum_1^l \alpha_i x_i$ and $x = \sum_1^l \beta_i x_i$ are two distinct affine combinations if and only if $0 = \sum_1^l (\beta_i - \alpha_i) x_i$ is a nontrivial linear combination with $\sum_1^l (\beta_i - \alpha_i) = 0$; this proves the first statement.

To prove the second statement, note that the conditions $\sum_1^l \alpha_i x_i = 0$ and $\sum_1^l \alpha_i = 0$ can be rewritten in the form $\sum_{i \neq j} \alpha_i (x_i - x_j) = 0$ with no restrictions on $\{\alpha_i\}_{i \neq j}$.

To prove the third statement, note that $L = A - x_j$. Thus, an arbitrary point $x \in A$ can be represented as an affine combination $x = \sum_1^m \alpha_i x_i$, $\sum_1^l \alpha_i x_i = 0$, if and only if the corresponding point $y =: x - x_j \in L$ has a representation as a linear combination $y = \sum_{i \neq j} \alpha_i (x_i - x_j)$, because

$$\begin{aligned} y = x - x_j &= \Big(\sum_1^m \alpha_i x_i\Big) - x_j = \sum_{i \neq j} \alpha_i (x_i - x_j) + \Big(\sum_1^m \alpha_i - 1\Big) x_j \\ &= \sum_{i \neq j} \alpha_i (x_i - x_j). \end{aligned}$$

The lemma is proved. □

Let $F : A \to B$ be a map between two affine sets A and B. It is easy to verify that F is an affine map if and only if its graph

$$\mathrm{gr}(F) := \{(x, F(x)) : x \in A\} \subseteq A \times B$$

is an affine set. If A and B are vector spaces and F is an affine map satisfying $f(0) = 0$, then it is also easy to verify, using Lemma 4.3, that F is a linear map. Consequently, a map between two vector spaces is linear if and only if its graph is a vector space. The following result elucidates the relationship between affine and linear maps.

Theorem 4.7. *Let $F : A \to B$ be an affine map between affine sets $A \subseteq E_1$ and $B \subseteq E_2$. Then F preserves affine combinations, that is,*

$$F\Big(\sum_{i=1}^{k} \alpha_i x_i\Big) = \sum_{i=1}^{k} \alpha_i F(x_i), \quad \text{where} \quad x_i \in A,\ \alpha_i \in \mathbb{R},\ \sum_{i=1}^{k} \alpha_i = 1.$$

Let $L = A - A$ and $M = B - B$ be the linear vector spaces associated with A and B, respectively. The function $\bar{F} : L \to M$ defined by

$$\bar{F}(x - a) := F(x) - F(a)$$

is a linear map that is independent of $a \in A$.

Consequently, if $F : A \to B$ is an affine map between two affine sets $A = a + L$ and $B = b + M$, then there exists a linear map $T : L \to M$ such that

$$F(x) = T(x - a) + F(a).$$

Proof. The graph $C = \{(x, y) : x \in A,\ y = F(x)\}$ is an affine set in $E_1 \times E_2$. It follows from Lemma 4.3 that if $\sum_1^k \alpha_i = 1$, then $\sum_1^k \alpha_i(x_i, F(x_i)) = (\sum_1^k \alpha_i x_i, \alpha_i F(x_i)) \in C$. This proves the statement that $F(\sum_1^k \alpha_i x_i) = \sum_1^k \alpha_i F(x_i)$. The set $K = C - \{(a, F(a)\}$ is a linear subspace of $E_1 \times E_2$. It is also the graph of the map $\bar{F} : L \to M$, which is a linear map. It follows from Theorem 4.4 that K and $\bar{F}$ are independent of $a \in A$. □

4.2 Convex Sets

In this section, convex sets will be defined within affine spaces, since this is their natural setting. However, a reader who prefers to work within vector spaces may assume that each affine set mentioned is a vector space.

Definition 4.8. *A subset C in an affine space E is called a* convex set *if for $x, y \in C$, the line segment*

$$[x, y] := \{(1 - \lambda)x + \lambda y : 0 \le \lambda \le 1\}$$

lies in C. In other words, if $x, y \in C$, then $(1-\lambda)x + \lambda y \in C$ for all $0 \le \lambda \le 1$.

In Figure 4.1, the sets A and B are convex. However, the set C is non-convex, since x and y are in C, but not all of the line segment connecting x and y.

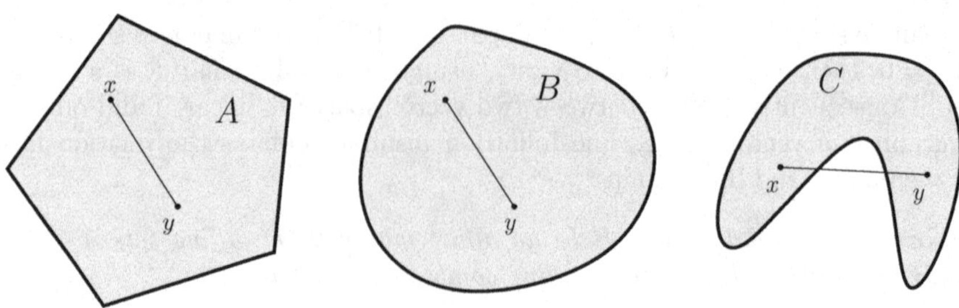

Fig. 4.1. Convex and nonconvex sets.

Lemma 4.9. *Let E be an affine space. The following statements are true.*

(a) Intersections of convex sets are convex: if $\{C_\gamma\}_{\gamma\in\Gamma}$ is a family of convex sets in E, then $\cap_{\gamma\in\Gamma}C_\gamma$ is a convex set.

(b) Minkowski sums of convex sets are convex: if $\{C_i\}_{i=1}^k$ is a set of convex sets, then their Minkowski sum

$$C_1 + \cdots + C_k := \{x_1 + \cdots + x_k : x_i \in C_i, i = 1, \ldots, k\}$$

is a convex set.

(c) An affine image of a convex set is convex: if $C \subseteq E$ is a convex set and $T : E \to F$ is an affine map from E into another affine space F, then $T(C) \subseteq F$ is also a convex set.

Proof. These statements are all easy to prove; we prove only (a). Let $x, y \in C := \cap_{\gamma\in\Gamma}C_\gamma$. For each $\gamma \in \Gamma$, we have $x, y \in C_\gamma$, and since C_γ is convex, $[x, y] \subseteq C_\gamma$; therefore, $[x, y] \subseteq C$ and C is a convex set. □

Definition 4.10. *Let $\{x_i\}_1^k$ be a finite set of points in an affine space E. A* convex combination *of $\{x_i\}_1^k$ is any point of the form*

$$\sum_{i=1}^k \lambda_i x_i, \quad \lambda_i \geq 0, \quad \sum_{i=1}^k \lambda_i = 1.$$

Let $A \subseteq E$ be a nonempty set. The convex hull *of A is the set of all convex combinations of points from A, that is,*

$$\mathrm{co}(A) := \left\{ \sum_{i=1}^k \lambda_i x_i \ : \ x_i \in A, \ \sum_{i=1}^k \lambda_i = 1, \ \lambda_i \geq 0, \ k \geq 1 \right\}.$$

Theorem 4.11. *Let $A \neq \emptyset$ be a subset of an affine space E. Then* co(A) *is a convex set; in fact,* co(A) *is the smallest convex set containing A.*

Proof. The proof is essentially a repeat of the proof of Lemma 4.2, but we now make the additional requirements that $0 < \alpha < 1$ and that $\{\lambda_i\}_1^k$, $\{\mu_j\}_1^l$ be nonnegative in that proof. It suffices to note that all the affine combinations now become convex combinations. □

The following result is an immediate consequence of the above theorem.

Corollary 4.12. *If C is a convex set in an affine space E, then* $\mathrm{co}(C) = C$, *that is, all convex combinations of elements from C lie in C,*

$$\lambda_i \geq 0,\ x_i \in C,\ i = 1, \ldots, k,\ \sum_{i=1}^{k} \lambda_i = 1 \quad \Longrightarrow \quad \sum_{i=1}^{k} \lambda_i x_i \in C.$$

The following theorem, due to Carathéodory [53], is a fundamental result in convexity in finite-dimensional vector spaces, and has many applications, including in optimization.

Theorem 4.13. (***Carathéodory***) *Let A be a nonempty subset of an affine space E. Every element of* $\mathrm{co}(A)$ *can be represented as a convex combination of affinely independent elements from A.*

Consequently, if $n = \dim(\mathrm{aff}(A)) < \infty$, *then every element of* $\mathrm{co}(A)$ *can be represented as a convex combination of at most $n+1$ elements from A; in other words,*

$$\mathrm{co}(A) = \Big\{ \sum_{i=1}^{n+1} \lambda_i x_i : x_i \in A,\ \lambda_i \geq 0,\ i = 1, \ldots, n+1,\ \sum_{i=1}^{n+1} \lambda_i = 1 \Big\}.$$

Proof. Let

$$x = \sum_{i=1}^{k} \lambda_i x_i \in \mathrm{co}(A), \quad \text{where } \sum_{i=1}^{k} \lambda_i = 1, \quad \lambda_i > 0. \tag{4.3}$$

If $\{x_i\}_1^k$ is affinely independent, then $\{x_i - x_1\}_2^k$ is linearly independent and $k - 1 \leq n$; thus, $k \leq n + 1$, and the theorem is proved.

Suppose that $\{x_i\}_1^k$ is affinely dependent. It follows from Lemma 4.6 that there exist scalars $\{\delta_i\}_1^k$ such that

$$\sum_{i=1}^{k} \delta_i x_i = 0, \quad \sum_{i=1}^{k} \delta_i = 0,\ (\delta_1, \ldots, \delta_k) \neq 0. \tag{4.4}$$

If we subtract from (4.3) ε times (4.4) ($\varepsilon > 0$), we obtain

$$x = (\lambda_1 - \varepsilon\delta_1)x_1 + \cdots + (\lambda_k - \varepsilon\delta_k)x_k, \quad \sum_{i=1}^{k} (\lambda_i - \varepsilon\delta_i) = 1. \tag{4.5}$$

Since $\sum_{i=1}^{k} \delta_i = 0$, there exist positive and negative scalars δ_i. If $\delta_i \leq 0$, then $\lambda_i - \varepsilon\delta_i \geq 0$ remains nonnegative for all $\varepsilon \geq 0$; however, if $\delta_i > 0$, then $\lambda_i - \varepsilon\delta_i \geq 0$ if and only if $\varepsilon \leq \lambda_i/\delta_i$. Therefore, if we set $\varepsilon = \min\{\lambda_i/\delta_i : \delta_i > 0\}$, then x remains a convex combination in (4.5), but has at least one fewer term. We can continue this process until the vectors $\{x_2 - x_1, \ldots, x_k - x_1\}$ in the representation (4.3) are linearly independent. When we halt, we will have $k \leq n + 1$. □

We immediately have the following.

Corollary 4.14. *If C is a nonempty subset of an n-dimensional vector space E, then every element of* $\operatorname{co}(A)$ *can be represented as a convex combination of at most $n+1$ elements from A.*

Corollary 4.15. *If C is a nonempty compact subset of a finite-dimensional affine space E, then so is the set* $\operatorname{co}(C)$.

Proof. It follows from Theorem 4.13 that

$$\operatorname{co}(C) = \left\{ \sum_{i=1}^{n+1} \lambda_i x_i : \lambda_i \geq 0,\ x_i \in C,\ i = 1, \ldots, n+1,\ \sum_{i=1}^{n+1} \lambda_i = 1 \right\},$$

where $n = \dim(C)$. Consider a sequence $\{x^k\}_1^\infty$ in $\operatorname{co}(C)$, where

$$x^k = \sum_{i=1}^{n+1} \lambda_i^k x_i^k.$$

Since C is compact, the sequence $\{x_1^k\}$ has a convergent subsequence $x_1^{k_j} \to x_1 \in C$. Next, let the sequence $\{x_2^{k_j}\}$ have convergent subsequence $x_2^{k_{j_l}} \to x_2 \in C$, and so on. Eventually, we can find a subsequence k_j such that

$$\lim_{j \to \infty} x_i^{k_j} = x_i \in C \quad \text{for all} \quad i = 1, \ldots, n+1.$$

Using the same arguments, we can assume that $\lim_{j \to \infty} \lambda_i^{k_j} = \lambda_i \geq 0$ for all $i = 1, \ldots, n+1$. Then we have $\sum_{i=1}^{n+1} \lambda_i = 1$, and

$$x^{k_j} \to \sum_{i=1}^{n+1} \lambda_i x_i \in \operatorname{co}(C).$$

This proves that $\operatorname{co}(C)$ is compact. □

An alternative proof runs as follows: Let

$$\Delta_n := \left\{ (\lambda_1, \ldots, \lambda_{n+1}) : \lambda_i \geq 0, i = 1, \ldots, n+1,\ \sum_{i=1}^{n+1} \lambda_i = 1 \right\}$$

be the standard unit simplex in $\mathbb{R}^{n+1}$, and consider the map

$$\Delta_n \times \underbrace{C \times \cdots \times C}_{n+1 \text{ times}} \to E,$$

given by

$$T(\lambda_1, \ldots, \lambda_{n+1}, x_1, \ldots, x_{n+1}) = \sum_{i=1}^{n+1} \lambda_i x_i.$$

Note that the image of T is $\mathrm{co}(C)$. Since the map T is continuous and the domain of T is compact (Δ_n and C are compact), we conclude that $\mathrm{co}(C)$ is compact.

We also record here the following elementary results.

Lemma 4.16. *Let E be an affine space in a normed vector space. If $C \subseteq E$ is a convex set, then its closure $\overline{C}$ is also a convex set. If $C_1, C_2 \subseteq E$ are convex sets, C_1 is compact, and C_2 is closed, then $C_1 + C_2$ is a closed, convex set.*

Proof. To prove the first statement, define the convex set

$$C_\epsilon := \{z : \|z - x\| < \epsilon,\, x \in C\} = \{x + u : x \in C,\ \|u\| < \epsilon\} = C + B_\epsilon(0).$$

Note that $\overline{C} := \cap_{\epsilon > 0} C_\epsilon$, because a point $z \in E$ lies in $\overline{C}$ if and only if given $\epsilon > 0$, there exists a point $x \in C$ such that $\|z - x\| < \epsilon$. It follows from Lemma 4.9 that $\overline{C}$ is a convex set.

To prove the second statement, let $\{z_k\}_{k=1}^{\infty}$ be a sequence in $C_1 + C_2$ converging to a point z. Write $z_k = x_k + y_k$ with $x_k \in C_1$ and $y_k \in C_2$. Since C_1 is compact, there exists a subsequence $x_{k_i} \to x \in C_1$. Since $z_{k_i} \to z$, y_{k_i} must converge to the point $y := z - x \in C_2$. Thus, $z = x + y \in C_1 + C_2$, proving that $C_1 + C_2$ is a closed set. □

4.2.1 Convex Cones

Definition 4.17. *A set K in a vector space E is called a* cone *if $tx \in K$ whenever $t > 0$ and $x \in K$. If K is also a convex set, then it is called a* convex cone.

Lemma 4.18. *A set K in a vector space E is a convex cone if and only if*

$$x, y \in K \text{ and } t > 0 \quad \Longrightarrow \quad tx \in K,\ \ x + y \in K. \tag{4.6}$$

Proof. Let $x, y \in K$. If K is a convex cone, then $(x + y)/2 \in K$, since K is convex, and $x + y = 2((x + y)/2) \in K$, since K is a cone. This proves (4.6). Conversely, if $0 < t < 1$ and (4.6) holds, then $(1 - t)x + ty \in K$, proving that K is a convex set. □

Many concepts and results for convex sets have analogues for convex cones.

Definition 4.19. *Let $\{x_i\}_1^k$ be a finite set of points in a vector space E. A* positive combination *of $\{x_i\}_1^k$ is any point of the form*

$$\sum_{i=1}^{k} \lambda_i x_i, \quad \lambda_i > 0,\ i = 1, \ldots, k.$$

Let $A \subseteq E$ be a nonempty set. The convex conical hull *of A is the set of all positive combinations of points from A, that is,*

$$\operatorname{cone}(A) := \left\{ \sum_{i=1}^{k} \lambda_i x_i \;:\; x_i \in X,\;\; \lambda_i > 0,\; k \geq 1 \right\}.$$

Theorem 4.20. *Let A be a nonempty set in a vector space E. Then* cone(A) *is the smallest convex cone containing A. If K is a convex cone, then* cone$(K) = K$.

This is proved in exactly the same way as Theorem 4.11. In fact, the proof here is somewhat simpler, since the weights $\{\lambda_i\}$ in a positive combination are not required to sum to one.

Theorem 4.21. (***Carathéodory***) *Let A be a nonempty subset of a vector space E. Every element of* cone(A) *can be represented as a positive combination of linearly independent elements from A. Consequently, if $n =$* dim(span$(A)) < \infty$, *then every element of* cone(A) *can be represented as a positive combination of at most n elements from A. In other words,*

$$\operatorname{cone}(A) = \left\{ \sum_{i=1}^{n} \lambda_i x_i : x_i \in A,\;\; \lambda_i > 0,\;\; i = 1, \ldots, n \right\}.$$

Proof. The proof is essentially the same as in the affine case. Let

$$x = \sum_{i=1}^{k} \lambda_i x_i \in \operatorname{cone}(A), \text{ where all } \lambda_i > 0. \tag{4.7}$$

If the vectors $\{x_i\}_1^k$ are linearly dependent, then there exist scalars $\{\delta_i\}_{i=1}^k$, not all zero, such that

$$\sum_{i=2}^{k} \delta_i x_i = 0.$$

If we subtract from (4.3) ε times this equation, we obtain

$$x = (\lambda_1 - \varepsilon\delta_1)x_1 + \cdots + (\lambda_k - \varepsilon\delta_k)x_k.$$

The rest of the proof is completed using the same arguments in the proof of Theorem 4.13. □

Corollary 4.22. *If C is a nonempty subset of an n-dimensional vector space E, then every element of* cone(A) *can be represented as a positive combination of at most n elements from A.*

4.3 Convex Functions

In this section, we define convex functions and discuss their most basic properties. Normally, a convex function $f : C \to \mathbb{R}$ is defined on a convex set C in a vector or affine space E. It will be sufficient for us to assume that E is a vector space. If C is a proper subset of E, we may extend f to the whole space E by defining $f(x) = \infty$ when $x \notin C$. This device is for convenience, since it relieves us from having to describe the domain of C each time a function is defined or mentioned. A related advantage is that a convex function is sometimes defined through a pointwise maximization process, which often introduces $+\infty$ as a function value. It is possible to consider convex functions taking the value $-\infty$; see [228, 89]. We do not consider them in this book, since they tend to be rather pathological.

Definition 4.23. *Let E be vector space. A function $f : E \to \mathbb{R} \cup \{\infty\}$ is called a* convex function *if*

$$f((1-t)x + ty) \le (1-t)f(x) + tf(y) \quad \textit{for all } x, y \in E,\ t \in [0,1]. \tag{4.8}$$

The set

$$\mathrm{dom}(f) = \{x : f(x) \in \mathbb{R}\}$$

is called the effective domain *of f. The function f* strictly convex *if*

$$f((1-t)x+ty) < (1-t)f(x)+tf(y) \quad \textit{for all } x \ne y \in \mathrm{dom}(f),\ t \in (0,1). \tag{4.9}$$

A function $f : E \to \mathbb{R} \cup \{-\infty\}$ is called a concave function *if $-f$ is a convex function, that is,*

$$f((1-t)x + ty) \ge (1-t)f(x) + tf(y) \quad \textit{for all } x, y \in E,\ t \in [0,1], \tag{4.10}$$

and strictly concave *if for all $x \ne y \in \mathrm{dom}(f)$ and $t \in (0,1)$,*

$$f((1-t)x + ty) > (1-t)f(x) + tf(y). \tag{4.11}$$

It is clear the function inequalities above need to be verified only when $f(x)$ and $f(y)$ are finite. Also, it is easy to show that $\mathrm{dom}(f)$ is a convex set of E. Thus, $f : \mathrm{dom}(f) \to \mathbb{R}$ is a convex or concave function in the usual sense, that is, it satisfies the relevant functional inequality above for all $x, y \in \mathrm{dom}(f)$.

If $f : C \to \mathbb{R}$ is a convex function on a convex set $C \subseteq E$ and $\alpha \in \mathbb{R}$ is a constant, then it is easy to show that the sublevel sets of f,

$$l_\alpha(f) := \{x \in C : f(x) \le \alpha\}$$

and $\{x \in C : f(x) < \alpha\}$, are convex. However, the converse is false, since the sublevel sets of any monotonic function in $\mathbb{R}$ are convex.

Recall that the epigraph of f is the set in $\mathbb{R}^{n+1}$ defined by the formula

$$\mathrm{epi}(f) := \{(x, \alpha) : x \in E, \alpha \in \mathbb{R},\ f(x) \le \alpha\}.$$

The following simple but important result makes it possible to view convex functions geometrically, an important theme in convex analysis.

Lemma 4.24. *Let $f : E \to \mathbb{R} \cup \{\infty\}$ be a function on a vector space E. The function f is convex if and only if* $\mathrm{epi}(f) \subseteq E \times \mathbb{R}$ *is a convex set in $E \times \mathbb{R}$.*

Proof. First, assume that f is convex. Let $(x_i, \alpha_i) \in \mathrm{epi}(f)$, $i = 1, 2$, and $0 \le \lambda \le 1$. We have $f(x_i) \le \alpha_i$ ($i = 1, 2$), and since f is a convex function, we obtain

$$f((1-\lambda)x_1 + \lambda x_2) \le (1-\lambda)f(x_1) + \lambda f(x_2) \le (1-\lambda)\alpha_1 + \lambda\alpha_2;$$

this implies $(1-\lambda)(x_1, \alpha_1) + \lambda(x_2, \alpha_2) \in \mathrm{epi}(f)$, and proves that $\mathrm{epi}(f)$ is a convex set.

Conversely, suppose that $\mathrm{epi}(f)$ is a convex set. Let $x_1, x_2 \in E$ and $0 \le \lambda \le 1$. We have $(x_i, f(x_i)) \in \mathrm{epi}(f)$, $i = 1, 2$, so that

$$(1-\lambda)(x_1, f(x_1)) + \lambda(x_2, f(x_2)) = ((1-\lambda)x_1 + \lambda x_2, (1-\lambda)f(x_1) + \lambda f(x_2))$$

lies in $\mathrm{epi}(f)$, that is,

$$f((1-\lambda)x_1 + \lambda x_2) \le (1-\lambda)f(x_1) + \lambda f(x_2),$$

thus proving that f is a convex function. □

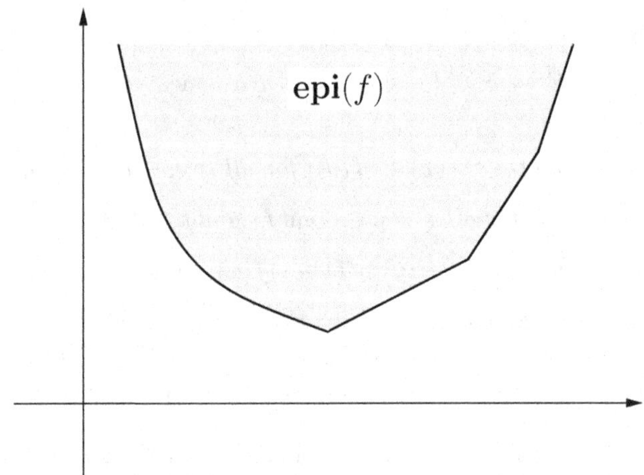

Fig. 4.2. Epigraph of a convex function.

Corollary 4.25. (***Jensen's inequality***) *If $f : E \to \mathbb{R} \cup \{\infty\}$ is a convex function, then*

$$f\Big(\sum_1^k \lambda_i x_i\Big) \le \sum_1^k \lambda_i f(x_i),$$

whenever $\lambda_i \ge 0$, $i = 1, \dots, k$, and $\sum_{i=1}^k \lambda_i = 1$.

Proof. Since the corollary clearly holds if $f(x_i) = \infty$ for some x_i, we assume $f(x_i) \in \mathbb{R}$ for all $i = 1, \ldots, k$. The points $(x_i, f(x_i))$ lie in epi(f), and since epi(f) is a convex set, we have

$$\sum_{i=1}^{k} \lambda_i (x_i, f(x_i)) = \Big(\sum_{i=1}^{k} \lambda_i x_i, \sum_{i=1}^{k} \lambda_i f(x_i)\Big) \in \text{epi}(f).$$

It follows that

$$f\Big(\sum_{1}^{k} \lambda_i x_i\Big) \leq \sum \lambda_i f(x_i).$$

□

Lemma 4.26. *The following functions are convex:*

(a) The affine function $f(x) = \langle a, x\rangle + b$, where $a \in E$, and $b \in \mathbb{R}$.
(b) A norm function $f(x) = \|x\|$ on a normed linear space E.
(c) If $\{f_i\}_{i=1}^{k}$ are convex functions, and $\alpha_i \geq 0$, then the function $\sum_{i=1}^{k} \alpha_i f_i$ is a convex function.
(d) If $\{f_\alpha\}_{\alpha\in A}$ is any family of convex function, $f_\alpha : E \to \mathbb{R} \cup \{\infty\}$, then the pointwise supremum of f_α, $f(x) := \sup_{\alpha\in A} f_\alpha(x)$, is a convex function $f : E \to \mathbb{R} \cup \{\infty\}$.

Proof. The proofs of (a) and (c) are trivial. To prove (b), note that if $x, y \in E$ and $0 < \lambda < 1$, then

$$\begin{aligned} f((1-\lambda)x + \lambda y) &= \|(1-\lambda)x + \lambda y\| \leq \|(1-\lambda)x\| + \|\lambda y\| \\ &= (1-\lambda)\|x\| + \lambda\|y\| = (1-\lambda)f(x) + \lambda f(y). \end{aligned}$$

To prove (d), we consider epi(f):

$$\begin{aligned} \text{epi}(f) &= \{(x,\lambda) : \sup_{\alpha\in A} f_\alpha(x) \leq \lambda\} = \{(x,\lambda) : f_\alpha(x) \leq \lambda \ \text{ for all } \ \alpha \in A\} \\ &= \bigcap_{\alpha\in A} \{(x,\alpha) : f_\alpha(x) \leq \lambda\} = \bigcap_{\alpha\in A} \text{epi}(f_\alpha). \end{aligned}$$

Since each epi(f_i) is convex, we see that epi(h) is convex, implying that $f(x)$ is a convex function. □

We remark that taking the pointwise supremum of convex functions as in (d) is a common operation in optimization and convex analysis. While this operation preserves convexity, it usually destroys differentiability.

4.4 Differentiable Convex Functions

In this section, we give characterizations of differentiable convex functions in terms of their gradients or Hessians. When applicable, these often provide the quickest means to establish the convexity of a function.

The following important result gives the differential characterization of convex functions.

Theorem 4.27. *Let C be a convex set in $\mathbb{R}^n$, and let f be a Gâteaux differentiable function on an open set containing C.*

Then f is convex on C if and only if the tangent plane at any point $x \in C$ lies below the graph of f, that is,

$$f(y) \geq f(x) + \langle \nabla f(x), y - x \rangle \quad \textit{for all} \quad x, y \in C. \tag{4.12}$$

Moreover, f is strictly convex on C if and only if the inequality above is strict when $y \neq x$.

Proof. First, assume that f is a convex function. If $t \in (0,1)$, then the inequality $f(x + t(y - x)) = f((1 - t)x + ty) \leq (1 - t)f(x) + tf(y)$ can be written as

$$\frac{f(x + t(y - x)) - f(x)}{t} \leq f(y) - f(x); \tag{4.13}$$

taking the limit of the left-hand side as $t \searrow 0$, we obtain

$$\langle \nabla f(x), y - x \rangle = \lim_{t \searrow 0} \frac{f(x + t(y - x)) - f(x)}{t} \leq f(y) - f(x).$$

Conversely, assume that (4.12) holds. Let $x, y \in C$, $x \neq y$, $t \in (0,1)$, and define $x_t := x + t(y - x) = (1 - t)x + ty$. The inequality (4.12) gives

$$\begin{aligned} f(x) &\geq f(x_t) + \langle \nabla f(x_t), x - x_t \rangle, \\ f(y) &\geq f(x_t) + \langle \nabla f(x_t), y - x_t \rangle. \end{aligned} \tag{4.14}$$

Multiplying these inequalities by $1 - t$ and t, and adding them, we obtain

$$\begin{aligned} (1 - t)f(x) + tf(y) &\geq f(x_t) + \langle \nabla f(x_t), (1 - t)x + ty - x_t \rangle \\ &= f(x_t) = f((1 - t)x + ty), \end{aligned} \tag{4.15}$$

which proves that f is a convex function.

Suppose that f is strictly convex. Let $x, y \in C$, $x \neq y$, and $t \in (0,1)$. Then, the inequality in (4.13) is strict. It is straightforward to verify that the difference quotient $(f(x + t(y - x)) - f(x))/t$ is a nonincreasing function of t, and strictly decreasing if f is strictly convex (see Exercise 9 on page 108); it follows that

$$\begin{aligned} \langle \nabla f(x), y - x \rangle &= \lim_{t \searrow 0} \frac{f(x + t(y - x)) - f(x)}{t} = \inf_{t > 0} \frac{f(x + t(y - x)) - f(x)}{t} \\ &< f(y) - f(x). \end{aligned}$$

Conversely, if the inequality (4.12) is strict for $x, y \in C$, $x \neq y$, then $x_t \in (x, y)$ and the inequalities in (4.14) and (4.15) are strict, proving that f is a strictly convex function. □

Exercise 10 offers a different insight into the above theorem.

The following important result provides the second-order differential characterizations of a convex function.

Theorem 4.28. *Let C be a convex set in $\mathbb{R}^n$, and let f be a twice Fréchet differentiable function on an open set containing C. Then,*

(a) The function f is convex on C if and only if the Hessian $Hf(x)$ is positive semidefinite at every point $x \in C$;
(b) If $Hf(x)$ is positive definite at every point $x \in C$, then f is strictly convex.

Proof. (a). First, assume that f is convex and consider an arbitrary direction $d \in E$. It follows from Theorem 4.27 and Taylor's formula that

$$\begin{aligned} f(x) + t\langle \nabla f(x), d\rangle &\leq f(x + td) \\ &= f(x) + t\langle \nabla f(x), d\rangle + \frac{t^2}{2}\langle Hf(x)d, d\rangle + o(t^2). \end{aligned}$$

This implies that $\langle Hf(x)d, d\rangle + o(t^2)/t^2 \geq 0$; letting $t \to 0$ proves that $Hf(x)$ is positive semidefinite.

Conversely, assume that $Hf(x)$ is positive semidefinite at every $x \in C$, and let $x, y \in C$, $x \neq y$. There exists a point $z \in (x, y)$ such that

$$\begin{aligned} f(y) &= f(x) + \langle \nabla f(x), y - x\rangle + \frac{1}{2}\langle Hf(z)(y - x), y - x\rangle \\ &\geq f(x) + \langle \nabla f(x), y - x\rangle, \end{aligned}$$

and Theorem 4.27 implies that f is convex.

The proof of (b) is similar to that of (a). □

It should be mentioned that the converse of statement (b) is false, as the function $f(x) = x^4$ shows: f is strictly convex, but $f''(0) = 0$.

Remark 4.29. Theorems 4.27 and 4.28, with straightforward modifications, are valid in much more general spaces.

Corollary 4.30. *Consider the quadratic function*

$$f(x) = \frac{1}{2}\langle Qx, x\rangle + \langle c, x\rangle,$$

where Q is a symmetric $n \times n$ matrix and $c \in \mathbb{R}^n$. Then:

(a) The function f is convex if and only if Q is positive semidefinite;
(b) The function $f(x)$ is strictly convex if and only if Q is positive definite;
(c) If f is bounded below on $\mathbb{R}^n$, then f is convex and achieves its minimum on $\mathbb{R}^n$;

(d) Let X^* be the set of the global minimizers of f; if $X^* \neq \emptyset$, then the values $\langle Qx^*, x^*\rangle$ and $\langle c, x^*\rangle$ are independent of $x^* \in X^*$; in fact,

$$\langle Qx^*, x^*\rangle = -\langle c, x^*\rangle = -2 \min_{\mathbb{R}^n} f \quad \textit{for all} \quad x^* \in X^*.$$

Proof. We have $\nabla f(x) = Qx + c$ and $Hf(x) = Q$; see Example 1.26. Thus, Theorem 4.28 proves (a).

If $f(x)$ is strictly convex and $h \neq 0$, then Theorem 4.27 gives

$$f(x) + \langle \nabla f(x), h\rangle < f(x+h) = f(x) + \langle \nabla f(x), h\rangle + \frac{1}{2}\langle Qh, h\rangle.$$

We have $\langle Qh, h\rangle > 0$ for all $0 \neq h \in \mathbb{R}^n$, that is, Q is positive definite. Conversely, if Q is positive definite and $h \neq 0$, then

$$f(x+h) = f(x) + \langle \nabla f(x), h\rangle + \frac{1}{2}\langle Qh, h\rangle > f(x) + \langle \nabla f(x), h\rangle,$$

and Theorem 4.27 implies that f is strictly convex. This proves (b).

If x^* is a global minimizer of f on $\mathbb{R}^n$, then $Hf(x^*) = Q$ is positive semidefinite by Theorem 2.12, and then it follows from (a) that f is a convex function.

Suppose that f is bounded from below on $\mathbb{R}^n$. Diagonalize Q in the form $Q = U\Lambda U^T$, where U is an orthogonal matrix and $\Lambda = \operatorname{diag}\{\lambda_1, \ldots, \lambda_n\}$, where $\lambda_i \neq 0$ for $i = 1, \ldots, k$ and $\lambda_i = 0$ for $i = k+1, \ldots, n$. Substituting $u = U^T x$ and setting $\overline{c} = U^T c$, we have

$$\begin{aligned} f(x) &= \frac{1}{2}\langle \Lambda U^T x, U^T x\rangle + \langle U^T c, U^T x\rangle = \frac{1}{2}\langle \Lambda u, u\rangle + \langle \overline{c}, u\rangle \\ &=: \overline{f}(u) = \sum_{i=1}^{k}\left(\frac{\lambda_i}{2}u_i^2 + \overline{c}_i u_i\right) + \sum_{i=k+1}^{n} \overline{c}_i u_i. \end{aligned}$$

It is clear that minimizing f over $x \in \mathbb{R}^n$ is equivalent to minimizing $\overline{f}$ over $\mathbb{R}^n$ and that $\overline{f}$ has a lower bound on $\mathbb{R}^n$ if and only if $\lambda_i > 0$ for $i = 1, \ldots, k$ and $\overline{c}_i = 0$ for $i = k+1, \ldots, n$. Moreover, if $\overline{f}$ is bounded from below, then it has a minimizer, since each function $\frac{1}{2}\lambda_i u_i^2 + \overline{c}_i u_i$ is minimized at $u_i^* = -\overline{c}_i/\lambda_i$ for $i = 1, \ldots, k$. This proves (c).

It remains to prove (d). If $x \in X^*$ is any global minimizer of f, then $\nabla f(x^*) = Qx^* + c = 0$ and

$$\begin{aligned} \min_{\mathbb{R}^n} f = f(x^*) &= \frac{1}{2}\langle Qx^*, x^*\rangle + \langle c, x^*\rangle = \frac{1}{2}\langle Qx^*, x^*\rangle - \langle Qx^*, x^*\rangle \\ &= -\frac{1}{2}\langle Qx^*, x^*\rangle = \frac{1}{2}\langle c, x^*\rangle. \end{aligned}$$

□

It is shown in Chapter 1 (see Exercise 9 on page 25) that Gâteaux differentiability is in general weaker than Fréchet differentiability, and additional conditions, such as the continuity of partial derivatives, are needed to improve Gâteaux differentiability to Fréchet differentiability; see Theorem 1.19. We end this section by showing that convexity collapses the distinction between Gâteaux and Fréchet differentiability.

Theorem 4.31. *Let C be a convex set in $\mathbb{R}^n$ such that* $\mathrm{int}(C) \neq \emptyset$, *and let* $f : C \to \mathbb{R}$ *be a convex function. If* $x \in \mathrm{int}(C)$, *and the partial derivatives* $\{\partial f(x)/\partial x_i\}_1^n$ *exist, then f is Fréchet differentiable at x.*

Consequently, if f is Gâteaux differentiable at x, then it is Fréchet differentiable at x.

Proof. Define the function

$$g(h) := f(x+h) - f(x) - \langle \nabla f(x), h \rangle.$$

Note that g is a convex function, $g(0) = 0$, and $\nabla g(0) = 0$. We have

$$g(h) = g(\frac{1}{n}\sum_{i=1}^{n} nh_i e_i) \leq \sum h_i \frac{g(nh_i e_i)}{nh_i} \leq \|h\| \cdot \sum \left| \frac{g(nh_i e_i)}{nh_i} \right|,$$

where the last two sums are over all i such that $h_i \neq 0$. Here the first inequality follows from Jensen's inequality, and the last inequality follows from the Cauchy–Schwarz inequality followed by the inequality $\|u\| \leq \|u\|_1 := \sum_{i=1}^n |u_i|$, $u \in \mathbb{R}^n$. The convexity of g also implies that $0 = 2g(0) \leq g(-h) + g(h)$. Using this and the above inequalities for g, we obtain

$$-\sum \left| \frac{g(-nh_i e_i)}{nh_i} \right| \leq \frac{-g(-h)}{\|h\|} \leq \frac{g(h)}{\|h\|} \leq \sum \left| \frac{g(nh_i e_i)}{nh_i} \right|.$$

The terms inside the sums above converge to $\partial g(0)/\partial x_i = 0$ as $h_i \to 0$. Thus, $g(h) = o(h)$, which is equivalent to the statement that f is Fréchet differentiable at x.

4.5 Optimization on Convex Sets

One of the most important and basic properties of convex functions is the fact that any local minimizer on a convex set is a global one.

Theorem 4.32. *Let $f : C \to \mathbb{R}$ be a convex function on a convex set C in a vector space E. Any local minimizer of f on C is a global minimizer of f on C. If f is strictly convex, then there exists at most one global minimizer of f on C.*

Proof. Let $x^* \in C$ be a local minimizer of f on C. If $x \in C$, then the line segment $[x^*, x]$ lies in C. For $t \in (0,1)$, the point $x_t := x^* + t(x - x^*) = (1-t)x^* + tx$ lies in C, and since x^* is a local minimizer, $f(x^*) \le f(x_t)$ if t is close to 0. We have

$$f(x^*) \le f(x_t) \le (1-t)f(x^*) + tf(x),$$

where the last inequality follows from the convexity of f. Consequently,

$$f(x^*) \le f(x) \quad \text{for all} \quad x \in C,$$

that is, x^* is a global minimizer of f on C.

If f is strictly convex and x_1^* and x_2^* are two global minimizers of f on C, then

$$f\left(\frac{x_1^* + x_2^*}{2}\right) < \frac{1}{2}f(x_1^*) + \frac{1}{2}f(x_2^*) = f(x_1^*) = f(x_2^*) = f^*,$$

a contradiction. The theorem is proved.

A slightly different proof runs as follows: If $x \in C$ satisfies $f(x) < f(x^*)$, then

$$f(x^* + t(x - x^*)) = f((1-t)x^* + tx) \le (1-t)f(x^*) + tf(x) < f(x^*)$$

for all $t \in (0,1]$, that is, $f(z) < f(x^*)$ for all $z \in (x^*, x]$. Since the segment $(x^*, x] \subseteq C$ contains points arbitrarily near x^*, this clearly contradicts the assumption that x^* is a local minimizer of f. □

Now we consider the minimization of a differentiable function f on a convex set C. We obtain an important first-order necessary condition, called a *variational inequality* (the inequality (4.16) below), that a local minimizer $x^* \in C$ of f must satisfy. If f a convex function, then the variational inequality is a sufficient condition as well, so that it provides a characterization of a global minimizer of the convex function f on C.

Theorem 4.33. *Let C be a convex set in $\mathbb{R}^n$, and let f be a Gâteaux differentiable function on an open set containing C.*

(a) ***(First-order necessary condition for a local minimizer)*** *If $x^* \in C$ is a local minimizer of f on C, then*

$$\langle \nabla f(x^*), x - x^* \rangle \ge 0 \quad \textit{for all} \quad x \in C. \tag{4.16}$$

(b) ***(First-order sufficient condition for a local minimizer)*** *If f is convex and* (4.16) *is satisfied at $x^* \in C$, then x^* is a global minimizer of f on C.*

Proof. To prove (a), pick a point $x \in C$. Since C is convex, $[x^*, x] \subseteq C$, and since x^* is a local minimizer of f on C, we have $f(x^* + t(x - x^*)) \ge f(x^*)$ when $t > 0$ is close to zero. Thus,

$$\langle \nabla f(x^*), x - x^* \rangle = \lim_{t \searrow 0} \frac{f(x^* + t(x - x^*)) - f(x^*)}{t} \geq 0.$$

To prove (b), suppose $x^* \in C$ satisfies the variational inequality (4.16). We have

$$f(x) \geq f(x^*) + \langle \nabla f(x^*), x - x^* \rangle \geq f(x^*) \text{ for all } x \in C$$

where the first inequality follows from the convexity of f, and the second one from (4.16). □

It is clear from the proof above that Theorem 4.33 is valid in very general spaces, including normed linear spaces.

4.5.1 Examples of Variational Inequalities

Example 4.34. Let f be a Gâteaux differentiable function in a neighborhood of a convex set $C \subseteq \mathbb{R}^n$. If C has nonempty interior, and $x^* \in \text{int}(C)$ is a local minimizer of f, then $\nabla f(x^*) = 0$, as we have seen in Chapter 2 (Theorem 2.7). This equation also follows from the variational inequality, since choosing $x = x^* - \epsilon \nabla f(x^*) \in C$ in (4.16) gives $\|\nabla f(x^*)\| \leq 0$.

Example 4.35. Consider a differentiable function $f : [a, b] \to \mathbb{R}$. If $x^* \in (a, b)$ is a local minimizer, then the preceding example above shows that $f'(x^*) = 0$, the familiar condition from elementary calculus. If $x^* = a$ is a local minimizer, then $x - x^* = x - a \geq 0$ in the variational inequality, so we can deduce only that $f'(a) \geq 0$. Thus, the condition $f'(a) \geq 0$ is the first-order necessary condition for a to be a local minimizer of f on $[a, b]$. A similar argument shows that if $x^* = b$ is a local minimizer of f on $[a, b]$, then $f'(b) \leq 0$. We see that the variational inequality gives something new, even in the one-dimensional case.

Example 4.36. Consider the minimization of a differentiable function on an affine subspace,

$$\begin{aligned} \min \quad & f(x) \\ \text{s.t.} \quad & Ax = b, \end{aligned}$$

where $f : \mathbb{R}^n \to \mathbb{R}$, A is an $m \times n$ matrix, and $b \in \mathbb{R}^m$. Define $C = \{x \in \mathbb{R}^n : Ax = b\}$. If $x^* \in C$ is a local minimizer of f on C, then it satisfies the variational inequality

$$\langle \nabla f(x^*), x - x^* \rangle \geq 0 \text{ for all } x \in C.$$

Since $\{z = x - x^* : x \in C\} = \{z : Az = 0\} = N(A)$, the variational inequality becomes

$$\langle \nabla f(x^*), z \rangle \geq 0 \text{ for all } z \in N(A).$$

If $z \in N(A)$, so is $-z \in L$, and the above inequality reduces to the equality

$$\langle \nabla f(x^*), z \rangle = 0 \ \text{ for all } \ z \in N(A).$$

We know from linear algebra that this is equivalent to the inclusion $\nabla f(x^*) \in N(A)^{\perp} = R(A^T)$. Consequently,

$$\nabla f(x^*) \in R(A^T) \tag{4.17}$$

is a necessary condition for x^* to be a local minimizer. If f is convex, then (4.17) is a necessary and sufficient condition for x^* to be a global minimum of f over C.

The condition (4.17) can be put in the form

$$\Pi_{N(A)} \nabla f(x^*) = 0,$$

which states that the component of $\nabla f(x^*)$ along the feasible set $C = \{x : Ax = b\}$ is zero. This resembles the first-order optimality condition in unconstrained optimization, and should make it easier to remember (4.17).

Example 4.37. Consider the *quadratic program*

$$\begin{aligned} \min \quad & f(x) := \frac{1}{2}\langle Qx, x \rangle + c^T x, \\ \text{s.t.} \quad & x \geq 0, \end{aligned}$$

where Q is an $n \times n$ symmetric and $c \in \mathbb{R}^n$.

If Q is positive definite, then the objective function $f(x)$ is coercive, and thus there exists a unique global minimizer x^* of f over the nonnegative orthant $\{x \in \mathbb{R}^n : x \geq 0\}$.

Let $x^* \geq 0$ be a local minimizer of f on the nonnegative orthant. Since $\nabla f(x^*) = Qx^* + c$, the variational inequality becomes

$$\langle Qx^* + c, x - x^* \rangle \geq 0 \ \text{ for all } \ x \geq 0 \text{ in } \mathbb{R}^n. \tag{4.18}$$

If we choose $x = 2x^*$ and then $x = 0$ in (4.18), we obtain $\langle Qx^* + c, x^* \rangle = 0$. Substituting this in (4.18) implies that $\langle Qx^* + c, x \rangle \geq 0$ for all $x \geq 0$, which in turn yields $Qx^* + c \geq 0$. Therefore, (4.18) implies the conditions

$$Qx^* + c \geq 0, \quad x^* \geq 0, \quad \text{and} \quad \langle Qx^* + c, x^* \rangle = 0. \tag{4.19}$$

Conversely, it is easy to verify that (4.19) implies (4.18).

Therefore, the two inequalities and the equation in (4.19) are the first-order necessary conditions for a local minimizer of a quadratic function f over the nonnegative orthant. If, moreover, f is a convex quadratic function, then (4.19) characterizes a global minimizer of f over the same orthant by virtue of Theorem 4.33.

Remark 4.38. The problem of finding a point x^* satisfying (4.19), where Q is an arbitrary $n \times n$ matrix Q, is called a *linear complementarity problem (LCP)*. Note that if Q is not symmetric, then (4.19) cannot be associated with an optimization problem, but it may come from a saddle point problem, for example.

Example 4.39. Consider the maximization problem

$$\begin{aligned} \max \quad & g(x) := x_1^{\alpha_1} \dots x_n^{\alpha_n}, \\ \text{s.t.} \quad & x_1 + \dots + x_n = 1 \\ & x_i \geq 0, \quad i = 1, \dots, n \end{aligned}$$

Since each x_i^* must clearly be positive at a local maximizer, we can reformulate the problem:

$$\begin{aligned} \min \quad & f(x) := -\alpha_1 \ln x_1 + \dots + (-\alpha_n) \ln x_n \\ \text{s.t.} \quad & x_1 + \dots + x_n = 1. \end{aligned}$$

We have $\nabla f(x) = (-\alpha_1/x_1, \dots, -\alpha_n/x_n)^T$, and the constraint set has the form $C = \{x : Ax = 1\}$, where $A = [1, \dots, 1]$; thus it follows from (4.17) that $\alpha_i/x_i^* = \lambda$ $(i = 1, \dots, n)$. Therefore, $x_i^* = \alpha_i/\lambda$ and

$$1 = \sum_{i=1}^n x_i^* = \sum_{i=1}^n \frac{\alpha_i}{\lambda},$$

giving $\lambda = \sum_{i=1}^n \alpha_i$ and

$$x_i^* = \frac{\alpha_i}{\sum_{k=1}^n \alpha_k}, \qquad i = 1, \dots, n.$$

Optimization is often a useful tool for proving inequalities. For example, if $\alpha_k = 1/n$ for all $k = 1, \dots, n$ in the above problem, then $x^* = (1/n, \dots, 1/n)$, and the optimal objective value of the maximization problem is $g(x^*) = 1/n$. This proves that $g(x) \leq 1/n$ whenever $x \geq 0$ and $x_1 + \dots + x_n = 1$. Since both the objective function and the function $h(x) := x_1 + \dots + x_n$ are homogeneous of first degree, that is, $g(tx) = tg(x)$ and $h(tx) = th(x)$ for $t \geq 0$, we have indeed proved the inequality

$$\sqrt[n]{x_1 x_2 \cdots x_n} \leq \frac{x_1 + x_2 + \dots + x_n}{n} \quad \text{for all} \quad x_i \geq 0,\ i = 1, \dots, n,$$

which is the precisely the arithmetic–geometric mean inequality. Moreover, since the maximum of g over the feasible set is unique, we see that the arithmetic–geometric mean inequality becomes an equality if and only if $x_1 = x_2 = \dots = x_n$.

Example 4.40. Finally, we consider the minimization of a differentiable function on a convex polyhedron,

$$\begin{aligned} \min \quad & f(x) \\ \text{s.t.} \quad & Ax \leq a, \\ & Bx = b, \end{aligned}$$

where $f : \mathbb{R}^n \to \mathbb{R}$, A and B are $m \times n$ and $p \times n$ matrices, respectively, $a \in \mathbb{R}^m$, and $b \in \mathbb{R}^p$. Define $C = \{x \in \mathbb{R}^n : Ax \leq a, Bx = b\}$. If $x^* \in C$ is a local minimizer of f on C, then it satisfies the variational inequality

$$\langle \nabla f(x^*), x - x^* \rangle \geq 0 \quad \text{for all} \quad x \in C,$$

or equivalently the implication

$$Ax \leq a, \; Bx = b \quad \Longrightarrow \quad \langle \nabla f(x^*), x \rangle \geq \langle \nabla f(x^*), x^* \rangle.$$

It is not a trivial matter to rewrite this system of potentially infinitely many conditions (one condition for each $x \in P$) in a compact, manageable form, but it is possible. Note that the implication above is equivalent to stating that the linear inequality system

$$Ax \leq a, \quad Bx = b, \quad \langle \nabla f(x^*), x \rangle < \langle \nabla f(x^*), x^* \rangle$$

is inconsistent. It follows from Theorem 3.17 that there exist multipliers $y \in \mathbb{R}^m$, $z \in \mathbb{R}^m$ such that

$$\nabla f(x^*) = A^T y + B^T z, \quad y \geq 0. \tag{4.20}$$

These optimality conditions are referred to as the *Karush–Kuhn–Tucker (KKT) conditions* for the problem of minimization of f over the set $C = \{x : Ax \leq a, Bx = b\}$. This topic will be discussed in great detail in Chapter 9.

If f is a convex function, then the KKT conditions (4.20) are, of course, necessary and sufficient conditions for a global minimizer of f over C.

4.6 Variational Principles on a Closed Convex Set

If we minimize a differentiable function $f : C \to \mathbb{R}$ over a convex set in $C \subseteq \mathbb{R}^n$ and f has a local minimizer on C, then the variational inequality may be applied. If f is bounded from below on C but has no minimizer on C, then Ekeland's ϵ-variational principle may be helpful.

Theorem 4.41. *Let $C \subseteq \mathbb{R}^n$ be a closed convex set, and f a lower semicontinuous, Gâteaux differentiable function in a neighborhood of C.*

If f is bounded from below on C, then there exists a sequence $\{x_k\}_{k=1}^{\infty}$ of points in C such that

$$f(x_k) \to \inf_C f, \quad \text{and} \quad \varliminf_{k \to \infty} \langle \nabla f(x_k), d \rangle \geq 0 \;\; \text{for all} \;\; d \in \operatorname{rec}(C),$$

where $\operatorname{rec}(C)$ *is the* recession cone *of* C*,*

$$\operatorname{rec} C := \{d \in \mathbb{R}^n : x + td \in C \;\; \text{for all } x \in C, \; t \geq 0\}.$$

Proof. Applying Theorem 3.2 with $\lambda = 1$ and $\epsilon = 1/k$, we obtain a point $x_k \in C$ satisfying the conditions

$$f(x_k) \le \inf_C f + \frac{1}{k} \quad \text{and} \quad f(x_k) \le f(x) + \frac{\|x - x_k\|}{k} \quad \text{for all } x \in C.$$

Let $d \in \operatorname{rec}(C)$ and $t > 0$. Since f is Gâteaux differentiable, we have

$$f(x_k + td) = f(x_k) + t\langle \nabla f(x_k), d \rangle + o(t);$$

substituting this in the second inequality above and simplifying, we obtain

$$0 \le \langle \nabla f(x_k), d \rangle + \frac{\|d\|}{k} + \frac{o(t)}{t}.$$

Letting $t \searrow 0$, we see that the sequence $\{x_k\}$ satisfies the required properties. □

4.7 Exercises

1. Let C be a nonempty set in a vector space E. Show that C is convex if and only if $sC + tC = (s + t)C$ for all positive numbers s, t.
2. Let A be a nonempty set in a vector space E, and define $\operatorname{co}_k(A)$ to be the set of all k-convex combinations of points from A, that is,

$$\operatorname{co}_k(A) := \Big\{ \sum_{i=1}^{k} \lambda_i x_i \ : \ x_i \in A,\ \lambda_i \ge 0,\ \sum_{i=1}^{k} \lambda_i = 1 \Big\}.$$

   Show that $\operatorname{co}_k(\operatorname{co}_l(A)) = \operatorname{co}_{kl}(A)$ for any positive integers k and l.
3. Let K_1 and K_2 be convex cones in a vector space E. Show that $K_1 + K_2$ is a convex cone, $K_1 + K_2 \subseteq \operatorname{co}(K_1 \cup K_2)$, and if both cones contain the origin, then $K_1 + K_2 = \operatorname{co}(K_1 \cup K_2)$.
4. Let C, D be two nonempty sets in $\mathbb{R}^n$.
   (a) If C is open, show that $C + D$ is open.
   (b) If C is open, then so is $\operatorname{co}(C)$.
   (c) Give an example of a closed set in $\mathbb{R}^2$ whose convex hull is not closed.
5. The convexity of a function is a one-dimensional concept in the following sense. Let $f : C \to \mathbb{R}$ be a convex function, where C is a convex subset of a vector space E. For $x, y \in C$, define $d = y - x$ and define the function $h(t) := f(x + td)$ on $[0, 1]$. Show that f is a convex function if and only if $h(t)$ is a convex function for every pair $x, y \in C$.
6. Let $f : C \to \mathbb{R}$ be a strictly convex function on a convex set C in a vector space E. Let $x_i \in C$ and $\lambda_i > 0$ for $i = 1, 2, \ldots, k$ such that $\sum_{i=1}^{k} \lambda_i = 1$. If

$$f(\lambda_1 x_1 + \cdots + \lambda_k x_k) = \lambda_1 f(x_1) + \cdots + \lambda_k f(x_k),$$

then show that $x_1 = x_2 = \cdots = x_k$.
Hint: Use induction on k.
This problem is important for applications of convexity to inequalities where it is used to characterize when the inequality becomes an equality.

7. Show that the function $f(x) := -\ln x$ is strictly convex on $(0, \infty)$. For n-tuples $(x_1, x_2, \ldots, x_n)$ and $(\alpha_1, \alpha_2, \ldots, \alpha_n)$ such that $x_k > 0$ and $\alpha_k > 0$ for all k, and $\sum \alpha_k = 1$, show that

$$x_1^{\alpha_1} x_2^{\alpha_2} \cdots x_n^{\alpha_n} \leq \alpha_1 x_1 + \alpha_2 x_2 + \cdots + \alpha_n x_n,$$

with equality holding if and only if $x_1 = x_2 = \cdots = x_n$.

8. Consider the *arithmetic–geometric–harmonic* mean inequality,

$$\frac{n}{\frac{1}{x_1} + \cdots + \frac{1}{x_n}} \leq (x_1 \cdots x_n)^{1/n} \leq \frac{x_1 + \cdots + x_n}{n},$$

where $x_1, \ldots, x_n$ are positive numbers.
   (a) Use Exercise 7 to prove the arithmetic–geometric mean inequality, and characterize when the inequality becomes an equality.
   (b) Prove the geometric–harmonic mean inequality, and characterize the equality case.

9. Let $f : I \to \mathbb{R}$ be a convex function on an interval $I = [a, b]$.
   (a) If $x_1 < x_2 < x_3$ are three points in I, then show that

$$\frac{f(x_2) - f(x_1)}{x_2 - x_1} \leq \frac{f(x_3) - f(x_1)}{x_3 - x_1} \leq \frac{f(x_3) - f(x_2)}{x_3 - x_2}.$$

   Moreover, show that the inequalities are strict when f is a strictly convex function.
   Hint: The inequalities may be obtained by purely algebraic means, but it may be helpful to draw a picture to visualize them.
   (b) Use (a) to prove that the one-sided derivatives exist: if x is an interior point of I, then show that the difference quotient $(f(x+t) - f(x))/t$ is an increasing function of $t \geq 0$. Use this to show that

$$f'_+(x) := \lim_{t \searrow 0} \frac{f(x+t) - f(x)}{t} = \inf_{t \searrow 0} \frac{f(x+t) - f(x)}{t}$$

   exists and is finite; similarly, show that

$$f'_-(x) := \lim_{t \nearrow 0} \frac{f(x+t) - f(x)}{t} = \sup_{t \nearrow 0} \frac{f(x+t) - f(x)}{t}$$

   exists and is finite.
   (c) If x is an interior point of I, show that $f'_-(x) \leq f'_+(x)$.
   (d) If x is not an interior point of I, show that one of the one-sided derivatives still makes sense, that it exists, but give an example to show that it may be infinite ($\mp\infty$).

(e) Let $f : C \to \mathbb{R}$ be a convex function on a convex set $C \subseteq \mathbb{R}^n$. Consider a point $x \in C$ and a direction $d \in \mathbb{R}^n$ such that $[x, x+\delta d] \in C$ for some $\delta > 0$. Recall that the directional derivative at x along d is given by

$$f'(x; d) = \lim_{t \searrow 0} \frac{f(x + td) - f(x)}{t}.$$

Show that the directional derivative exists by proving that

$$f'(x; d) = \inf_{t>0} \frac{f(x + td) - f(x)}{t},$$

but it may be infinite. Show that if $[x - \delta d, x + \delta d] \in C$ for some $\delta > 0$, then $f'(x; d)$ is finite.
Hint: Use parts (a)–(c).
This proves the important result that a convex function always has directional derivatives in the interior of its domain, regardless of whether it is differentiable in any other sense.

10. This problem provides a condition equivalent to the convexity inequality

$$f((1 - t)x + ty) \leq (1 - t)f(x) + tf(y), \quad 0 \leq t \leq 1.$$

(a) Let $f : I \to \mathbb{R}$ be a function, where $I \subseteq \mathbb{R}$ is an interval. Let $x, y \in I$ and let l be the line connecting $(x, f(x))$ and $(y, f(y))$. Recall that the function f is convex if and only if the segment of l between $(x, f(x))$ and $(y, f(y))$ lies *above* the graph of the function f. Show that this is equivalent to the statement that the remaining part of the line l lies *below* the graph, that is,

$$f((1 - t)x + ty) \geq (1 - t)f(x) + tf(y), \quad t \geq 1, \ (1 - t)x + ty \in I.$$

(b) Let $f : \mathbb{R} \to \mathbb{R}$ be a convex differentiable function. Use part (a) to give a more geometric proof of the convexity inequality

$$f(y) \geq f(x) + f'(x)(y - x)$$

given in Theorem 4.27 on page 98.

(c) Let $f : \mathbb{R} \to \mathbb{R}$ be a convex function satisfying $f(0) = 0$ and $f(x) > 0$ for some $x > 0$. Show that $\lim_{x\to\infty} f(x) = \infty$. If, moreover, $f(y) > 0$ for some $y < 0$, then show that f is a coercive function.

(d) Generalize parts (b) and (c) to a function $f : C \to \mathbb{R}$, where $C \subseteq \mathbb{R}^n$ is a convex set.

11. Let $f : C \to \mathbb{R}$ be a twice Fréchet differentiable function on a convex open set $C \subseteq \mathbb{R}^n$. The following statements are known to be equivalent:
(a) f is convex.
(b) $f(y) \geq f(x) + \langle \nabla f(x), y - x \rangle$ for all $x, y \in C$.
(c) $\langle \nabla f(y) - \nabla f(x), y - x \rangle \geq 0$ for all $x, y \in C$.

(d) $Hf(x)$ is positive semidefinite at every $x \in C$.

In fact, we have proved that (a), (b), and (d) are equivalent conditions. Give direct proofs of

(i) (b) implies (c).

(ii) (c) implies (b).

Hint: First, show that the function $g(t) := f(x+t(y-x)) - t\langle \nabla f(x), y-x\rangle$ is differentiable and nondecreasing.

(iii) (c) implies (d).

(iv) (d) implies (c).

12. (a) Let $f : \mathbb{R} \to \mathbb{R}$ be a differentiable function such that the graph of the function f at every point $x \in \mathbb{R}$ is supported by a line, that is, there exists $\alpha_x \in \mathbb{R}$ satisfying

$$f(y) \geq f(x) + \alpha_x(y - x).$$

Show that $\alpha_x = f'(x)$, and consequently, that f is a convex function.

(b) Suppose that $f : \mathbb{R}^n \to \mathbb{R}$ is a differentiable function such that the graph of f at every point $x \in \mathbb{R}^n$ is supported by a hyperplane, that is, there exists $\alpha_x \in \mathbb{R}^n$ satisfying

$$f(y) \geq f(x) + \langle \alpha_x, y - x\rangle.$$

Show that $\alpha_x = \nabla f(x)$, so that f is a convex function.

(c) Prove (a) and (b) without assuming that f is differentiable.

13. *(Infimal convolution of convex functions)* Let $f, g : \mathbb{R}^n \to \mathbb{R} \cup \{\infty\}$ be convex functions. The function

$$(f \,\Box\, g)(x) := \inf\{f(x - y) + g(y) : y \in \mathbb{R}^n\}$$

is called the *infimal convolution* of f and g.

Show that $f \,\Box\, g$ is a convex function.

14. Let $C \subseteq \mathbb{R}^n$ be a convex set. Show that the following functions are convex:

(a) The *indicator function* of C defined by

$$\delta_C(x) := \begin{cases} 0, & \text{if } x \in C, \\ +\infty, & \text{otherwise.} \end{cases}$$

(b) The *distance function* to C defined by

$$d_C(x) := \inf\{\|z - x\| : z \in C\}.$$

(c) The *support function* of C defined by

$$\sigma_C(x) := \sup\{\langle z, x\rangle : z \in C\}.$$

Hint: Consider the epigraph of the function.

Furthermore,

(d) Show that $d_C = \|\cdot\| \,\square\, \delta_C$, where $\|\cdot\|(x) = \|x\|$ is the norm function on $\mathbb{R}^n$.

15. Let the function $f : C \to \mathbb{R}$ given by $f(x) = \langle a, x\rangle^2/\langle b, x\rangle$ be defined on a convex set $C \subseteq \mathbb{R}^n$ such that $\langle b, x\rangle > 0$ on C. Show that f is a convex function.
16. (a) Show that if $g : \mathbb{R} \to \mathbb{R}$ is a concave function with $g(x) > 0$, then $f = 1/g$ is a convex function.
    (b) Show that a function $f : \mathbb{R}^n \to \mathbb{R}$ is convex if and only if for any given $x, y \in \mathbb{R}^n$ the function $h : \mathbb{R} \to \mathbb{R}$ given by $h(t) = f(x + ty)$ is convex.
    (c) Use (a) and (b) to show that if a function $g : \mathbb{R}^n \to \mathbb{R}$ is concave and $g(x) > 0$, then the function $f = 1/g$ is a convex function.
17. Suppose that $f : \mathbb{R}^n \to \mathbb{R}$ is a convex function.
    (a) Show that if $f(0) = 0$ and f is an even function, that is, $f(-x) = f(x)$, then $f(x) \geq 0$ for all $x \in \mathbb{R}^n$.
    Hint: Compare $f(x), f(-x), f(0)$ using the convexity of f.
    (b) Suppose that f is homogeneous of degree $p \neq 1$, that is, $f(tx) = t^p f(x)$ for all $x \in \mathbb{R}^n$ and for all $t \geq 0$. Show that $f(0) = 0$ and $f(x) \geq 0$ for all $x \in \mathbb{R}^n$.
    Hint: Use the convexity inequality for f on the three points $-x, tx, 0$, and vary $t > 0$.
    (c) Show that if f is not a constant function, then there exists $0 \neq x_0 \in \mathbb{R}^n$ such that
    $$\lim_{t\to\infty} f(tx_0) = \infty.$$
    Hint: We may assume that $f(0) = 0$. Show that there exists a point x_0 with $f(x_0) > 0$ and apply the convexity inequality for f at the points $0, x_0, tx_0$, where $t > 1$. (If x_0 does not exist, then one would get a contradiction by applying the convexity inequality at $x, -x, 0$ for any $x \neq 0$.)
18. Let D be a compact, convex set in $\mathbb{R}^n$, and let C be its projection onto $\mathbb{R}^{n-1}$, that is, onto the hyperplane $x_n = 0$. Show that there exist two convex functions f and g on C such that
    $$D = \{(x, \lambda) : x \in C,\ -g(x) \leq \lambda \leq f(x)\}.$$
19. *(Logarithmically convex functions)* A positive-valued function $f : C \to \mathbb{R}$, where $C \subseteq \mathbb{R}^n$ is a convex set, is called *logarithmically convex* if $g(x) := \ln f(x)$ is a convex function.
    (a) Let $f : [a, b] \to \mathbb{R}$. Show that f is a convex function if and only if
    $$\sup_{t\in I} \{f(t) - \alpha t\} \leq \max_{t\in\partial I} \{f(t) - \alpha t\}$$
    for all $\alpha \in \mathbb{R}$ and all closed subintervals I of $[a, b]$.

(b) Let f be a positive-valued function on $[a,b]$. Use part (a) to show that the function $\ln f(t)$ is convex if and only if $e^{ct}f(t)$ is a convex function for every $c \in \mathbb{R}$.

(c) Let $\{f_i(t)\}_1^k$ be positive-valued functions on $[a,b]$ such that each $\ln f_i(t)$ is a convex function. Show that $\ln\bigl(\sum_{i=1}^k f_i(t)\bigr)$ is a convex function.

(d) Show that $\ln\bigl(\sum_{i=1}^k e^{f_i(t)}\bigr)$ is a convex function if the functions $\{f_i\}_1^k$ are all convex.
This proves that if $\{f_i\}_1^k$ are convex functions, then $\sum_{i=1}^k e^{f_i(t)}$ is a logarithmically convex function.

(e) Generalize (d) to show that $\int e^{f(t,s)}ds$ is logarithmically convex under appropriate assumptions on $f(t,s)$.

(f) Generalize parts (c)–(e) to functions on a convex set $C \subseteq \mathbb{R}^n$.

20. *(Finite difference characterization of convex functions)* Let $f : I = [a,b] \to \mathbb{R}$, and let $x_1, x_2, x_3 \in I$ be three distinct points. Define the *finite difference*

$$[x_1,x_2]f := \frac{f(x_1)-f(x_2)}{x_1-x_2},$$

and the second-order finite difference

$$[x_1,x_2,x_3]f := \frac{[x_1,x_2]f - [x_2,x_3]f}{x_1-x_3}.$$

Show that f is a convex function if and only if $[x_1,x_2,x_3]f \geq 0$ for all distinct $x_1, x_2, x_3 \in I$.

21. Jensen's inequality for convex functions has integral versions. The following is an example. Let $f : C \to \mathbb{R}$ be a continuous convex function on a convex set $C \subseteq \mathbb{R}^n$. If $x : [0,1] \to C$ is any continuous function with component functions x_i, that is, $x(t) = (x_1(t), \ldots, x_n(t))$, then prove that

$$f\Bigl(\int_0^1 x(t)dt\Bigr) \leq \int_0^1 f(x(t))dt.$$

Here

$$\int_0^1 x(t)dt = \Bigl(\int_0^1 x_1(t)dt, \ldots, \int_0^1 x_n(t)dt\Bigr).$$

Hint: Approximate the integral with Riemann sums,

$$\int_0^1 f(x(t))dt \approx \sum f(x_i)(t_{i+1}-t_i),$$

and do the same with the integral $\int_0^1 x(t)dt$. Use Jensen's inequality on the sums and pass to the limit.

22. *(Convexity of the function* $-\ln\det X$*)* Consider the function $f(X) = -\ln\det X$ on the positive definite cone P_n, the set of symmetric, positive definite $n \times n$ matrices. The purpose of this exercise is to prove that f is strictly convex, in three different ways.

(a) In Example 1.27 on page 18, it is shown that $Hf(X) = X^{-1} \otimes X^{-1}$. If D is a symmetric $n \times n$ matrix and $X \in P_n$, show that

$$\langle Hf(X)(D), D\rangle = \operatorname{tr}(X^{-1}DX^{-1}D),$$

and if $D \neq 0$, prove that $\langle Hf(X)(D), D\rangle > 0$.

(b) First, show by direct computation that f is strictly convex on the set of diagonal matrices with positive entries. Next, let $X, Y \in P_n$, and use Theorem 2.21 on page 42 to write $X = Z^T \Lambda Z$ and $Y = Z^T \Sigma Z$, where Z is an $n \times n$ matrix Z and Λ, Σ are diagonal matrices. Combine the two results to prove that f is strictly convex on P_n.

(c) Let $X \in P_n$ and let D be an $n \times n$ symmetric matrix. Show that the function

$$p(t) := \det(X + tD) = \det X + \det(I + tX^{-1/2}DX^{-1/2})$$

is a polynomial that can be written as $p(t) = c\prod_1^k(1 - tt_i)$, where $t_i \neq 0$ (in fact, t_i^{-1} is an eigenvalue of $-X^{-1/2}DX^{-1/2}$, thus real). Show that if $D \neq 0$, then

$$\langle Hf(X)D, D\rangle = (-\ln p)''(0) = \sum_1^k t_i^{-2} > 0.$$

This approach comes from the theory of *hyperbolic polynomials*; see [101, 102, 119, 26].

23. Consider the function $F(x) = (x_1x_2\cdots x_n)^{1/n}$ on the nonnegative orthant $\mathbb{R}^n_+ := \{x \in \mathbb{R}^n : x_i \geq 0,\ i = 1, \ldots, n\}$.

(a) Show that F is a concave function by computing its Hessian $Hf(x)$.

(b) Use (a) to show that

$$(y_1y_2\cdots y_n)^{1/n} \leq (x_1x_2\cdots x_n)^{1/n}\frac{1}{n}\sum_{i=1}^n \frac{y_i}{x_i},$$

and prove that this implies the arithmetic–geometric mean inequality.

(c) Prove the inequality

$$\prod_1^n(x_i + y_i)^{1/n} \geq \Big(\prod_1^n x_i^{1/n}\Big) + \Big(\prod_1^n y_i^{1/n}\Big), \tag{4.21}$$

and show that equality holds if and only if $x, y \in \mathbb{R}^n_+$ are proportional, that is, x and y lie on a line through the origin.

24. **(Oppenheim)**; see [127]. The inequality (4.21) has interesting applications to matrices.

(a) Let A, B be two $n \times n$ symmetric, positive definite matrices. Show that there exists a nonsingular $n \times n$ matrix X such that

$$X^TAX = \Lambda = \operatorname{diag}\{\lambda_1, \ldots, \lambda_n\} \text{ and } X^TBX = \Delta = \operatorname{diag}\{\delta_1, \ldots, \delta_n\}.$$

Hint: Use Theorem 2.21.

(b) Use the inequality (4.21) on $(\lambda_1, \ldots, \lambda_n)$ and $(\delta_1, \ldots, \delta_n)$ to prove that

$$(\det(A+B))^{1/n} \geq (\det A)^{1/n} + (\det B)^{1/n},$$

with equality holding if and only if A and B are multiples of each other.

(c) If $A = \left[\begin{smallmatrix} A_{11} & A_{12} \\ A_{21} & A_{22} \end{smallmatrix}\right]$ is a partition of A such that A_{11} and A_{22} are $r \times r$ and $(n-r) \times (n-r)$ matrices, respectively, then

$$\det A \leq \det A_{11} \cdot \det A_{22},$$

with equality holding if and only if A is block diagonal, that is, $A_{12} = 0$ and $A_{21} = 0$.

Hint: Use (b) with $B = \left[\begin{smallmatrix} A_{11} & -A_{12} \\ -A_{21} & A_{22} \end{smallmatrix}\right]$. Why is B positive definite?

(d) Show that

$$\det A \leq \prod_{i=1}^{n} a_{ii},$$

with equality holding if and only if A is a diagonal matrix.

(e) Prove *Hadamard's inequality*: if $C = [c_1, c_2, \ldots, c_n]$ is a nonsingular $n \times n$ matrix with columns $\{c_i\}_1^n$, then

$$|\det C| \leq \prod_{i=1}^{n} \|c_i\|,$$

with equality holding if and only if $\{c_i\}$, the columns of C, are mutually orthogonal.

Hint: Apply (d) to $A = C^T C$.

25. **(Gauss–Lucas)** Let $p(z) = \sum_0^n a_j z^j$ be a polynomial with complex coefficients a_j. Show that the roots of the derivative $p'(z)$ are contained in the convex hull of the roots of $p(z)$.
Hint: Let $\{z_j\}_1^n$ be the roots of $p(z)$. If z is a root of p' but not of p, show that

$$0 = \frac{p'(z)}{p(z)} = \sum_{j=1}^{n} \frac{1}{z - z_j} = \sum_{j=1}^{n} \frac{1}{\overline{z} - \overline{z_j}} = \sum_{j=1}^{n} \frac{z - z_j}{\|z - z_j\|^2}.$$

26. Let $p(z) = \sum_0^n a_j z^j$ be a polynomial with complex coefficients a_j and let $C \subset \mathbb{C} = \mathbb{R}^2$ be a closed convex set. Show that

$$D := \{w \in \mathbb{C} : \text{ all solutions } z \text{ of } p(z) = w \text{ lie in } C\}$$

is a convex set.
Hint: For given positive integers n_1, n_2 and points $w_1, w_2 \in D$, define the polynomial

$$q(z) = (p(z) - w_1)^{n_1} (p(z) - w_2)^{n_2},$$

and use Exercise 25 to show that all roots of q' lie in C. Use the known fact (which you may assume) that the roots of a polynomial depend continuously on its coefficients to finish the proof.

27. **(Carleman's inequality)** Let $\sum_{i=1}^{\infty} a_i < \infty$, where $a_i > 0$. Show that

$$\sum_{i=1}^{\infty} (a_1 a_2 \cdots a_n)^{1/n} \leq e \sum_{i=1}^{\infty} a_i.$$

Hint: Let $\{c_i\}_1^\infty$ be a positive, yet unspecified, sequence. Use the arithmetic–geometric mean inequality to get

$$\begin{aligned}(a_1 a_2 \cdots a_n)^{1/n} &\leq (c_1 c_2 \cdots c_n)^{-1/n} (c a_1 c a_2 \cdots c a_n)^{1/n} \\ &\leq (c_1 c_2 \cdots c_n)^{-1/n} \frac{c a_1 + c a_2 + \cdots + c a_n}{n}.\end{aligned}$$

Choose $c_m = \frac{(m+1)^m}{m^{m-1}}$, and use the fact that

$$\sum_{n \geq m} \frac{1}{n(n+1)} = \sum_{n \geq m} \left(\frac{1}{n} - \frac{1}{n+1}\right) = \frac{1}{m}$$

and $(1 + 1/m)^m < e$ to complete the proof of the inequality.

28. This problem gives an integral representation for a twice differentiable convex function $f : I \to \mathbb{R}$ on an interval $I = [a, b]$. (In fact, a suitable generalization gives an integral representation for any convex function, but this involves more general Stieltjes integrals, which we do not consider here.)

(a) Let $I = [0, 1]$, and define $K : I \times I \to \mathbb{R}$ by

$$K(x, y) = \begin{cases} x(1 - y), & \text{if } x \leq y, \\ (1 - x)y, & \text{if } y \leq x. \end{cases}$$

($K(x, y)$ is the *Green's function* for the differential equation $u'' = 0$.) Show that if $f(0) = f(1) = 0$, then

$$f(x) = -\int_0^1 K(x, y) f''(y) dy.$$

(b) Show that in general,

$$f(x) = x f(0) + (x - 1) f(1) - \int_0^1 K(x, y) f''(y) dy.$$

(c) What is $K(x, y)$ if $I = [a, b]$ is a general interval, and what is the analogue of the equation in (b)?

5

Structure of Convex Sets and Functions

This chapter is devoted to investigating the deeper properties of convex sets and convex functions on fairly general affine and vector spaces. The separation properties of convex sets are very important in optimization, especially in duality theory, with more sophisticated separation theorems leading to better duality results; see for example Theorem 11.15, on the strong duality in convex programming. In turn, sophisticated separation theorems are obtained by a careful study of the properties of interior points of convex sets. Since a convex set is not necessarily full-dimensional, it is important to study the "relative interior" of convex sets. It turns out that the relative interior of a convex set can be studied in a purely algebraic setting, without any use of topological notions, and this leads to a rich theory of the relative algebraic interior of convex sets. If the space under consideration has a topology, then it turns out that there is a strong connection between the algebraic and topological notions of relative interior, and the topological results can be obtained from the corresponding algebraic ones with relative ease.

The boundary structure of convex sets, especially the properties of extreme points and directions of convex sets, are useful in many applications, since it is possible to represent convex sets in terms of these. For example, a theorem of Minkowski states that a compact convex set in $\mathbb{R}^n$ is the convex hull of its extreme points.

These are among the many reasons that make it worthwhile to undertake a close study of the finer structure of convex sets.

5.1 Algebraic Interior and Algebraic Closure of Convex Sets

Definition 5.1. *Let C be a convex subset of an affine space A. The* algebraic interior *of C, denoted by $\mathrm{ai}_A(C)$ ($\mathrm{ai}(C)$ if A is understood from context), is the set of all points $x \in C$ such that every line $\ell \subseteq A$ through x contains a line segment in C having x in its interior, that is,*

O. Güler, *Foundations of Optimization*, Graduate Texts in Mathematics 258,
DOI 10.1007/978-0-387-68407-9_5,

$$\begin{aligned}\mathrm{ai}_A(C) &:= \{x \in C \,:\, \forall z \in A,\ \exists \delta > 0,\ [x - \delta(z-x), x + \delta(z-x)] \subseteq A\} \\ &= \{x \in C \,:\, \forall z \in A,\ \exists \delta > 0,\ [x, x + \delta(z-x)) \subseteq A\}.\end{aligned}$$

A convex set $C \subseteq A$ is called a convex algebraic body *if* $\mathrm{ai}_A(C) \neq \emptyset$.

If the affine set A is $\mathrm{aff}(C)$, *the affine hull of C, then* $\mathrm{ai}_A(C)$ *is called the* relative algebraic interior *of C, and is denoted by* $\mathrm{rai}(C)$.

The algebraic closure *of C, denoted by* $\mathrm{ac}(C)$, *is the set of all points $z \in A$ such that $[x, z) \subseteq C$ for some $x \in C$,*

$$\mathrm{ac}(C) := \{z \in A \,:\, \exists x \in C,\ [x, z) \subseteq C\}.$$

Lemma 5.2. *If C is a convex set in an affine space A, then* $\mathrm{ai}(C)$ *and* $\mathrm{ac}(C)$ *are also convex sets in A.*

Proof. Let $x, y \in \mathrm{ai}(C)$. If $u \in A$, then there exists $\delta > 0$ such that $x + \delta(u - x)] =: [x, p] \subset C$ and $[y, y + \delta(u - y)] =: [y, q] \subset C$; see the first figure in Figure 5.1. If $z := (1-t)x + ty$ for some $t \in (0, 1)$, then we have that

$$r := z + \delta(u - z) = (1-t)(x + \delta(u - x)) + t(y + \delta(u - y)) = (1-t)p + tq$$

lies in C; thus $[z, r] \subset C$, meaning that $z \in \mathrm{ai}(C)$.

Let $u, v \in \mathrm{ac}(C)$, where and $[x, u) \subseteq C$ and $[y, v) \subseteq C$; see the middle figure in Figure 5.1. Let $t \in (0, 1)$ and define $w := (1-t)u + tv$ and $z := (1-t)x + ty \in C$. We claim that $[z, w) \subseteq C$. Let $a := (1-\delta)z + \delta w$ for some $\delta \in (0, 1)$. We have

$$\begin{aligned}a &= (1-\delta)z + \delta w = (1-\delta)((1-t)x + ty) + \delta((1-t)u + tv) \\ &= (1-t)((1-\delta)x + \delta u) + t((1-\delta)y + \delta v) \in C,\end{aligned}$$

proving the claim and the lemma. □

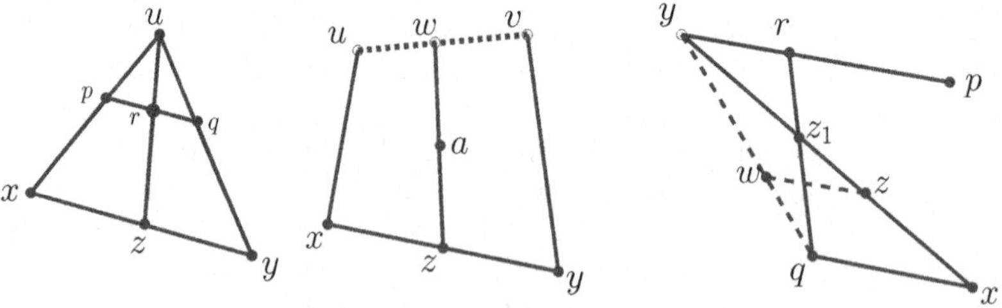

Fig. 5.1. Algebraic interior and algebraic closure of a convex set.

Lemma 5.3. *If C is a convex set in a finite-dimensional affine space A, then* $\mathrm{rai}(C) \neq \emptyset$.

Proof. Let $\{x_i\}_1^k \subseteq C$ be an affine basis of $\text{aff}(C)$, and let $x := (\sum_1^k x_i)/k$ be the center of the simplex defined by $\{x_i\}_1^k$. We claim that $x \in \text{rai}(C)$. Indeed, if $u = \sum_1^k t_i x_i$, where $\sum_1^k t_i = 1$ is an arbitrary point in $\text{aff}(C)$, then $x + \delta(u - x) = (1 - \delta)x + \delta u = \sum_1^k [(1 - \delta)/k + \delta t_i] x_i$. If $\delta > 0$ is sufficiently small, then each $(1 - \delta)/k + \delta t_i$ is positive, and we have $x + \delta(u - x) \in C$. □

Remark 5.4. Every infinite-dimensional affine space A contains a convex set C such that $\text{rai}(C) = \emptyset$. Indeed, let $X = \{x_i\}_1^\infty$ be a set of affinely independent points in A and consider the convex hull $C = \text{co}(X)$. Suppose that $x \in \text{rai}(C)$. Then $x \in C_m := \text{co}(\{x_i\}_1^m)$ for some integer m. If $y \in C$ and $y \neq x$, then there exists $z \in C$ such that $x \in (y, z)$. But then $y, z \in C_n$ for some $n > m$, and it is easy to prove that C_m is a face of C_n. It follows that $y, z \in C_m$ (see Definition 5.24 and Lemma 5.25), and this implies that $C = C_m$, a contradiction.

Lemma 5.5. *Let C be a convex set in an affine space A. If $y \in \text{ac}(C)$ and $x \in \text{ai}(C)$, then $[x, y) \subset \text{ai}(C)$.*

Proof. First, assume that $y \in C$. Let $z := tx + (1 - t)y = y + t(x - y) \in C$, $t \in (0, 1)$; see the last figure in Figure 5.1. We claim that $z \in \text{ai}(C)$. Let $d := u - x$ be an arbitrary direction in A, where $u \in A$. Since $x \in \text{ai}(C)$, there exists $\delta > 0$ such that $[x, x + \delta(u - x)] =: [x, q] \subset C$. We have that

$$w := z + t\delta(u - x) = z + t(q - x) = y + t(x - y) + t(q - x) = y + t(q - y)$$

lies in C; thus $[z, w] = [z, z + t\delta(u - x)] \subset C$, proving the claim.

Now assume that $y \in \text{ac}(C) \setminus C$. There exists $p \in C$ such that $[p, y) \subset C$. We claim that $[x, y) \in \text{ai}(C)$. Let $z \in (x, y)$. If $p = x$, pick $z_1 \in (x, y)$ such that $z \in (z_1, x)$. The first paragraph of the proof shows that $z \in \text{ai}(C)$. Finally, suppose that $p \neq x$. There exists $\delta > 0$ such that $[x, x + \delta(y - p)] =: [x, q] \subset C$. Pick a point $r \in (y, p)$ such that $[r, q]$ intersects $[y, x]$ at a point z_1 such that $z \in (z_1, x)$. It follows again from the first paragraph that $z \in \text{ai}(C)$. □

5.2 Minkowski Gauge Function

Definition 5.6. *Let C be a convex set in a vector space E such that $0 \in \text{rai}(C)$. The* (Minkowski) gauge function *of C is the function p_C defined on E by the formula*

$$p_C(x) := \inf\{t > 0 : x \in tC\} = \inf\{t \geq 0 : x \in tC\}.$$

If C is a convex set in an affine space A and $x_0 \in \text{rai}(C)$, then the gauge function of C with respect to x_0 is the function $p(x) := p_{C - x_0}(x - x_0)$ defined on A, that is,

$$p(x) := \inf\{t > 0 : x \in x_0 + t(C - x_0)\}, \quad \textit{where} \quad x \in A.$$

Theorem 5.7. *Let C be a convex set in a vector space E such that $0 \in \operatorname{rai}(C)$. The gauge function p_C is a nonnegative extended-valued function, $p_C : E \to \mathbb{R} \cup \{+\infty\}$, that is finite-valued precisely on the linear subspace* $\operatorname{span}(C)$.

Moreover, p_C is a sublinear function, that is, for all $x, y \in E$ and for all $t \geq 0$,

$$p_C(tx) = tp_C(x) \quad \text{and} \quad p_C(x+y) \leq p_C(x) + p_C(y).$$

Thus, p_C is a convex homogeneous function of degree one.

Proof. Evidently, $p(x) := p_C(x) \geq 0$ for all $x \in E$. Since $0 \in \operatorname{rai}(C)$, there exists $t > 0$ such that $x \in tC$ if and only if $x \in L := \operatorname{span}(C)$. Thus $p(x)$ is finite if and only if $x \in L$.

The homogeneity of p is obvious from its definition, and the convexity of p is thus equivalent to its subadditivity. To prove the former, let $x_1, x_2 \in L$ and $0 < t < 1$. Given an arbitrary $\epsilon > 0$, note that $x_i \in (p(x_i) + \epsilon)C$, $i = 1, 2$. We have

$$\begin{aligned}(1-t)x_1 + tx_2 &\in (1-t)[(p(x_1) + \epsilon)C)] + t[(p(x_2) + \epsilon)C] \\ &= [(1-t)p(x_1) + tp(x_2) + \epsilon]C,\end{aligned}$$

where the equality follows since C is a convex set. Since $\epsilon > 0$ is arbitrary, we have

$$p((1-t)x_1 + tx_2) \leq (1-t)p(x_1) + tp(x_2).$$

□

Theorem 5.8. *If C is a convex set in a vector space E such that $0 \in \operatorname{rai}(C)$, then*

$$\begin{aligned}\operatorname{rai}(C) &= \operatorname{rai}(\operatorname{rai}(C)) = \operatorname{rai}(\operatorname{ac}(C)) = \{x \in E : p(x) < 1\}, \\ \operatorname{ac}(C) &= \operatorname{ac}(ac(C)) = \operatorname{ac}(\operatorname{rai}(C)) = \{x \in E : p(x) \leq 1\}.\end{aligned} \tag{5.1}$$

Proof. Note that without any loss of generality, we may assume that $E = \operatorname{span}(C)$. Then p_C is finite-valued, and relative algebraic interiors become algebraic interiors.

First, it is clear from the definition of p that

$$\{x \in E : p(x) < 1\} \subseteq C \subseteq \{x : p(x) \leq 1\}.$$

Let $x \in E$ be such that $p(x) < 1$. If $y \in E$ is arbitrary, then for small enough $t > 0$, we have $p(x+ty) \leq p(x) + tp(y) < 1$; thus $[x, x+ty] \subset C$, which proves that $x \in \operatorname{ai}(C)$. Lemma 5.5 then implies that $x + sy \in \operatorname{ai}(C)$ for $s \in [0, t)$, and we have $x \in \operatorname{ai}(\operatorname{ai}(C))$. Therefore,

$$\{x \in E : p(x) < 1\} \subseteq \operatorname{ai}(\operatorname{ai}(C)), \tag{5.2}$$

and consequently,

$$\operatorname{ai}(\operatorname{ai}(C)) \subseteq \operatorname{ai}(C) \subseteq \{x \in E : p(x) < 1\} \overset{(5.2)}{\subseteq} \operatorname{ai}(\operatorname{ai}(C)) \subseteq \operatorname{ai}(\operatorname{ac}(C)).$$

Here the second inclusion follows because if $x \in \mathrm{ai}(C)$, then there exists $t > 1$ such that $tx \in C$, implying $tp(x) = p(tx) \le 1$, and hence $p(x) < 1$. Note that the first three inclusions above are actually equalities, so that the first line of (5.1) is proved, except for the inclusion $\mathrm{ai}(\mathrm{ac}(C)) \subseteq \{x \in E : p(x) < 1\}$.

To prove the second line of (5.1), assume that $x \in \mathrm{ac}(C)$, with $[u, x) \subset C$. If $t \in [0, 1)$, then

$$\begin{aligned} p(x) &= p(u + t(x-u) + (1-t)(x-u)) \le p(u + t(x-u)) + (1-t)p(x-u) \\ &\le 1 + (1-t)p(x-u). \end{aligned}$$

Letting $t \nearrow 1$, we conclude that $p(x) \le 1$; this proves that

$$\mathrm{ac}(C) \subseteq \{x : p(x) \le 1\}.$$

Now if $x \in \mathrm{ac}(\mathrm{ac}(C))$, then since $0 \in \mathrm{ai}(C) \subseteq \mathrm{ai}(\mathrm{ac}(C))$, Lemma 5.5 implies that $tx \in \mathrm{ai}(ac(C)) \subseteq \mathrm{ac}(C)$ for $t \in (0, 1)$; thus $tp(x) = p(tx) \le 1$, and letting $t \nearrow 1$ leads to $p(x) \le 1$, proving

$$\mathrm{ac}(\mathrm{ac}(C)) \subseteq \{x \in E : p(x) \le 1\}. \tag{5.3}$$

Also, if $p(x) = 1$, we have $p(z) < 1$ for every $z \in [0, x)$, hence $[0, x) \subset \mathrm{ai}(\mathrm{ai}(C)) \subset \mathrm{ai}(C)$, proving $x \in \mathrm{ac}(\mathrm{ai}(C))$, so that

$$\{x \in E : p(x) = 1\} \subseteq \mathrm{ac}(\mathrm{ai}(C)). \tag{5.4}$$

These give

$$\mathrm{ac}(C) \subseteq \mathrm{ac}(\mathrm{ac}(C)) \overset{(5.3)}{\subseteq} \{x \in E : p(x) \le 1\} \overset{(5.2),(5.4)}{\subseteq} \mathrm{ac}(\mathrm{ai}(C)) \subseteq \mathrm{ac}(C),$$

and hence all inclusions are equalities, proving the second line in (5.1).

Finally, it remains to prove that $\mathrm{ai}(\mathrm{ac}(C)) \subseteq \{x \in E : p(x) < 1\}$. If $x \in \mathrm{ai}(\mathrm{ac}(C))$, then there exists $t > 1$ such that $tx \in \mathrm{ac}(C)$. Then (5.3) implies $tp(x) = p(tx) \le 1$, and thus $p(x) < 1$. The theorem is proved. □

Corollary 5.9. *Let C be a convex algebraic body in an affine space A. Then*

$$\begin{aligned} \mathrm{rai}(C) &= \mathrm{rai}(\mathrm{rai}(C)) = \mathrm{rai}(\mathrm{ac}(C)) = \{x \in A : p(x) < 1\}, \\ \mathrm{ac}(C) &= \mathrm{ac}(\mathrm{ac}(C)) = \mathrm{ac}(\mathrm{rai}(C)) = \{x \in A : p(x) \le 1\}, \end{aligned}$$

where $p(x)$ is the gauge function with respect to any point $x_0 \in \mathrm{rai}(C)$.

Proof. Let $x_0 \in \mathrm{rai}(C)$. Evidently, $0 \in \mathrm{rai}(C - x_0) = \mathrm{rai}(C) - x_0$, $\mathrm{ac}(C - x_0) = \mathrm{ac}(C) - x_0$, and $p_C(x) = p_{(C-x_0)}(x - x_0)$. The corollary follows immediately from Theorem 5.8. □

5.3 Calculus of Relative Algebraic Interior and Algebraic Closure of Convex Sets

Lemma 5.10. *Let $\{C_i\}_1^m$ be convex sets in a vector space E. If $\cap_1^m \operatorname{rai}(C_i) \neq \emptyset$, then*

(a) $\operatorname{aff}(\cap_1^m C_i) = \cap_1^m \operatorname{aff}(C_i)$,
(b) $\operatorname{rai}(\cap_1^m C_i) = \cap_1^m \operatorname{rai}(C_i)$.

Proof. Write $C := \cap_1^m C_i$ and $D := \cap_1^m \operatorname{aff}(C_i)$. Since $C \subseteq D$ and D is an affine set, it follows that $\operatorname{aff}(C) \subseteq D$. To prove the reverse inclusion, let $x_0 \in \cap_1^m \operatorname{rai}(C_i)$. If $x \in D$, then x lies in each $\operatorname{aff}(C_i)$, and since $x_0 \in \operatorname{rai}(C_i)$, the line l passing through x and x_0 contains an open segment around x_0 that is contained in C_i, hence in C. It follows that $x \in \operatorname{aff}(C)$, completing the proof of (a).

We next claim the inclusion $\operatorname{rai}(C) \subseteq \cap_1^m \operatorname{rai}(C_i)$ in (b). Let $x \in \operatorname{rai}(C)$ and pick a point $x_0 \in \cap_1^m \operatorname{rai}(C_i)$. Since $x_0 \in C$ and $x \in \operatorname{rai}(C)$, Lemma 5.5 implies that there exists $z \in C$ such that $x \in (x_0, z)$. But $z \in C_i$ and $x_0 \in \operatorname{rai}(C_i)$ for each i, and Lemma 5.5 implies $x \in \operatorname{rai}(C_i)$, proving the claim.

To prove the reverse inclusion $\cap_1^m \operatorname{rai}(C_i) \subseteq \operatorname{rai}(C)$ in (b), let $x \in \cap_1^m \operatorname{rai}(C_i)$ and pick a point $y \in \operatorname{rai}(C)$, $y \neq x$. (If there is no $y \neq x$, then $\operatorname{rai}(C) = \cap_1^m \operatorname{rai}(C_i) = \{x\}$ and we are done.) Since $x \in \operatorname{rai}(C_i)$ and $y \in C_i$, there exists a point $z_i \in C_i$ such that $x \in s_i := (z_i, y)$. Then the open interval $s := \cap_1^m s_i$ has the form $s = (z, y)$ where $z \in C$. Since $y \in \operatorname{rai}(C)$ and $z \in C$, Lemma 5.5 implies that $x \in \operatorname{rai}(C)$. □

The condition $\cap_1^m \operatorname{rai}(C_i) \neq \emptyset$ is needed for the validity of the lemma. For example, both (a) and (b) fail for the sets $C_1 = [-1, 0]$, $C_2 = [0, 1]$.

Lemma 5.11. *Let C and D be two convex sets in a vector space E. If $\operatorname{rai}(C) \neq \emptyset$ and $\operatorname{rai}(D) \neq \emptyset$, then $\operatorname{rai}(C + D) = \operatorname{rai}(C) + \operatorname{rai}(D)$.*

Proof. Let $x \in \operatorname{rai}(C)$ and $y \in \operatorname{rai}(D)$. Given arbitrary points $u \in C$ and $v \in D$, there exist $u_1 \in C$ and $v_1 \in D$ such that $x \in (u, u_1)$ and $y \in (v, v_1)$. We may assume that $x = (1 - \lambda)u + \lambda u_1$ and $y = (1 - \lambda)v + \lambda v_1$. Then $x + y = (1 - \lambda)(u + v) + \lambda(u_1 + v_1)$, that is, $x + y \in (u + v, u_1 + v_1)$ with $u + v$ and $u_1 + v_1$ lying in $C + D$. Since $u + v \in C + D$ is arbitrary, we have $x + y \in \operatorname{rai}(C + D)$. This proves the inclusion $\operatorname{rai}(C) + \operatorname{rai}(D) \subseteq \operatorname{rai}(C + D)$.

To prove the reverse inclusion, fix $x_0 \in \operatorname{rai}(C)$ and $y_0 \in \operatorname{rai}(D)$. Then $x_0 + y_0 \in \operatorname{rai}(C) + \operatorname{rai}(D)$, and since $\operatorname{rai}(C) + \operatorname{rai}(D) \subseteq \operatorname{rai}(C + D)$ as we proved above, $x_0 + y_0 \in \operatorname{rai}(C + D)$. If $z \neq x_0 + y_0$ is an arbitrary element of $\operatorname{rai}(C + D)$, then $z \in (x_0 + y_0, u + v)$ for some $u + v \in C + D$, where $u \in C$ and $v \in D$, say $z = (1 - \lambda)(x_0 + y_0) + \lambda(u + v)$ for some $\lambda \in (0, 1)$. Thus, $z = x' + y'$, where $x' = (1 - \lambda)x_0 + \lambda u \in (x_0, u)$ and $y' = (1 - \lambda)y_0 + \lambda v \in (y_0, v)$. It follows from Lemma 5.5 that $x' \in \operatorname{rai}(C)$ and $y' \in \operatorname{rai}(D)$; consequently, we have $z = x' + y' \in \operatorname{rai}(C) + \operatorname{rai}(D)$. □

Definition 5.12. *Let E and F be two arbitrary sets. A nonempty subset of $G \subsetneq E \times F$ may be considered the* graph *of a* multivalued map *A from E to F. The* domain *and* range *of A are defined by*

$$\operatorname{dom}(A) := \{x \in E : \exists y \in F, (x, y) \in \operatorname{gr}(A)\},$$
$$R(A) := \{y \in F : \exists x \in E, (x, y) \in \operatorname{gr}(A)\},$$

respectively. If $x \in E$, the set $A(x) := \{y \in F : (x, y) \in \operatorname{gr}(A)\}$ is the value *of x; thus $A(x) \neq \emptyset$ if and only if $x \in \operatorname{dom}(A)$.*

If E and F are affine sets and $\operatorname{gr}(A) \subseteq E \times F$ is also an affine set, then A is called a multivalued affine map*; then $\operatorname{dom}(A)$ and $R(A)$ are both affine sets.*

Lemma 5.13. *Let E and F be affine spaces, and $A : E \to F$ a multivalued map whose graph $\operatorname{gr}(A) = C$ is a nonempty convex set in $E \times F$. Then a point (x, y) belongs to $\operatorname{rai}(\operatorname{gr}(A))$ if and only if $x \in \operatorname{rai}(\operatorname{dom}(A))$ and $y \in \operatorname{rai}(A(x))$.*

Proof. Let us first prove that if $(x, y) \in \operatorname{rai}(\operatorname{gr}(A))$, then $x \in \operatorname{rai}(\operatorname{dom}(A))$ and $y \in \operatorname{rai}(A(x))$. If $\bar{x} \in \operatorname{dom}(A)$, $\bar{x} \neq x$, then there exists $\bar{y} \in F$ such that $(\bar{x}, \bar{y}) \in \operatorname{gr}(A)$; see Figure 5.2. Since $(\bar{x}, \bar{y}) \neq (x, y)$ and $(x, y) \in \operatorname{rai}(\operatorname{gr}(A))$, there exists $(x_1, y_1) \in \operatorname{gr}(A)$ such that $(x, y) \in ((\bar{x}, \bar{y}), (x_1, y_1))$ in $E \times F$. This proves that $x \in \operatorname{rai}(\operatorname{dom}(A))$, since we have $x_1 \in \operatorname{dom}(A)$ and $x \in (\bar{x}, x_1)$. To prove that $y \in \operatorname{rai}(A(x))$, let $y_2 \in A(x)$, $y_2 \neq y$. Then $(x, y) \neq (x, y_2) \in \operatorname{gr}(A)$, and since $(x, y) \in \operatorname{rai}(\operatorname{gr}(A))$, there exists a point $(x, y_3) \in \operatorname{gr}(A)$ such that $(x, y) \in ((x, y_2), (x, y_3))$. We clearly have $y \in (y_2, y_3)$, and so $y \in \operatorname{rai}(A(x))$.

Let us now prove the converse statement that if $x \in \operatorname{rai}(\operatorname{dom}(A))$ and $y \in \operatorname{rai}(A(x))$, then $(x, y) \in \operatorname{rai}(\operatorname{gr}(A))$. Let $(\bar{x}, \bar{y}) \in \operatorname{gr}(A)$, $(\bar{x}, \bar{y}) \neq (x, y)$. We first consider the case $\bar{x} = x$. Then $\bar{y} \neq y$, and since $y \in \operatorname{rai}(A(x))$, there exists some $\hat{y} \in A(x)$ such that $y \in (\bar{y}, \hat{y})$. But then $(x, y) \in ((x, \bar{y}), (x, \hat{y}))$. If $\bar{x} \neq x$, then there exists $x_4 \in \operatorname{dom}(A)$ such that $x \in (\bar{x}, x_4)$. Let $(x_4, y_4) \in \operatorname{gr}(A)$; see again Figure 5.2. We need to prove that there exists a point $(x_1, y_1) \in \operatorname{gr}(A)$ such that the line segment $((\bar{x}, \bar{y}), (x_1, y_1))$ contains (x, y). If the line segment $((\bar{x}, \bar{y}), (x_4, y_4))$ already contains (x, y), then the point (x_4, y_4) can serve as the point (x_1, y_1). Otherwise, there exists a point (x, y_3) on the line segment between $(\bar{x}, \bar{y})$ and (x_4, y_4). Since $y_3 \neq y$ and $y \in \operatorname{rai}(A(x))$, there exists a point $y_2 \in A(x)$ such that $y \in (y_2, y_3)$. Since $(x, y_3) \in ((\bar{x}, \bar{y}), (x_4, y_4))$ and $(x, y) \in ((x, y_2), (x, y_3))$, (x, y) is in the relative interior of the convex hull of the triangle with vertices $\{(\bar{x}, \bar{y}), (x_4, y_4), (x, y_2)\}$, that is, (x, y) is a convex combination of these three points with positive weights. But then (x, y) can be obtained first by taking a nontrivial convex combination of the vertices $\{(x, y_2), (x_4, y_4)\}$ to obtain a point (x_1, y_1), and then taking a nontrivial convex combination of the points $\{(x_1, y_1), (\bar{x}, \bar{y})\}$. Consequently, $(x, y) \in ((\bar{x}, \bar{y}), (x_1, y_1))$. This completes the proof that $(x, y) \in \operatorname{rai}(\operatorname{gr}(A))$.

□

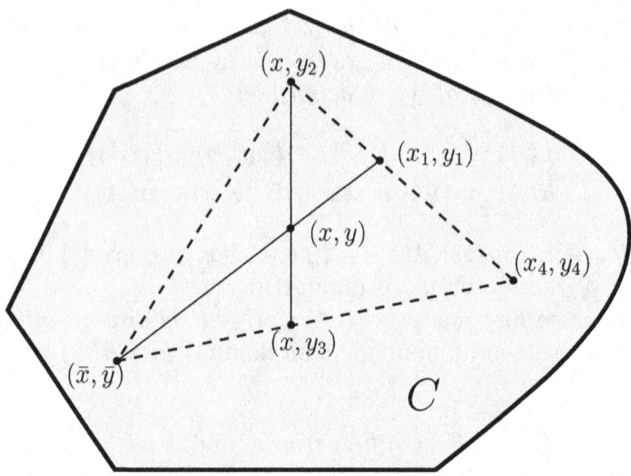

Fig. 5.2. Relative algebraic interior of a convex set in a product space.

Corollary 5.14. *Let E and F be affine spaces, and $A : E \to F$ a multivalued map whose graph* $\mathrm{gr}(A) = C$ *is a nonempty convex set in $E \times F$. Then*

$$\mathrm{dom}\ \mathrm{rai}(\mathrm{gr}(A)) \subseteq \mathrm{rai}\ \mathrm{dom}(A).$$

If $\mathrm{rai}\, A(x) \neq \emptyset$ *for all* $x \in \mathrm{rai}\ \mathrm{dom}(A)$*, then*

$$\mathrm{dom}\ \mathrm{rai}(\mathrm{gr}(A)) = \mathrm{rai}\ \mathrm{dom}(A).$$

In particular, the above equality holds when F has finite dimension.

Proof. The inclusion follows immediately from Lemma 5.13. If $x \in \mathrm{rai}\ \mathrm{dom}(A)$ and $y \in \mathrm{rai}\, A(x) \neq \emptyset$, then Lemma 5.13 implies that $(x, y) \in \mathrm{rai}\, \mathrm{gr}(A)$, and we have $x \in \mathrm{dom}\ \mathrm{rai}(\mathrm{gr}(A))$. If $x \in \mathrm{rai}\ \mathrm{dom}(A)$ and F is finite-dimensional, then Lemma 5.3 implies that $\mathrm{rai}\, A(x) \neq \emptyset$. □

Corollary 5.15. *Let C_i be a convex set in an affine space A_i, $i = 1, \ldots, k$. If each* $\mathrm{rai}(C_i)$ *is nonempty, then*

$$\mathrm{rai}(C_1 \times C_2 \times \cdots \times C_k) = \mathrm{rai}(C_1) \times \mathrm{rai}(C_2) \times \cdots \times \mathrm{rai}(C_k).$$

Proof. The proof is trivial for $k = 1$ and follows immediately from Lemma 5.13 for $k = 2$. The proof is easily completed by induction on k. □

It is also possible to give an independent easy proof of the corollary from scratch.

Lemma 5.16. *Let $A : E \to F$ be a multivalued affine map between two affine spaces E and F, and $C \subseteq E$ a convex set such that* $\mathrm{rai}(C) \cap \mathrm{dom}(A) \neq \emptyset$.

Then we always have

$$A(\mathrm{rai}(C)) \subseteq \mathrm{rai}(A(C)).$$

Moreover, if E is finite-dimensional, or more generally if $\mathrm{rai}(C \cap A^{-1}(y)) \neq \emptyset$ *for all $y \in \mathrm{rai}\ A(C)$, then*

$$A(\mathrm{rai}(C)) = \mathrm{rai}(A(C)).$$

Proof. Consider the multivalued map $B : F \to E$ whose graph is the convex set

$$\mathrm{gr}(B) := \mathrm{gr}(A) \cap (C \times F),$$

and note that

$$\mathrm{dom}(B) = \{y : \exists x \in C,\ y \in A(x)\} = A(C)$$

and

$$B(y) = C \cap A^{-1}(y) \quad \text{for } y \in \mathrm{dom}(B).$$

We have

$$\mathrm{rai}(\mathrm{gr}(B)) = \mathrm{rai}(\mathrm{gr}(A)) \cap (\mathrm{rai}(C \times F)) = \mathrm{gr}(A) \cap (\mathrm{rai}(C) \times F) \neq \emptyset,$$

where the last relation follows from the assumption $\mathrm{rai}(C) \cap \mathrm{dom}(A) \neq \emptyset$, and the second equation from the equality $\mathrm{rai}(\mathrm{gr}(A)) = \mathrm{gr}(A)$ and Corollary 5.15. Then the first equality follows from Lemma 5.10. Consequently, we have

$$\mathrm{dom}\ \mathrm{rai}(\mathrm{gr}(B)) = \{y : \exists x \in \mathrm{rai}(C), y \in A(x)\} = A(\mathrm{rai}(C)).$$

With these preparations, the lemma follows immediately from Corollary 5.14. □

5.4 Topological Interior and Topological Closure of Convex Sets

In this section we compare the algebraic and topological concepts of interior, relative interior, and closure for convex sets. As Theorem 5.20 and Corollary 5.21 show, the algebraic and topological concepts agree to a remarkable degree.

Let us recall some basic topological notions; see [232] for a quick introduction to general topology, and [161, 45, 46] for comprehensive treatments. Let X be a set and $\mathcal{T}$ a set of subsets of X. Then $(X, \mathcal{T})$ is called a *topological space* if $\emptyset \in \mathcal{T}$, $X \in \mathcal{T}$, and $\mathcal{T}$ is closed under unions and finite intersections, that is, any union of sets in $\mathcal{T}$ is in $\mathcal{T}$, and the intersection of two sets in $\mathcal{T}$ is in $\mathcal{T}$. The sets in $\mathcal{T}$ are the *open sets* of the topological space $(X, \mathcal{T})$. A set $F \subseteq X$ is called *closed* if $X \backslash F$ is open. A *neighborhood* of a point $x \in X$ is a set $V \subseteq X$ that contains an open set $U \in \mathcal{T}$ such that $x \in U \subseteq V$. The *interior* of a set A in X, denoted by $\mathrm{int}(A)$, is the set of points $x \in A$ such that x has a neighborhood that lies entirely in A. The *closure* of a set $A \subseteq X$, denoted by $\overline{A}$, is the intersection of all the closed sets containing A. Alternatively, a point $x \in \overline{A}$ if and only if every neighborhood of x intersects A.

If Y is a subset of X, then $(Y, \mathcal{S})$ inherits the *relative topology* from $(X, \mathcal{T})$: the open sets in $\mathcal{S}$ are simply the sets of the form $U \cap Y$, where $U \in \mathcal{T}$.

A real vector space E is called a *topological vector space* if there exists a topology $\mathcal{T}$ on E such that the linear operations $(x, y) \mapsto x+y$ and $(\alpha, x) \mapsto \alpha x$ are continuous maps from the product topological spaces $E \times E$ and $\mathbb{R} \times E$ to E, respectively. We refer the reader to any book on functional analysis, for example [177, 233, 44], for more details.

Let $(E, \mathcal{T})$ be a topological vector space, and $A \subset E$ an affine subset of E. Then $(A, \mathcal{S})$, $\mathcal{S}$ the relative topology inherited from $\mathcal{T}$, is called a *topological affine space*.

Definition 5.17. *Let $C \subseteq A$ be a convex set in a topological affine space A. The* relative interior *of C, denoted by* $\mathrm{ri}(C)$, *is the interior of C in the relative topology of the affine space* $\mathrm{aff}(C)$.

If the topology on A is given by a norm, for example, we have

$$\mathrm{ri}(C) := \{x \in C : \exists \varepsilon > 0, B_\varepsilon(x) \cap \mathrm{aff}(C) \subseteq C\}.$$

Lemma 5.18. *Let C be a convex set in a topological affine space A with a nonempty interior. If $x \in \mathrm{int}(C)$ and $y \in \overline{C}$, then $[x, y) \subseteq \mathrm{int}(C)$. Consequently,* $\mathrm{int}(C)$ *is a convex set.*

Moreover, $\mathrm{int}(C) \subseteq \mathrm{ai}(C)$.

Proof. Let $z := y + t(x - y)$, $t \in (0, 1)$. We claim that $z \in \mathrm{int}(C)$. Let $U \subset C$ be a neighborhood of x; see Figure 5.3. Since $y = (z - tx)/(1 - t) \in (z - tU)/(1 - t) =: V$ and $v \mapsto (z - tv)/(1 - t)$ is a homeomorphism, V is a neighborhood of y. Pick $p \in C \cap V$, and define $u \in U$ by the equation $p = (z - tu)/(1 - t)$, that is, $z = p + t(u - p)$. Then z lies in the open set $p + t(U - p)$, which is a subset of C by the convexity of C. This proves the claim.

The convexity of $\mathrm{int}(C)$ follows from this: if $x, y \in \mathrm{int}(C)$, then $[x, y) \subseteq \mathrm{int}(C)$. Since $y \in \mathrm{int}(C)$ as well, the whole segment $[x, y]$ lies in $\mathrm{int}(C)$.

Let $x \in \mathrm{int}(C)$ such that $x \in U \subseteq C$, where U is a neighborhood of x. If $u \in A$, then the map $t \mapsto x + t(u - x)$ is continuous; thus there exists $\delta > 0$ such that $x + t(u - x) \in U$ for all $|t| \leq \delta$. This proves that $x \in \mathrm{ai}(C)$. □

Lemma 5.19. *Let C be a nonempty convex set in a topological affine space A. Then $\overline{C}$ is a convex set, and* $\mathrm{ac}(C) \subseteq \overline{C}$.

Proof. Let $x, y \in \overline{C}$ and $z := y + t(x - y) = tx + (1 - t)y$, $t \in (0, 1)$. We claim that $z \in \overline{C}$. Let U_z be a neighborhood of z. Since the map $(u, v) \mapsto tu + (1 - t)v$ is continuous, there exist neighborhoods $U_x \ni x$ and $U_y \ni y$ such

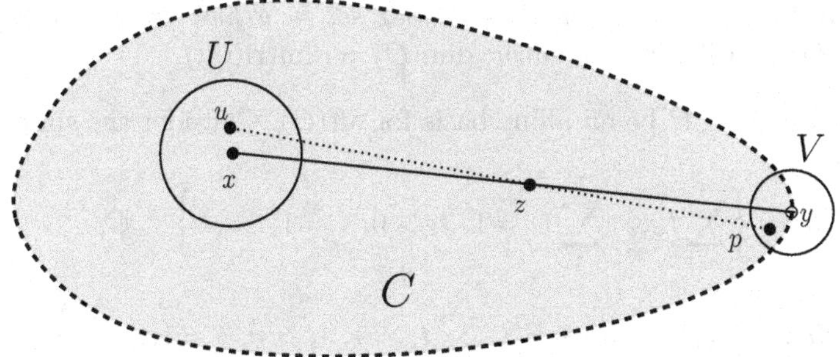

Fig. 5.3. Interior of a convex set.

that $tU_x + (1-t)U_y \subset U_z$. Pick $p_x \in C \cap U_x$ and $p_y \in U_y$. Then the point $p := tp_x + (1-t)p_y$ belongs to U_z, and lies in C as well, due to the convexity of C. This proves the claim.

Let $y \in \mathrm{ac}(C)$ with $[x, y) \subset C$, and let U be a neighborhood of y. Since the map $t \mapsto y + t(x-y)$ is continuous, there exists $\delta > 0$ such that $y + t(x-y) \in U \cap C$ for all $0 < t \leq \delta$. This proves that $x \in \overline{C}$. □

Theorem 5.20. *If C is a convex body in a topological affine space A, that is, $\mathrm{int}(C) \neq \emptyset$, then*

$$\mathrm{int}(C) = \mathrm{int}(\overline{C}) = \mathrm{ai}(C) \quad \textit{and} \quad \overline{C} = \overline{\mathrm{int}(C)} = \mathrm{ac}(C).$$

Proof. To prove the equality $\mathrm{int}(C) = \mathrm{ai}(C)$, it suffices to prove the claim that $\mathrm{ai}(C) \subseteq \mathrm{int}(C)$, since the reverse inclusion is already proved in Lemma 5.18. Let $x \in \mathrm{ai}(C)$ and $y \in \mathrm{int}(C)$. There exists a point $z \in C$ such that $x \in (y, z)$. Then Lemma 5.18 implies that $x \in \mathrm{int}(C)$, proving the claim.

To prove the equality $\mathrm{int}(C) = \mathrm{int}(\overline{C})$, it clearly suffices to prove the inclusion $\mathrm{int}(\overline{C}) \subseteq \mathrm{int}(C)$. Let $x \in \mathrm{int}(C)$ and $y \in \mathrm{int}(\overline{C})$. There exists a point $z \in \overline{C}$ such that $y \in (x, z)$. Then Lemma 5.18 implies that $y \in \mathrm{int}(C)$.

Next, let us prove the equality $\overline{C} = \overline{\mathrm{int}(C)}$: let $x \in \mathrm{int}(C)$ and $y \in \overline{C}$. Lemma 5.18 implies that $[x, y) \subset \mathrm{int}(C)$, which in turn implies $y \in \overline{\mathrm{int}(C)}$. This proves the inclusion $\overline{C} \subseteq \overline{\mathrm{int}(C)}$, and hence the equality $\overline{C} = \overline{\mathrm{int}(C)}$, because the reverse inclusion is trivial.

Finally, we consider the equality $\mathrm{ac}(C) = \overline{C}$. The inclusion $\mathrm{ac}(C) \subseteq \overline{C}$ is already proved in Lemma 5.19. To prove the reverse inclusion, let $y \in \overline{C}$ and pick $x \in \mathrm{int}(C)$. We have $[x, y) \subseteq \mathrm{int}(C) \subseteq C$, where the first inclusion follows from Lemma 5.18; this gives $y \in \mathrm{ac}(C)$, proving the equality $\mathrm{ac}(C) = \overline{C}$. □

By considering the relative topology on $\mathrm{aff}(C)$, we immediate obtain the following result.

Corollary 5.21. *Let C be a convex set in a topological affine space A. If $\mathrm{ri}(C) \neq \emptyset$, then*

$$\mathrm{ri}(C) = \mathrm{ri}(\overline{C}) = \mathrm{rai}(C) \quad \textit{and} \quad \overline{C} = \overline{\mathrm{ri}(C)} = \mathrm{ac}(C).$$

Lemma 5.22. *If C is a nonempty convex set in a finite-dimensional affine space A, then* $\mathrm{ri}(C) \neq \emptyset$. *Moreover,* $\dim(C) = \dim(\mathrm{ri}(C))$.

Proof. Let $\{x_i\}_1^k \subset C$ be an affine basis for $\mathrm{aff}(C)$. Consider the simplices

$$S := \left\{ \sum_{i=1}^{k} t_i x_i : \sum_{1}^{k} t_i = 1,\ t_i > 0,\ i = 1, \ldots, k \right\} \subset C,$$

$$S_k := \left\{ (t_1, \ldots, t_k) \in \mathbb{R}^k : \sum_{1}^{k} t_i = 1,\ t_i > 0,\ i = 1, \ldots, k \right\}.$$

Since $\{x_i\}_1^k$ is affinely independent, the map

$$T(t_1, \ldots, t_k) = \sum_{i}^{k} t_i x_i$$

is a one-to-one onto affine transformation from the affine space $\mathrm{aff}(C)$ to the hyperplane $H := \{t \in \mathbb{R}^k : \sum_1^k t_i = 1\}$, hence a homoeomorphism between the two affine spaces. Evidently, S_k is open in H, and thus $S = T(S_k)$ is open in $\mathrm{aff}(C)$. This proves that $\mathrm{ri}(C) \neq \emptyset$.

Note that we also have $\dim(C) = k - 1 = \dim(\mathrm{ri}(C))$. □

Combining Corollary 5.21 and Lemma 5.22, we immediately obtain the following important result for finite-dimensional convex sets.

Theorem 5.23. *If C is a nonempty finite-dimensional convex set, then* $\mathrm{ri}(C) \neq \emptyset$, *and*

$$\mathrm{ri}(C) = \mathrm{ri}(\overline{C}) = \mathrm{rai}(C) \quad \textit{and} \quad \overline{C} = \overline{\mathrm{ri}(C)} = \mathrm{ac}(C).$$

5.5 Facial Structure of Convex Sets

In this section, we decompose a closed convex set C in a vector space E as a Minkowski sum

$$C = L + K + B,$$

where L is a linear subspace of E, K is a closed convex cone containing no lines, and B is a bounded convex set that is the convex hull of the set of extreme points of a set related to C. We give a further decomposition of K and B in terms of their extreme directions and points, respectively. The precise meaning of this decomposition is contained in Theorem 5.37 below. These decompositions find many applications in optimization and elsewhere.

Definition 5.24. *Let C be a nonempty convex set in a vector space E. A* face *of C is convex subset $F \subseteq C$ such that if $x, y \in C$ and the line segment (x, y) intersects F, then $[x, y] \subseteq F$.*

A point $x \in C$ is called an extreme point *of C if $\{x\}$ is a face of C. We denote the set of extremal points of C by* ext(C).

A vector $d \in E$ is called an extreme direction *of C if there is a point $p \in C$ such that the ray $p + \mathbb{R}_+ d := \{p + td : t \geq 0\}$ is a face of C.*

A vector $d \in E$ is called a recession direction *of C if there exists a point $p \in C$ such that the ray $p + \mathbb{R}_+ d$ stays in C. The set of all recession directions of C is called the* recession cone *of C,*

$$\operatorname{rec} C := \{d \in \mathbb{R}^n : \textit{there exists } p \in C \textit{ such that } p + td \in C \ \textit{ for all } t \geq 0\}.$$

Lemma 5.25. *Let F and C be two convex sets such that $F \subset C$. Then F is a face of C if and only if $C \setminus F$ is a convex set.*

This is an easy consequence of the definition of a face, and can be used as an alternative definition of a face.

Remark 5.26. It is easy to see that the union of a nested set of faces of C is a face (the nestedness condition is needed only to ensure the convexity of the union), and the same is true for the intersection of any set of faces. The face of a face is a face, that is, if $F_2 \subseteq F_1 \subseteq C$ with F_2 a face of F_1 and F_1 a face of C, then F_2 is a face of C. Also, if F is a face of C, then $\operatorname{ext}(F) = \operatorname{ext}(C) \cap F$.

A face actually satisfies a stronger property given below. This can be used to show, among other things, that a convex set C can be written as a disjoint union of relative interiors of different faces of C; see Rockafellar [228], Theorem 18.2.

Lemma 5.27. *Let C be a convex set in a vector space E, F a face of C, and D a convex subset of C. If $\operatorname{ri}(D) \cap F \neq \emptyset$, then $D \subseteq F$.*

Proof. Pick $z \in \operatorname{ri}(D) \cap F$. If $x \in D$, then $z \in \operatorname{ri}(D)$ implies that there exists $y \in D$ such that $z \in (x, y)$. Since F is a face of C, we have $x \in F$. □

As a first step toward the decomposition of a closed convex set, we characterize its affine faces.

Lemma 5.28. *Let C be a closed convex set in a vector space E. If $0 \neq d \in E$ is a recession direction of C, then $q + \mathbb{R}_+ d \subseteq C$ for every $q \in C$.*

Consequently, $C + \operatorname{rec} C = C$.

Proof. Suppose that $p + \mathbb{R}_+ d \subseteq C$, and let $q \in C$, $q \neq p$. It is easy to see geometrically that the convex hull of q and $p + R$ is the union of sets $\{q\}$ and $[p, q) + R$, whose closure is the set $[p, q] + R$; see Figure 5.4. Since C is closed, we have $q + R \subseteq C$. □

Corollary 5.29. *Let C be a closed, convex set in a vector space E. If C contains an affine subspace $K := p + L$, where $p \in C$ and L is a linear subspace of E, then $C = C + L$.*

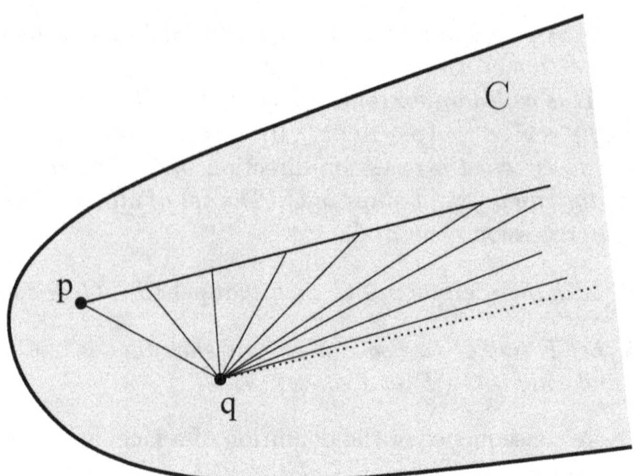

Fig. 5.4. Recession direction in a convex set.

Theorem 5.30. *Let C be a nonempty, closed, convex set in a finite-dimensional linear space E. Then C contains an affine face F.*

The affine face F is unique, is a maximal affine subspace of C, and has the form $F = p + L_C$, where $p \in C$ is arbitrary and

$$L_C := \{x \in E : C + x = C\}$$

is a linear subspace of E.

Moreover, $C = C + L_C$.

Proof. We first demonstrate the existence of an affine face by induction on the dimension of C. Suppose $\dim(C) = n$ and that we have proved the existence of affine faces for convex sets with dimension less than n. Clearly, we may assume that C is full-dimensional, that is, E has dimension n. We may also assume that $\partial C \neq \emptyset$; otherwise $C = E$ (C is both closed and open), in which case the lemma is obviously true. Let $p \in \partial C$. By Theorem 6.8, there exists a support hyperplane $H := H_{(a,\alpha)}$ at p such that $C \subseteq \bar{H}^-$. Then $D := C \cap H$ is a convex set in E with dimension less than n, so by the induction hypothesis, D contains an affine face F. We claim that F is a face of C. By Remark 5.26, this will follow if we can show that D is a face of C. Suppose that $x, y \in C$ with $z \in (x, y) \cap D$. We have $\langle a, x\rangle \leq \alpha$, $\langle a, y\rangle \leq \alpha$, but $\langle a, z\rangle = \alpha$; consequently, $\langle a, x\rangle = \alpha = \langle a, y\rangle$, which proves that D is a face of C.

Next, we show that an affine face $F \subseteq C$ is a maximal affine subset of C. Let M be a maximal affine subset of C containing F. If $x \in M$, pick $z \in F$ and consider the line passing through x and z. This line is contained in M, so that $z \in (x, y)$ for some $y \in M$; since F is a face of C, we have $x \in F$. This proves that $F = M$.

If $F = p + L$ is an affine face of C, Corollary 5.29 implies that $C = C + L$. Consider the set L_C defined in the statement of the theorem. Obviously, $L \subseteq$

L_C. We claim that L_C is a linear subspace. It will then follow that $L = L_C$. It is easily seen that L_C is convex, closed under addition, and that $L_C = -L_C$. Thus, if $x \in L_C$, then $kx \in L_C$ for every integer $k > 0$. The convexity of L_C implies that $tx \in L_C$ for every $t > 0$, and consequently for every $t \in \mathbb{R}$. $\square$

The linear subspace L_C defined above is called the *lineality subspace* of C, and its dimension the *lineality* of C.

Since affine faces are points if and only if $L_C = \{0\}$, we have the following corollary.

Corollary 5.31. *A nonempty, closed, convex set in a finite-dimensional linear space contains an extreme point if and only if it is line-free, that is, it contains no whole lines.*

Definition 5.32. *Two linear subspaces L, M of a vector space E are called* complementary *if $E = L + M$ and $L \cap M = \{0\}$, and we denote this by the notation*

$$E = L \oplus M.$$

If $C \subseteq E$ is a convex set such that $C = C_1 + C_2$, where C_1 and C_2 are convex subsets of L and M, respectively, we write

$$C = C_1 \oplus C_2.$$

Lemma 5.33. *Let C be a nonempty, closed, convex set in a finite-dimensional linear space E, and let M be a linear subspace of E complementary to L_C.*

The set C can decomposed as

$$C = \hat{C} \oplus L_C,$$

where $\hat{C}$ is a line-free, closed convex set in M.

Proof. Define $\hat{C} := C \cap M$. If $x \in C$, we can write $x = l + m$, where $l \in L_C$ and $m \in M$; then we have $m \in \hat{C}$, since $m \in M$, and $m = x - l \in C + L_C = C$ by virtue of Theorem 5.30. This shows that $\hat{C} \neq \emptyset$, $x = m + l \in \hat{C} + L_C$, and hence $C \subseteq L_C + \hat{C}$. The reverse inclusion also holds, since $L_C + \hat{C} \subseteq L_C + C = C$, so we have $C = L_C + \hat{C}$.

Suppose that $\hat{C}$ contains a line $q + \{td : t \in \mathbb{R}\} = q + \mathbb{R}d$, where $q \in \hat{C}$ and $0 \neq d \in M$. Then

$$q + L_C \subset q + (\mathbb{R}d + L_C) \subseteq C + L_C = C,$$

where $q + (\mathbb{R}d + L_C)$ is an affine subset of C strictly containing $q + L_C$, which by Theorem 5.30 contradicts the maximality of the affine subspace $q + L_C$. This proves that $\hat{C}$ is line-free. $\square$

Lemma 5.34. *Let C be a closed convex set C in a finite-dimensional linear space E. The relative boundary* $\operatorname{rbd}(C) := C \setminus \operatorname{ri}(C)$ *of C is convex if and only if C is either an affine subspace of E or the intersection of an affine subspace with a closed half-space.*

Proof. If C is either an affine subset or the intersection of an affine subspace with a closed half-space, then it is clear that $\operatorname{rbd}(C)$ is convex.

To prove the converse, assume without loss of generality that C is full-dimensional. If the boundary of C is empty, then $C = E$ (C is both open and closed), and we are done. Otherwise, Theorem 6.16 implies that there exists a support hyperplane H such that $\partial C \subseteq H$, C does not lie entirely on H, and $C \subseteq \bar{H}^+$, where $\bar{H}^+$ is one of the two closed half-spaces bounding H.

We claim that $C = \bar{H}^+$. Pick $p \in C \setminus H$; then $p \in H^+ := \operatorname{int}(\bar{H}^+)$. If x is in H^+ but not in C, then the line segment $[x, p]$ must contain a point $w \in \partial C$, a contradiction because $w \in H^+$; consequently, $\operatorname{int}(C) = H^+$ and $C = \bar{H}^+$. □

Lemma 5.35. *Let C be a nonempty, closed, convex set in a finite-dimensional linear space E. If C is not an affine subspace or the intersection of an affine subspace with a closed half-space, then every point in the relative interior of C lies on a line segment whose endpoints lie on the relative boundary of C; consequently $C = \operatorname{co}(\operatorname{rbd}(C))$.*

Proof. Again, we may assume that C is full-dimensional. Since ∂C is not convex by Lemma 5.34, there exist two points $x, y \in \partial C$ such that $[x, y]$ intersects $\operatorname{int}(C)$. It follows from Lemma 5.28 that the line passing through any point of $\operatorname{int}(C)$ and parallel to $[x, y]$ must intersect C in a line segment. □

Theorem 5.36. *A nonempty, line-free, closed, convex set C in a finite-dimensional linear space E is the convex hull of its extreme points and extreme rays, that is, any point $x \in C$ has a representation*

$$x = \sum_{i=1}^{k} \lambda_i v_i + \sum_{j=1}^{l} \mu_j d_j, \tag{5.5}$$

where $\{v_i\}_1^k$ and $\{d_j\}_1^l$ are extreme points and extreme directions of C, respectively, $\{\lambda_i\}_1^k$, $\{\mu_j\}_1^l$ are nonnegative, and $\sum_{i=1}^k \lambda_i = 1$.

Proof. We use induction on the dimension of C. Suppose $\dim(C) = n$ and that we have proved the theorem for convex sets with dimension less than n.

First, suppose that $x \in \operatorname{rbd}(C) = C \setminus \operatorname{ri}(C)$. By Theorem 6.8, there exists a support hyperplane H at x such that $C \subseteq \bar{H}^+$. Then $D := C \cap H$ is a convex set in E with dimension less than n, so by the induction hypothesis, x has a representation (5.5), where $\{v_i\}_1^k$ and $\{d_j\}_1^l$ are extreme points and extreme directions of D, respectively. Since D is a face of C (see the proof of Theorem 5.30), $\{v_i\}_1^k$ and $\{d_j\}_1^l$ are also extreme points and directions of C, respectively.

If $x \in \operatorname{ri}(C)$, then Lemma 5.35 implies that $x \in (y, z)$ for two points $y, z \in \partial C$. Since y and z both have representations in the form (5.5), so does x. □

Theorem 5.37. *Let C be a closed convex set C in a finite-dimensional vector space E. The set C can be decomposed as*

$$C = L_C \oplus \hat{C},$$

where L_C is a linear subspace and $\hat{C}$ is a line-free convex set lying in a linear subspace complementary to L_C. The set of extreme points of $\hat{C}$ is nonempty, and

$$\hat{C} = \mathrm{rec}(\hat{C}) + \mathrm{co}(\mathrm{ext}(\hat{C})).$$

Moreover, we also have the decompositions

$$\mathrm{rec}\, C = L_C \oplus \mathrm{rec}(\hat{C}) \quad \textit{and} \quad C = \mathrm{co}(\mathrm{ext}(\hat{C})) + \mathrm{rec}(C).$$

Proof. The theorem follows immediately from Lemma 5.33 and Theorem 5.36. □

We also have the following classical result.

Theorem 5.38. **(*Minkowski*)** *A compact, convex set in a finite-dimensional linear space is the convex hull of its extreme points.*

Finally, we include the following two interesting results relating the faces of a sum set to the faces of its summands.

Lemma 5.39. *Let $C = C_1 + C_2$, where C_1 and C_2 are nonempty convex sets. If F is a face of C, then there exist faces F_i of C_i, $i = 1, 2$, such that $F = F_1 + F_2$.*

Proof. Define the sets

$$F_1 = \{x \in C_1 : \exists y \in C_2,\ x + y \in F\}, \quad F_2 = \{x \in C_2 : \exists x \in C_1,\ x + y \in F\}.$$

It is easy to verify that F_1 and F_2 are convex sets. We claim that F_1 is a face of C_1. Let $x \in F_1$ such that $x + y \in F$ for some $y \in C_2$. If $x \in (u_1, u_2)$ for some $u_1, u_2 \in C_1$, then $x + y \in (u_1 + y, u_2 + y)$, and since F is a face of C, we have $[u_1 + y, u_2 + y] \subseteq F$. It follows from the definition of F_1 that $u_1, u_2 \in F_1$, proving that F_1 is a face of C_1. Similarly, F_2 is a face of C_2.

It follows from the definition of F_1 and F_2 that $F \subseteq F_1 + F_2$. To prove the reverse inclusion $F_1 + F_2 \subseteq F$, let $x_1 \in F_1$ and $y_2 \in F_2$; it suffices to show that $x_1 + y_2 \in F$. If $y_1 \in C_2$ and $x_2 \in C_1$ are such that $x_i + y_i \in F$, $i = 1, 2$, then $(x_1 + y_1)/2 + (x_2 + y_2)/2 = (x_1 + y_2)/2 + (x_2 + y_1)/2 \in F$; since F is a face of C, we have $x_1 + y_2 \in F$. □

Lemma 5.40. *Let $C = C_1 + C_2$, where C_1 and C_2 are nonempty convex sets. If z is an extreme point of C, then z has a unique representation $z = x + y$, where $x \in C_1$, $y \in C_2$. Moreover, in this representation x is an extreme point of C_1 and y is an extreme point of C_2.*

Proof. It follows from Lemma 5.39 that $z = x + y$, where x and y are extreme points of C_1 and C_2, respectively. If $z = \overline{x} + \overline{y}$, where $\overline{x} \in C_1$, $\overline{y} \in C_2$, then we have $z = (x + \overline{y})/2 + (\overline{x} + y)/2$, and because z is an extreme point of C, $x + \overline{y} = \overline{x} + y = z$. Comparing this equation with $z = x + y$ gives $x = \overline{x}$ and $y = \overline{y}$. □

5.6 Homogenization of Convex Sets

It is often easier to establish results for convex cones than for the more general class of convex sets. One may desire to extend an already established result for convex cones to convex sets. This is usually possible, thanks to a procedure called *homogenization* that manufactures a convex cone out of a convex set. Homogenization is thus a very useful technique for translating results between convex sets and convex cones.

Let C be a nonempty convex set in a vector space E. We can naturally identify C with the convex set

$$\hat{C} := \{(x, z) \in E \times \mathbb{R} : x \in C,\ z = 1\}$$

lying on the hyperplane $H = \{(x, 1) : x \in E\}$ of the vector space $E \times \mathbb{R}$, and then form the set

$$K(C) := \{t(x, 1) : x \in C,\ t > 0\} = \operatorname{cone}(\hat{C});$$

see Figure 5.5. This process is called the *homogenization* of C. The intersection of $K(C)$ with the hyperplane $x_{n+1} = 1$ is $\hat{C}$, so that C may be considered a cross section of $K(C)$.

It is clear that $K(C)$ is a cone from the above description. The cone $K(C)$ is convex, because if $u_i = (x_i, 1) \in K(C)$ with $x_i \in C$ $(i = 1, 2)$, then

$$u_1 + u_2 = 2\left(\frac{x_1 + x_2}{2}, 1\right) \in K(C).$$

When dealing with a closed convex set C, one often desires to homogenize C so as to obtain a *closed* convex cone. The set $K(C)$ defined above does not contain 0, so it is never closed. If one adds 0 to $K(C)$, then it is easily seen that the resulting set is $\{t(x, 1) : x \in C, t \geq 0\}$, which can be shown to be closed if C is a compact convex set. However, if C is an unbounded closed convex set, then $K(C) \cup \{0\}$ is not closed. For example, if $C = \mathbb{R}_+$ is the nonnegative real line, then $K(C) \cup \{0\} = \mathbb{R}^2_+ \setminus \{(x, 0) : x > 0\}$ is not closed.

Therefore, the following result is of interest.

Lemma 5.41. *If $C \neq \emptyset$ is a closed convex set in a finite-dimensional vector space E, then*

$$\overline{K(C)} = K(C) \cup \{(d, 0) : d \in \operatorname{rec}(C)\}.$$

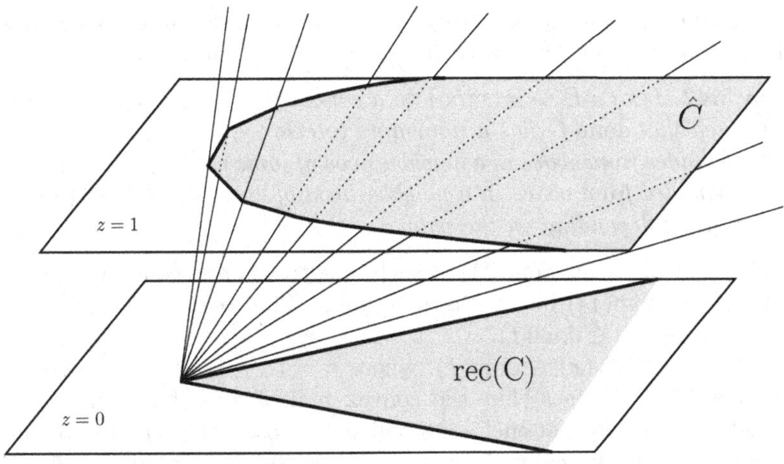

Fig. 5.5. Homogenization of a convex set.

Proof. Denote the set on the right-hand side by D. Let $d \in \text{rec}(C)$. If $x \in C$, then $x_n := x + nd \in C$, $(x_n, 1)/n \in K(C)$, and $(x_n, 1)/n \to (d, 0) \in \overline{K(C)}$, proving $D \subseteq \overline{K(C)}$.

To prove the reverse inclusion $\overline{K(C)} \subseteq D$, let $(d, t) \in \overline{K(C)}$ be such that $t_n(x_n, 1) \to (d, t)$. If $t > 0$, then letting $d = tx$, we have $t_n(x_n, 1) \to t(x, 1)$. Thus, $x_n \to x \in C$, since C is closed, and consequently $(d, t) = t(x, 1) \in K(C) \subseteq D$.

If, however, $t = 0$, then $t_n x_n \to d$ and $t_n \to 0$; we claim that $d \in \text{rec}(C)$. Let $x \in C$ and $s > 0$ be arbitrary. We have $d_n := t_n(x_n - x) \to d$, $x_n = x + d_n/t_n \in C$, and for large enough n such that $0 \leq st_n \leq 1$,

$$x + sd_n = (1 - st_n)x + st_n\left(x + \frac{d_n}{t_n}\right) = (1 - st_n)x + st_n x_n \in C.$$

This gives $x + sd_n \to x + sd \in C$, and proves the claim. □

5.7 Continuity of Convex Functions

There are no continuity assumptions made in the definition of a convex function. However, it turns out that a convex function is continuous in the relative interior of its domain, even Lipschitz continuous under a mild boundedness assumption. In finite dimensions, this boundedness assumption is not needed, because it is automatically satisfied. These constitute some of the most basic results on convex functions.

Let $f : E \to \mathbb{R} \cup \{\infty\}$ be a convex function on a normed linear space E, and define $L = \text{aff}(\text{dom}(f))$. Since f is always $+\infty$ off L, in questions regarding the continuity of f, it makes sense to consider f only on L, since otherwise f

can never be continuous at any point when L is a proper subset of E. Thus, in this section, we consider convex functions with full-dimensional domains.

Lemma 5.42. *Let $f : E \to \mathbb{R} \cup \{\infty\}$ be a convex function in a normed linear space E such that* $\operatorname{dom}(f)$ *has a nonempty interior.*

If f is bounded from above in a neighborhood of some point $x_0 \in \operatorname{int}(\operatorname{dom}(f))$, then f is bounded from above in a neighborhood of every point of $\operatorname{int}(\operatorname{dom}(f))$ *(with the bound depending on the point).*

Proof. Let $B_{r_0}(x_0) \in \operatorname{int}(\operatorname{dom}(f))$ be such that $f(x) \leq a < \infty$ for $x \in B_{r_0}(x_0)$, and let $x \in \operatorname{int}(\operatorname{dom}(f))$ be an arbitrary point. There exists $\gamma > 1$ such that $w := x_0 + \gamma(x - x_0) \in \operatorname{dom}(f)$.

We claim that $B_r(x) \in \operatorname{dom}(f)$, where $r = (\gamma - 1)r_0/\gamma$. Geometrically, this follows from the fact that the convex hull of w and $B_{r_0}(x_0)$ forms a truncated conic region that encloses a ball at x whose radius can be computed using similarity. Analytically, if $\|v - x\| \leq r$, consider the point u defined by the equation $w =: u + \gamma(v - u)$. Then $0 = (1 - \gamma)(u - x_0) + \gamma(v - x)$ and $\|u - x_0\| \leq r\gamma/(\gamma - 1) = r_0$, so that $u \in \operatorname{dom}(f)$. This proves the claim.

Since $v = w/\gamma + ((\gamma - 1)/\gamma)u \in \operatorname{dom}(f)$, using the convexity of f, we obtain

$$f(v) = f\left(\frac{1}{\gamma}w + \frac{\gamma - 1}{\gamma}u\right) \leq \frac{1}{\gamma}f(w) + \frac{\gamma - 1}{\gamma}f(u) \leq \frac{1}{\gamma}f(w) + \frac{\gamma - 1}{\gamma}a =: b.$$

Thus f is bounded from above by the constant b on $B_r(x)$. □

Theorem 5.43. *Let $f : E \to \mathbb{R} \cup \{\infty\}$ be a convex function on a normed linear space E. Then f is continuous in* $\operatorname{ri}(\operatorname{dom}(f))$ *if and only if there exists a point $x_0 \in \operatorname{ri}(\operatorname{dom}(f))$ such that f is bounded from above in a relative neighborhood of x_0 (that is, a neighborhood of x_0 in the relative topology of* $\operatorname{aff}(\operatorname{dom}(f))$*).*

Proof. Define $C = \operatorname{dom}(f)$ and assume that $\operatorname{int}(\operatorname{dom}(f)) \neq \emptyset$, by restricting f to $\operatorname{aff}(\operatorname{dom}(f))$ if necessary. If f is continuous at $x_0 \in \operatorname{int}(C)$, then there exists a neighborhood $x_0 \in N \subseteq C$ such that $|f(x) - f(x_0)| \leq 1$ on N, and f is bounded from above on N by the constant $f(x_0) + 1$.

Conversely, assume that $f(x) \leq a < \infty$ for x in a neighborhood N of x_0. We may assume, if necessary by considering the function $x \mapsto f(x + x_0) - f(x_0)$ on the set $C - x_0$, that $x_0 = 0 \in \operatorname{int}(C)$ and $f(0) = 0$. Consider the symmetric neighborhood $S = N \cap (-N)$ of 0, and pick any $0 < \varepsilon < 1$. If $z \in \varepsilon S$, then $\pm z/\varepsilon \in S$, and we have

$$-\varepsilon a \leq f(z) \leq (1 - \varepsilon)f(0) + \varepsilon f(z/\varepsilon) \leq \varepsilon a,$$

where the second inequality follows from the convexity of f and the first inequality from

$$\begin{aligned} 0 = f(0) &= f\left(\frac{1}{1 + \varepsilon}z + \frac{\varepsilon}{1 + \varepsilon}(-z/\varepsilon)\right) \\ &\leq \frac{1}{1 + \varepsilon}f(z) + \frac{\varepsilon}{1 + \varepsilon}f(-z/\varepsilon) \leq \frac{1}{1 + \varepsilon}f(z) + \frac{\varepsilon}{1 + \varepsilon}a. \end{aligned}$$

This proves the continuity of f at $x_0 = 0$. By virtue of Lemma 5.42, f is bounded from above on a neighborhood of every point of $x \in \text{int}(C)$, so that f is continuous on $\text{int}(C)$. □

It is possible to strengthen the continuity result above.

Definition 5.44. *A function $f : X \to \mathbb{R}$ on a metric space X is called Lipschitz continuous if there exists a constant $K > 0$ such that*

$$|f(x) - f(y)| \leq Kd(x, y) \quad \textit{for all} \quad x, y \in X.$$

The function f is called locally Lipschitz continuous if each point $x \in X$ has a neighborhood on which f is Lipschitz continuous.

Theorem 5.45. *Let $f : E \to \mathbb{R} \cup \{\infty\}$ be a convex function on a normed linear space E. If there exists a point $x_0 \in \text{ri}(\text{dom}(f))$ such that f is bounded from above in a relative neighborhood of x_0, then the function f is locally Lipschitz continuous on* $\text{ri}(\text{dom}(f))$.

Proof. We may again assume that $\text{int}(\text{dom}(f)) \neq \emptyset$. By virtue of Theorem 5.43, f is continuous in a neighborhood $\overline{B}_{r_0}(x_0)$. Suppose that

$$m \leq f(x) \leq M \quad \text{on} \quad \overline{B}_{r_0}(x_0),$$

and pick $0 < r < r_0$. We claim that f is Lipschitz continuous in $\overline{B}_r(x_0)$. Let $v_1, v_2 \in \overline{B}_r(x_0)$, and assume for now that $\|v_2 - v_1\| \leq r_0 - r$. The function $g(w) := f(v_1 + w) - f(v_1)$ is convex and finite-valued, $g(0) = 0$, and g is bounded from above in the ball $N := \overline{B}_{r_0 - r}(0)$ by the constant $M - m$. It follows from the proof of Theorem 5.43 that $|g(w)| \leq \varepsilon(M - m)$ on εN. Consequently, since $v_2 - v_1 \in (\|v_2 - v_1\|/(r_0 - r))N$, we have

$$|f(v_2) - f(v_1)| = |g(v_2 - v_1)| \leq \frac{M - m}{r_0 - r}\|v_2 - v_1\|. \tag{5.6}$$

Now if $v_1, v_2 \in \overline{B}_r$ are arbitrary, we can partition the interval $[v_1, v_2]$ into N points $\{u_k\}_1^N$ in such a way that $u_1 = v_1$, $u_N = v_2$, and $\|u_k - u_{k-1}\| \leq r_0 - r$. Applying (5.6) to each pair (u_{k-1}, u_k), $k = 2, \ldots, N$, and adding the results gives

$$\begin{aligned} |f(v_2) - f(v_1)| &\leq \sum_{k=2}^{N} |f(u_k) - f(u_{k-1})| \leq \sum_{k=2}^{N} \frac{M - m}{r_0 - r}\|u_k - u_{k-1}\| \\ &= \frac{M - m}{r_0 - r}\|v_2 - v_1\|, \end{aligned}$$

where the equality follows since each vector $u_k - u_{k-1}$ has the same direction $v_2 - v_1$. This proves the claim and the theorem. □

In infinite-dimensional vector spaces, the boundedness assumption on the function f is really necessary, where even the continuity of linear functionals is tied up with boundedness. However, the situation is different in the finite-dimensional case.

Lemma 5.46. *Let $f : E \to \mathbb{R} \cup \{\infty\}$ be a convex function on a finite-dimensional normed linear space E. If $x \in \operatorname{ri}(\operatorname{dom}(f))$, then f is bounded from above in a relative neighborhood of x.*

Proof. Let $n = \dim(E)$. As in the proof of Theorem 5.43, we may assume that $\operatorname{int}(\operatorname{dom}(f)) \neq \emptyset$. If $x \in \operatorname{int}(\operatorname{dom}(f))$, then

$$x \in N = \operatorname{int}(\Delta) = \left\{ \sum_{i=1}^{n+1} \lambda_i u_i : \lambda_i > 0, \sum_{i=1}^{n+1} \lambda_i = 1 \right\},$$

where $\{u_i\}_1^{n+1}$ are affinely independent vectors in $\operatorname{dom}(f)$, so that $\Delta = \operatorname{co}(\{u_i\}_1^{n+1})$ is an n-simplex contained in $\operatorname{dom}(f)$. If $y = \sum_{i=1}^{n+1} \lambda_i u_i \in \Delta$, then

$$f(y) = f\Big(\sum_{i=1}^{n+1} \lambda_i u_i\Big) \leq \sum_{1=1}^{n+1} \lambda_i f(u_i) \leq \max\{f(u_1), \ldots, f(u_{n+1})\},$$

proving our claim. □

Corollary 5.47. *Let $f : E \to \mathbb{R} \cup \{\infty\}$ be a convex function on a finite-dimensional normed linear space E. If we consider f as a function $f : \operatorname{aff}(C) \to \mathbb{R} \cup \{\infty\}$, then f is locally Lipschitz continuous on $\operatorname{ri}(C)$.*

5.8 Exercises

1. (a) Show that

$$\operatorname{co}(\{x_i\}_1^k) + \operatorname{co}(\{y_j\}_1^l) = \operatorname{co}\big(\{x_i + y_j := 1, \ldots, k, j = 1, \ldots, l\}\big).$$

   (b) Let C_1 and C_2 be two nonempty sets in a vector space E. Use (a) to show that $\operatorname{co}(C_1 + C_2) = \operatorname{co}(C_1) + \operatorname{co}(C_2)$.
2. Let C be a k-dimensional convex set in $\mathbb{R}^n$. Given any basis $\{e_i\}_1^n$ for $\mathbb{R}^n$, show that there exists a linear subspace $L \subseteq \mathbb{R}^n$ spanned by k of the basis vectors $\{e_i\}$ such that the projection of C onto L has dimension k.
3. Let C be a convex set and $\{A_i\}_1^k$ affine subspaces in a vector space E. If $C \subseteq A_1 \cup A_2 \cup \cdots \cup A_k$, show that $C \subseteq A_i$ for some i, $1 \leq i \leq k$.
   Hint: Use induction on k.
4. The purpose of this problem is to establish a connection between the Minkowski gauge function and the support function.

Let $C \subseteq E$ be a convex body in a finite-dimensional vector space E containing zero in its interior, and let

$$C^\circ := \{x : \langle x, z\rangle \le 1 \text{ for all } z \in C\}$$

be its *polar set*, and the function

$$\sigma_C(x) := \sup_{z \in C} \langle z, x\rangle$$

the *support function* of C.

(a) Show that the gauge function of C° is the support function of C, that is,

$$p_{C^\circ}(x) = \sigma_C(x) \text{ for all } x \in E.$$

(b) If C is closed, then show that

$$p_C(x) = \sigma_{C^\circ}(x) \text{ for all } x \in E.$$

Hint: you may assume that $(C^\circ)^\circ = C$, a fact that requires an appropriate separation argument.

5. Let $C \subseteq E$ be a convex set in a finite-dimensional vector space E such that C is a relatively open convex in the algebraic sense, that is, $\text{rai}(C) = C$. Give a direct proof of a result proved in Theorem 5.23, namely, that C is a relative open convex set in the topological sense, that is, $\text{ri}(C) = C$:
   (a) Using the properties of the Minkowski gauge function.
   (b) Using an elementary approach, starting with a basis $\{d_i\}_1^k$ of directions in E at a point $x_0 \in \text{rai}(C)$, and then using the convexity of C.
6. Let $f(x, y) = y^2/x$, where $x > 0$.
   (a) Show that f is a convex function.
   (b) Show that f cannot be made continuous at zero, that is, we cannot prescribe a value to $f(0, 0)$ so that f becomes continuous at $(0, 0)$.

6

Separation of Convex Sets

Separation theorems involving convex sets is an important topic in optimization. The duality theory of convex programming depends on them, and the various optimality conditions in optimization theory, from the Fritz John and Karush–Kuhn–Tucker (KKT) conditions to the Pontryagin maximum principle, can be obtained from them. In addition, the familiar Hahn–Banach theorem, which is one of the cornerstones in functional analysis with an enormous number of applications in the field, is an analytic form of a separation theorem.

Most separation theorems in the literature deal with the separation of two convex sets. However, it is sometimes advantageous to prove theorems dealing with separation of several convex sets. We will deal with both types of separation theorems in this chapter.

It is a fact that it is technically easier to prove separation theorems in finite-dimensional (and Hilbert) spaces than in general vector spaces. Since our emphasis is more on finite-dimensional spaces, we find it expedient to first prove separation theorem in the simpler setting of finite-dimensional spaces (which for the most part can be extended to Hilbert spaces). This is done in Sections 6.1–6.5. Our proofs are valid verbatim in any finite-dimensional Euclidean space E equipped with an inner product $\langle \cdot, \cdot \rangle$, but we assume for convenience that $E = \mathbb{R}^n$ equipped with the usual inner product $\langle x, y \rangle = x^T y$. Then in Sections 6.6 and 6.7, we prove separation theorems (involving both two and several convex sets) in general vector spaces. Our treatment emphasizes the algebraic approach, which accounts for its generality. The topological separation theorems are then obtained with ease as corollaries; see for example the proof of Theorem 6.39. Finally, Section 6.8 deals with the Hahn–Banach theorem in a general setting.

A reader who is interested in only finite-dimensional spaces may skip Sections 6.6–6.8 without any loss of continuity.

O. Güler, *Foundations of Optimization*, Graduate Texts in Mathematics 258,
DOI 10.1007/978-0-387-68407-9_6, © Springer Science+Business Media, LLC 2010

6.1 Projection of a Point onto a Finite-Dimensional Closed Convex Set

Let $C \subseteq \mathbb{R}^n$ be a closed convex set. If x is a point in $\mathbb{R}^n$, then the point in C that is closest to x is called the *projection* of x onto C, and is denoted by $\Pi_C(x)$.

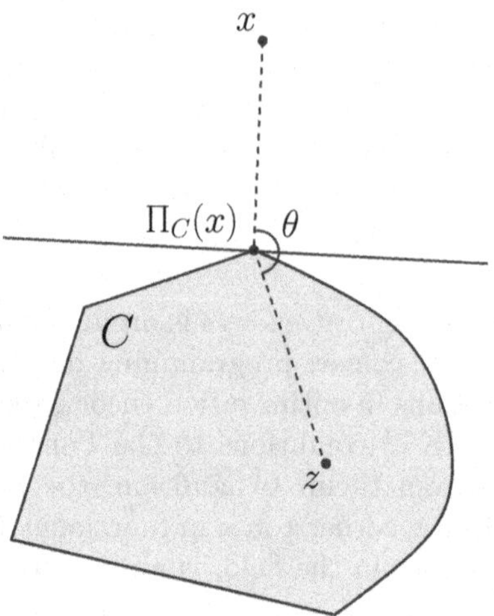

Fig. 6.1. Projecting a point onto a convex set.

Figure 6.1 illustrates the situation. We may convince ourselves that for any $z \in C$, the angle θ between the vectors $x - \Pi_C(x)$ and $z - \Pi_C(x)$ must be obtuse. Since $\langle x - x^*, z - x^* \rangle = \|x - x^*\| \cdot \|z - x^*\| \cos\theta$, and since $\cos\theta \le 0$ when θ is obtuse, we may expect that the following theorem holds.

Theorem 6.1. *Let $C \subseteq \mathbb{R}^n$ be a nonempty closed convex set. The projection $\Pi_C(x)$ of x onto C is characterized by the variational inequality*

$$\langle x - \Pi_C(x), z - \Pi_C(x) \rangle \le 0 \quad \textit{for all} \quad z \in C. \tag{6.1}$$

Proof. Consider the minimization problem

$$\min_{z \in C} f(z) := \frac{1}{2}\|z - x\|^2.$$

The function f is clearly coercive on $\mathbb{R}^n$, so that any sublevel set $l_\alpha(f) := \{z \in C : f(z) \le \alpha\}$ is a compact set. It follows from Theorem 2.2 that there exists a global minimizer of f on C. Since f is a strictly convex function, the minimizer is unique. Finally, it follows from Theorem 4.33 that the minimizer $x^* := \Pi_C(x)$ is characterized by the variational inequality

$$\langle \nabla f(x^*), z - x^* \rangle \ge 0 \text{ for all } z \in C.$$

Since $\nabla f(x^*) = x^* - x$, this is equivalent to the condition

$$\langle x - x^*, z - x^* \rangle \leq 0 \ \text{ for all } \ z \in C.$$

The theorem is proved. □

Remark 6.2. The above theorem holds even in Hilbert spaces; notice that we need to show only that $\Pi_C(x)$ exists. Let $\{x_n\}_1^\infty$ be a minimizing sequence, that is, $x_n \in C$ such that

$$\|x - x_n\| \to d = d_C(x) := \inf\{\|x - z\| : z \in C\}.$$

The Euclidean distance satisfies the parallelogram equality $\|u - v\|^2 + \|u + v\|^2 = 2(\|u\|^2 + \|v\|^2)$; therefore

$$\begin{aligned} \|x_m - x_n\|^2 &= 2\|x - x_m\|^2 + 2\|x - x_n\|^2 - 4\left\|x - \frac{x_m + x_n}{2}\right\|^2 \\ &\leq 2\|x - x_m\|^2 + 2\|x - x_n\|^2 - 4d \to 0, \end{aligned}$$

where the inequality follows from the fact $(x_m + x_n)/2 \in C$. This shows that $\{x_m\}$ is a Cauchy sequence and thus converges to a point $x^* \in C$ satisfying the equality $\|x - x^*\| = d$.

Corollary 6.3. *The function $\Pi_C : \mathbb{R}^n \to C$ is nonexpansive, that is,*

$$\|\Pi_C(x_2) - \Pi_C(x_1)\| \leq \|x_2 - x_1\| \ \text{ for all } \ x_1, x_2 \in \mathbb{R}^n.$$

Consequently, Π_C is a continuous mapping.

Proof. The variational inequality (6.1) gives

$$\begin{aligned} \langle x_1 - \Pi_C(x_1), \Pi_C(x_2) - \Pi_C(x_1) \rangle &\leq 0, \\ \langle x_2 - \Pi_C(x_2), \Pi_C(x_1) - \Pi_C(x_2) \rangle &\leq 0; \end{aligned}$$

rearranging and adding these inequalities, we obtain

$$\langle x_1 - x_2 + \Pi_C(x_2) - \Pi_C(x_1), \Pi_C(x_2) - \Pi_C(x_1) \rangle \leq 0,$$

or

$$\begin{aligned} \|\Pi_C(x_2) - \Pi_C(x_1)\|^2 &\leq \langle x_2 - x_1, \Pi_C(x_2) - \Pi_C(x_1) \rangle \\ &\leq \|x_2 - x_1\| \cdot \|\Pi_C(x_2) - \Pi_C(x_1\|, \end{aligned}$$

where the last inequality follows from the Cauchy–Schwarz inequality. □

6.2 Separation of Convex Sets in Finite-Dimensional Vector Spaces

We start by defining several relevant concepts.

Definition 6.4. *A* hyperplane H *in* $\mathbb{R}^n$ *is an* $(n-1)$*-dimensional affine subset of* $\mathbb{R}^n$*, that is,* $H = \{x \in \mathbb{R}^n : \ell(x) = \alpha\}$ *is the level set of a nontrivial linear function* $\ell : \mathbb{R}^n \to \mathbb{R}$*. If* ℓ *is given by* $\ell(x) = \langle a, x\rangle$ *for some* $a \neq 0$ *in* $\mathbb{R}^n$*, then*

$$H = H_{(a,\alpha)} := \{x \in \mathbb{R}^n : \langle a, x\rangle = \alpha\}.$$

A hyperplane H partitions $\mathbb{R}^n$ into two half-spaces.

Definition 6.5. *Let* $H = H_{(a,\alpha)}$ *be a hyperplane in* $\mathbb{R}^n$*. The closed half-spaces associated with* H *are the two closed sets*

$$\bar{H}^+_{(a,\alpha)} = \{x \in E : \langle a, x\rangle \geq \alpha\},$$
$$\bar{H}^-_{(a,\alpha)} = \{x \in E : \langle a, x\rangle \leq \alpha\}.$$

Similarly, the open half-spaces associated with H *are the two open sets*

$$H^+_{(a,\alpha)} = \{x \in E : \langle a, x\rangle > \alpha\},$$
$$H^-_{(a,\alpha)} = \{x \in E : \langle a, x\rangle < \alpha\}.$$

Definition 6.6. *Let* C *and* D *be two nonempty sets and* $H := H_{(a,\alpha)}$ *a hyperplane in* $\mathbb{R}^n$*.*

H *is called a* separating hyperplane *for the sets* C *and* D *if* C *is contained in one of the closed half-spaces determined by* H *and* D *in the other, say* $C \subseteq \bar{H}^+_{(a,\alpha)}$ *and* $D \subseteq \bar{H}^-_{(a,\alpha)}$*.*

H *is called a* strictly separating hyperplane *for the sets* C *and* D *if* C *is contained in one of the open half-spaces determined by* H *and* D *in the other, say* $C \subseteq H^+_{(a,\alpha)}$ *and* $D \subseteq H^-_{(a,\alpha)}$*.*

H *is called a* strongly separating hyperplane *for the sets* C *and* D *if there exist* β *and* γ *satisfying* $\beta > \alpha > \gamma$*,* $C \subseteq \bar{H}^+_{(a,\beta)}$*, and* $D \subseteq \bar{H}^-_{(a,\gamma)}$*.*

H *is called a* properly separating hyperplane *for the sets* C *and* D *if* H *separates* C *and* D *and* C *and* D *are not* both *contained in the hyperplane* H*.*

H *is called a* support hyperplane *of* C *at a point* $x \in \overline{C}$ *if* $x \in H$ *and* $C \subseteq \bar{H}^+$*.*

If there exists a hyperplane H *separating the sets* C *and* D *in one of the senses above, we say that* C *and* D *can be* separated, strictly separated, strongly separated, properly separated, *respectively.*

We are ready to study the separation properties of convex sets in $\mathbb{R}^n$. We begin by considering the separation of a single point from a convex set.

Theorem 6.7. *If $C \subset \mathbb{R}^n$ is a nonempty convex set and $\overline{x} \notin \mathrm{ri}(C)$, then there exists a hyperplane $H_{(a,\alpha)}$ such that $\overline{x} \in H_{(a,\alpha)}$ and $C \subseteq \bar{H}^+_{(a,\alpha)}$, that is,*

$$\langle a, x\rangle \geq \langle a, \overline{x}\rangle \quad \textit{for all} \quad x \in C.$$

Proof. We first assume that $\overline{x} \notin \overline{C}$. The variational inequality (6.1) gives

$$\langle \Pi_{\overline{C}}\overline{x} - \overline{x}, x - \Pi_{\overline{C}}\overline{x}\rangle \geq 0 \quad \text{for all} \quad x \in \overline{C};$$

defining $a := \Pi_C(\overline{x}) - \overline{x} \neq 0$, and writing

$$x - \Pi_{\overline{C}}\overline{x} = x - \overline{x} + (\overline{x} - \Pi_{\overline{C}}\overline{x}) = x - \overline{x} - a,$$

we get $\langle a, x - \overline{x} - a\rangle \geq 0$, or $\langle a, x - \overline{x}\rangle \geq \|a\|^2 > 0$. Therefore,

$$\langle a, x\rangle \geq \langle a, \overline{x}\rangle \quad \text{for all} \quad x \in C,$$

and the theorem is proved in this case.

If $\overline{x} \in \overline{C} \setminus \mathrm{ri}(C)$, then there exists a sequence $\{x_k\}$ of points not in $\overline{C}$ such that $x_k \to \overline{x}$. It follows from (6.1) that

$$\langle \Pi_{\overline{C}}(x_k) - x_k, x - \Pi_{\overline{C}}(x_k)\rangle \geq 0 \quad \text{for all} \quad x \in \overline{C}.$$

Since $x_k \notin \overline{C}$ and $\Pi_{\overline{C}}(x_k) \in \overline{C}$, we have $\Pi_{\overline{C}}(x_k) \neq x_k$; defining

$$a_k := \frac{\Pi_{\overline{C}}(x_k) - x_k)}{\|\Pi_{\overline{C}}(x_k) - x_k\|},$$

we have

$$\langle a_k, x - \Pi_{\overline{C}}(x_k)\rangle \geq 0 \quad \text{for all} \quad x \in \overline{C}. \tag{6.2}$$

Since the sequence $\{a_k\}$ is bounded, it has a convergent subsequence; to avoid cumbersome notation, we assume that the sequence $\{a_k\}$ itself converges, say $a_k \to a$, where $\|a\| = 1$. Since $x_k \to \overline{x}$ and $\Pi_{\overline{C}}$ is continuous,

$$\Pi_{\overline{C}}(x_k) \to \Pi_{\overline{C}}(\overline{x}) = \overline{x};$$

thus letting $k \to \infty$ in (6.2) gives

$$\langle a, x\rangle \geq \langle a, \overline{x}\rangle \quad \text{for all} \quad x \in C,$$

and the theorem is proved. □

Theorem 6.7 immediately implies the following.

Theorem 6.8. (***Support hyperplane theorem***) *If $C \subseteq \mathbb{R}^n$ is a nonempty convex set and $x \in \overline{C} \setminus \mathrm{ri}(C)$, then there exists a support hyperplane to C at x.*

The following theorem provides the weakest separation result for two convex sets. It follows easily from Theorem 6.7, because a useful trick reduces the problem of separating two convex sets to the separation of a point from a suitably defined convex set.

Theorem 6.9. *Let C and D be two nonempty convex sets in $\mathbb{R}^n$. If C and D are disjoint, then there exists a hyperplane $H_{(a,\alpha)}$ that separates C and D, that is,*

$$\langle a, x\rangle \le \alpha \le \langle a, y\rangle \quad \textit{for all} \quad x \in C,\ y \in D.$$

Proof. The trick is to define the set

$$A := C - D = \{x - y : x \in C,\ y \in D\},$$

and note that $0 \notin A$, due to $C \cap D = \emptyset$. The set A is convex, because $A = C + (-D)$ is the Minkowski sum of the convex sets C and $-D$; it follows from Theorem 6.7 that there exists a hyperplane $H_{(a,\alpha)}$ with $\alpha = \langle a, 0\rangle = 0$ such that $\langle a, u\rangle \le 0$ for all $u \in A$. Since $A = C - D$, this means that $\langle a, x - y\rangle \le 0$ for all $x \in C$ and all $y \in D$, or

$$\langle a, x\rangle \le \langle a, y\rangle \quad \text{for all} \quad x \in C,\ y \in D.$$

Then, any hyperplane $H_{(a,\alpha)}$ with α satisfying

$$\sup_{x \in C} \langle a, x\rangle \le \alpha \le \inf_{y \in D} \langle a, y\rangle$$

separates the sets C and D. □

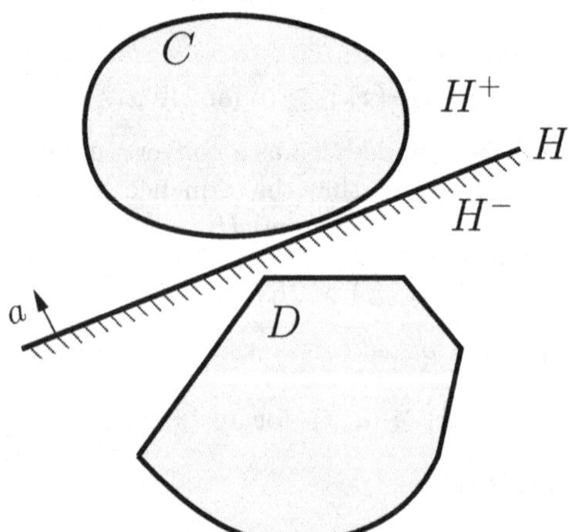

Fig. 6.2. Separation of two convex sets by a hyperplane.

Theorem 6.10. (***Strong separation theorem***) *Let C, D be two nonempty, disjoint, closed, convex sets in $\mathbb{R}^n$. If one of them is compact, then C and D can be strongly separated.*

Proof. Let us assume for definiteness that D is compact. We note that the theorem is equivalent to the existence of a hyperplane $H_{(a,\alpha)}$ satisfying the condition

$$\inf_{x\in C}\langle a,x\rangle > \alpha > \max_{y\in D}\langle a,y\rangle,$$

or equivalently the condition

$$\langle a,x\rangle > \alpha > \langle a,y\rangle \quad \text{for all } x\in C,\ y\in D.$$

Define $A := C - D$. The set A is convex, and we claim that it is closed. Let $u_k \to u$ be a convergent sequence with $u_k \in A$; we will show that $u \in A$. Write $u_k = x_k - y_k$, where $x_k \in C$ and $y_k \in D$. Since D is compact, we can extract a convergent subsequence from $\{y_k\}$. To avoid cumbersome notation, assume that $y_k \to y \in D$. Since $x_k - y_k \to u$ and $y_k \to y$, we see that $x_k \to x := u + y \in C$. Thus, $u = x - y \in A$, and the claim is proved.

Since $C \cap D = \emptyset$, we have $0 \notin A = \overline{A}$, and it follows from the first part of the proof of Theorem 6.7 that there exists a hyperplane $H_{(a,0)}$ such that $\langle a,u\rangle \geq \|a\|^2 > 0$ for all $u \in A$, or $\langle a, x-y\rangle \geq \|a\|^2$ for all $x \in C$ and $y \in D$. This implies

$$\langle a,x\rangle \geq \langle a,x\rangle - \frac{\|a\|^2}{2} \geq \langle a,y\rangle + \frac{\|a\|^2}{2} > \langle a,y\rangle \quad \text{for all } x\in C,\ y\in D;$$

it is easy to see that the theorem holds with $\alpha = \|a\|^2/2 + \max_{y\in D}\langle a,y\rangle$. □

Remark 6.11. The strong separation theorem breaks down if neither C nor D is compact. For example, consider the closed convex sets $C = \{(0,y)\}$ (the y-axis) and $D = \{(x,y) : y \leq \ln x\}$. Neither set is compact, and it is easy to see that the only hyperplane separating C and D is the y-axis. Since C coincides with the separating hyperplane, there exists no hyperplane separating C and D strictly, let alone strongly.

Theorem 6.12. *If $C \subseteq \mathbb{R}^n$ is a nonempty closed convex set, then C is the intersection of all the closed half-spaces containing it, that is,*

$$C = \bigcap_{(a,\alpha)} \{\bar{H}^+_{(a,\alpha)} : C \subseteq \bar{H}^+_{(a,\alpha)}\}.$$

Proof. Denote by D the intersection set above. It is clear that D is a closed, convex set containing C, so it remains to show that $D \subseteq C$.

If this is not true, then there exists a point $x_0 \in D$ that does not lie in C. Applying Theorem 6.10 to the convex sets $\{x_0\}$ and C, we see that there exists a hyperplane $H := H_{(a,\alpha)}$ such that $x_0 \in H^-$ and $C \subseteq H^+$; but then $\bar{H}^+$ is one of the closed half-spaces intersected to obtain D, and so $D \subseteq \bar{H}^+$. Since $x_0 \in D$, we obtain $x_0 \in \bar{H}^+$, which contradicts $x_0 \in H^-$. The theorem is proved. □

Remark 6.13. The result above provides an "external" or "from outside" characterization of closed convex sets as intersections of closed half-spaces. That is, all closed convex sets are obtained from half-spaces using the *intersection* operation. In contrast, the convex hull operation generates a convex set by enlargement, "from inside." This is an instance of a "duality," which is a common phenomenon in convexity.

The most useful separation result for two convex sets in a vector space is perhaps the *proper separation theorem* (Theorem 6.15 below and its general version Theorem 6.33 on page 161). It is here that the properties of the relative interior developed in Chapter 5 prove most useful.

We need the following result in its proof.

Lemma 6.14. *Two nonempty convex sets C and D in $\mathbb{R}^n$ can be properly separated if and only if the origin and the convex set $K := C - D$ can be properly separated.*

Proof. Let $H := H_{(a,\alpha)}$ be a hyperplane properly separating C and D such that $C \subseteq \bar{H}^+$, $D \subseteq \bar{H}^-$, and assume without loss of generality that C does not lie on H. Then

$$\langle a, x\rangle \geq \alpha \geq \langle a, y\rangle \quad \text{for all} \quad x \in C,\ y \in D,$$

and $\langle a, x_0\rangle > \alpha$ for some $x_0 \in C$; it follows that $\langle a, z\rangle \geq 0$ for all $z \in K$, with strict inequality holding for some $z_0 \in K$. This proves that the hyperplane $H_{(a,0)}$ properly separates the sets $\{0\}$ and K.

Conversely, suppose that the sets $\{0\}$ and K are properly separated by a hyperplane $H_{(a,\alpha)}$ such that $K \subseteq \bar{H}^+_{(a,\alpha)}$. Then $\langle a, x - y\rangle \geq \alpha \geq 0$ for all $x \in C$ and $y \in D$, or

$$\langle a, x\rangle \geq \alpha + \langle a, y\rangle \quad \text{for all} \quad x \in C,\ y \in D,$$

and either the first inequality is strict for some $x_0 \in C$ and $y_0 \in D$, or else $\alpha > 0$. In the first case, any hyperplane $H_{(a,\gamma)}$ with $\gamma \in \mathbb{R}$ satisfying

$$\inf_{x\in C} \langle a, x\rangle \geq \gamma \geq \alpha + \sup_{y\in D} \langle a, y\rangle$$

properly separates C and D; in the second case we have $\alpha > 0$ and $\langle a, x\rangle = \alpha + \langle a, y\rangle$ for any $x \in C$ and $y \in D$, so the hyperplane $H_{(a,\gamma)}$ with $\gamma = \alpha/2 + \langle a, y\rangle$ properly separates C and D. □

Theorem 6.15. (***Proper separation theorem***) *Two nonempty convex sets C and D in $\mathbb{R}^n$ can be properly separated if and only if* $\mathrm{ri}(C)$ *and* $\mathrm{ri}(D)$ *are disjoint.*

Proof. Define the convex set $K := C - D$. It follows from Lemma 5.11 and Corollary 5.21 that $\mathrm{ri}(K) = \mathrm{ri}(C - D) = \mathrm{ri}(C) - \mathrm{ri}(D)$; thus, $\mathrm{ri}(C) \cap \mathrm{ri}(D) = \emptyset$

and $0 \notin \text{ri}(K)$ are equivalent statements. Consequently, Lemma 6.14 reduces the proof of the theorem to proving that the sets $\{0\}$ and K are properly separable if and only if $0 \notin \text{ri}(K)$.

Suppose that the origin and K are properly separated by a hyperplane H, such that $0 \in \bar{H}^-$ and $K \subseteq \bar{H}^+$. We claim that $0 \notin \text{ri}(K)$. If $0 \notin H$, then $\text{ri}(K) \subseteq \bar{H}^+$, so that $0 \notin \text{ri}(K)$. Otherwise, $0 \in H$ and there exists a point $x \in K \setminus H$. If we had $0 \in \text{ri}(K)$, there would exist a point $y \in K$ such that $0 \in (x, y)$, giving the contradiction $y \in C \cap H^- = \emptyset$. This proves the claim.

Conversely, suppose that $0 \notin \text{ri}(K)$. Write

$$L := \text{aff}(K) = u_0 + \text{span}\{u_1, \ldots, u_k\},$$

where $\{u_i\}_1^k$ is linearly independent. If $0 \notin L$, then $\{u_i\}_0^k$ is linearly independent, and we can extend it to a basis $\{u_i\}_0^{n-1}$ of $\mathbb{R}^n$. Then the hyperplane $H := u_0 + \text{span}\{u_1, \ldots, u_{n-1}\}$ does not contain the origin, so it properly separates $\{0\}$ and K.

If $0 \in L$, we apply Theorem 6.9 *within* the vector space L to the sets $\{0\}$ and $\text{ri}(K)$, and obtain a hyperplane P in L separating 0 and K such that $\text{ri}(K) \subseteq \overline{P}^+$; see Figure 6.3. We may assume that $0 \in P$; otherwise the translation of P so that it passes through the origin also satisfies the same separation properties. Extending P to the hyperplane $H = \text{span}\{P, u_{k+1}, \ldots, u_{n-1}\}$, it is evident that H properly separates $\{0\}$ and K. □

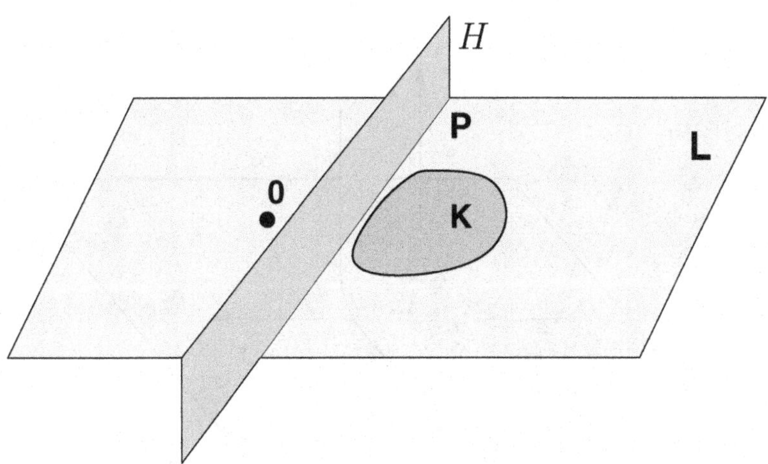

Fig. 6.3. Proper separation of convex sets.

We use Theorem 6.15 to obtain an improved version of Theorem 6.8.

Theorem 6.16. *Let C be a nonempty convex set in $\mathbb{R}^n$. If D is a nonempty convex subset of the relative boundary of C ($D \subseteq \bar{C} \setminus \text{ri}(C)$), then there exists a support hyperplane to C containing D but not all points of C.*

Proof. Since $\mathrm{ri}(C) \cap \mathrm{ri}(D) \subseteq \mathrm{ri}(C) \cap D = \emptyset$, Theorem 6.15 implies that there exists a hyperplane $H := H_{(a,\alpha)}$ properly separating C and D, say

$$\langle a, x \rangle \leq \alpha \leq \langle a, y \rangle \quad \text{for all} \quad x \in C, \; y \in D.$$

If $y \in D$, then $\langle a, y \rangle \geq \alpha$, but since $y \in \bar{C}$, we also have $\langle a, y \rangle \leq \alpha$. This means that $\langle a, y \rangle = \alpha$ for all $y \in D$, that is, $D \subseteq H$. Since H properly separates C and D, we must have $C \not\subseteq H$. □

Finally, we present a proper separation theorem involving two convex sets one of which is an affine set.

Theorem 6.17. *Let $C \subset \mathbb{R}^n$ be a nonempty convex set. If M is an affine set such that* $\mathrm{ri}(C)$ *and M are disjoint, then M can be extended to a hyperplane H such that* $\mathrm{ri}(C)$ *and H are disjoint.*

Proof. Since M is affine, $\mathrm{ri}(M) = M$, and Theorem 6.15 implies that there exists a hyperplane H properly separating C and M. If $M \subseteq H$, we are done; if not, there exists a point $x \in M \setminus H \neq \emptyset$. The affine sets H and M must be disjoint; otherwise, there would exist a point $u \in M \cap H$, and the line passing through x and u would intersect both half-spaces $\bar{H}^+$ and $\bar{H}^-$, a contradiction because the line stays in M, which is included in one of these half-spaces.

If the hyperplane H is shifted parallel to itself so that the new hyperplane $\tilde{H}$ includes M, then $\tilde{H} \cap \mathrm{ri}(C) = \emptyset$. □

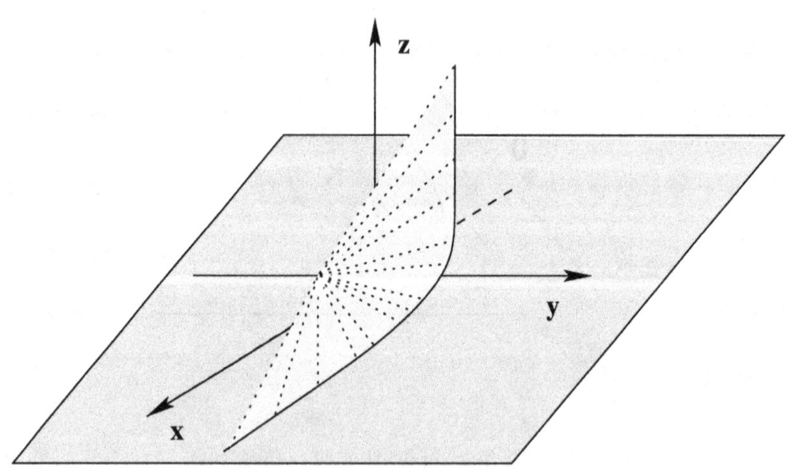

Fig. 6.4. Separation of an affine set and a convex set.

It is perhaps surprising that even when the affine set M and the convex set C in Theorem 6.17 are disjoint, the hyperplane $H \supset M$ may contain points of C. The example in Figure 6.4 is such a case in which $M \subset \mathbb{R}^3$ is the x-axis and C is the convex hull of the point $(0,-1,0)$ and the hyperbola $\{(x,y,z) : xz = 1, x > 0, y = 1\}$. Even though C and M have no common points, the separating hyperplane H is the xy-plane and contains the point $(0,-1,0)$.

6.3 Two Applications of Separation Theorems

6.3.1 Dual Cone

Duality is a very important aspect of convex analysis. In this subsection, we deal with a particular case of duality.

Definition 6.18. *If $K \subseteq \mathbb{R}^n$ is a nonempty set, the set*

$$K^* := \{y \in \mathbb{R}^n : \langle x, y\rangle \le 0 \ \text{ for all } \ x \in K\}$$

is called the polar cone *of K, or sometimes the* dual cone *of K.*

The cone K^* is always a closed convex cone, because

$$K^* = \bigcap_{x\in K} \{y \in \mathbb{R}^n : \langle x, y\rangle \le 0\} = \bigcap_{x\in K} \bar{H}^-_{(x,0)},$$

and each $\bar{H}^-_{(x,0)}$ is a closed convex cone.

Theorem 6.19. *If $K \subseteq \mathbb{R}^n$ is a closed convex cone, then $K = K^{**}$.*

Proof. From the definition of K^*, we see that if $x \in K$, then $\langle x, y\rangle \le 0$ for all $y \in K^*$; this proves that $K \subseteq K^{**}$. Suppose that the reverse inclusion $K^{**} \subseteq K$ is not true, and pick a point $x \in K^{**} \setminus K$. It follows from Theorem 6.10 that there exists a nonzero vector $a \in \mathbb{R}^n$ such that

$$\langle a, x\rangle > \langle a, z\rangle \ \text{ for all } \ z \in K. \tag{6.3}$$

On the one hand, setting $z = 0$ in (6.3) gives $\langle a, x\rangle > 0$; on the other hand, if $z \in K$ is a fixed point, then $tz \in K$ for all $t > 0$, and (6.3) gives $\langle a, x\rangle > \langle a, tz\rangle$, or $\langle a, z\rangle < \langle a, x\rangle/t$. Letting $t \to \infty$, we obtain

$$\langle a, z\rangle \le 0 \ \text{ for all } \ z \in K,$$

which implies that $a \in K^*$. However, since $z \in K^{**}$, we must have $\langle a, x\rangle \le 0$, which contradicts the fact $\langle a, x\rangle > 0$ proved above. ☐

Definition 6.18 and Theorem 6.19, with appropriate modifications, hold in much more general settings.

6.3.2 A Convex Barrier Function on an Open Convex Set

Theorem 6.20. *Let $C \subset \mathbb{R}^n$ be a nonempty open convex set. The distance function*

$$d_C(x) := d(x, \partial C) = \min\{\|x - z\| : z \in \partial C\}$$

is a concave function on C that vanishes on the boundary of C.

Furthermore, the function $-\ln d_C(x)$ is a convex barrier function on C, that is,

$$-\ln d_C(x) \to \infty \qquad \text{as } x \to \partial C.$$

Proof. Let $x \in C$. If H^+ is an open half-space containing C, then it is a simple consequence of the definition of d_C that $d_C(x) \leq d_{H^+}(x)$; thus

$$d_C(x) \leq \inf_{C \subseteq H^+} d_{H^+}(x),$$

where the infimum is taken over the set of open half-spaces H^+ containing C. However, $d_C(x) = \|x - z_x\|$ for some $z_x \in \partial C$, and it follows from Theorem 6.8 that there exists $0 \neq a \in \mathbb{R}^n$ such that $H(x)^+ := \{x : \langle a, x\rangle > \langle a, z_x\rangle\}$ contains C. Clearly

$$d_{H(x)^+}(x) = d_C(x),$$

and this proves that

$$d_C(x) = \inf_{C \subseteq H^+} d_H(x) \quad \text{for} \quad x \in C. \tag{6.4}$$

Now, if $H^+ = H^+_{(a,\alpha)}$, where $\|a\| = 1$ and $x \in H^+$, then it is easily shown that $d_H(x) = \langle a, x\rangle - \alpha$, a linear function. Consequently, $d_C(x)$ is a concave function, being the pointwise infimum of a set of linear functions.

The same argument shows that

$$-\ln d_C(x) = \sup_{C \subseteq H} -\ln d_H(x), \quad x \in C.$$

Since the function $-\ln d_H(x)$ is convex, we see that $-\ln d_C(x)$ is a convex function on C, and approaches $+\infty$ on ∂C. □

The converse statement is also true, that is, if C is an open set and d_C is a concave function (or $-\ln d_C$ is a convex function), then C is a convex set. We will not prove this here; the interested reader may consult [47] or [141], pp. 57–60, for details.

6.4 Proper Separation of a Convex Set and a Convex Polyhedron

If, in separation theorems, one of the convex sets has a special character, then it is possible to prove more powerful separation results; see [175, 176] for examples of such results. Rockafellar proves a particular separation theorem when one set is a convex polyhedron; see his Theorem 20.2 in [228]. He uses it to prove two important results, a convex transposition theorem and as a corollary the strong duality theorem of convex programming, which more or less correspond to our Theorems 11.14 and 11.15, respectively.

We here give a simpler proof of Rockafellar's separation theorem using a device in [147].

Theorem 6.21. *Let C and P be nonempty convex sets in $\mathbb{R}^n$ such that P is a convex polyhedron, that is, P is the intersection of finitely many closed half-spaces in $\mathbb{R}^n$. There exists a hyperplane separating C and P properly and not containing C if and only if* $\mathrm{ri}(C)$ *and P are disjoint.*

Proof. Suppose that there exists a hyperplane H that separates C and P and does not contain C. If $x \in \mathrm{ri}(C) \cap P \neq \emptyset$ as well, then we have $x \in H$, and if $y \in C \setminus H$, then the line segment $[y, x]$ can be extended beyond x without leaving C (since $x \in \mathrm{ri}(C)$), intersecting both half-paces H^+ and H^-; this contradicts the fact that H separates C and P.

To complete the proof, we need to show that if $\mathrm{ri}(C) \cap P = \emptyset$, then we can find a hyperplane H separating C and P and not containing C. We will do this by a suitable separation argument. Suppose that P has the representation

$$P = \{x \in \mathbb{R}^n : h_j(x) \leq 0, \ j = 1, \ldots, m\},$$

where each h_j is an affine function.

A crucial part of the argument involves the introduction of the *maximal* set of indices $I \subseteq \{1, \ldots, m\}$ (which could be the empty set) such that there exist positive multipliers $\alpha_j > 0$ satisfying the conditions

$$\sum_{j \in I} \alpha_j h_j(x) = 0 \text{ for all } x \in \mathrm{ri}(C).$$

An important property of the set I (which is easy to see) is that if a point $x \in \mathrm{ri}(C)$ satisfies the linear inequalities $h_j(x) \leq 0$ for all $j \in I$, then it satisfies them with equality, that is, $h_j(x) = 0$ for all $j \in I$.

We define a set of "right-hand-side vectors" of the affine functions on $\mathrm{ri}(C)$,

$$R := \bigl\{z \in \mathbb{R}^m : \exists x \in \mathrm{ri}(C), \ \ h_j(x) \leq z_j, \ j \notin I, \ \ h_j(x) = z_j, \ j \in I\bigr\}.$$

which is clearly a nonempty convex set. We have $0 \notin R$ (since any $x \in \mathrm{ri}(C)$ must satisfy $h_j(x) = 0$ for $j \in I$), and by Theorem 6.17 there exists a hyperplane H through the origin such that $\mathrm{ri}(R) \cap H = \emptyset$. Consequently, R lies in one of the closed half-spaces defined by H, but not entirely on H. Thus, there exists a nonzero vector $\mu \in \mathbb{R}^m$ such that $\langle \mu, z \rangle \geq 0$ for all $z \in R$, and the strict inequality is satisfied for some $\overline{z} \in R$. This means that

$$\sum_{j \notin I} \mu_j (h_j(x) + t_j) + \sum_{j \in I} \mu_j h_j(x) \geq 0 \ \text{ for all } x \in \mathrm{ri}(C), \ t \geq 0, \tag{6.5}$$

and the inequality is strict for some $\overline{x} \in \text{ri}(C)$, $\overline{t} \geq 0$.

If $\mu_j < 0$ for some $j \notin I$, then we obtain a contradiction to (6.5) by letting $t_j \to \infty$; thus $\mu_j \geq 0$ for $j \notin I$. Letting $t_j \to 0$, we obtain

$$\sum_{j=1}^{m} \mu_j h_j(x) \geq 0 \quad \text{for all} \quad x \in \text{ri}(C).$$

Some components of the multiplier vector $\mu_I := (\mu_j, j \in I)$ may be negative, but adding to it a positive multiple of the vector $\alpha_I := (\alpha_j, j \in I)$ makes all components of μ_I positive without changing the above inequality.

The affine function

$$\ell(x) := \sum_{j=1}^{m} \mu_j h_j(x)$$

is nonnegative on $\text{ri}(C)$ and clearly nonpositive on P. Thus, $H := \{x : \ell(x) = 0\}$ is a hyperplane separating $\text{ri}(C)$ and P.

The function $\ell(x)$ does not vanish identically on $\text{ri}(C)$: otherwise, in the case that $\mu_j > 0$ for some $j \notin I$, the maximality of I is violated; and in the remaining case in which $\mu_j = 0$ for all $j \notin I$, (6.5) implies that $\ell(\overline{x}) > 0$ for some $\overline{x} \in \text{ri}(C)$.

Thus far, we have proved that H separates C and P, and that such $\text{ri}(C)$ does not lie on H. It remains only to verify the claim that $\ell(x) \geq 0$ for all $x \in C$. If there is a point $x \in C$ such that $\ell(x) < 0$, then the line segment $[x_0, x]$, where $x_0 \in \text{ri}(C)$, contains a point $x_1 \in (x_0, x)$ such that $\ell(x_1) < 0$. Since $x_1 \in \text{ri}(C)$ by Lemma 5.18, this gives a contradiction, proving the claim. □

6.5 Dubovitskii–Milyutin Theorem in Finite Dimensions

Dubovitskii and Milyutin [80] devise a general scheme to derive optimality conditions in very diverse optimization problems, ranging from mathematical programming to optimal control. In this theory, the optimality conditions are reduced to the condition that the intersection of a certain set of convex sets is empty. Then, the *Dubovitskii–Milyutin theorem* below is invoked to write the optimality conditions in a more convenient, analytical, form. Two general versions of the theorem are given in Section 6.7.

We first prove a generalization of this result in finite dimensions following [124].

Lemma 6.22. *Let $\{C_i\}_1^{k+1}$, $k \geq 1$, be convex sets in $\mathbb{R}^n$ such that $0 \in \overline{C_i}$ for $i = 1, \ldots, k+1$. Consider the conditions*

(a) $\cap_{i=1}^{k+1} C_i = \emptyset$.

(b) There exists $l := (l_1, \ldots, l_{k+1}) \neq 0$ *such that*

$$\langle l_i, x_i \rangle \leq 0 \quad \textit{for all} \quad x_i \in C_i, \quad \sum_{i=1}^{k+1} l_i = 0.$$

Then (a) implies (b). Moreover, if k of the sets, say $\{C_i\}_1^k$*, are open, then (b) implies (a) as well, and the two conditions are equivalent.*

Proof. *(a) implies (b)*: The important idea here is to define the sets

$$K_1 := \{(x_{k+1}, \ldots, x_{k+1}) : x_{k+1} \in C_{k+1}\},$$
$$K_2 := C_1 \times \cdots \times C_k,$$

and to note that $K_1 \cap K_2 = \emptyset$, and $0 \in \overline{K_i}$, $i = 1, 2$. Theorem 6.9 implies that the sets K_1 and K_2 can be separated, that is, there exists $0 \neq (l_1, \ldots, l_k)$, $l_i \in \mathbb{R}^n$, such that

$$\sum_{i=1}^{k} \langle l_i, x_i \rangle \leq \langle \sum_{i=1}^{k} l_i, y \rangle \text{ for all } x_i \in C_i, \ i = 1, \ldots, k, \ y \in C_{k+1}. \tag{6.6}$$

Fix i, $1 \leq i \leq k$, and let the remaining $x_j \to 0$ and $y \to 0$. This gives

$$\langle l_i, x_i \rangle \leq 0 \text{ for all } x_i \in C_i, \ i = 1, \ldots, k.$$

Similarly, letting all $x_i \to 0$ in (6.6) shows that $l_{k+1} := -\sum_{i=1}^{k} l_i$ satisfies the inequality $\langle l_{k+1}, y \rangle \leq 0$ for all $y \in C_{k+1}$.

(b) implies (a): We must have $l_i \neq 0$ for some $i \leq k$, since otherwise $l_{k+1} = 0$ as well, and $l = 0$. If $x \in \cap_{i=1}^{k+1} C_i \neq \emptyset$, then $0 = \sum_{i=1}^{k+1} \langle l_i, x \rangle$, and because $\langle l_i, x \rangle \leq 0$, we actually have $\langle l_i, x \rangle = 0$ for each $i \leq k$. Let $l_i \neq 0$. Since C_i is open, there exists $\varepsilon > 0$ such that $x + \varepsilon l_i \in C_i$; but then we have a contradiction, since

$$0 \geq \langle l_i, x + \varepsilon l_i \rangle = \varepsilon \|l_i\|^2 > 0.$$

□

Theorem 6.23. (***Dubovitskii–Milyutin***) *Let* $\{K_i\}_1^k$ *be open convex cones, and* K_{k+1} *a convex cone in* $\mathbb{R}^n$*. Then*

$$\cap_{i=1}^{k+1} K_i = \emptyset \tag{6.7}$$

if and only if there exist

$$l_i \in K_i^*, \ \{l_i\}_{i=1}^{k+1} \textit{ not all zero, such that } \ l_1 + l_2 + \cdots + l_{k+1} = 0. \tag{6.8}$$

This finite-dimensional version of the Dubovitskii–Milyutin theorem follows immediately from Lemma 6.22. The theorem is also true in infinite-dimensional topological vector spaces; see Section 6.7.

6.6 Separation of Convex Sets in General Vector Spaces

In this section and the next, we prove separation theorems involving two or more convex sets in arbitrary vector spaces over $\mathbb{R}$. We first prove separation theorems using a synthetic, algebraic framework, suggested in the works [167, 168, 169, 177, 17, 18], and especially in the charming book [63], because this approach brings out the basic ideas behind the separation theorems most clearly, and gives the most general results. Moreover, this approach effectively isolates the role of topological considerations in separation theorems, and makes it possible to prove topological separation theorems with relative ease.

Readers who are not interested infinite-dimensional vector spaces may skip this section and the next two without any loss of continuity.

We start by defining several relevant concepts.

Definition 6.24. *Let E be a real vector space. A* hyperplane *H in E is the level set of a nontrivial linear functional $\ell : E \to \mathbb{R}$, that is,*

$$H = \{x \in E : \ell(x) = \alpha\}$$

for some $\alpha \in \mathbb{R}$.

The hyperplane H partitions E into two half-spaces; in the definitions below, “closed” and “open” are algebraic concepts and do not refer to any topology of E.

Definition 6.25. *An algebraically closed half-space in E is a set either of the form*

$$\bar{H}^{+}_{(\ell,\alpha)} := \{x \in E : \ell(x) \geq \alpha\},$$

or of the form

$$\bar{H}^{-}_{(\ell,\alpha)} := \{x \in E : \ell(x) \leq \alpha\},$$

where ℓ is a nonzero linear functional on E and $\alpha \in \mathbb{R}$.

Similarly, an algebraically open half-space in E is a set either of the form

$$H^{+}_{(\ell,\alpha)} := \{x \in E : \ell(x) > \alpha\},$$

or of the form

$$H^{-}_{(\ell,\alpha)} := \{x \in E : \ell(x) < \alpha\}.$$

Definition 6.26. *Let C and D be two nonempty sets, and $H := H_{(\ell,\alpha)}$ a hyperplane in a vector space E.*

H is called a separating hyperplane *for the sets C and D if C is contained in one of the algebraically closed half-spaces determined by H and D in the other, say $C \subseteq \bar{H}^{+}_{(\ell,\alpha)}$ and $D \subseteq \bar{H}^{-}_{(\ell,\alpha)}$.*

H is called a strictly separating hyperplane *for the sets C and D if C is contained in one of the algebraically open half-spaces determined by H and D in the other, say $C \subseteq H^{+}_{(\ell,\alpha)}$ and $D \subseteq H^{-}_{(\ell,\alpha)}$.*

H is called a strongly separating hyperplane *for the sets C and D if there exist β and γ satisfying $\gamma < \alpha < \beta$, such that $C \subseteq \bar{H}^+_{(\ell,\beta)}$, and $D \subseteq \bar{H}^-_{(\ell,\gamma)}$.*

H is called a properly separating hyperplane *for the sets C and D if H separates C and D, and C and D are not* both *contained in the hyperplane H.*

If there exists a hyperplane H separating the sets C and D in one of the senses above, we say that C and D can be separated, strictly separated, strongly separated, properly separated, *respectively.*

As in the finite-dimensional case, hyperplanes are proper, maximal affine subsets. However, when E is a topological vector space, it is no longer true that every hyperplane is necessarily closed.

Lemma 6.27. *Let E be a real vector space. A set $H \subset E$ is a hyperplane if and only if H is a proper maximal affine subset of E.*

Moreover, if E is a topological vector space, then the hyperplane $H_{(\ell,\alpha)}$ is closed if and only if ℓ is a continuous linear functional.

Proof. Clearly, a hyperplane $H_{(\ell,\alpha)}$ is a proper affine subset of E. The maximality of H holds: if $a \in E \setminus H$, then $\ell(a) \neq 0$, so that if $x \in E$, we have $\ell(x) = \ell((\ell(x)/\ell(a))a)$, that is, $x - \ell(x)/\ell(a)a \in H$, proving that $E = \text{span}\{H, a\}$.

Conversely, suppose that H is a proper maximal affine subset of E. Assume without loss of generality that H is a linear subspace of E. If $a \in E \setminus H$, then $E = \text{span}\{H, a\}$, so that every $x \in E$ has a representation $x = u + ta$, where $u \in H$ and $t \in \mathbb{R}$. This representation is unique, since $x = u_1 + t_1 a = u_2 + t_2 a$ implies that $u_2 - u_1 = (t_1 - t_2)a \in H \cap \text{span}(\{a\}) = \{0\}$, that is, $u_2 = u_1$ and $t_2 = t_1$. Define

$$\ell(x) = t, \quad \text{where} \quad x = u + ta, \ u \in H, \ t \in \mathbb{R},$$

which is easily shown to be a linear functional. Clearly, $H = H_{(\ell,0)}$, proving that H is a hyperplane.

Now suppose that E is a topological vector space. If ℓ is continuous, it is clear that $H_{(\ell,\alpha)}$ is a (topologically) closed set. Conversely, if $H := H_{(\ell,\alpha)}$ is closed, we claim that ℓ is continuous. Pick a point x in the complement of H, which is an open set. There exists an open neighborhood N of the origin such that $x + N \subseteq E \setminus H$. We may assume that N is a symmetric neighborhood, that is, $N = -N$: since $(t, x) \mapsto tx$ is continuous, there exist $\delta > 0$ and a neighborhood W of the origin such that $tV \subseteq N$ for all $|t| < \delta$. The set $\bar{N} := \cup_{|t|<\delta} tW \subseteq N$ is clearly a symmetric neighborhood of the origin. If $\ell(N)$ is unbounded, then it is easy to see that $\ell(N) = \mathbb{R}$, so that there exists $y \in N$ satisfying $\ell(y) = \alpha - \ell(x)$, which gives the contradiction $x + y \in (x + N) \cap H = \emptyset$. Therefore, $\ell(N)$ is bounded, say $|\ell(x)| \leq M$ for $x \in N$. The continuity of ℓ follows, because given $\epsilon > 0$, $|\ell(x)| < \epsilon$ for every $x \in (\epsilon/M)N$. □

Note that the proof above also establishes the following result.

Corollary 6.28. *Let E be a real topological vector space, and $H := H_{(\ell,\alpha)}$ a hyperplane. The linear functional ℓ is continuous if and only if one of the half-spaces H^+, H^- contains an open set.*

Some form of Zorn's lemma is needed to prove separation theorems in general vector spaces. Recall that a *partial order* $\preceq$ on a set X is a reflexive, antisymmetric, and transitive relation on X, that is, for $x, y, z \in X$, we have

(a) $x \preceq x$,
(b) $x \preceq y,\ y \preceq x \implies x = y$,
(c) $x \preceq y,\ y \preceq z \implies x \preceq z$.

A subset $Y \subseteq X$ is called *totally ordered* if any two elements $x, y \in Y$ can be compared, that is, either $x \preceq y$ or $y \preceq x$. An *upper bound* of any set $Z \subseteq X$ is a point $x \in X$ such that $z \preceq x$ for every $z \in Z$. A *maximal element* of a partially ordered set X is a point $x \in X$ such that $x \preceq z$ implies that $z = x$.

Lemma 6.29. (***Zorn's lemma***) *A partially ordered set has a maximal element if every totally ordered subset of it has an upper bound.*

Zorn's lemma is a basic axiom of set theory equivalent to the axiom of choice or the well-ordering principle; see for example [125] for more details.

A pair of nonempty convex sets C and D satisfying $C \cap D = \emptyset$ and $C \cup D = E$ are called *complementary* convex sets. The following result essentially goes back to [152] and [248].

Lemma 6.30. *If A and B are two nonempty, disjoint convex sets in a vector space E, then there exist complementary convex sets C and D in E such that $A \subseteq C$ and $B \subseteq D$.*

Proof. We introduce a relation $\preceq$ on the set $\mathcal{C}$ of disjoint convex subsets $(C, D) \subseteq E \times E$ such that $A \subseteq C$ and $B \subseteq D$ by the inclusion relation, that is, we declare $(C, D) \preceq (C', D')$ if $C \subseteq C'$ and $D \subseteq D'$. It is evident that $\preceq$ is a partial order relation on $\mathcal{C}$. Moreover, if $\mathcal{D} \subset \mathcal{C}$ is any totally ordered subset, then the union of sets in $\mathcal{D}$ is a pair of disjoint convex sets that is an upper bound for $\mathcal{D}$. Thus, Zorn's lemma applies, and there exists a maximal element $(C, D) \in \mathcal{C}$, that is, C and D are convex sets satisfying $A \subseteq C$ and $B \subseteq D$, and whenever C' and D' are convex sets satisfying $C \subseteq C'$ and $D \subseteq D'$, then $C' = C$ and $D' = D$.

We claim that $C \cup D = E$. If this is not true, pick a point $x \in E \setminus (C \cup D)$. Since (C, D) is a maximal pair, we have $\operatorname{co}(\{x\} \cup C) \cap D \neq \emptyset$ and $\operatorname{co}(\{x\} \cup D) \cap C \neq \emptyset$. Let $y_1 \in \operatorname{co}(\{x\} \cup D) \cap C$ and $y_2 \in \operatorname{co}(\{x\} \cup C) \cap D$; then there exist $x_2 \in D$ such that $y_1 \in (x, x_2)$, and $x_1 \in C$ such that $y_2 \in (x, x_1)$; see Figure 6.5. But the intersection point z of the line segments $[x_1, y_1]$ and $[x_2, y_2]$ belongs to both C and D, a contradiction. This proves the claim and the lemma. □

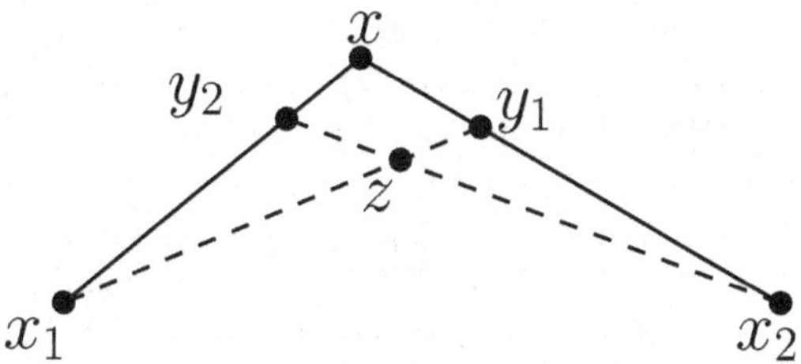

Fig. 6.5.

Lemma 6.31. *Let* (C, D) *be complementary convex sets in a vector space* E*. The set*

$$L := \mathrm{ac}(C) \cap \mathrm{ac}(D)$$

is either a hyperplane in E *or the whole space* E*.*

Moreover,

(a) $L = E$ *if and only if* $\mathrm{ai}(C) = \mathrm{ai}(D) = \emptyset$*, or equivalently if and only if* $\mathrm{ac}(C) = \mathrm{ac}(D) = E$*.*
(b) If L *is a hyperplane, then the sets* $\mathrm{ai}(C)$ *and* $\mathrm{ai}(D)$ *are both nonempty, and the pairs* $(\mathrm{ai}(C), \mathrm{ai}(D))$ *and* $(\mathrm{ac}(C), \mathrm{ac}(D))$ *are the algebraically open and closed half-spaces associated with* L*, respectively.*

Proof. The set L is convex, because $\mathrm{ac}(C)$ and $\mathrm{ac}(D)$ are convex sets. The set L is not empty: let $x \in C$ and $y \in D$; there exists a point $w \in (x, y)$ such that $[x, w) \subseteq C$ and $[y, w) \subseteq D$, implying $w \in \mathrm{ac}(D) \cap \mathrm{ac}(D) \neq \emptyset$.

First, we claim that

$$\mathrm{ac}(C) = E \setminus \mathrm{ai}(D).$$

If $x \notin \mathrm{ai}(D)$, then there exists $u \in E$ such that any point $v \in E$ satisfying $x \in (u, v)$ has the property that $(x, v] \subseteq E \setminus D = C$; thus, $x \in \mathrm{ac}(C)$, and we have proved that $\mathrm{ac}(C) \cup \mathrm{ai}(D) = E$. The sets $\mathrm{ac}(C)$ and $\mathrm{ai}(D)$ are disjoint: if $x \in \mathrm{ac}(C) \cap \mathrm{ai}(D)$, then there exists $u \in C$ such that $[u, x) \subseteq C$. Let v be a point such that $x \in (u, v)$; either $v \in C$, in which case $[u, v] \subset C$ and $x \in \mathrm{ai}(D) \subseteq D$, which is impossible, or $v \in D$ and then $[u, x)$ intersects D, since $x \in \mathrm{ai}(D)$, which is also impossible. Our claim is proved, and we have

$$\mathrm{ac}(C) = E \setminus \mathrm{ai}(D), \text{ and } \mathrm{ac}(D) = E \setminus \mathrm{ai}(C). \tag{6.9}$$

It follows immediately that $L = E$ if and only if $\mathrm{ai}(C) = \mathrm{ai}(D) = \emptyset$, or equivalently if and only if $\mathrm{ac}(C) = \mathrm{ac}(D) = E$, proving (a).

It is now easy to show that L is an affine set. If $x, y \in L$ and z satisfies $y \in (x, z)$ and $z \notin L = \mathrm{ac}(C) \cap \mathrm{ac}(D)$, then $z \notin \mathrm{ac}(C)$, say; but then $z \in \mathrm{ai}(D)$, and since $x \in \mathrm{ac}(D)$, we must have $y \in \mathrm{ai}(D)$ by Lemma 5.5, which contradicts the assumption that $y \in \mathrm{ac}(C)$. Thus, $z \in L$, and L is an affine set.

Finally, assume that $L \neq E$. Pick $p \in \mathrm{ai}(C) = E \setminus \mathrm{ac}(D)$, so that $p \notin L$; see Figure 6.6. To prove that L is a hyperplane, it suffices to show that $E =$

$\text{aff}(\{p\} \cup L)$. Pick a point $r \in L$ and consider a point $q = \alpha r - p$ with $\alpha > 1$, that is, $r \in (p, q)$. We must have $q \in \text{ai}(D)$, because otherwise $q \in \text{ac}(C)$ and $r \in \text{ai}(C) = E \setminus \text{ac}(D)$ by Lemma 5.5, contradicting $r \in L$. If $x \in C \setminus L$ is an arbitrary point, then the segment $[x, q]$ must intersect L; in fact, the point $w \in (x, q)$ satisfying $[x, w) \subseteq C$ and $[q, w) \subseteq \text{ai}(D)$ must lie on L, because $w \in \text{ac}(C)$ and $w \in \text{ac}(\text{ai}(D)) = \text{ac}(D)$. This proves that $x \in \text{aff}(\{p\} \cup L)$. Similarly, if $y \in D \setminus L$ is an arbitrary point, then $y \in \text{aff}(\{p\} \cup L)$. Altogether, we have proved that $E = \text{aff}(\{p\} \cup L)$, that is, L is a hyperplane.

It follows from (6.9) that the sets $\text{ai}(C), L, \text{ai}(D)$ are disjoint and their union is E. This proves (b). □

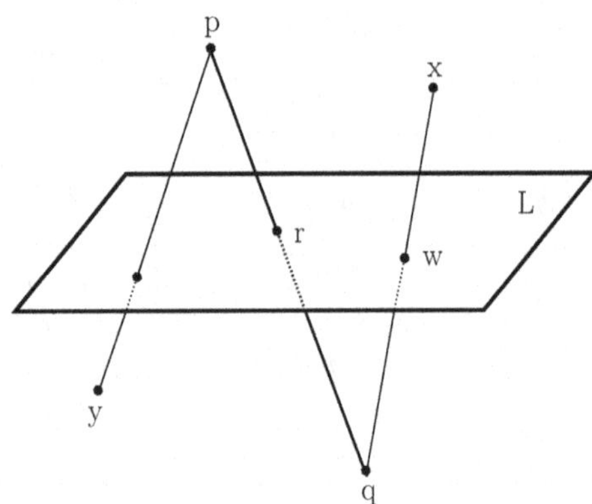

Fig. 6.6.

Theorem 6.32. *Let C and D be nonempty convex sets in a vector space E such that $\text{ai}(C) \neq \emptyset$. Then there exists a hyperplane H separating C and D if and only if $\text{ai}(C) \cap D = \emptyset$, in which case $\text{ai}(C)$ lies in one of the algebraically open half-spaces associated with H.*

Proof. Suppose that the hyperplane H separates C and D, such that $C \subseteq \bar{H}^+$ and $D \subseteq \bar{H}^-$. The set C cannot lie on H, since $\text{aff}(C) = E$; hence there exists a point $y \in C \cap H^+$. We must have $\text{ai}(C) \subseteq H^+$, because if there is a point $x \in \text{ai}(C) \cap H$, then there exists a point $z \in C$ such that $x \in (y, z)$; since $z \in C \cap H^-$, this gives a contradiction.

Conversely, if $\text{ai}(C) \cap D = \emptyset$, then Lemma 6.30 implies that there exist complementary convex sets $(\tilde{C}, \tilde{D})$ such that $\text{ai}(C) \subseteq \tilde{C}$ and $D \subseteq \tilde{D}$. We claim that $\text{ai}(C) \subseteq \text{ai}(\tilde{C}) \neq \emptyset$. If $x \in \text{ai}(C)$, then for any $y \in E$, there exists $u \in C$ such that $x \in (u, y)$. Since $[x, u) \subseteq \text{ai}(C)$ by Lemma 5.5, we may assume that $u \in \text{ai}(C) \subseteq \tilde{C}$; this proves the claim. Lemma 6.31 implies that $H := \text{ac}(\tilde{C}) \cap \text{ac}(\tilde{D})$ is a hyperplane that separates $\tilde{C}$ and $\tilde{D}$, hence $\text{ai}(C)$ and D. Supposing $\text{ai}(C) \subseteq \bar{H}^+$ and $D \subseteq \bar{H}^-$, we must have $C \subseteq \bar{H}^+$, because if

$x \in C \cap H^-$ and $y \in \mathrm{ai}(C) \subseteq \bar{H}^+$, then Lemma 5.5 implies that (y, x) contains a point $z \in \mathrm{ai}(C) \cap H^- = \emptyset$, a contradiction.

We have proved that H separates C and D. □

Theorem 6.33. (***Proper separation theorem***) *Let C and D be nonempty convex sets in a vector space E, such that* $\mathrm{rai}(C) \neq \emptyset$ *and* $\mathrm{rai}(D) \neq \emptyset$.

Then, there exists a hyperplane H properly separating C and D if and only if $\mathrm{rai}(C) \cap \mathrm{rai}(D) = \emptyset$.

Proof. The proof is partly the same as the proof Theorem 6.15. Define the convex set $K := C - D$. It follows from Lemma 5.11 that $\mathrm{rai}(K) = \mathrm{rai}(C - D) = \mathrm{rai}(C) - \mathrm{rai}(D)$; thus, $\mathrm{rai}(C) \cap \mathrm{rai}(D) = \emptyset$ and $0 \notin \mathrm{rai}(K)$ are equivalent statements. Lemma 6.14 holds in arbitrary vector spaces, as is evident from its proof. Thus, the proof of the theorem reduces to establishing the fact that the sets $\{0\}$ and K are properly separable if and only if $0 \notin \mathrm{rai}(K)$.

Suppose that the origin and K are properly separated by a hyperplane H, such that $0 \in \bar{H}^-$ and $K \subseteq \bar{H}^+$. We claim that $0 \notin \mathrm{rai}(K)$. If $0 \notin H$, then $\mathrm{rai}(K) \subseteq \bar{H}^+$, so that $0 \notin \mathrm{rai}(K)$. Otherwise, $0 \in H$ and there exists a point $x \in K \setminus H$. If we had $0 \in \mathrm{rai}(K)$, there would exist a point $y \in K$ such that $0 \in (x, y)$, giving the contradiction $y \in C \cap H^- = \emptyset$. This proves the claim.

To prove the converse implication, suppose that $0 \notin \mathrm{rai}(K)$, and let $L := \mathrm{aff}(K)$ be the affine hull of K. If $0 \notin L$, we will show that $\{0\}$ and L can be properly separated; this will imply that $\{0\}$ and K can be properly separated. Consider the set of all affine subsets of E that contain L but not 0, partially ordered by set inclusion. Zorn's lemma guarantees the existence of a maximal affine subspace H containing L but not 0. We claim that H is a hyperplane; otherwise, $H' = \mathrm{aff}(\{0\} \cup H) \neq E$, and if we pick $x \in E \setminus H'$, then the affine set $\mathrm{aff}(\{x\} \cup H)$ strictly includes H but does not include 0, contradicting the maximality of H. This proves the claim. It is clear that H properly separates $\{0\}$ and L.

If $0 \in L$, we apply Theorem 6.32 *within* the vector space L to the sets $\{0\}$ and K, and obtain a hyperplane P in L separating 0 and K such that $\mathrm{rai}(K) \subseteq P^+$; see Figure 6.3. We may again assume that $0 \in P$ (otherwise the translation of P so that it passes through the origin also satisfies the same separation properties). Zorn's lemma implies that there exists a maximal linear subspace H of E extending P and satisfying $P = H \cap L$. We claim that H is a hyperplane in E. Otherwise, pick $x \in E \backslash H$ and form the linear subspace $H' := \mathrm{span}\{x, H\}$, which strictly contains H. We have $H' \cap L = P$: any $y \in H'$ can be written as $y = \alpha x + h$ with $\alpha \in \mathbb{R}$ and $h \in H$; if $y \in L$, then $y \in H$, and consequently $\alpha x \in H$, which implies that $\alpha = 0$ and $y \in H \cap L = P$. This proves that $H' \cap L = P$ (the inclusion $P \subseteq H' \cap L$ is trivial), contradicting the maximality of H.

We have proved that H is a hyperplane; clearly H properly separates $\{0\}$ and K. □

Theorem 6.34. *Let C be a nonempty convex set in a vector space E. If M is an affine set such that* $\text{rai}(C) \cap M = \emptyset$, *then there exists a hyperplane H extending M such that* $\text{rai}(C) \cap H = \emptyset$.

Proof. The proof of the theorem is the same as the proof of Theorem 6.17 except that we replace $\text{ri}(C)$ in that proof by $\text{rai}(C)$ and invoke Theorem 6.33 instead of Theorem 6.15. □

6.7 Separation of Several Convex Sets

In this section, we deal with separation of several convex sets, in both the algebraic and topological senses. The algebraic separation theorems give the most general results, being valid under very mild conditions; they also make it fairly easy to obtain topological separation from them, since the only thing that needs proof is the continuity of the linear functional defining the hyperplane; see for example the proof of Theorem 6.39 below.

The major results of this section are Theorem 6.37, which characterizes, under mild conditions, when several convex sets can be properly separated in the algebraic sense, and Theorems 6.38 and 6.39, which generalize the Dubovitskii–Milyutin theorem, Theorem 6.23, in the algebraic and topological senses, respectively.

Readers who are not interested infinite-dimensional vector spaces may skip this section without any loss of continuity.

Definition 6.35. *Let $\{C_i\}_1^k$ be a family of nonempty proper convex sets in a vector space E. The sets $\{C_i\}_1^k$ are called* separable *if there exist hyperplanes $\{H_i\}_1^k$ such that $C_i \subseteq \bar{H}_i^-$ for all i and the algebraically open half-spaces $\{H_i^-\}_1^k$ have an empty intersection, that is, $\cap_1^k H_i^- = \emptyset$.*

The sets $\{C_i\}_1^k$ are called properly separable *if they are separable and at least one set C_i does not lie on the corresponding hyperplane H_i.*

Theorem 6.37 below generalizes Theorem 6.33 to more than two convex sets. We need the following preliminary result in its proof.

Lemma 6.36. *Let E be a real vector space, and $\{\ell_i\}_1^k$ a set of linear functionals on E. The system of strict linear inequalities*

$$\ell_i(x) < \alpha_i, \quad i = 1, \ldots, k, \tag{6.10}$$

is inconsistent if and only if there exist nonnegative scalars $\{\lambda_i\}_1^k$, not all zero, such that

$$\sum_{i=1}^k \lambda_i \ell_i = 0, \quad \sum_{i=1}^k \lambda_i \alpha_i \leq 0.$$

Proof. Both sets of conditions cannot hold simultaneously, because if x satisfies (6.10), then we have the contradiction

$$0 < \sum_{i=1}^{k} \lambda_i(\alpha_i - \ell_i(x)) = \sum_{i=1}^{k} \lambda_i \alpha_i \leq 0.$$

Suppose that (6.10) is inconsistent, and define the sets

$$L := \{u \in \mathbb{R}^k : \exists x \in E,\ u_i = \ell_i(x) - \alpha_i,\ i = 1, \ldots, k\},$$
$$K := \{u \in \mathbb{R}^k : u < 0\};$$

L is affine, and K is an open convex set. Note that $K \cap L = \emptyset$; it follows from Theorem 6.34 (in fact Theorem 6.17 suffices) that there exists a hyperplane $H = H_{(\lambda,\gamma)} \subset \mathbb{R}^k$ such that $L \subseteq H$ and $K \subseteq H^-$. On the one hand, the fact $K \subseteq H^-$ gives

$$\langle \lambda, v \rangle < \gamma \text{ for all } v \in \mathbb{R}^k,\ v < 0,$$

which implies that $\lambda \geq 0$ and $\gamma \geq 0$; on the other hand, the fact $L \subseteq H$ means that

$$\Big(\sum_{i=1}^{k} \lambda_i \ell_i\Big)(x) - \sum_{i=1}^{k} \lambda_i \alpha_i = \sum_{i=1}^{k} \lambda_i(\ell_i(x) - \alpha_i) = \gamma \text{ for all } x \in E,$$

which implies that $\sum_{i=1}^{k} \lambda_i \ell_i = 0$ and $\sum_{i=1}^{k} \lambda_i \alpha_i = -\gamma \leq 0$. □

An independent proof of the lemma can be found in Appendix A.

Theorem 6.37 gives both analytic and geometric characterizations of the disjointness condition $\cap_1^k \operatorname{rai} C_i = \emptyset$ of k proper nonempty sets $\{C_i\}_1^k$ under the mild conditions $\operatorname{rai} C_i \neq \emptyset$ for all $i = 1, \ldots, k$, which are always satisfied in the finite-dimensional case.

Theorem 6.37. *Let $\{C_i\}_1^k$, $k > 1$, be proper convex sets in a vector space E such that $\operatorname{rai} C_i \neq \emptyset$ for all $i = 1, \ldots, k$.*

The following conditions are equivalent:

(a) $\cap_1^k \operatorname{rai} C_i = \emptyset$.
(b) There exist linear functionals $\{\ell_i\}_1^k$ on E, not all identically zero, and scalars $\{\alpha_i\}_1^k$ such that

$$\ell_i(x_i) \leq \alpha_i \ \text{ for all } x_i \in C_i,\ i = 1, \ldots, k, \quad \sum_1^k \ell_i = 0, \quad \sum_1^k \alpha_i \leq 0,$$

and there exists a point x belonging to a set C_i with corresponding $\ell_i \neq 0$ such that $\ell_i(x_i) < \alpha_i$.
(c) The sets $\{C_i\}_1^k$ are properly separable.

Proof. We first prove the equivalence of parts (a) and (b). We start by proving (a) implies (b), the longest part of the whole theorem. Suppose that (a) is true. We prove (b) by induction on k. If $k = 2$, Theorem 6.33 implies that there exists a hyperplane $H = H_{(\ell,\alpha)}$ such that $\ell(x_1) \leq \alpha$ for all $x_1 \in C_1$, $\ell(x_2) \geq \alpha$ for all $x_2 \in C_2$, and $C_1 \cup C_2 \not\subseteq H$; (b) is clearly satisfied with the choices $\ell_1 = \ell$, $\ell_2 = -\ell$, $\alpha_1 = \alpha$, and $\alpha_2 = -\alpha$.

Assume that $k > 2$, and that we have proved (a) implies (b) for all integers smaller than k. If the relative algebraic interiors of $k-1$ of the sets have empty intersection, say $\cap_2^k \operatorname{rai} C_i = \emptyset$, then by the induction hypothesis, there exist $\{(\ell_i, \alpha_i)\}_2^k$ satisfying (b). If we define $\ell_1 = 0$ and $\alpha_1 = 0$, then (b) is satisfied with $\{(\ell_i, \alpha_i)\}_1^k$.

Thus, we may assume that $\cap_2^k \operatorname{rai} C_i \neq \emptyset$. Define the sets

$$\begin{aligned} K_1 &:= \{(x_1, x_1, \ldots, x_1) : x_1 \in C_1\}, \\ K_2 &:= C_2 \times \cdots \times C_k. \end{aligned}$$

We have, by elementary arguments,

$$\begin{aligned} \operatorname{rai} K_1 &= \{(x_1, x_1, \ldots, x_1) : x_1 \in \operatorname{rai} C_1\} \neq \emptyset, \\ \operatorname{rai} K_2 &= \operatorname{rai} C_2 \times \cdots \times \operatorname{rai} C_k \neq \emptyset. \end{aligned}$$

Since $\cap_1^k \operatorname{rai} C_i = \emptyset$, we see that $\operatorname{rai} K_1 \cap \operatorname{rai} K_2 = \emptyset$. Theorem 6.33 implies that there exists $(\ell, \alpha) = ((\ell_2, \ldots, \ell_k), \alpha)$, $\ell \neq 0$, such that

$$\ell_2(x_2) + \cdots + \ell_k(x_k) \leq \alpha \leq \Big(\sum_2^k \ell_i\Big)(x_1) \quad \text{for all } x_i \in C_i,\ i = 1, \ldots, k, \tag{6.11}$$

and strict inequality holds in one of the inequalities above for some choice of $x_1, \ldots, x_k$.

Define $\alpha_i := \sup_{x_i \in C_i} \ell_i(x_i)$ for $i = 2, \ldots, k$, and $\ell_1 := -\sum_2^k \ell_i$, $\alpha_1 := -\alpha$. It follows from (6.11) that $\alpha_2 + \cdots + \alpha_k \leq \alpha$. Since we have $\ell_i \neq 0$ for some $2 \leq i \leq k$, (b) holds except possibly when $\ell_i(x_i) = \alpha_i$ for all $x_i \in C_i$, $i = 2, \ldots, k$, and $\sum_2^k \alpha_i = \alpha$. However, in this last case we have $\alpha < \ell_1(\bar{x}_1)$ for some $\bar{x}_1 \in C_1$ and $\alpha \leq \ell_1(x_1)$ for all $x_1 \in C_1$. We must have $\ell_1 \neq 0$, because otherwise letting $x_i = x \in \cap_2^k \operatorname{rai}(C_i)$ gives

$$0 = \Big(\sum_1^k \ell_i\Big)(x) = \sum_2^k \ell_i(x) = \alpha < \ell_1(\bar{x}_1) = 0,$$

a contradiction. Thus, (b) holds in this case as well.

Conversely, let us prove that (b) implies (a). Suppose that (b) is true but (a) is false, that is, there exists a point $x \in \cap_1^k \operatorname{rai} C_i \neq \emptyset$. If the set C_i is such that $\ell_i \neq 0$ and C_i does not lie on the hyperplane $H_i := H_{(\ell_i, \alpha_i)}$, we have $\operatorname{rai}(C_i) \subseteq H_i^-$; otherwise, there exist $w \in \operatorname{rai}(C_i) \cap H_i$ and $u \in C_i$ such that $w \in (x_i, u)$, and this gives a contradiction because $u \in C \cap H_i^+ = \emptyset$. Therefore, we have

$$0 > \sum_{i=1}^{k} (\ell_i(x) - \alpha_i) = -\sum_{i=1}^{k} \alpha_i \geq 0,$$

a contradiction, which proves that (a) must be true.

It remains to demonstrate that parts (b) and (c) are equivalent. It follows immediately from Lemma 6.36 that (c) implies (b). Conversely, suppose that (b) is true. Define $I := \{i : 1 \leq i \leq k,\, \ell_i \neq 0\}$ and $H_i := H_{(\ell_i, \alpha_i)}$ for $i \in I$. We have $C_i \subseteq \bar{H}_i^-$. The open half-spaces $\{H_i^-\}_{i \in I}$ must already have an empty intersection, because $x \in \cap_{i \in I} H_i^- \neq \emptyset$ gives the contradiction

$$0 > \sum_{i=1}^{k} (\ell_i(x) - \alpha_i) = -\sum_{i=1}^{k} \alpha_i \geq 0.$$

It remains to show that every C_i, $i \notin I$, is contained in a half-space. Pick a point $x \notin C_i$ and invoke Theorem 6.34 to obtain a hyperplane H_i such that $\text{rai}(C_i) \subseteq H_i^-$. The hyperplanes $\{H_i\}_1^k$ properly separate the sets $\{C_i\}_1^k$, proving (c). □

The next theorem is a vast generalization of the Dubovitskii–Milyutin theorem, whose finite-dimensional versions were given in Lemma 6.22 and Theorem 6.23. The cone version of the theorem, with a different proof, is given in [82].

Theorem 6.38. ***(Dubovitskii–Milyutin)*** *Let $\{C_i\}_1^k$, $k > 1$, be nonempty convex sets in a vector space E, such that $\{C_i\}_1^{k-1}$ are algebraically open, that is, $\text{ai}(C_i) = C_i$, $i = 1, \ldots, k-1$.*

The following conditions are equivalent:

(a) $\cap_1^k C_i = \emptyset$.
(b) There exist linear functionals $\{\ell_i\}_1^k$ on E, not all identically zero, and scalars $\{\alpha_i\}_1^k$ such that

$$\ell_i(x_i) \leq \alpha_i \;\; \textit{for all} \;\; x_i \in C_i,\; i = 1, \ldots, k, \quad \sum_1^k \ell_i = 0, \quad \sum_1^k \alpha_i \leq 0.$$

Proof. We first prove that (a) implies (b). Define the sets

$$K_1 := \{(x_k, x_k, \ldots, x_k) : x_k \in C_k\},$$
$$K_2 := C_1 \times \cdots \times C_{k-1}.$$

We have, by elementary arguments,

$$\text{ai}\, K_2 = \text{ai}\, C_1 \times \cdots \times \text{ai}\, C_{k-1} = C_1 \times \cdots \times C_{k-1} = K_2,$$

so that K_2 is algebraically open. Clearly, $K_1 \cap K_2 = \emptyset$, so by Theorem 6.32 there exists $(\ell, \alpha) = ((\ell_1, \ldots, \ell_{k-1}), \alpha)$, $\ell \neq 0$, such that

$$\ell_1(x_1) + \cdots + \ell_{k-1}(x_{k-1}) < \alpha \le \Big(\sum_1^{k-1} \ell_i\Big)(x_k) \quad \text{for all } x_i \in C_i,\ i = 1, \ldots, k.$$

Define $\alpha_i := \sup_{x_i \in C_i} \ell_i(x_i)$ for $i = 1, \ldots, k-1$, so that $\alpha_1 + \cdots + \alpha_{k-1} \le \alpha$, and $\ell_k := -\sum_1^{k-1} \ell_i$, $\alpha_k := -\alpha$; this proves (b).

Conversely, let us assume (b) and prove (a). If (a) is false, then there exists a point $x \in \cap_1^k C_i \neq \emptyset$ that satisfies (b). On the one hand, we have

$$0 = \sum_1^k \ell_i(x) \le \sum_1^k \alpha_i \le 0,$$

so that $\ell_i(x) = \alpha_i$ for all $i = 1, \ldots, k$, and $\sum_1^k \alpha_i = 0$. On the other hand, if $z \in E$ is an arbitrary point, then there exists $\epsilon > 0$ such that $[x - \epsilon z, x + \epsilon z] \subseteq C_i$ for $i = 1, \ldots, k-1$, because C_i is algebraically open. This gives $\ell_i(x \mp \epsilon z) \le \alpha_i$, so that

$$\alpha_i \mp \epsilon \ell_i(z) = \ell_i(x) \mp \epsilon \ell_i(z) = \ell_i(x \mp \epsilon z) \le \alpha_i,$$

which means that $\ell_i(z) = 0$, that is, the linear functions $\{\ell_i\}_1^{k-1}$ are identically zero, and hence all the $\{\ell_i\}_1^k$ are identically zero, which contradicts (b). This proves that (a) must be true. □

The topological version of Theorem 6.38 is now easy to establish. The following proof should serve as a model for obtaining a topological separation theorem from an algebraic one.

Theorem 6.39. ***(Dubovitskii–Milyutin)*** *Let $\{C_i\}_1^k$, $k > 1$, be nonempty convex sets in a topological vector space E, such that $\{C_i\}_1^{k-1}$ are open, that is, $\operatorname{int}(C_i) = C_i$, $i = 1, \ldots, k-1$.*

The following conditions are equivalent:

(a) $\cap_1^k C_i = \emptyset$.
(b) There exist continuous linear functionals $\{\ell_i\}_1^k$ on E, not all identically zero, and scalars $\{\alpha_i\}_1^k$ such that

$$\ell_i(x_i) \le \alpha_i \quad \text{for all } x_i \in C_i,\ i = 1, \ldots, k, \quad \sum_1^k \ell_i = 0, \quad \sum_1^k \alpha_i \le 0.$$

Proof. By virtue of Theorem 5.20, $\operatorname{int}(C_i) = \operatorname{ai}(C_i)$ for $i = 1, \ldots, k-1$, so it follows from Theorem 6.38 that we need to prove only that if (a) is true, then the linear functionals $\{\ell_i\}_1^k$ in (b) are continuous; in fact, since $\ell_k = -\sum_1^{k-1} \ell_i$, it suffices to prove the continuity of $\ell := (\ell_1, \ldots, \ell_{k-1})$.

Suppose that (a) holds. As shown in the proof of Theorem 6.38, there is a hyperplane $H := H_{(\ell,\alpha)}$ such that the open convex set $C_1 \times \cdots \times C_{k-1}$ lies in the algebraically open half-space H^-; ℓ is continuous by Corollary 6.28. □

6.8 Hahn–Banach Theorem

The Hahn–Banach theorem is a cornerstone of functional analysis. Its proof is given in every book in functional analysis, following almost verbatim its original proof given by Banach [20]. Earlier results of Helly [129] and Hahn [123] were important in the development of the Hahn–Banach theorem. The article [52] gives a survey of the Hahn–Banach theorem and related results and includes 351 references.

It will be apparent in this section that the Hahn–Banach theorem is intimately related to Theorem 6.34; it is in fact an analytic formulation of it.

Here is an algebraic version of the Hahn–Banach theorem, which we prove using Theorem 6.34.

Theorem 6.40. ***(Hahn–Banach theorem)*** *Let E be a vector space, and $p : E \to \mathbb{R}$ a sublinear functional,*

$$p(x+y) \le p(x) + p(y), \quad p(\lambda x) = \lambda p(x) \quad \text{for all} \quad x, y \in E,\ \lambda \ge 0.$$

If $L \subset E$ is a linear subspace, and $g : L \to \mathbb{R}$ is a linear functional majorized by p, that is,

$$g(y) \le p(y) \quad \text{for all} \quad y \in L,$$

then there exists a linear functional $f : E \to \mathbb{R}$ that extends g and that is majorized by p, that is,

$$f(y) = g(y) \ \text{for all}\ y \in L \quad \text{and} \quad f(x) \le p(x) \ \text{for all}\ x \in E.$$

Moreover, if E is a real topological vector space and p is a continuous function, then f is also continuous.

Proof. Define the following sets in the vector space $E \times \mathbb{R}$:

$$\begin{aligned} A &:= \operatorname{epi}(p) = \{(x,\alpha) \in E \times \mathbb{R} : p(x) \le \alpha\}, \\ B &:= \operatorname{gr}(g) \ = \{(x,\alpha) \in L \times \mathbb{R} : g(x) = \alpha\}. \end{aligned}$$

Since p is a convex function and g is a linear functional, A is a convex set (in fact a convex cone) and B is a linear subspace of $E \times \mathbb{R}$. By Lemma 5.13,

$$\operatorname{rai}(A) = \{(x,\alpha) \in E \times R : p(x) < \alpha\}.$$

We have $\operatorname{rai}(A) \cap B = \emptyset$, because if $(y,\alpha) \in L \times \mathbb{R}$ is in the intersection set, then we have the contradiction $p(y) < \alpha = g(y)$.

It follows from Theorem 6.34 that there exists a hyperplane H extending B, and disjoint from $\operatorname{rai}(A)$. Let

$$H := H_{(\ell,m,0)} = \{(x,\alpha) : \ell(x) + m\alpha = 0\},$$

where the nonzero linear functional (ℓ, m) on $E \times \mathbb{R}$ is given by

$$(\ell, m)(x, \alpha) := \ell(x) + m\alpha,$$

and suppose that $A \subseteq \bar{H}^-$. Then

$$\begin{aligned} \ell(y) + mg(y) &= 0 \quad \text{for all} \quad y \in L, \\ \ell(x) + m\alpha &< 0 \quad \text{for all} \quad (x, \alpha) \in E \times \mathbb{R}, \;\; p(x) < \alpha. \end{aligned} \tag{6.12}$$

We must have $m \leq 0$, since otherwise we get a contradiction in the inequality above by letting $\alpha \to \infty$. If $m = 0$, then $\ell(x) < 0$ for all $x \in E$, which is impossible, since $\ell(0) = 0$. Thus, $m < 0$, and we may assume that $m = -1$; then the linear functional $f : E \to \mathbb{R}$ defined by

$$f(x) := \ell(x)$$

satisfies $f(y) = g(y)$ for $y \in L$, that is, f extends g. Finally, the inequality in (6.12) gives $f(x) < \alpha$ for all $\alpha > p(x)$; it follows that $f(x) \leq p(x)$ for all $x \in E$, meaning that p dominates f.

If E is a real topological vector space and p is continuous, then $f(x) \leq p(x)$ implies that

$$\{x \in E : p(x) < \alpha\} \subseteq \{x \in E : f(x) < \alpha\} = H^-_{(f,\alpha)},$$

so that $H^-_{(f,\alpha)}$ contains an open set; Corollary 6.28 implies that f is continuous. □

Next, we give an independent, analytic proof of a general algebraic form of the Hahn–Banach theorem, which is in the spirit of the original proof by Banach.

Theorem 6.41. (***Extended Hahn–Banach theorem***) *Let E be a vector space, $p : E \to \mathbb{R} \cup \{\infty\}$ a convex function, $L \subset E$ a linear subspace, and $g : L \to \mathbb{R}$ is a linear functional.*

If

$$g(y) \leq p(y) \quad \text{for all} \quad y \in L \quad \text{and} \quad \operatorname{rai}(\operatorname{dom} p) \cap L \neq \emptyset,$$

then there exists a linear function $f : E \to \mathbb{R}$ that extends g and that is majorized by p,

$$f(y) = g(y) \;\; \text{for all} \;\; y \in L \quad \text{and} \quad f(x) \leq p(x) \;\; \text{for all} \;\; x \in E.$$

Proof. Fix a point $x \in E \setminus L$. For $u, v \in L$ and scalars $\alpha > 0$, $\beta < 0$, we have

$$\begin{aligned} \alpha g(u) - \beta g(v) &= g(\alpha u - \beta v) \\ &= (\alpha - \beta) g\Big(\frac{\alpha}{\alpha - \beta} u - \frac{\beta}{\alpha - \beta} v\Big) \\ &\leq (\alpha - \beta) p\Big(\frac{\alpha}{\alpha - \beta} u - \frac{\beta}{\alpha - \beta} v\Big) \\ &= (\alpha - \beta) p\Big(\frac{\alpha}{\alpha - \beta}(u + x/\alpha) - \frac{\beta}{\alpha - \beta}(v + x/\beta)\Big) \\ &\leq \alpha p(u + x/\alpha) - \beta p(v + x/\beta). \end{aligned}$$

Consequently,

$$\beta[p(v+x/\beta)-g(v)] \le \alpha[p(u+x/\alpha)-g(u)] \quad \forall\, u,v\in L,\ \alpha>0,\ \beta<0,$$

so that

$$-\infty < \sup_{v\in L,\,\beta<0} \beta[p(v+x/\beta)-g(v)] \le \inf_{u\in L,\,\alpha>0} \alpha[p(u+x/\alpha)-g(u)].$$

We claim that there exists $c\in\mathbb{R}$ such that

$$\beta \sup_{v\in L,\,\beta<0} [p(v+x/\beta)-g(v)] \le c \le \inf_{u\in L,\,\alpha>0} \alpha[p(u+x/\alpha)-g(u)]. \qquad (6.13)$$

Clearly, the claim holds unless both the infimum and supremum above are equal to $+\infty$, in which case picking $u\in \operatorname{rai}(\operatorname{dom} p)\cap L$, we have $p(u+\lambda_0 x)\in\mathbb{R}$ for some $\lambda_0<0$ but $p(u+\lambda x)=\infty$ for all $\lambda>0$, in contradiction to the fact that $u\in\operatorname{rai}(\operatorname{dom} p)$.

Define $f(x):=c$, where c is given in (6.13); this determines f on all of the linear subspace $M=\operatorname{span}\{L,x\}$: any $w\in M$ has the form $w=u+\lambda x$ for some $\lambda\in\mathbb{R}$, and

$$f(w)=f(u+\lambda x)=f(u)+\lambda f(x)=g(u)+\lambda c.$$

If $\lambda<0$, (6.13) implies $[p(u+\lambda x)-g(u)]/\lambda\le c$, and if $\lambda>0$, $[p(u+\lambda x)-g(u)]/\lambda\ge c$; in either case, we have

$$f(u+\lambda x)=g(u)+\lambda c\le p(u+\lambda x),$$

which proves that the linear functional f can be extended from the linear subspace L to a linear subspace M strictly containing L in such a way that f is still dominated by p on M.

Consider the set $\mathcal{A}=\{(f,W)\}$, where W is a linear subspace of E containing L and $f:W\to\mathbb{R}$ is a linear functional that agrees with g on L and is dominated by p on W, and define the relation $\preceq$ on $\mathcal{A}$ by declaring $(f_1,W_1)\preceq(f_2,W_2)$ if $W_1\subseteq W_2$ and g_2 is an extension of f_1 to W_2. Clearly, $\preceq$ is a partial order, and it is easy to see that if $\{(f_\alpha,W_\alpha)\}$ is a chain in the partial order, then the pair (f,W), where $W=\cup W_\alpha$ and f agrees with f_α on W_α, is an upper bound to the chain. It follows from Zorn's lemma that the partial order has a maximal element (f,W). We must have $W=E$, because otherwise g can be extended from W to a strictly larger subspace as shown above, contradicting the maximality of W. □

A proof of Theorem 6.41 can be given that employs a separation argument; see Exercise 13.

To complete the proof of the equivalence of Theorems 6.41 and 6.34, we now deduce the latter theorem from the former.

Corollary 6.42. *Let C be a nonempty convex set in a vector space E, such that* $\operatorname{rai}(C) \neq \emptyset$. *If $M \subset E$ is an affine set satisfying* $\operatorname{rai}(C) \cap M = \emptyset$, *then there exists a hyperplane $H \supseteq M$ extending M such that* $\operatorname{rai}(C) \cap H = \emptyset$.

Proof. We assume without any loss of generality that $0 \in \operatorname{rai}(C)$. Define the linear subspace $L := \operatorname{span} M$; then M is a hyperplane in L given by the formula $M = \{x \in L : g(x) = 1\}$, where g is a linear functional on L. Recall that the Minkowski function $p_C : E \to \mathbb{R} \cup \{\infty\}$ is a convex function satisfying

$$\operatorname{dom} p_C = \operatorname{aff}(C) = \operatorname{span} C$$

and

$$\operatorname{rai}(C) = \{x \in E : p_C(x) < 1\}.$$

Since $\operatorname{rai}(C)$ and M are disjoint, we have $g(y) \leq p_C(y)$ for all $y \in M$.

Let $x = ty$ be an arbitrary point in L, where $y \in M$ and $t \in \mathbb{R}$. If $t > 0$, then by the homogeneity of g and p_C, we have $g(ty) = tg(y) \leq tp_C(y) = p_C(ty)$, so that $g(x) \leq p_C(x)$, and if $t \leq 0$, then $g(x) = tg(y) \leq 0 \leq p_C(x)$; therefore

$$g(x) \leq p_C(x) \quad \text{for all} \quad x \in L.$$

We have

$$0 \in \operatorname{rai}(\operatorname{dom} p_C) \cap L = \operatorname{span}(C) \cap P \neq \emptyset,$$

and Theorem 6.41 implies that there exists a linear functional $f : E \to \mathbb{R}$ satisfying

$$f(x) \leq p_C(x) \quad \text{for all} \quad x \in E.$$

The hyperplane $H = \{x \in E : f(x) = 1\}$ is clearly an extension of M, and we have $\operatorname{rai} C \cap H = \{x \in E : p_C(x) < 1 = f(x)\} = \emptyset$. □

6.9 Exercises

1. Let $f : \mathbb{R}^n \to \mathbb{R}$ be a differentiable convex function, $C \subset \mathbb{R}^n$ a closed convex set, and $\alpha \geq 0$. Prove that $x^* \in C$ solves the problem $\min_{x \in C} f(x)$ if and only if $x^* = \Pi_C(x^* - \alpha \nabla f(x^*))$.
2. Let $C \subseteq \mathbb{R}^n$ be a closed convex set, $x \in \mathbb{R}^n \setminus C$, $\Pi_C(x)$ the projection of x onto C, and $d_C(x) = \|x - \Pi_C(x)\|$ the distance from x to C.
   (a) Show that if D is a closed convex set containing C, then $d_D(x) \leq d_C(x)$. Conclude that if the half-space
   $$H = \bar{H}^-_{(d,\alpha)} := \{z \in \mathbb{R}^n : \langle d, z \rangle \leq \alpha\}, \quad d \neq 0,$$
   contains C, then $d_H(x) \leq d_C(x)$.
   (b) Consider the particular half-space $H = \bar{H}^-_{(d,\alpha)}$, where $d = x - \Pi_C(x)$ and $\alpha = \langle d, \Pi_C(x) \rangle$. Show that the variational inequality characterization of $\Pi_C(x)$ implies that
   $$(i)\ C \subseteq H, \qquad (ii)\ d_C(x) = d_H(x).$$

(c) Show that these results lead to the following geometrically appealing "duality" result: *the minimum distance from a point x to a convex set C not containing x is equal to the maximum among the distances from x to the closed half-spaces containing C.*

3. Let C, D be nonempty convex sets in $\mathbb{R}^n$ such that a hyperplane H properly separates C and D, $C \subseteq \bar{H}^+$, but $C \not\subseteq H$. Show that $\mathrm{ri}(C) \subseteq H^+$.
Hint: Pick $x_0 \in C \setminus H$. Show that the assumption $x \in \mathrm{ri}(C) \cap H$ leads to a contradiction.
4. Let $f : \mathbb{R}^n \to \mathbb{R}$ be a concave function such that $f(0) = 0$ and
$$C = \{x \in \mathbb{R}^n : f(x) > 0\} \neq \emptyset.$$
Let $\langle l, x\rangle$ be a linear functional on $\mathbb{R}^n$. Show that the following statements are equivalent:
(a) $f(x) > 0 \quad$ implies $\quad \langle l, x\rangle > 0$,
(b) $\exists \lambda > 0, \;\; \langle l, x\rangle \geq \lambda f(x)$ for all $x \in \mathbb{R}^n$.
Moreover, show that equality holds above if f is a linear functional.
Hint: Define
$$D = \{(\langle l, x\rangle, f(x) - t) : x \in \mathbb{R}^n, t \geq 0\} \subseteq \mathbb{R}^2;$$
show that D is a convex set, and that $0 \notin \mathrm{ri}(D)$ when (a) holds; then use an appropriate separation argument.
5. Let $\{K_i\}_1^m$ be closed cones in $\mathbb{R}^n$ such that the conical hull of their union is not closed, that is, $K := \mathrm{cone}(\cup_1^m K_i)$ is not closed. Show that there exist vectors $x_i \in K_i$, not all zero, such that
$$x_1 + x_2 + \cdots + x_m = 0.$$
Hint: Pick $\overline{x} \in \overline{K} \setminus K$, and let $y^k \in K$ converge to $\overline{x}$. Write $y^k = y_1^k + \cdots + y_m^k$, where $y_i^k \in K_i$, and define $a_k := \max\{\|y_1^k\|, \ldots, \|y_m^k\|\}$. Argue that $a_k \to \infty$. Finally, consider the convergence behavior of the sequence $x^k := y^k / a_k$ as $k \to \infty$.
6. (***Stiemke's theorem***) Prove that the system
$$\sum_{i=1}^{m} x_i a_i = 0, \; x_i > 0, \; i = 1, \ldots, m,$$
has no solution if and only if the system
$$\langle a_i, y\rangle \leq 0, \; i = 1, \ldots, m, \quad \text{not all zero},$$
has a solution.
Hint: Use the finite-dimensional version of the Dubovitskii–Milyutin theorem.

7. Let A, B, C be compact convex sets in $\mathbb{R}^n$ such that $A + C = B + C$. The purpose of this problem is to prove that $A = B$. Define the support function
$$\sigma_A(x) := \max\{\langle x, u\rangle : u \in A\}.$$
   (a) Show that σ_A is a convex function defined for all $x \in \mathbb{R}^n$.
   (b) Show that
$$\sigma_{A+B} = \sigma_A + \sigma_B.$$
   (c) Show that if $F, G \subseteq E$ are compact convex sets such that $\sigma_F = \sigma_G$, then $F = G$ (this requires a separation argument).
   (d) Prove that $A = B$.

   How far can we relax the assumptions that the sets A, B, C are compact? Can we remove the compactness hypothesis altogether?
8. Let $C \subseteq \mathbb{R}^n$ be a compact convex set and assume that $0 \in \operatorname{int}(C)$. Define the polar body
$$C^\circ := \{y \in \mathbb{R}^n : \langle x, y\rangle \leq 1 \text{ for all } x \in C\}.$$
   (a) Show that $\overline{B}_r(0)^\circ = \overline{B}_{1/r}(0)$.
   (b) Show that if $C_1 \subseteq C_2$, then $C_2^\circ \subseteq C_1^\circ$.
   (c) Show that C° is a compact, convex body with $0 \in \operatorname{int}(C^\circ)$.
   (d) Show that $(C^\circ)^\circ = C$.
   Hint: This will involve a separation argument.
   (e) Show that the polar body of the unit cube $\{x \in \mathbb{R}^n : |x_i| \leq 1, i = 1, \ldots, n\}$ is the *cross polytope* $\{x \in \mathbb{R}^n : \sum_{i=1}^n |x_i| \leq 1\}$.
9. Let K be a nonempty convex cone in $\mathbb{R}^n$.
   (a) Show that $\operatorname{ri}(K)$ and $-\operatorname{ri}(K^*)$ must intersect, that is, $\operatorname{ri}(K) \cap (-\operatorname{ri}(K^*)) \neq \emptyset$.
   Hint: use a separation argument.
   (b) Show that $K \cap (-K^*) = \{0\}$ if and only if K (hence K^*) is a linear subspace of $\mathbb{R}^n$.
   Hint: to prove the harder part of the statement, show that if $K \cap (-K^*) = \{0\}$, then the origin lies in $\operatorname{ri}(K)$. Consequently, prove that if x is in K, then so is $-x$.
10. Let ℓ be a linear functional on a topological vector space. Show that if ℓ is nonnegative on an open set, then ℓ is continuous.
11. ***(Krein's theorem)*** Let K be a convex cone in a topological vector space E, containing interior points. Let $L \subset E$ be a linear subspace such that $L \cap \operatorname{int}(K) \neq \emptyset$, and $\overline{f} : L \to \mathbb{R}$ a linear functional that is nonnegative on $L \cap K$, that is, $\overline{f}(x) \geq 0$ for all $x \in L \cap K$.
   Krein's theorem states that *there exists a continuous linear functional $f : E \to \mathbb{R}$ extending $\overline{f}$ such that f is nonnegative on K.*

   Prove the theorem by completing the following steps:

(a) Define $M := \{x \in L : \overline{f}(x) = 0\}$. If $M = L$, show that $f = 0$ satisfies the requirements of Krein's theorem. Assume that $M \neq L$, that is, $\overline{f}$ is not identically zero on L.
(b) Show that $\text{int}(K) \cap M = \emptyset$.
(c) Use a separation argument to prove that there exists a hyperplane $H \subset E$ extending M such that $\text{int}(K) \cap H = \emptyset$.
(d) Let $g : E \to \mathbb{R}$ be a linear functional such that $H = \{x \in E : g(x) = 0\}$. Show that g is either positive or negative on $\text{int}(K)$. Assume the first possibility.
(e) Show that g is a continuous linear functional.
(f) Notice that $\{x \in L : \overline{f}(x) = 0\} \subseteq \{x \in L : g(x) = 0\}$ (because the first set is M and g extends $\overline{f}$). Show that the two sets are either equal or g is identically zero on L (recall that M is a hyperplane in L). Show that the second possibility is impossible. Deduce that there exists λ such that $f(x) = \lambda g(x)$ for all $x \in L$. Show that $\lambda > 0$.
(g) Show that the functional $f := \lambda g$ on E satisfies the required properties.

12. Prove Lemma 6.36 using Theorem 6.38. (Notice that the proof of the theorem does not depend on the lemma, so there is no circular reasoning involved.)
Is it possible to prove the results of Appendix A (Theorem A.3 and its affine version) using Theorem 6.38?
13. Give a separation proof of Theorem 6.41 by mimicking the proof of Theorem 6.40.

7

Convex Polyhedra

In this chapter, we develop the basic results of the theory of convex polyhedra. This is a large area of research that has been studied from many different points of view. Within optimization, it is very important in linear programming, especially in connection with the simplex method for solving linear programs. The choice of the topics we treat in this chapter is dictated mostly by the needs of optimization. However, we do not have space to treat the extensive body of work concerning the combinatorial theory of convex polyhedra, some of which is intimately related to the simplex method and its variants. The interested reader may consult the books [115, 50, 274] for more information on this topic and the book [5] for differential-geometric questions regarding convex polyhedra.

In this chapter, E will be a finite-dimensional vector space.

7.1 Convex Polyhedral Sets and Cones

Definition 7.1. *Let $\{a_j\}_{j=1}^k$ be a given set of vectors in E. A cone $K \subseteq E$ is called a* convex polyhedral cone *if it has the form*

$$K = \{x : \langle a_j, x\rangle \le 0, \quad j = 1, \ldots, k\} = \cap_{j=1}^k \bar{H}^-_{a_j,0},$$

that is, a polyhedral cone is the intersection of finitely many half-spaces passing through the origin.

A convex cone K is called finitely generated *if it has the form*

$$K := \Big\{\sum_{j=1}^k ta_j : t_j \ge 0, j = 1, \ldots, k\Big\},$$

that is, a finitely generated cone is a finite sum of rays, $K = \mathbb{R}_+ a_1 + \cdots + \mathbb{R}_+ a_k$.

We first prove an important result on finitely generated cones.

O. Güler, *Foundations of Optimization*, Graduate Texts in Mathematics 258,
DOI 10.1007/978-0-387-68407-9_7, © Springer Science+Business Media, LLC 2010

Lemma 7.2. *A finitely generated cone is a closed set.*

Proof. Let K be a finitely generated cone:

$$K = \Big\{\sum_{j=1}^{k} t_j a_j : t_j \geq 0, j = 1, \ldots, k\Big\}.$$

By Carathéodory's theorem (Theorem 4.21, p. 94), any point $x \in K$ can be written as

$$x = \sum_{j=1}^{k} \delta_j b_j, \quad \delta_j \geq 0,$$

where $\{b_j\}_1^p$ is a linearly independent subset of $\{a_i\}_1^k$. It follows that $x \in \{\sum_1^p \delta_j b_j : \delta_j \geq 0\}$, a simplical cone that is the image of the nonnegative orthant $\mathbb{R}_+^p$ under the linear map

$$T(\delta) = \sum_{j=1}^{k} \delta_j b_j.$$

Since $\{b_j\}_1^p$ is linearly independent, T is a homeomorphism, and since $\mathbb{R}_+^p$ is closed, so is the simplical cone. The cone K is a union of such simplical cones, which are finitely many in number, so must be closed. □

It follows from this lemma that $\{\sum_1^k \delta_i a_i : \delta_i \geq 0\} = \text{cl cone}(a_1, \ldots, a_k)$. We will denote this set by $\overline{\text{cone}}(a_1, \ldots, a_k)$; thus

$$\overline{\text{cone}}(a_1, \ldots, a_k) := \text{cl cone}(a_1, \ldots, a_k).$$

Lemma 7.3. *The dual of finitely generated cone $K = \overline{\text{cone}}(a_1, \ldots, a_k)$ is the polyhedral cone $L = \cap_{j=1}^k \{x : \langle a_j, x\rangle \leq 0\}$.*

Proof. Clearly, we have

$$K^* = \left\{x : \Big\langle \sum_{j=1}^{k} t_j a_j, x\Big\rangle \leq 0 \ \text{ for all } \ t_j \geq 0\right\} \supseteq \cap_1^k \{x : \langle a_j, x\rangle \leq 0\}.$$

If $x \in K^*$, choosing $t_j = 1$ and all other $t_i = 0$ implies $\langle a_j, x\rangle \leq 0$, proving that $K^* \subseteq L$. □

7.1.1 Convex Polyhedral Cones

Theorem 7.4. *Let $a_1, \ldots, a_k \in E$. The finitely generated cone*

$$K = \overline{\text{cone}}(a_1, \ldots, a_k)$$

and the polyhedral cone

$$L = \{x : \langle a_j, x\rangle \leq 0, j = 1, \ldots, k\}$$

are polars of each other, that is, $K^* = L$ and $L^* = K$.

Proof. It follows from Lemma 7.3 that $K^* = L$. Since K is closed by Lemma 7.2, Theorem 6.19 implies that $K = (K^*)^* = L^*$. □

If E is endowed with a basis, the above theorem takes the following form.

Corollary 7.5. *Let A be an $n \times k$ matrix. Then the cones $K = \{Av : v \geq 0\}$ and $L = \{x : A^T x \leq 0\}$ are polars of each other.*

Proof. Let $A = [a_1 \ldots, a_k]$, where $\{a_i\}$ are the columns of A. Then $K = \overline{\text{cone}}(a_1, \ldots, a_k)$ and $L = \{x : \langle a_j, x\rangle \leq 0, j = 1, \ldots, k\}$. □

Theorem 7.6. (*Farkas's lemma, homogeneous version*) *Let $a_1, \ldots, a_k$ be given vectors in E. The following statements are equivalent:*

(a) If $x \in E$ satisfies the inequalities $\langle a_i, x\rangle \leq 0$, $i = 1, \ldots, k$, then it also satisfies the inequality $\langle b, x\rangle \leq 0$. In other words,

$$[\langle a_i, x\rangle \leq 0, i = 1, \ldots, k] \quad \Longrightarrow \quad [\langle b, x\rangle \leq 0]\,.$$

(b) The vector b is a nonnegative linear combination of $\{a_i\}_1^k$, that is, $b = \sum_{i=1}^k t_i a_i$ for some $t_i \geq 0$, $i = 1, \ldots, k$.

Proof. This is essentially a restatement of Theorem 7.4. Define

$$K = \{x : \langle a_i, x\rangle \leq 0, i = 1, \ldots, k\}.$$

Part (a) is equivalent to the statement, $b \in K^*$, whereas part (b) states that $b \in \overline{\text{cone}}(a_1, \ldots, a_k)$. We have $K^* = \overline{\text{cone}}(a_1, \ldots, a_k)$ by Theorem 7.4. □

The general, affine version of Farkas's lemma is given in Theorem 7.20 on page 185.

Corollary 7.7. *Let $c_1, \ldots, c_k$, $a_1, \ldots, a_l$ be given vectors in E. The following statements are equivalent:*

(a)

$$[\langle c_i, x\rangle = 0, i = 1, \ldots, k,\ \langle a_j, x\rangle \leq 0, j = 1, \ldots, l] \quad \Longrightarrow \quad [\langle b, x\rangle \leq 0]\,.$$

(b) There exist $t_i \in \mathbb{R}$ $(i = 1, \ldots, k)$ and $s_j \geq 0$ $(j = 1, \ldots, l)$ such that

$$b = \sum_{i=1}^k t_i c_i + \sum_{j=1}^l s_j a_j.$$

Proof. The equality $\langle c_i, x\rangle = 0$ is equivalent to the inequalities $\langle c_i, x\rangle \leq 0$ and $\langle -c_i, x\rangle \leq 0$. By Farkas's lemma, part (a) is equivalent to

$$b \in \overline{\text{cone}}(c_1, \ldots, -c_1, \ldots, -c_k, a_1, \ldots, a_l) := L.$$

An arbitrary element of $x \in L$ can be written as $x = \sum_{i=1}^k (\alpha_i - \beta_i)c_i + \sum_{j=1}^l s_j a_j$ with $\alpha_i \geq 0$, $\beta_i \geq 0$ $(i = 1, \ldots, k)$, and $s_j \geq 0$ $(j = 1, \ldots, l)$. Since $t_i : \alpha_i - \beta_i$ can be any real number, we see that parts (a) and (b) are equivalent. □

The following important result establishes the *equivalence of finitely generated and polyhedral cones.*

Theorem 7.8. *Every finitely generated cone K is a convex polyhedral cone, and vice versa.*

Proof. We first show that every finitely generated cone K is a polyhedral cone. Let $K = \overline{\text{cone}}(a_1, \ldots, a_k) \subseteq E$ be a finitely generated cone. We claim that K is a polyhedral cone using induction on k. If $k = 1$, then $K = \{ta : t \geq 0\}$. If $a = 0 \in E$, then

$$K = \{0\} = \{x : x_i = 0, i = 1, \ldots, n\} = \{x : \langle e_i, x\rangle = 0, i = 1, \ldots, n\},$$

where $\{e_i\}_1^n$ is a basis of E, and each equation $\langle e_i, x\rangle = 0$ can be written as two inequalities $\langle e_i, x\rangle \leq 0$ and $\langle -e_i, x\rangle \leq 0$. This proves that K is a polyhedral cone. If $0 \neq a \in E$, then K is a half-line. In this case, pick a basis $\{e_i\}_1^n$ of E such that $e_1 = a$ and $\{e_i\}_2^n$ is a basis of $\{a\}^\perp$. Then we can write K in the form

$$K = \{x : \langle -e_1, x\rangle \leq 0, \langle e_i, x\rangle = 0, i = 2, \ldots, n\},$$

proving that K is again a polyhedral cone.

Supposing that the claim is proved for $k - 1$, we will prove it for k. Let $K = \overline{\text{cone}}(a, a_1, \ldots, a_{k-1})$. Define $K_1 := \overline{\text{cone}}(a_1, \ldots, a_{k-1})$. Any $x \in K$ can be written as $x = y + ta$, where $y \in K$ and $t \geq 0$. This means that

$$K = \{x \in E : \exists t \geq 0, x - ta \in K_1\}.$$

By the induction hypothesis, there exist $\{b_j\}_1^m$ such that

$$K_1 = \{x \in E : \langle b_j, x\rangle \leq 0, j = 1, \ldots, m\}.$$

Consequently, we have

$$\begin{aligned} K &= \{x \in E : \exists t \geq 0, \langle b_j, x - ta\rangle \leq 0, j = 1, \ldots, m\} \\ &= \{x \in E : \exists t \geq 0, \langle b_j, x\rangle \leq t\langle b_j, a\rangle, j = 1, \ldots, m\}. \end{aligned} \tag{7.1}$$

We will write K as a polyhedral cone by "eliminating" the variable t in these inequalities. Define the index sets $I^+ := \{j : \langle b_j, a\rangle > 0\}$, $I^- := \{j : \langle b_j, a\rangle < 0\}$, and $I^0 := \{j : \langle b_j, a\rangle = 0\}$. The conditions in (7.1) can then be written as $\langle b_i, x\rangle / \langle b_i, a\rangle \leq t$ for $i \in I^+$, $\langle b_j, x\rangle / \langle b_j, a\rangle \geq t$ for $j \in I^-$, and $\langle b_l, x\rangle \leq 0$ for $j \in I^0$. Therefore,

$$K = \left\{x \in E : \exists t \geq 0, \frac{\langle b_i, x\rangle}{\langle b_i, a\rangle} \leq t \leq \frac{\langle b_j, x\rangle}{\langle b_j, a\rangle}, \langle b_l, x\rangle \leq 0, i \in I^+, j \in I^-, l \in I^0\right\}.$$

Clearly, a variable $t \geq 0$ exists above if and only if

$$\max_{i \in I^+} \frac{\langle b_i, x\rangle}{\langle b_i, a\rangle} \leq \min_{j \in I^-} \frac{\langle b_j, x\rangle}{\langle b_j, a\rangle}, \quad \min_{j \in I^-} \frac{\langle b_j, x\rangle}{\langle b_j, a\rangle} \geq 0.$$

Since $\langle b_j, a\rangle < 0$ for $j \in I^-$, the second inequality above is equivalent to the condition that $\langle b_j, a\rangle \le 0$ for all $j \in I^-$. It follows that

$$K = \left\{ x \in E : \langle b_l, x\rangle \le 0, \frac{\langle b_i, x\rangle}{\langle b_i, a\rangle} \le \frac{\langle b_j, x\rangle}{\langle b_j, a\rangle},\ l \in I^- \cup I^0, i \in I^+, j \in I^- \right\},$$

which proves that K is a polyhedral cone.

Conversely, suppose that K is a polyhedral cone. It follows from Theorem 7.4 that K^* is finitely generated. The above argument shows that K^* is polyhedral. Theorem 7.4 again implies that $K = (K^*)^*$ is finitely generated. □

Remark 7.9. The method of elimination of the variable t from (7.1) is called the *Fourier–Motzkin elimination method*. It is a powerful tool that can be used to derive most theoretical results for systems of linear equalities and inequalities, including the derivation of (various forms of) Farkas's lemma; see [180]. The elimination method can also be used to solve systems of linear equalities and inequalities numerically. However, it is a very inefficient tool in this respect, since the elimination of a single variable typically leads to the creation of many additional equations and inequalities.

Remark 7.10. A more general version of the elimination of variables idea applies to systems of polynomial equations and inequalities, and goes by the name *Tarski–Seidenberg principle*; see [34]. This is an indispensable theoretical tool in real algebraic geometry. Unfortunately, the Tarski–Seidenberg principle is also a very inefficient computational tool for solving systems of polynomial equations and inequalities, for the same reasons.

Remark 7.11. Another elimination procedure is at work in multilinear algebra. Let V and W be vector spaces over $\mathbb{R}$ and consider bilinear maps $f(v, w)$ from $V \times W$ into an arbitrary vector space Z. Thus, f is a map that is linear in each of the variables separately, that is, $f(\alpha_1 v_1 + \alpha_2 v_2, w) = \alpha_1 f(u_1, w) + \alpha_2 f(u_2, w)$ and $f(v, \beta_1 w_1 + \beta_2 w_2) = \beta_1 f(u,, w_1) + \beta_2 f(u, w_2)$. It is well known in multilinear algebra that the condition

$$\begin{aligned} &\alpha_1 f(v_1, w_1) + \alpha_2 f(v_2, w_2) + \cdots + \alpha_n f(v_n, w_n) = 0 \\ &\text{for all bilinear maps } f : V \times W \to Z \end{aligned} \tag{7.2}$$

is equivalent to the condition that

$$\alpha_1(v_1 \otimes w_1) + \alpha_2(v_2 \otimes w_2) + \cdots + \alpha_n(v_n \otimes w_n) = 0.$$

Consequently, the elimination of the quantifier "for all f" in (7.2) leads to the concept of tensor products.

7.2 Convex Polyhedra

Definition 7.12. *A nonempty set $P \subseteq E$ is called a* convex polyhedron *if P is the intersection of finitely many closed half-spaces, that is,*

$$P = \{x \in E : \langle a_j, x\rangle \leq \alpha_j, j = 1, \dots, m\}, \tag{7.3}$$

where $\{a_j\}_1^m$ are given vectors in E and $\{\alpha_j\}_1^m$ are given scalars.

The polyhedron P is called a convex polytope *if P is a bounded set.*

We note that

$$P = \bigcap_{j=1}^{m} \{x : \langle a_j, x\rangle \leq \alpha_j\} = \bigcap_{j=1}^{m} \bar{H}^-_{a_j,\alpha_j}.$$

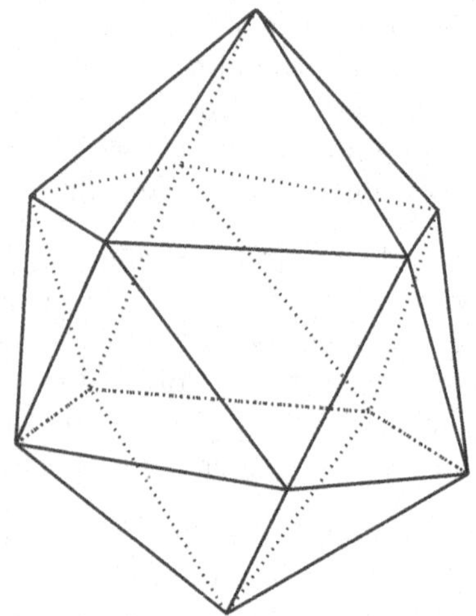

Fig. 7.1. A convex polytope (icosahedron).

Theorem 7.13 below, due to Minkowski [198] and Weyl [266, 267], is the *fundamental theorem of convex polyhedral theory,* giving a basic decomposition of a convex polyhedron in terms of vertices and directions.

Theorem 7.13. (***Minkowski–Weyl***) *A nonempty set $P \subseteq E$ is a convex polyhedron if and only if there exist vectors $\{v_i\}_1^k$ and $\{d_j\}_1^l$ such that*

$$\begin{aligned} P &= \mathrm{co}(v_1, \dots, v_k) + \overline{\mathrm{cone}}(d_1, \dots, d_l) \\ &= \Big\{ \sum_{i=1}^{k} \lambda_i v_i + \sum_{j=1}^{l} \mu_j d_j : \sum_{i=1}^{k} \lambda_i = 1, \lambda_i \geq 0, \mu_j \geq 0 \Big\}. \end{aligned} \tag{7.4}$$

Proof. Let P be a convex polyhedron, say in the form (7.3). We show that P has the form (7.4). Define the polyhedral cone

$$\begin{aligned} K &:= \{(x,t) \in E \times \mathbb{R} : \langle a_j, x\rangle \le \alpha_j t, t \ge 0, j = 1,\dots,m\} \\ &= \{(x,t) \in E \times \mathbb{R} : \langle a_j, x\rangle - \alpha_j t \le 0, t \ge 0, j = 1,\dots,m\}. \end{aligned}$$

Theorem 7.8 implies that K is a finitely generated cone, say in the form

$$K = \overline{\operatorname{cone}}\,\{(v_i,1),(d_j,0), i = 1,\dots,k, j = 1,\dots,l\}.$$

But $P = \{x : (x,1) \in K\}$, and a simple calculation shows that P has the form (7.4).

Conversely, let P have the form given in (7.4). Define the cone

$$\begin{aligned} K &:= \Big\{\sum_{i=1}^{k} \lambda_i (v_i,1) + \sum_{j=1}^{l} \mu_j (d_j,0) : \lambda_i \ge 0, \mu_j \ge 0, i = 1,\dots,k, j = 1,\dots,l.\Big\} \\ &= \overline{\operatorname{cone}}((v_i,1),(d_j,0), i = 1,\dots,k, j = 1,\dots,l). \end{aligned}$$

Theorem 7.8 implies that K is a polyhedral cone, say in the form

$$K = \{(x,t) : x \in E, t \ge 0, \langle a_j, x\rangle \le t\alpha_j, j = 1,\dots,m\}.$$

We have

$$P = \{x : (x,1) \in K\} = \{x \in E : \langle a_j, x\rangle \le \alpha_j, j = 1,\dots,m\}.$$

The theorem is proved. □

It should be possible to prove Theorem 7.13 using the techniques in Section 5.5. In particular, it should be possible to prove that a convex polyhedron $P = \{x : Ax \le b\}$ has finitely many vertices and finitely many extreme directions. The duality arguments as above would then prove the complete Theorem 7.13.

7.2.1 Homogenization of Convex Polyhedra

For completeness, we include here descriptions of $\overline{K(P)}$, the closure of the homogenization of a convex polyhedron P, when P is given either in the form (7.3) in terms of linear inequalities or in the form (7.4) in terms of vertices and directions. It should be noted that the cones $\overline{\operatorname{cone}}\, K(P)$ in Lemmas 7.14 and 7.15 are precisely the cones attached to the polyhedron P in the proof of Theorem 7.13. This extra information was not used for the sake of a simple proof.

Lemma 7.14. *If*

$$P = \{x \in E : \langle a_j, x\rangle \le \alpha_j, j = 1,\dots,m\}$$

is a nonempty polyhedron, then

$$\overline{K(P)} = \{(x,t) : x \in E, t \ge 0, \langle a_j, x\rangle \le t\alpha_j, j = 1,\dots,m\}.$$

Proof. We claim that

$$\operatorname{rec}(P) = \{d : \langle a_j, d\rangle \le 0, j = 1, \ldots, m\}.$$

Let $x_0 \in P$. If $d \in \operatorname{rec} P$, then $x_0 + td \in P$ for $t > 0$, that is, $\langle a_j, x_0 + td\rangle \le \alpha_j$. Dividing both sides by t and letting $t \to \infty$ proves that $\langle a_j, d\rangle \le 0$. Conversely, if this inequality is satisfied, then $x_0 + td \in P$, since

$$\langle a_j, x_0 + td\rangle \le \langle a_j, x_0\rangle + t\langle a_j, d\rangle \le \alpha_j.$$

Consequently, Lemma 5.41 (p. 134) implies that

$$\begin{aligned}\overline{K(P)} &= \{t(x,1) : x \in E, t > 0\} \cup \{(x,0) : x \in \operatorname{rec}(P)\}\\ &= \{(x,t) : \langle a_j, x\rangle \le t\alpha_j, t > 0, j = 1, \ldots, m\}\\ &\quad \cup \{(x,0) : \langle a_j, x\rangle \le 0, j = 1, \ldots, m\}\\ &= \{(x,t) : \langle a_j, x\rangle \le t\alpha_j, t \ge 0, j = 1, \ldots, m\}.\end{aligned}$$

□

Lemma 7.15. *Let*

$$\begin{aligned}P &= \operatorname{co}(v_1, \ldots, v_k) + \overline{\operatorname{cone}}(d_1, \ldots, d_l)\\ &= \Big\{\sum_{i=1}^k \lambda_i v_i + \sum_{j=1}^l \mu_j d_j : \sum_{i=1}^k \lambda_i = 1, \lambda_i \ge 0, \mu_j \ge 0\Big\}.\end{aligned}$$

Then

$$\overline{K(P)} = \overline{\operatorname{cone}}\,\{(v_i, 1), (d_j, 0), i = 1, \ldots, k, j = 1, \ldots, l\}. \tag{7.5}$$

Proof. We claim that $\operatorname{rec}(P) = \overline{\operatorname{cone}}\{d_j : j = 1, \ldots, l\} =: L$. Clearly, we have that each d_j belongs to $\operatorname{rec}(P)$, proving the inclusion $L \subseteq \operatorname{rec}(P)$. Conversely, if $d \in \operatorname{rec}(P)$, $t > 0$, and $x_0 \in P$, then $x_0 + td \in P$, so that

$$x_0 + td = w_t + \sum_{j=1}^l \mu_j(t) d_j,$$

where $w_t \in \operatorname{co}(v_1, \ldots, v_k)$. Now w_t is bounded, since $\|w_t\| \le \max\{\|v_i\| : 1 \le i \le k\}$. Also, $d + \frac{x_0 - w_t}{t}$ is in L and converges to d, and we have $d \in L$, since L is closed. This proves $\operatorname{rec}(P) \subseteq L$, and hence the claim.

By Lemma 5.41, we have

$$\begin{aligned}\overline{K(P)} &= \{t(x,1) : x \in P, t > 0\} \cup \{(x,0) : x \in \operatorname{rec}(P)\}\\ &= \Big\{(\sum_1^k \lambda_i v_i + \sum_1^l \mu_j d_j, t) : \sum_1^k \lambda_i = t > 0, \lambda_i \ge 0, \mu_j \ge 0\Big\}\\ &\quad \cup \Big\{(\sum_1^l \delta_j d_j, 0) : \delta_j \ge 0\Big\}\\ &= \overline{\operatorname{cone}}\,\{(v_i, 1), (d_j, 0), i = 1, \ldots, k, j = 1, \ldots, l.\}.\end{aligned}$$

□

7.3 Linear Inequalities

In this section, we prove the basic results on linear inequalities using separation theorems of convex analysis. These include *Gordan's lemma*, the affine version of Farkas's lemma, and the affine version of *Motzkin's transposition theorem*. These are important results in their own right; our interest in them stems from their utility in deriving optimality conditions in various constrained optimization problems. These results were proved in Section 3.3 using Ekeland's ϵ-variational principle. Another, elementary, approach to proving these results is given in Appendix A. Each approach is independent of the other two, which provides more insight into the subject and gives more flexibility in covering the material.

The central result on linear inequalities is the homogeneous version of Motzkin's transposition theorem, Theorem 7.17 below. It will be seen that all other results on linear inequalities are more or less straightforward applications of it.

We first need a preliminary result.

Lemma 7.16. *For a given vector $0 \neq a \in E$, consider the open half-space*

$$C := \{d \in E : \langle a, d\rangle < 0\}.$$

The dual cone C^* is given by

$$C^* = \overline{\text{cone}}(a) = \{ta : t \geq 0\}.$$

Proof. It is easily verified that $C^* = (\overline{C})^*$ for any set C, and it is equally easy to see that $\overline{C} = \{d \in E : \langle a, d\rangle \leq 0\}$. It then follows from Theorem 7.4 that $C^* = \overline{\text{cone}}(a)$. □

Theorem 7.17. **(*Motzkin's transposition theorem, homogeneous version*)** *Let A, B, C be matrices with the same number of rows. Then either the system*

$$(a) \qquad A^T x < 0, \quad B^T x \leq 0, \quad C^T x = 0,$$

is consistent, or the system

$$(b) \qquad Ay + Bz + Cw = 0, \quad y \geq 0, \quad y \neq 0, \quad z \geq 0,$$

is consistent, but not both.

Proof. Let $A = [a_1, \ldots, a_l]$, $B = [b_1, \ldots, b_m]$, $C = [c_1, \ldots, c_p]$, where $\{a_i\}_1^l$, $\{b_j\}_1^m$, $\{c_k\}_1^p$, are the columns of A, B, and C, respectively.

We first show that (a) and (b) cannot both be consistent. If (b) is consistent and x satisfies (a), we have the contradiction

$$\begin{aligned} 0 &= \langle x, Ay + Bz + Cw\rangle = \langle A^T x, y\rangle + \langle B^T x, z\rangle + \langle C^T x, w\rangle \\ &= \sum_1^l y_i\langle a_i, x\rangle + \sum_1^m z_j\langle b_j, x\rangle < 0, \end{aligned}$$

since $\langle b_j, x\rangle \le 0$ and $z_j \ge 0$ for all j, $\langle a_i, x\rangle < 0$ and $y_i \ge 0$ for all i, and some $y_i > 0$.

It remains to prove the claim that if (a) is inconsistent, then (b) must be consistent. First, suppose that $C_1 := \{x : Ax < 0\} = \emptyset$. Applying Theorem 6.23 on page 155 to the open cones $K_i := \{x : \langle a_i, x\rangle < 0\}$, $C_1 = \cap_1^l K_i$, and using Lemma 7.16, we see that there exists a nonnegative vector $0 \le y \in \mathbb{R}^l$, $y \ne 0$, such that $\sum_1^l y_i a_i = Ay = 0$. Then (b) is satisfied with $(y, z, w) = (y, 0, 0)$.

In the remaining case $C_1 \ne \emptyset$, and C_1 is disjoint from the cone $C_2 := \{x : B^T x \le 0, C^T x = 0\} \ne \emptyset$. Theorem 6.9 implies that there exist $\ell \ne 0$ and α such that

$$\langle \ell, u\rangle \le \alpha \le \langle \ell, v\rangle \quad \text{for all} \quad u \in C_1,\ v \in C_2.$$

Letting $u \to 0$ and setting $v = 0$ gives $\alpha = 0$. Thus, we have $\ell \in C_1^*$ and $-\ell \in C_2^*$.

The inclusion $\ell \in C_1^*$ is equivalent to the assertion that the system

$$\langle -\ell, u\rangle < 0, \ \ \langle a_i, u\rangle < 0, \ i = 1, \dots, l$$

is inconsistent. By the argument above in the first case, there exist multipliers $\{y_i\}_0^l$, not all zero, such that $-y_0\ell + \sum_1^l y_i a_i = 0$. If $y_0 = 0$, then $\sum_1^l y_i a_i = 0$, and Theorem 6.23 implies that the system $Ax < 0$ is inconsistent, a contradiction. Therefore, $y_0 > 0$, and we may assume that $y_0 = 1$ by scaling. This gives $\ell = \sum_1^l y_i a_i \ne 0$, and $y := (y_1, \dots, y_l) \ne 0$. The second inclusion $-\ell \in C_2^*$ is precisely the statement that if v satisfies all the inequalities $B^T v \le 0$ and $C^T v = 0$, then $\langle -\ell, v\rangle \le 0$. Farkas's lemma (Theorem 7.6) implies that there exist multipliers $z \ge 0$ and w such that $-\ell = \sum_1^m z_j b_j + \sum_1^p w_k c_k$. Summarizing, we have

$$\sum_1^l y_i a_i + \sum_1^m z_j b_j + \sum_1^p w_k c_k = 0, \quad y \ge 0,\ y \ne 0,\ z \ge 0.$$

The theorem is proved. □

We obtain *Gordan's lemma* as an easy corollary.

Corollary 7.18. (*Gordan's lemma*) *Let $\{a_i\}_{i=1}^k$, $\{b_j\}_{j=1}^l$ be given vectors in E. Then*

$$\{d : \langle a_i, d\rangle < 0, i = 1, \dots, k, \langle b_j, d\rangle = 0, j = 1, \dots, l\} = \emptyset$$

if and only if there exist vectors $\lambda := (\lambda_1, \dots, \lambda_k) \ge 0$ and $\mu := (\mu_1, \dots, \mu_l)$ satisfying $\lambda \ne 0$, and

$$\sum_{i=1}^k \lambda_i a_i + \sum_{j=1}^l \mu_j b_j = 0.$$

Theorem 7.19. (***Motzkin's transposition theorem, affine version***) *Let A, B, and C be matrices with the same number of rows. The linear system*

$$A^T x < a, \; B^T x \le b, \; C^T x = c \tag{7.6}$$

is inconsistent *if and only if the system*

$$\begin{aligned} &Ay + Bz + Cw = 0, \quad \langle a, y\rangle + \langle b, z\rangle + \langle c, w\rangle + y_0 = 0, \\ &(y_0, y, z) \ge 0, \quad (y_0, y) \ne 0 \end{aligned} \tag{7.7}$$

is consistent.

Proof. The system (7.6) is consistent if and only if the homogeneous system

$$t > 0, A^T x < ta, \; B^T x \le tb, \; C^T x = tc$$

in the variables (x, t), that is, the system

$$(0, -1)\begin{pmatrix} x \\ t \end{pmatrix} < 0, \; [A^T, -a]\begin{pmatrix} x \\ t \end{pmatrix} < 0, \; [B^T, -b]\begin{pmatrix} x \\ t \end{pmatrix} \le 0, \; [C^T, -c]\begin{pmatrix} x \\ t \end{pmatrix} = 0,$$

is consistent. The theorem follows immediately from Theorem 7.17. □

7.4 Affine Version of Farkas's Lemma

The following, affine, version of Farkas's lemma is essentially equivalent to the strong duality theorem of linear programming; see Theorem 8.6.

Theorem 7.20. (***Farkas's lemma, affine version***) *Let $\{a_i\}_1^m$, $a_i \in E$, $\{\alpha_i\}_1^m$, $\alpha_i \in \mathbb{R}$, be given vectors and scalars. Suppose that the linear inequalities*

$$\langle a_i, x\rangle \le \alpha_i, \quad i = 1, \ldots, m,$$

are consistent. Then the following statements are equivalent:

$$(a) \quad [\langle a_i, x\rangle \le \alpha_i, i = 1, \ldots, m,] \quad \Longrightarrow \quad [\langle c, x\rangle \le \gamma],$$

$$(b) \quad \exists (y_1, \ldots, y_m) \ge 0 \text{ such that } \sum_{i=1}^m a_i y_i = c, \; \sum_{i=1}^m y_i \alpha_i \le \gamma.$$

Proof. Define $A = [a_1, \ldots, a_m]$, $a = (\alpha_1, \ldots, \alpha_m)^T$. Then (a) is equivalent to the inconsistency of the system

$$A^T x \le a, \quad -c^T x < -\gamma.$$

By the affine version of Motzkin's transposition theorem (Theorem 7.19), this is equivalent to the consistency of the system

$$Ay - z_1 c = 0, \; \langle a, y\rangle - z_1\gamma + z_0 = 0, \; y \ge 0, \; (z_0, z_1) \ge 0, \; (z_0, z_1) \ne 0.$$

Since $Ax \le a$ is consistent, we cannot have $z_1 = 0$. Thus, z_0 is positive, and we may assume that it is one. The theorem is proved. □

7.4.1 An Example of Farkas's Lemma

Here we work out a numerical example in order to develop an intuitive understanding of Farkas's lemma. Consider the system (L) of linear inequalities and equalities

$$\begin{aligned} x_1 - x_2 - x_3 + 3x_4 &\le 1, \\ 5x_1 + x_2 + 3x_3 + 8x_4 &\ge 55, \\ -x_1 + 2x_2 + 3x_3 - 5x_4 &= 3, \\ x_1 &\ge 0, \\ x_3 &\ge 0, \\ x_4 &\le 0. \end{aligned} \tag{7.8}$$

A linear inequality (M) of the form

$$c_1x_1 + c_2x_2 + c_3x_3 + c_4x_4 \le \beta \qquad (C)$$

is called a *consequence* of L if every point $x = (x_1, x_2, x_3, x_4)$ satisfying L must satisfy C. Obviously, any of the linear inequalities (or the equality) in (7.8) is a consequence of (7.8). The inequality $-2x_1 + 2x_2 + 2x_3 - 6x_4 \ge -2$ is also a consequence of (7.8), since it is obtained from the second inequality in (7.8) by multiplying both sides by -2 and reversing the direction of the inequality.

Another way to obtain a consequence inequality is by *aggregation*. For example, multiplying the first three inequalities by 5, the equality by -4, and the sign inequality $x_1 \ge 0$ by -10, we obtain the valid consequence inequality

$$-x_1 - 13x_2 - 17x_3 + 35x_4 \le -7.$$

Of course, to obtain an inequality of the form (C) with direction $\le$, we need to multiply a $\le$ inequality by a nonnegative number, a $\ge$ inequality by a nonpositive number. To obtain a consequence inequality with the direction $\ge$, the signs on the multipliers are reversed. An equality constraint may be multiplied by any number.

We can also *relax* a consequence inequality to obtain a valid consequence inequality; for example, the inequality

$$-x_1 - 13x_2 - 17x_3 + 35x_4 \le 3$$

is a consequence inequality.

Now note that Farkas's lemma (Theorem 7.20) can be paraphrased as stating that any valid consequence inequality to a linear inequality/equality system such as (7.8) must have been obtained by aggregation and then possibly a relaxation. For example, the inequality

$$2x_1 - 45x_2 - 78x_3 + 90x_4 \le -100$$

is a consequence of the system (7.8), although it may not be easy to see this by inspection. In fact, this inequality is obtained from (7.8) by aggregating it using the multipliers 2, −3, −20, −5, −7, 8, respectively, and then relaxing the obtained right-hand side constant −223 to −100. The validity of this inequality can be numerically verified by solving the *linear programming* problem of maximizing the objective function $2x_1 - 45x_2 - 78x_3 + 90x_4$ subject to the constraint that x satisfies the linear system (7.8). One then needs only to check that the optimal objective value of the linear program is at most −100.

Remark 7.21. In a vector space endowed with a coordinate system with respect to a basis, Farkas's lemma can be stated in a more compact way. Let n be the dimension of E. Then E can be identified with $\mathbb{R}^n$. Let A be the $m \times n$ matrix having a_i as ith row, and define $a = (\alpha_1, \ldots, \alpha_m)^T$ and $y = (y_1, \ldots, y_m)^T$. Then (a) and (b) become

$$\begin{array}{ll} (a) & \langle c, x\rangle \le \beta \text{ for all } x \text{ satisfying } Ax \le a, \\ (b) & \text{there exists } y \in \mathbb{R}^m,\ y \ge 0 \text{ such that } A^T y = c \text{ and } \langle a, y\rangle \le \beta. \end{array}$$

7.4.2 Application of Farkas's Lemma to Optimization

Recall that the variational inequality

$$\langle \nabla f(x^*), x - x^*\rangle \ge 0 \text{ for all } x \in C,$$

for minimizing a function f over a closed convex set C, is a set of conditions for every $x \in C$, hence infinitely many in number. In the case that C is a polyhedron, Farkas's lemma is a useful tool to turn these infinitely many conditions into a set of much more manageable, finitely many conditions.

Theorem 7.22. *Let f be a Gâteaux differentiable function. Consider the optimization problem*

$$\begin{array}{ll} \min & f(x) \\ \text{s.t.} & \langle a_i, x\rangle \ge \beta_i, \quad i = 1, \ldots, m. \end{array}$$

If x^* is a local minimizer of f, then there exist nonnegative multipliers $\{\lambda_i\}_1^m$ such that

$$\sum_{i=1}^m \lambda_i a_i = \nabla f(x^*), \qquad \sum_{i=1}^m \lambda_i b_i \ge \langle \nabla f(x^*), x^*\rangle. \tag{7.9}$$

If f is a convex function, then conditions (7.9) are also sufficient for x^* to be a global minimizer of f.

Proof. The variational inequality for this problem is

$$[\langle a_i, x\rangle \ge \beta_i, \quad i = 1, \ldots, m] \qquad \Longrightarrow \qquad [\langle \nabla f(x^*), x\rangle \ge \langle \nabla f(x^*), x^*\rangle].$$

It follows from Farkas's lemma (Theorem 7.20) that there exist nonnegative multipliers $\{\lambda_i\}_1^m$ such that (7.9) holds. Since the variational inequality is a sufficient condition for optimality in case f is convex, the rest of the theorem follows. □

7.5 Tucker's Complementarity Theorem

Theorem 7.23. (***Tucker*** [255]) *If $L \subseteq \mathbb{R}^n$ is a linear subspace and $L^\perp$ its orthogonal complement, then there exist vectors $x^* \in L$ and $y^* \in L^\perp$ such that x^* and y^* are* strictly complementary, *that is, $x^* \geq 0$, $y^* \geq 0$, and for each index $1 \leq i \leq n$, either $x_i^* > 0$ and $y_i^* = 0$ or $x_i^* = 0$ and $y_i^* > 0$. Moreover, the indices $I(x^*) = \{i : x_i^* > 0\}$, $J(y^*) = \{i : y_i^* > 0\}$ are independent of (x^*, y^*), and are uniquely determined by L.*

Proof. Write $L = \{x : Ax = 0\}$, where A is an $m \times n$ matrix. Then $L^T = A^T(\mathbb{R}^m)$ is the range of A^T. For each index i, $1 \leq i \leq n$, it follows from the homogeneous version of Motzkin's transposition theorem (Theorem 7.17) that exactly one of the systems

$$Ax = 0, x \geq 0, x_i > 0; \quad A^T\delta = \mu + \lambda e_i, \mu \geq 0, \lambda > 0,$$

is consistent. Define $y := A^T\delta \in L^\perp$, and denote the sets of indices such that the first or the second system is consistent by I and J, respectively, and the corresponding solution to the consistent system by $(x^i)^*$ or $(y^i)^*$. The vectors

$$x^* := \sum_{i \in I} (x^i)^* \text{ and } y^* := \sum_{i \in J} (y^i)^*$$

satisfy the requirements of the theorem. The independence of the indices on x^* and y^* is also clear. □

A closely related result states that if a linear program $\min\{c^Tx : Ax = b\}$ and its dual $\max\{b^Ty : A^Ty + s = c, s \geq 0\}$ both have optimal solutions, then there exists an optimal solution pair $(x^*, (y^*, s^*))$ such that (x^*, s^*) are strictly complementary, that is, $x^* + s^* > 0$; see Section 8.5. This fact has important applications in interior point methods for linear programming.

7.6 Exercises

1. Using Farkas's lemma or otherwise, determine the polar of the cone

$$K = \{(x, y, z) : x + y - z \leq 0, -2x + 3y \geq 0, x + 2y - 4z = 0, x \leq 0, y \geq 0\}.$$

2. Consider the cone

$$\begin{aligned} x_1 + x_2 - x_3 &\le 0, \\ x_3 &\ge 0, \\ 2x_1 - x_2 + x_3 &\ge 0. \end{aligned}$$

Determine the polar cone. Find the extreme directions of the polar cone.
3. Determine the polar cone K^*, where $K = \{x \in R^n : Ax = 0, x \ge 0\}$.
4. If $C \subseteq E$ is a closed convex set, define

$$C^\circ = \cap_{x \in C}\{u \in E : \langle u, x\rangle \le 1\}.$$

Since C° is the intersection closed half-spaces, it is closed and convex. Prove the following:
   (a) If

$$C = \text{co}(v_1, \ldots, v_k) + \overline{\text{cone}}(d_1, \ldots, d_l),$$

   describe C° in terms of $\{v_i, d_j\}$. Consequently, show that C° is a polyhedral set. If C is a polytope, determine the conditions under which C° is also a polytope.
   (b) If C is cone, show that $C^\circ = C^*$, the polar cone.
   (c) Determine C° explicitly (in terms of A, b) if $C = \{x : Ax \le b\}$.
   (d) Show that the polar of the cube $\{x \in \mathbb{R}^3 : |x_i| \le 1, i = 1, 2, 3\}$ is the octahedron $\{x \in \mathbb{R}^3 : |x_1| + |x_2| + |x_3| \le 1\}$.

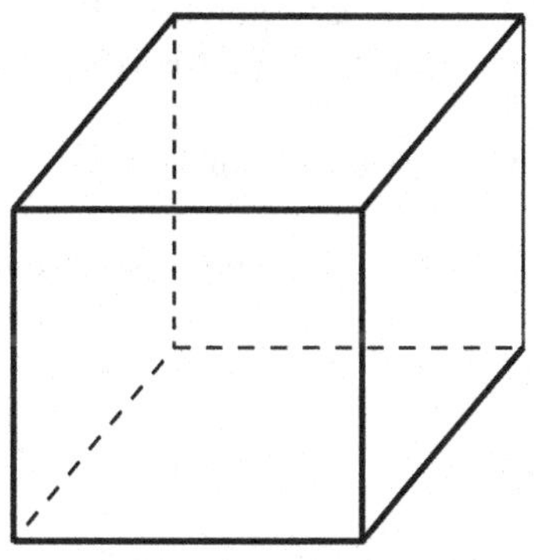
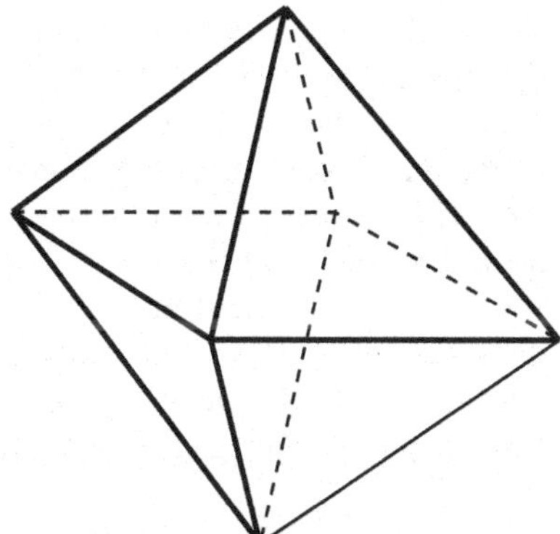

Fig. 7.2. A pair of polar polytopes (cube and octahedron).

5. If $C = \{x : x^T Qx \le 1\}$, where Q is a symmetric, positive semidefinite $n \times n$ matrix, describe C°.
6. Let K_1, K_2 be polyhedral convex cones. Show that their Minkowski sum $K = K_1 + K_2$ is also a polyhedral convex cone.

7. Consider the convex polytope

$$P = \{x \in E : \langle a_i, x\rangle \leq b_i, i = 1, \ldots, k\}. \qquad (I)$$

 (a) Theorem 7.13 on page 180 gives a representation of P. Show that the cone C in that proposition is null in our situation, that is, $C = \{0\}$. Thus, P can also be written as

$$P = \left\{x : x = \sum_{j=1}^{m} \mu_j v_j, \sum_{j=1}^{m} \mu_j = 1, \mu_j \geq 0, j = 1, \ldots, m\right\}. \qquad (II)$$

 Use either the representation (I) or (II), whichever is more convenient, to prove the following statements. (The point of this problem is that the proof of each of these statements is simplified greatly if one uses the appropriate representation.)
 (b) Every intersection of a polytope with an affine subspace is a polytope. Thus, a cross section of a polytope is a polytope.
 (c) Every intersection of a polytope with a polyhedron is a polytope.
 (d) The Minkowski sum of two polytopes is a polytope, that is, if P_1 and P_2 are polytopes, then the sum

$$P = P_1 + P_2 = \{x_1 + x_2 : x_i \in P_i, i = 1, 2\}$$

 is a polytope.
 (e) An affine image of a polytope is a polytope.
 (f) A projection of a polytope onto an affine subspace is a polytope.
 (g) If $A : \mathbb{R}^n \to \mathbb{R}^m$ is an affine mapping and P is a polytope in $\mathbb{R}^m$, then $A^{-1}(P)$ is a polyhedron in $\mathbb{R}^n$. Show that the result is true even if P is a polyhedron.

8. If P is a polytope and A, B nonempty convex sets such that $A + B = P$, show that A and B are polytopes.
 Hint: Let $\{p_i\}_1^k$ be the vertices of P and $p_i = a_i + b_i$, where $a_i \in A$ and $b_i \in B$, $i = 1, \ldots, k$. Define $A' = \text{co}(a_1, \ldots, a_k)$, $B' = \text{co}(b_1, \ldots, b_k)$, and show that $A' + B' = P$. Finally, argue that $A = A'$ and $B = B'$.

9. Find explicitly the projection of a point $x \in \mathbb{R}^n$
 (a) onto the nonnegative orthant

$$K = \{x \in \mathbb{R}^n : x = (x_1, \ldots, x_n)^T, x_i \geq 0, i = 1, ..., n\}.$$

 (Of course, the projection $\Pi_K(x)$ is the unique solution to the minimization problem $\min\{\|z - x\| : z \in K\}$.)
 (b) onto the unit cube

$$C = \{x \in \mathbb{R}^n : x = (x_1, \ldots, x_n)^T, 0 \leq x_i \leq 1, i = 1, \ldots, n\}.$$

10. *(Projection onto a simplex)* Find the projection of a point $a \in \mathbb{R}^n$ onto the standard unit simplex Δ_{n-1} in $\mathbb{R}^n$. (Recall that $\Delta_{n-1} := \{x =$

$(x_1, \dots, x_n) \in \mathbb{R}^n : x_1 + \cdots + x_n = 1, x_i \geq 0, \dots, x_n \geq 0\}$.) There is no simple, explicit formula for the projection $x^* = \Pi_{\Delta_{n-1}}(a)$. Nevertheless, Farkas's lemma simplifies the problem to a manageable form.

(a) Show that Farkas's lemma applied to the variational inequality for the projection problem gives a system that is equivalent to the system

$$x^* - \lambda = a - \mu e, \ x^* \geq 0, \ \lambda \geq 0, \ \text{and} \ \langle x^*, \lambda \rangle = 0,$$

where $\lambda \in \mathbb{R}^n$ and $\mu \in \mathbb{R}$.

(b) Show that

$$x^* - \lambda = (a - \mu e)^+, \ \text{and} \ \lambda = -(a - \mu e)^-;$$

x^* and λ are thus the positive and negative parts of the vector $a - \mu e$, respectively, that is, $x_i^* = (a_i - \mu)^+ := \max\{0, a_i - \mu\}$ and $\lambda_i = -(a_i - \mu)^- := -\min\{0, a_i - \mu\}$, $i = 1, \dots, n$.

(c) Consequently, show that x^* is characterized by the condition

$$\exists \mu \in \mathbb{R}, \ \sum_1^n (a_i - \mu)^+ = 1.$$

(d) Devise a search routine to compute μ and hence the projection $x^* = \Pi_{\Delta_{n-1}}(a)$.

11. Let $\{v_i\}_1^m$ be given vectors in $\mathbb{R}^n$. Show that the following conditions are equivalent:
 (a) If $x \neq 0$, then $\langle v_i, x \rangle > 0$ for some i.
 (b) $\overline{\text{cone}}(v_1, \dots, v_m) = \mathbb{R}^n$.
 (c) $0 \in \text{int}(\text{co}(v_1, \dots, v_m))$.

 Hint: Let $\{e_i\}_1^n$ be the standard unit vectors in $\mathbb{R}^n$. To prove that (b) implies (c), show that some positive multiples of the vectors $\mp e_k$ lie in $\text{co}(v_1, \dots, v_m)$.
12. Let A be an $m \times n$ matrix. Recall that Gordan's lemma states that the system

$$\{d : Ad < 0\}$$

is inconsistent if and only if the system

$$\lambda \geq 0 \in \mathbb{R}^m, \ \lambda \neq 0, \ A^T \lambda = 0$$

is consistent. Prove this instead using the following direct approach: let C be the convex hull of the rows of A, that is, $C = \text{co}(a_1, \dots, a_m) \subseteq E$, where a_i is the ith row of A.

 (a) Show that the consistency of the second system is equivalent to the statement that $0 \in C$.
 (b) Use a separation argument to prove that $\{d : Ad < 0\} \neq \emptyset$ if and only if $0 \notin C$, thereby proving Gordan's lemma.

13. Show that exactly one of the following two systems has a solution:

$$\begin{aligned} &(a) \quad 0 \neq Bx \geq 0 \quad \text{for some } x \in \mathbb{R}^n, \\ &(b) \quad B^T y = 0 \quad \text{for some } 0 < y \in \mathbb{R}^m, \end{aligned}$$

where B is an $m \times n$ matrix.

14. Prove the following version of Farkas's lemma: for compatible matrices A, B, C and vectors u, v, w, *either* there exists a solution vector x satisfying

$$Ax = u, \ Bx \geq v, \ Cx \leq w,$$

or there exist solution vectors a, b, c with

$$A^T a + B^T b + C^T c = 0, \ b \leq 0, \ c \geq 0, \ \langle a, u \rangle + \langle b, v \rangle + \langle c, w \rangle < 0.$$

Hint: Reformulate the problem so that it reduces to a well-known version of Farkas's lemma.

15. This problem shows how to obtain Gordan's lemma directly from Farkas's lemma. Recall that Gordan's lemma states that either the system

$$Ax < 0$$

or the system

$$A^T u = 0, \ u \geq 0, \ u \neq 0$$

has a solution but never both. Prove this by reducing it to Farkas's lemma. *Hint:* Let a^i be the ith row of A. First, show that $Ax < 0$ has no solution x if and only if

$$\big[\langle a^i, x \rangle \leq t \ \text{ for all } \ i\big] \quad \Longrightarrow \quad t \geq 0.$$

Then, apply Farkas's lemma to this system in the variables (x, t).

16. Assume that the linear system involving $\{a_i\}_1^m$ is consistent in each of the following systems. In each case, characterize (c, γ) that makes the corresponding implication true:

$$\begin{aligned} &(a) \quad \langle a_i, x \rangle < \alpha_i, \ i = 1, \ldots, m \quad &\Longrightarrow \quad &\langle c, x \rangle < \gamma. \\ &(b) \quad \langle a_i, x \rangle \leq \alpha_i, \ i = 1, \ldots, m \quad &\Longrightarrow \quad &\langle c, x \rangle < \gamma. \\ &(c) \quad \langle a_i, x \rangle < \alpha_i, \ i = 1, \ldots, m \quad &\Longrightarrow \quad &\langle c, x \rangle \leq \gamma. \\ &(d) \quad \langle a_i, x \rangle = \alpha_i, \ i = 1, \ldots, m \quad &\Longrightarrow \quad &\langle c, x \rangle < \gamma. \\ &(e) \quad \langle a_i, x \rangle \leq \alpha_i, \ i = 1, \ldots, m \quad &\Longrightarrow \quad &\langle c, x \rangle = \gamma. \end{aligned}$$

17. Prove *Carver's theorem*: the linear system $Ax < a$ is consistent if and only if $y = 0$ is the only solution for the system $y \geq 0$, $A^T y = 0$, and $\langle a, y \rangle \leq 0$.

18. Prove by a direct separation argument involving two convex sets that the following two statements are equivalent:

$$(i)\quad Ax \geq a \text{ is consistent;}$$
$$(ii)\quad \lambda \geq 0,\ A^T\lambda = 0] \quad \Longrightarrow \quad \langle a, \lambda\rangle \leq 0.$$

Hint: The harder part is to prove that (ii) implies (i). Assume that (i) is inconsistent and let A be an $m \times n$ matrix. Form the affine set $S = \{Ax - a : x \in \mathbb{R}^n\}$ and write it in the form $S = \{y : By = b\}$. Show that the set $C := \{y : y \in S,\ y \geq 0\}$ is a closed convex set and does not contain b. Then use an appropriate separation theorem to prove the existence of a λ that violates the conditions of (ii).

19. (***Von Neumann–Morgenstern***) The following "theorem of alternative" plays an important role in game theory in the famous book [212]. Let A be an $m \times n$ matrix. Let Δ_{n-1} and Δ_{m-1} denote the standard unit simplices in $\mathbb{R}^n$ and $\mathbb{R}^m$, respectively. Prove that either there exists a vector $x \in \mathbb{R}^n$ satisfying

$$Ax \leq 0,\ x \in \Delta_{n-1}, \qquad (I)$$

or there exists a vector $y \in \mathbb{R}^m$ satisfying

$$A^T y > 0,\ y \in \Delta_{m-1}, \qquad (II)$$

but not both.
Hint: Use Motzkin's transposition theorem.

20. (***Strong separation of two disjoint convex polyhedra***) Let $P_1 = \{x : A^T x \leq a\}$ and $P_2 = \{x : B^T x \leq b\}$ be two nonempty convex polyhedra in $\mathbb{R}^n$. Show that P_1 and P_1 can be strongly separated by completing the following steps:
   (a) Use Motzkin's transposition theorem (Theorem 7.19) to prove that there exist nonnegative, nonzero multipliers y and z such that

$$Ay + Bz = 0,\ \ \langle a, y\rangle + \langle b, z\rangle < 0.$$

   (b) Show that in fact, $Ay \neq 0$ and $Bz \neq 0$.
   (c) Define $l := Ay = -Bz \neq 0$, and show that a suitable hyperplane $H_{(l,\gamma)} = \{x : \langle l, x\rangle = \gamma\}$ separates P_1 and P_2 strongly.

8

Linear Programming

A *linear program* is a problem of minimization (or maximization) of a linear function subject to linear equalities or inequalities, that is, the optimization of a linear function over a polyhedron. It is probably the most practical and important class of practical optimization problems, having wide applications in industry.

We have developed enough theory by now to give the fundamental existence and duality results in this subject. There is a huge literature on linear programming and its applications. For more information on linear programming, the simplex method for solving it, and its wide-ranging applications, the reader can profitably consult the books [67, 58, 238] and the references therein. For interior-point methods for solving linear programs; see the book [270].

8.1 Fundamental Theorems of Linear Programming

In this section, we present the fundamental results in linear programming, concerning the existence of solutions and the duality theory of linear programming.

For concreteness, we focus on the linear program

$$\begin{aligned} \max \quad & \langle c, x \rangle \\ \text{s.t.} \quad & \langle a_i, x \rangle \leq b_i, \ i = 1, \ldots, m, \end{aligned} \qquad (P)$$

but our results here can be generalized to any linear program.

Let us first consider the existence of solutions.

Lemma 8.1. *Assume that the linear program (P) has a nonempty feasible region. Then (P) has a solution if and only if its objective function $\langle c, x \rangle$ is bounded from above on the constraint set.*

O. Güler, *Foundations of Optimization*, Graduate Texts in Mathematics 258,
DOI 10.1007/978-0-387-68407-9_8, © Springer Science+Business Media, LLC 2010

Proof. Denote by

$$C := \{x : \langle a_i, x\rangle \le b_i,\ i = 1, \dots, m\}$$

the constraint set of (P), and suppose that the objective function is bounded from above by a constant $M < \infty$. By Theorem 7.13, C has a representation

$$C = \mathrm{co}(v_1, \dots, v_k) + \overline{\mathrm{cone}}(d_1, \dots, d_l).$$

Since $v_1 + td_j \in C$ for all $t \ge 0$, we have $\langle c, v_1 + td_j\rangle \le M$. Dividing both sides of this inequality by t and letting $t \to \infty$ gives $\langle c, d_j\rangle \le 0$ for $j = 1, \dots, l$. Now, the supremum of the objective function over C equals

$$\sup\left\{\sum_{i=1}^{k} \lambda_i \langle c, v_i\rangle + \sum_{j=1}^{l} \delta_j \langle c, d_j\rangle : \lambda \in \Delta_{m-1},\ \delta_j \ge 0,\ j = 1, \dots, l\right\}$$

$$= \sup\left\{\sum_{i=1}^{k} \lambda_i \langle c, v_i\rangle : \lambda \in \Delta_{m-1}\right\} = \max\{\langle c, v_i\rangle : i = 1, \dots, m\},$$

where Δ_{m-1} is the standard unit simplex in $\mathbb{R}^m$. This proves that the supremum of the objective function is attained, and attained at some vertex v_i. □

The next result points to the existence of a duality theory for linear programming.

Theorem 8.2. *Let the linear program (P) have a solution. A feasible point x^* is a solution to (P) if and only if there exist multipliers $\{y_i^*\}_1^m$ such that*

$$\sum_{i=1}^{m} y_i^* a_i = c, \quad y_i^* \ge 0, \quad \text{and} \quad \langle b, y^*\rangle = \langle c, x^*\rangle. \tag{8.1}$$

Proof. Let x^* and y^* satisfy (8.1). If x is feasible, then

$$\langle c, x\rangle = \left\langle \sum_{i=1}^{m} y_i^* a_i, x\right\rangle = \sum_{i=1}^{m} y_i^* \langle a_i, x\rangle \le \sum_{i=1}^{m} y_i^* b_i = \langle c, x^*\rangle,$$

proving that x^* is a solution of (P).

Conversely, let x^* be a solution of (P). Since every feasible point x satisfies $\langle c, x\rangle \le \langle c, x^*\rangle$, that is,

$$\langle a_i, x\rangle \le b_i,\ i = 1, \dots, m \quad \Longrightarrow \quad \langle c, x\rangle \le \langle c, x^*\rangle,$$

it follows from Lemma 7.20 (Farkas's lemma) that there exist nonnegative $\{y_i^*\}_1^m$ satisfying $\sum_{i=1}^m y_i^* a_i = c$ and $\langle b, y^*\rangle \le \langle c, x^*\rangle$. Moreover,

$$\langle c, x^*\rangle - \langle b, y^*\rangle = \sum_{i=1}^{m} y_i^* \langle a_i, x^*\rangle - \langle b, y^*\rangle = \sum_{i=1}^{m} y_i^* (\langle a_i, x^*\rangle - b_i) \le 0,$$

where the inequality follows since $y_i^* \ge 0$ and $\langle a_i, x^*\rangle \le b_i$. This gives $\langle c, x^*\rangle \le \langle b, y^*\rangle$, and we conclude that $\langle c, x^*\rangle = \langle b, y^*\rangle$. □

Remark 8.3. The multipliers $\{y_i^*\}_1^m$ serve as "certificates of optimality" for x^*: to convince a skeptic that x^* is really an optimal solution of (P), all we need to do is to verify is that x^* is feasible for (P) and y^* satisfies (8.1). This amounts to verifying that (x^*, y^*) satisfies some linear equations and inequalities, a trivial task in comparison to computing a solution to a linear program from scratch.

Corollary 8.4. *Suppose that the linear program (P) has a solution. A feasible point x^* is a solution to (P) if and only if there exist multipliers $\{y_i^*\}_1^m$ satisfying the conditions*

$$\sum_{i=1}^{m} y_i^* a_i = c, \quad y_i^* \geq 0, \; i = 1, \ldots, m,$$

and the complementarity *conditions*

$$\textit{for each } i = 1, \ldots, m, \; \textit{either } y_i^* = 0 \textit{ or } \langle a_i, x^* \rangle = b_i.$$

This follows immediately from the proof of the theorem above, since the condition $\langle c, x^* \rangle = \langle b, y^* \rangle$ is equivalent to

$$\langle b, y^* \rangle - \langle c, x^* \rangle = \sum_{i=1}^{m} y_i^* (\langle a_i, x^* \rangle - b_i) = 0; \tag{8.2}$$

and since $y_i^* \geq 0$ and $\langle a_i, x^* \rangle - b_i \leq 0$, each term in the sum above is zero.

The optimality condition

$$\langle c, x \rangle \leq \langle c, x^* \rangle = \langle b, y^* \rangle, \quad \text{for all } \; x \in C,$$

makes it intriguing to consider the linear program

$$\begin{aligned} \min \quad & \langle b, y \rangle \\ \text{s. t.} \quad & \sum_{i=1}^{m} y_i a_i = c, \qquad\qquad (D) \\ & y_i \geq 0, \; i = 1, \ldots, m. \end{aligned}$$

Concerning the linear program pair (P) and (D), we have the following fundamental results.

Theorem 8.5. (***Weak duality theorem for linear programming***) *If x is a feasible solution to (P) and y is a feasible solution to (D), then*

$$\langle c, x \rangle \leq \langle b, y \rangle.$$

Proof. This follows from the observation, already used above, that

$$\langle b, y \rangle - \langle c, x \rangle = \langle b, y \rangle - \sum_{i=1}^{m} y_i \langle a_i, x \rangle = \sum_{i=1}^{m} y_i (b_i - \langle a_i, x \rangle) \geq 0,$$

where the inequality follows from $y_i \geq 0$ and $\langle a_i, x \rangle \leq b_i$. □

The expression

$$\langle b, y\rangle - \langle c, x\rangle$$

is called the *duality gap*. Thus, the weak duality theorem states that if x and y are feasible, then their duality gap is nonnegative.

Theorem 8.6. (***Strong duality theorem for linear programming***) *If (P) has an optimal solution x^*, then (D) has an optimal solution y^* and the optimal objective values of (P) and (D) are the same, that is,*

$$\langle c, x^*\rangle = \langle b, y^*\rangle.$$

This follows immediately from Theorem 8.2.

The linear programs (P) and (D) can be written in more compact form using matrices. Denoting by A the $m \times n$ matrix whose rows are a_i^T, that is, $A^T = [a_1, a_2, \ldots, a_m]$, we have

$$\begin{aligned} \max \quad & \langle c, x\rangle \\ \text{s.t.} \quad & Ax \le b, \end{aligned} \qquad (P)$$

$$\begin{aligned} \min \quad & \langle b, y\rangle \\ \text{s.t.} \quad & A^T y = c \qquad (D) \\ & y \ge 0, \end{aligned}$$

8.2 An Intuitive Formulation of the Dual Linear Program

In this section, we provide an intuitive method to formulate the dual of a linear program. Consider the linear program

$$\begin{aligned} \max \quad & 4x_1 + x_2 + 5x_3 + 3x_4 \\ \text{s.t.} \quad & x_1 - x_2 - x_3 + 3x_4 \le 1 \\ & 5x_1 + x_2 + 3x_3 + 8x_4 \ge 15 \qquad\qquad (P) \\ & -x_1 + 2x_2 + 3x_3 - 5x_4 = 3 \\ & x_1 \ge 0,\ x_2 \text{ free},\ x_3 \ge 0,\ x_4 \le 0. \end{aligned}$$

Let us try to estimate the optimal objective value z^* of (P) by finding *lower* and *upper* bounds for it. To find a lower bound, it suffices to find a feasible point of (P). For example, the feasible point $x = (3, 0, 2, 0)^T$ allows us to obtain a lower bound of $4x_1 + 2x_3 = 4 \times 3 + 5 \times 2 = 2$; thus $z^* \ge 22$.

Determining an upper bound for z^* relies on a different strategy, and this will *motivate* the formulation of the dual linear program (D) corresponding to (P). We proceed as follows: multiply each of the three constraints of (P) by some numbers y_1, y_2, y_3, respectively, and add the resulting inequalities. We would like the resulting constraint to have the form

$$\text{lhs} \le \text{rhs},$$

where lhs and rhs denote its left-hand side and right-hand side, respectively. We then *compare* lhs with the objective function $4x_1 + x_2 + 5x_3 + 3x_4$: *we would like to have the inequality* $4x_1 + x_2 + 5x_3 + 3x_4 \leq$ lhs. Since lhs $\leq$ rhs, this will imply that

$$4x_1 + x_2 + 5x_3 + 3x_4 \leq \text{rhs} \quad \text{for all feasible } x = (x_1, x_2, x_3, x_4)^T.$$

Clearly, rhs will then be an upper bound for the objective value $z^* = \max(P)$.

Since we would like to have lhs $\leq$ rhs and the first constraint $x_1 - x_2 - x_3 + 3x_4 \leq 1$ has the same direction $\leq$, this forces us to choose $y_1 \geq 0$. However, the second inequality has the direction $\geq$, so we multiply it by $y_2 \leq 0$ to convert it to the form $\leq$. Since the third constraint is an equality, y_3 is unconstrained, or free. Adding the resulting constraints leads to the inequality

$$\begin{aligned} &y_1(x_1 - x_2 - x_3 + 3x_4) + y_2(5x_1 + x_2 + 3x_3 + 8x_4) \\ &\quad + y_3(-x_1 + 2x_2 + 3x_3 - 5x_4) \\ &\leq y_1 + 15y_2 + 3y_3, \end{aligned}$$

which we rewrite by collecting each x_i term separately,

$$\begin{aligned} \text{lhs} := {} & (y_1 + 5y_2 - y_3)x_1 + (-y_1 + y_2 + 2y_3)x_2 \\ & + (-y_1 + 3y_2 + 3y_3)x_3 + (3y_1 + 8y_2 - 5y_3)x_4 \\ \leq {} & y_1 + 15y_2 + 3y_3 \\ =: {} & \text{rhs}\,. \end{aligned}$$

Now, to enforce the condition $4x_1 + x_2 + 5x_3 + 3x_4 \leq$ lhs, that is, the condition

$$\begin{aligned} 4x_1 + x_2 + 5x_3 + 3x_4 \ \leq {} & (y_1 + 5y_2 - y_3)x_1 + (-y_1 + y_2 + 2y_3)x_2 \\ & + (-y_1 + 3y_2 + 3y_3)x_3 + (3y_1 + 8y_2 - 5y_3)x_4, \end{aligned}$$

we compare, for each i, the two x_i terms on both sides of the inequality. First, let us see how to ensure the inequality

$$4x_1 \leq (y_1 + 5y_2 - y_3)x_1.$$

Since $x_1 \geq 0$, we can guarantee this condition only when $4 \leq y_1 + 5y_2 - y_3$. Next, we would like to have

$$x_2 \leq (-y_1 + y_2 + 2y_3)x_2;$$

since x_2 is free, this can be guaranteed only when

$$1 = -y_1 + y_2 + 2y_3.$$

Finally, we would like to have

$$3x_4 \leq (3y_1 + 8y_2 - 5y_3)x_4;$$

since $x_4 \leq 0$, we can guarantee this inequality only when

$$3 \geq 3y_1 + 8y_2 - 5y_3.$$

Summarizing, we see that as long as y_1, y_2, y_3 satisfy the inequalities

$$\begin{aligned} y_1 + 5y_2 - y_3 &\geq 4, \\ -y_1 + y_2 + 2y_3 &= 1, \\ -y_1 + 3y_2 + 3y_3 &\geq 5, \\ 3y_1 + 8y_2 - 5y_3 &\leq 3, \\ y_1 \geq 0,\ y_2 \leq 0,\ y_3 &\text{ free,} \end{aligned}$$

we have an upper bound for (P), since for any feasible point $x = (x_1, x_2, x_3, x_4)^T$, we have

$$4x_1 + x_2 + 5x_3 + 3x_4 \leq \text{lhs} \leq y_1 + 15y_2 + 3y_3.$$

In other words, the value $y_1 + 15y_2 + 3y_3$ is an upper bound for z^*.

Suppose we want as *tight* an upper bound as possible, meaning that we would like to *minimize* the upper bound $y_1 + 15y_2 + 3y_3$. This leads us to the dual linear program

$$\begin{aligned} \min\quad & y_1 + 15y_2 + 3y_3 \\ \text{s.t.}\quad & y_1 + 5y_2 - y_3 \geq 4 \\ & -y_1 + y_2 + 2y_3 = 1 \\ & -y_1 + 3y_2 + 3y_3 \geq 5 \\ & 3y_1 + 8y_2 - 5y_3 \leq 3 \\ & y_1 \geq 0,\ y_2 \leq 0,\ y_3 \text{ free.} \end{aligned} \qquad (D)$$

8.3 Duality Rules in Linear Programming

The rules above for formulating the dual of a linear program can be summarized as follows:

Primal Linear Program	Dual Linear Program
Objective: Maximize	Objective: Minimize
ith constraint: $\leq$	ith variable: ≥ 0
ith constraint: $=$	ith variable: unconstrained
ith constraint: $\geq$	ith variable: ≤ 0
jth variable: ≥ 0	jth constraint: $\geq$
jth variable: unconstrained	jth constraint: $=$
jth variable: ≤ 0	jth constraint: ≤ 0

We emphasize that a linear program and its dual form a *primal–dual pair*. Either of the two linear programs may be designated as the primal one; then the other one is called its dual. The choice is arbitrary. What is important is the fact that the direction and the sign conventions *depend* on whether the LP is a maximization or a minimization problem.

It may be more convenient to remember the following table, which shows the usual situation:

Primal Linear Program	Dual Linear Program
Objective: Maximize	Objective: Minimize
ith constraint: $\le$	ith variable: ≥ 0
jth variable: ≥ 0	jth constraint: $\ge$

In other words, *in a maximization problem, the usual constraint has the direction* $\le$ *and the usual variable is nonnegative, and in a minimization problem, the usual constraint has the direction* $\ge$ *and the usual variable is nonnegative.* If a constraint (variable) has the opposite direction (sign), then the corresponding dual variable (constraint) has the opposite sign (direction). If a constraint is an equality, then the corresponding dual variable is unconstrained, and vice versa.

A programming pair with the usual constraints and variables is written in matrix notation in the form

$$\begin{aligned} \max\quad & \langle c, x\rangle \\ \text{s.t.}\quad & Ax \le b \quad (P) \\ & x \ge 0, \end{aligned} \qquad\qquad \begin{aligned} \min\quad & \langle b, y\rangle \\ \text{s.t.}\quad & A^T y \ge c \quad (D) \\ & y \ge 0, \end{aligned}$$

where A is an $m \times n$ matrix, $c, x \in \mathbb{R}^n$ are n-vectors, and $b, y \in \mathbb{R}^m$ are m-vectors. This primal–dual pair of linear programs is said to be in *symmetric form.* One may verify, using the above duality rules, that the following programming problems also form a primal–dual pair:

$$\begin{aligned} \min\quad & \langle c, x\rangle \\ \text{s.t.}\quad & Ax = b \quad (P) \\ & x \ge 0, \end{aligned} \qquad\qquad \left.\begin{aligned} \max\quad & \langle b, y\rangle \\ \text{s.t.}\quad & A^T y \le c. \end{aligned}\right. \quad (D)$$

Such a pair of primal–dual linear programs is said to be in *standard form.* The simplex method for linear programming is applied to a linear program (P) in standard form. If a linear program is not in standard form, then slack and surplus variables must be added to transform it into a standard-form linear program.

8.4 Geometric Formulation of Linear Programs

In the primal–dual linear programming pairs discussed above, the two linear programs do not appear to be symmetric. For example, in the last standard linear programming pair above, the primal variable x lies in $\mathbb{R}^n$, while the dual variable y lies in $\mathbb{R}^m$. We will show in this section that the asymmetry of the two programs is due to their particular *representation*. When properly viewed geometrically, the two programs will have exactly the same form. This holds for any pair of primal–dual programs, but we will demonstrate it only for the standard linear programming pair above.

We introduce a slack variable s in (D) and rewrite the linear programs in the form

$$\begin{array}{lll} \min & \langle c, x\rangle & \\ \text{s.t.} & Ax = b & (P) \\ & x \geq 0, & \end{array} \qquad \begin{array}{lll} \max & \langle b, y\rangle & \\ \text{s.t.} & A^T y + s = c & (D) \\ & s \geq 0. & \end{array}$$

Suppose that the linear equations in (P) and (D) are feasible, and pick x_0 and (y_0, s_0) satisfying them,

$$Ax_0 = b \text{ and } A^T y_0 + s_0 = c.$$

The constraint $Ax = b$ is then equivalent to $A(x - x_0) = 0$, or $x \in x_0 + L$, where $L := N(A)$. We can rewrite the objective function as well:

$$\langle c, x\rangle = \langle A^T y_0 + s_0, x\rangle = \langle y_0, Ax\rangle + \langle s_0, x\rangle = \langle b, y_0\rangle + \langle s_0, x\rangle.$$

Similarly, in the dual program, the constraint $A^T y + s = c = A^T y_0 + s_0$ has the form

$$s - s_0 = A^T(y_0 - y) \in R(A^T) = L^\perp,$$

and the objective function the form

$$\langle b, y\rangle = \langle Ax_0, y\rangle = \langle x_0, A^T y\rangle = \langle x_0, c - s\rangle = \langle c, x_0\rangle - \langle x_0, s\rangle. \tag{8.3}$$

With these changes, the linear programs take on their geometric form

$$\begin{array}{lll} \min & \langle s_0, x\rangle & \\ \text{s.t.} & x \in x_0 + L & (GP) \\ & x \geq 0, & \end{array} \qquad \begin{array}{lll} \min & \langle x_0, s\rangle & \\ \text{s.t.} & s \in s_0 + L^\perp & (GD) \\ & s \geq 0, & \end{array}$$

in which the primal and dual programs are in exactly the same form.

Several remarks should be made here. First of all, note that the original variable y no longer appears in (GD). This shows that the duality is really between the primal variables x and the dual slack variables s, with y playing a secondary role.

Secondly, while the feasible and optimal solution sets of (P) and (GP) are the same, we have

$$\min(P) = \langle b, y_0\rangle + \min(GP).$$

The differences between (D) and (GD) are a bit more substantial. The sets of dual feasible (optimal) slack variables remain the same; hence the set of dual feasible (optimal) y can be recovered from (GD), if desired. Moreover, the dual program (GD) is now a *minimization* problem by virtue of (8.3), and we have

$$\max(D) = \langle c, x_0 \rangle - \min(GD).$$

Since $\min(P) = \max(D)$, and $c = A^T y_0 + s_0$, $b = Ax_0$, we have

$$\min(GP)+\min(GD) = \langle c, x_0 \rangle - \langle b, y_0 \rangle = \langle A^T y_0 + s_0, x_0 \rangle - \langle Ax_0, y_0 \rangle = \langle x_0, s_0 \rangle,$$

which is the duality gap of the feasible points x_0 and s_0.

8.5 Strictly Complementary Optimal Solutions

Theorem 8.7. *If both programs in the primal–dual pair of linear programs*

$$\begin{array}{llll} \min & \langle c, x \rangle & & \\ \text{s.t.} & Ax = b & (P) & \\ & x \geq 0, & & \end{array} \qquad \begin{array}{llll} \max & \langle b, y \rangle & & \\ \text{s.t.} & A^T y + s = c & (D) & \\ & s \geq 0, & & \end{array}$$

have solutions, then they have a strictly complementary solution, that is, there exists a solution x^* to (P) and (y^*, s^*) to (D) such that $x^* + s^* > 0$.

Proof. Let A be an $m \times n$ matrix and $z^* := \langle c, x^* \rangle = \langle b, y^* \rangle$ the common optimal value of (P) and (D). Fix an index i, $1 \leq i \leq n$. We claim that either there exists a solution x^* to (P) such that $x_i^* > 0$, or there exists a solution (y^*, s^*) to (D) such that $s_i^* > 0$. Consider the systems

$$(a) \qquad Ax = b,\ x \geq 0,\ \langle c, x \rangle \leq z^*,\ \langle -e_i, x \rangle < 0,$$

$$(b) \qquad \begin{array}{c} -A^T y - \mu + \mu_0 c - \lambda_1 e_i = 0,\ -\langle b, y \rangle + \mu_0 z^* + \lambda_0 = 0, \\ (\mu_0, \mu, \lambda_0, \lambda_1) \geq 0,\ (\lambda_0, \lambda_1) \neq 0. \end{array}$$

It follows from Theorem 3.17 that exactly one of the systems is consistent.

If (a) is consistent and has a solution x^*, then x^* is clearly a solution of (P) with $x_i^* > 0$, and the claim is proved.

Otherwise, (b) is consistent, with a solution $(y, \lambda_0, \lambda_1, \mu_0, \mu)$. We need to show that there exists a solution (y^*, s^*) to (D) such that $s_i^* > 0$. We examine separately the cases $\mu_0 > 0$ and $\mu_0 = 0$.

If $\mu_0 > 0$, we can assume that $\mu_0 = 1$. Then $A^T y + (\mu + \lambda_1 e_i) = c$, and $\langle b, y \rangle = z^* + \lambda_0 \geq z^*$. We conclude that $(y^*, s^*) := (y, \mu + \lambda_1 e_i) \geq 0$ is a solution to (D). Moreover, we have in fact $\langle b, y^* \rangle = z^*$ and $\lambda_0 = 0$ and hence $\lambda_1 > 0$. This proves that $s_i^* > 0$, proving the claim in this case, too.

If $\mu_0 = 0$, then $A^T y = -\mu - \lambda_1 e_i \leq 0$ and $\langle b, y \rangle = \lambda_0$. If x is feasible, then

$$0 \geq \langle x, A^T y \rangle = \langle Ax, y \rangle = \langle b, y \rangle = \lambda_0 \geq 0,$$

and we conclude again that $\lambda_0 = 0$, and hence $\lambda_1 > 0$. Now if $(\overline{y}, \overline{s})$ is a solution to (D), then we have

$$A^T(\overline{y} + y) + (\overline{s} + \mu + \lambda_1 e_i) = c, \quad \langle b, \overline{y} + y \rangle = z^*,$$

so that $(y^*, s^*) := (\overline{y} + y, \overline{s} + \mu + \lambda_1 e_i)$ is an optimal solution to (D) with $s_i^* > 0$.

We have established that for a given index i, there exists an optimal solution pair (x^i, y^i, s^i) to the pair (P) and (D) such that either $x^i > 0$ or $s^i > 0$ (but not both by complementarity). Then the optimal solution pair

$$x^* := \sum_{i=1}^{n} x^i, \quad (y^*, s^*) = \sum_{i=1}^{n} (y^i, s^i)$$

satisfies $x^* + s^* > 0$. □

This result is due to Goldman and Tucker [105]. One of its consequences is that interior-point methods for linear programming generally produce strictly complementary optimal solutions [122]. This result has many important applications in interior-point methods; see Ye [270].

8.6 Exercises

1. Formulate the dual to the linear program

$$\begin{aligned} \min \quad & x_1 - 3x_2 - x_3 \\ \text{s.t.} \quad & 3x_1 - x_2 + 2x_3 \geq 1 \\ & -2x_1 + 4x_2 \leq 12 \\ & -4x_1 + 3x_2 + 3x_3 = 14 \end{aligned}$$

without transforming the linear program in any way.

2. Formulate the dual to the linear program

$$\begin{aligned} \min \quad & 2x_1 + 3x_2 \\ \text{s.t.} \quad & x_1 \geq 125 \\ & x_1 + x_2 \geq 350 \\ & 2x_1 + x_2 \leq 600 \\ & x_1 \geq 0, \; x_2 \geq 0. \end{aligned}$$

3. Consider the linear program

$$\begin{aligned} \min \quad & -x_2 \\ \text{s.t.} \quad & x_1 + x_2 \leq 4 \\ & x_1 - x_2 \geq -2 \\ & 2x_1 + x_2 \leq 4 \\ & x_1 \geq 0, \; x_2 \geq 0. \end{aligned}$$

(a) Convert the linear program into one in standard form.
(b) Determine the dual linear program to the standard form LP in (a).

4. Let $A = [a_1, \ldots, a_n]$ be an $m \times n$ matrix such that the linear program

$$\begin{aligned} \min \quad & c^T x \\ \text{s.t.} \quad & Ax = b, \qquad (P) \\ & x \geq 0, \end{aligned}$$

is feasible. It is well known that the simplex method is an iterative method that moves from a vertex (extreme point) of the polyhedron $X := \{x \in \mathbb{R}^n : Ax = b, x \geq 0\}$ to an adjacent one, along the edges of X.
This problem provides an algebraic characterization of the vertices of X. The *support* of a point $x \in X$ is the set of indices of the positive components of x, that is,

$$\text{supp}(x) := \{i : x_i > 0\}.$$

Prove that a point x satisfies

$$x \in X \text{ is a vertex of } X \text{ if and only if}$$
$$\{a_i\}_{i \in \text{supp}(x)} \text{ is linearly independent,}$$

by completing the following steps:
(a) If x is a vertex of X, show that $\{a_i\}_{i \in \text{supp}(x)}$ is linearly independent.
Hint: Otherwise, pick a nontrivial solution $\delta \in \mathbb{R}^n$, $\delta_j = 0$ for $j \notin \text{supp}(x)$, to the equation $\sum_{i=1}^n \delta_i a_i = 0$. Show that if $\epsilon > 0$ is small enough, then the vectors $y = x + \epsilon\delta$ and $y = x - \epsilon\delta$ belong to X, $y \neq z$, and satisfy $x = (y + z)/2$.
(b) Conversely, suppose that x is a point in X such that $\{a_i\}_{i \in \text{supp}(x)}$ is linearly independent. Suppose that $x = (1-t)y + tz$ for some $0 < t < 1$, $x \in X$, $y \in X$. Prove that

$$\text{supp}(x) = \text{supp}(y) \cup \text{supp}(z),$$

$$\sum_{i \in \text{supp}(x)} x_i a_i = \sum_{i \in \text{supp}(y)} y_i a_i = \sum_{i \in \text{supp}(z)} z_i a_i = 0,$$

and show that these imply $x = y = z$.

5. Consider the primal–dual linear programming pair

$$\begin{aligned} \min \quad & c^T x \\ \text{s.t.} \quad & Ax = b \quad (P) \\ & x \geq 0, \end{aligned} \qquad\qquad \begin{aligned} \max \quad & b^T y \\ \text{s.t.} \quad & A^T y \leq c, \end{aligned} \quad (D)$$

where A is an $m \times n$ matrix.
(a) Suppose that (D) has a bounded, nonempty, feasible solution set. Show that $A(\{x : x \geq 0\}) = R^m$. Thus, in particular, A has linearly independent rows.
Hint: Show that (D) has an optimal solution when $b \in R^m$ is arbitrary, and then use the strong duality theorem of linear programming.

(b) What can you conclude about A when (P) has a bounded feasible solution set?

(c) Suppose that both (P) and (D) are feasible linear programs, that is, their feasible solution sets are nonempty. Introduce slack variables $0 \le s \in R^m$ in (D) such that $A^T y + s = c$. Consider the pair (x_i, s_i), where x_i is the ith coordinate of a feasible x for (P) and s_i is the ith coordinate of a feasible s for (D).
Prove that for each i, one of x_i, s_i must be bounded, the other unbounded.
Hint: Suppose, say, x_i is bounded. Then (P), but with the objective function $-x_i$, has an optimal solution. Then use LP duality. Use a similar argument when s_i is bounded.

6. Let A be an $m \times n$ matrix and $p \in \mathbb{R}^n$. Consider the linear program

$$\begin{aligned} \min \quad & t \\ \text{s.t.} \quad & Ay = 0, \\ & p^T y - t = -1, \\ & y \ge 0,\ t \ge 0. \end{aligned} \qquad (P)$$

(a) Formulate the dual program (D) as an explicitly written linear program.

(b) Show that both (P) and (D) have optimal solutions.
Hint: Use LP duality.

(c) Let v^* be the common optimal objective value of (P) and (D). Show that it follows from (P) and (a) that $0 \le v^* \le 1$. In fact, show that v^* is equal to either zero or one.
Hint: Use complementarity for the last part.

7. **(Klain and Rota [166])** Let $c_1 \ge c_2 \ge \cdots \ge c_n > 0$ be given positive constants. Consider the linear program

$$\begin{aligned} \min \quad & \sum_{k=1}^{n} \frac{x_k}{c_k} \\ & 0 \le x_k \le c_k, \\ & \sum_{k=1}^{n} x_k \ge \sum_{i=1}^{r} c_i. \end{aligned} \qquad (P)$$

Show that the optimal objective value of (P) is r, that is, $\min(P) = r$; moreover, if $c_1 > c_2 > \cdots > c_n > 0$, then the optimal solution to (P) is unique, and is given by $x_i = c_i$ for $i = 1, \ldots, r$ and $x_i = 0$ for $i > r$.
Note that this proves the inequality that $\sum_{k=1}^{n} \frac{x_k}{c_k} \ge r$ for all x_k satisfying the conditions $0 \le x_k \le c_k$ and $x_1 + x_2 + \cdots + x_n \ge c_1 + c_2 + \cdots + c_r$.
Hint: Formulate the dual linear program. Define the index sets $I_1 = \{i : x_i = c_i, i = 1, \ldots, n\}$, $I_2 = \{i : 0 < x_i < c_i, i = 1, \ldots, n\}$, $I_3 = \{i : x_i = 0, i = 1, \ldots, n\}$, and the dual variables y_i, λ corresponding to the

constraints $x_i \leq c_i$ and $x_1 + x_2 + \cdots + x_n \geq c_1 + \cdots + c_r$, respectively. Show that $\lambda \neq 0$ and determine y_i. Show that we can always choose $I_1 \subseteq \{1, \ldots, r\}$, and that we have equality if $I_2 = \emptyset$.

9

Nonlinear Programming

A *nonlinear program*, or a *mathematical program*, is a constrained optimization (say minimization) problem having the form

$$\begin{aligned} \min \quad & f(x) \\ \text{s.t.} \quad & g_i(x) \le 0, \quad i = 1, \dots, r, \qquad (P) \\ & h_j(x) = 0, \quad j = 1, \dots, m, \end{aligned} \tag{9.1}$$

where f, $\{g_i\}_1^r$, and $\{h_j\}_1^m$ are real-valued functions defined on some subsets of $\mathbb{R}^n$. The function f is called the *objective function* of (P), and the inequalities and equalities involving g_i and h_j, respectively, are called the *constraints* of the problem. The *feasible region (or constraint set)* of (P) is the set of all points satisfying all the constraints,

$$\mathcal{F}(P) = \{x \in \mathbb{R}^n \ : \ g_i(x) \le 0, i = 1, \dots, r, \ h_j(x) = 0, j = 1, \dots, m\}.$$

Definition 9.1. *A feasible point $x^* \in \mathcal{F}(P)$ is called a* local minimizer *of (P) if x^* is a minimizer of f on a feasible neighborhood of x^*, that is, there exists $\epsilon > 0$ such that*

$$f(x^*) \le f(x) \text{ for all } x \in \mathcal{F}(P) \cap \overline{B}_\epsilon(x^*).$$

The point x^* is called a global minimizer *of (P) if*

$$f(x^*) \le f(x) \ \text{ for all } \ x \in \mathcal{F}(P).$$

Local and global maximizers are defined similarly, by changing the directions of the above inequalities.

The geometry of the feasible set $\mathcal{F}(P)$ around a local minimizer $x^* \in \mathcal{F}(P)$ dictates the optimality conditions that x^* has to satisfy. For example, it should be clear that if $g_i(x^*) < 0$, then the constraint function g_i plays no role in determining whether x^* is a local minimizer of (P). We call such a constraint *inactive*. More formally, if $x \in \mathcal{F}(P)$, we denote by

O. Güler, *Foundations of Optimization*, Graduate Texts in Mathematics 258,
DOI 10.1007/978-0-387-68407-9_9, © Springer Science+Business Media, LLC 2010

$$I(x) := \{i : g_i(x) = 0\}$$

the index set of the *active constraints* at x. If $i \notin I(x)$, then g_i is an *inactive constraint* at x.

The purpose of this section is to give necessary and sufficient conditions for a feasible point x^* to be a local minimizer of (P).

Toward this goal, we first define some relevant concepts. Recall (Definition 2.28) that a vector $d \in \mathbb{R}^n$ is called a *tangent direction* of a nonempty set $M \subseteq \mathbb{R}^n$ at the point $x \in M$ if there exist a sequence $x_n \in M$ converging to x and a nonnegative sequence α_n such that $\lim_{n\to\infty} \alpha_n(x_n - x) = d$. We also say that d is a tangent direction of the sequence $\{x_n\}$.

Definition 9.2. *Let x^* be a feasible point of (P) in (9.1). A tangent direction of $\mathcal{F}(P)$ at x^* is called a* feasible direction *of (P) at x^*. We denote the set of feasible directions of (P) at x^* by $\mathcal{FD}(x^*)$.*

A vector $d \in \mathbb{R}^n$ is called a descent direction *for f at x^* if there exists a sequence of points $x_n \to x^*$ in $\mathbb{R}^n$ (not necessarily feasible) with tangent direction d such that $f(x_n) \le f(x^*)$ for all n. If $f(x_n) < f(x^*)$ for all n, we call d a* strict descent direction *for f at x^*. We denote the set of strict descent directions at x^* by $\mathcal{SD}(f; x^*)$.*

Lemma 9.3. *If $x^* \in \mathcal{F}(P)$ is a local minimum of (P), then*

$$\mathcal{FD}(x^*) \cap \mathcal{SD}(f; x^*) = \emptyset.$$

Proof. The lemma is obvious: if the intersection is not empty, then there exists a sequence of feasible points $x_n \to x^*$ such that $f(x_n) < f(x^*)$, which contradicts our assumption that x^* is a local minimizer of (P). □

Although this lemma is very important from a conceptual point of view, it is hard to extract meaningful results from it, since the sets $\mathcal{FD}(x^*)$ and $\mathcal{SD}(f; x^*)$ are difficult to describe in a useful way, unless the functions f, g_i, h_j appearing in (P) have additional useful properties, such as differentiability.

9.1 First-Order Necessary Conditions (Fritz John Optimality Conditions)

The Fritz John (FJ) conditions are first-order necessary conditions for a local minimizer in the nonlinear program (P) in (9.1) when all the functions involved, f, g_i, h_j, are continuously differentiable in an open neighborhood of the feasible region $\mathcal{F}(P)$.

Let us define the *linearized* versions of the feasible and strict descent directions defined above,

$$\begin{aligned}\mathcal{LFD}(x^*) &:= \{d : \langle \nabla g_i(x^*), d\rangle < 0, \quad i = 1, \ldots, r,\\ &\qquad \langle \nabla h_j(x^*), d\rangle = 0, \quad j = 1, \ldots, m, \},\\ \mathcal{LSD}(f; x^*) &:= \{d : \langle \nabla f(x), d\rangle < 0\}.\end{aligned}$$

To motivate these definitions, let F be a differentiable function defined on an open set in $\mathbb{R}^n$. If $d \in \mathcal{LSD}(F; x)$, then

$$F(x + td) = F(x) + t\left[\langle \nabla F(x), d\rangle + \frac{o(t)}{t}\right] < F(x),$$

for all $t > 0$ small, since $\langle \nabla F(x), d\rangle < 0$ and $\lim_{t\to 0} o(t)/t = 0$, so that the term inside the brackets is negative. Thus, if $d \in \mathcal{LFD}(x^*) \cap \mathcal{LSD}(f; x^*)$ and $t > 0$ is small enough, then $f(x^* + td) < f(x^*)$ and $g_i(x^* + td) < g_i(x^*) = 0$ for an active constraint function g_i. The requirement that $\langle \nabla h_j(x), d\rangle = 0$ is more delicate and will be handled below.

Theorem 9.4. (Fritz John) *If a point x^* is a local minimizer of (P), then there exist multipliers $(\lambda, \mu) := (\lambda_0, \lambda_1, \ldots, \lambda_r, \mu_1, \ldots, \mu_m)$, not all zero, $(\lambda_0, \lambda_1, \ldots, \lambda_r) \geq 0$, such that*

$$\lambda_0 \nabla f(x^*) + \sum_{i=1}^{r} \lambda_i \nabla g_i(x^*) + \sum_{j=1}^{m} \mu_j \nabla h_j(x^*) = 0, \tag{9.2}$$

$$\lambda_i \geq 0,\ g_i(x^*) \leq 0,\ \lambda_i g_i(x^*) = 0,\ \ i = 1, \ldots, r. \tag{9.3}$$

Proof. Because of the "complementarity conditions" (9.3), we may write (9.2) in the form

$$\lambda_0 \nabla f(x^*) + \sum_{j\in I(x^*)} \lambda_j \nabla g_j(x^*) + \sum_{j=1}^{m} \mu_j \nabla h_j(x^*) = 0.$$

If the vectors $\{\nabla h_i(x^*)\}_1^m$ are linearly *dependent*, then there exist multipliers $\mu := (\mu_1, \ldots, \mu_m) \neq 0$ such that $\sum_{j=1}^m \mu_j \nabla h_j(x^*) = 0$. Then setting $\lambda := (\lambda_0, \ldots, \lambda_r) = 0$, we see that the theorem holds with the multipliers $(\lambda, \mu) \neq 0$.

Assume now that $\{\nabla h_j(x^*)\}_{j=1}^m$ are linearly *independent*. We claim that

$$\begin{aligned}\Big\{d : \langle \nabla f(x^*), d\rangle < 0,\ \ \langle \nabla g_i(x^*), d\rangle < 0,\ i \in I(x^*),&\\ \langle \nabla h_j(x^*), d\rangle = 0,\ j = 1, \ldots, m\Big\} = \emptyset.&\end{aligned} \tag{9.4}$$

Suppose that (9.4) is false, and pick a direction d, $\|d\| = 1$, in the set above. Since $\{\nabla h_j(x^*)\}_1^m$ is linearly independent, it follows from Lyusternik's theorem (see Theorem 2.29 or Theorem 3.23) that there exists a sequence $x_n \to x^*$ that has tangent direction d and satisfies the equations $h_j(x_n) = 0$, $j = 1, \ldots, m$. We also have

$$f(x_n) = f(x^*) + \left[\langle \nabla f(x^*), \frac{x_n - x^*}{\|x_n - x^*\|}\rangle + \frac{o(x_n - x^*)}{\|x_n - x^*\|}\right] \cdot \|x_n - x^*\|,$$

where $(x_n - x^*)/\|x_n - x^*\| \to d$, and $o(x_n - x^*)/\|x_n - x^*\| \to 0$ as $n \to \infty$. It follows that the term inside the brackets is negative and thus $f(x_n) < f(x^*)$ for sufficiently large n. The same arguments show that if g_i is an active constraint function at x^*, then $g_i(x_n) < g_i(x^*) = 0$ for sufficiently large n. We conclude that $\{x_n\}_1^\infty$ is a feasible sequence for (P) such that $f(x_n) < f(x^*)$ for large enough n. This contradicts our assumption that x^* is a local minimizer of (P), and proves (9.4).

The theorem follows immediately from (9.4) using the homogeneous version of Motzkin's transposition theorem; see Theorem 3.15, Theorem 7.17, or Theorem A.3. □

Definition 9.5. *The function*

$$L(x; \lambda, \mu) := \lambda_0 f(x) + \sum_{i=1}^{r} \lambda_i g_i(x) + \sum_{j=1}^{m} \mu_j h_j(x) \quad (\lambda_i \geq 0, i = 0, \ldots, r)$$

is called the weak Lagrangian function *for* (P).

If $\lambda_0 > 0$*, then we may assume without loss of generality that* $\lambda_0 = 1$*, and the resulting function,*

$$L(x, \lambda, \mu) = f(x) + \sum_{i=1}^{r} \lambda_i g_i(x) + \sum_{j=1}^{m} \mu_j h_j(x), \quad \lambda_i \geq 0,\ i = 1, \ldots, r,$$

is called the Lagrangian function.

The Lagrangian function is named in honor of Lagrange, who first introduced an analogue of the function L in the eighteenth century in order investigate optimality conditions in calculus of variations problems.

We remark that the equality (9.3) in the FJ conditions can be written as

$$\nabla_x L(x, \lambda, \mu) = 0.$$

The conditions expressed in (9.3) are called complementarity conditions, since

$$\lambda_i g_i(x^*) = 0, \quad \lambda_i \geq 0, \quad g_i(x^*) \leq 0,$$

imply that either $\lambda_i = 0$ or $g_i(x^*) = 0$. In particular, if $g_i(x^*) < 0$, that is, g_i is inactive at x^*, then $\lambda_i = 0$. It is possible that a constraint is active and the corresponding multiplier is zero. For example, in Exercise 17 on page 245, this happens at every KKT point. Otherwise, we say that *strict complementarity* holds at x^*. In every linear program with an optimal solution, there always exists a strictly complementary solution; see Theorem 8.7 on page 203. This fact, first proved by Goldman and Tucker [105], plays an important role in *interior-point methods* for linear programming.

The Fritz John theorem is remarkable, since it always holds at a local minimizer. However, we will see later (Example 9.8, page 218) an example of an abnormal nonlinear program in which $\lambda_0 = 0$. This is an awkward situation, since the objective function f is not involved in the first necessary optimality conditions! Additional assumptions on (P) must be made in order to rule out this possibility. Such assumptions that ensure $\lambda_0 > 0$ (in fact $\lambda_0 = 1$) are called *constraint qualifications*, and the resulting optimality conditions are called *Karush–Kuhn–Tucker (KKT) conditions*.

Corollary 9.6. *If the vectors*

$$\{\nabla g_i(x^*),\ i \in I(x^*),\ \ \nabla h_j(x^*),\ j = 1, \ldots, m\}$$

are linearly independent, then $\lambda_0 > 0$ *and we have*

$$\nabla f(x^*) + \textstyle\sum_{i=1}^{r} \lambda_i \nabla g_i(x^*) + \sum_{j=1}^{m} \mu_j \nabla h_j(x^*) = 0, \tag{9.5}$$

$$\lambda_i \geq 0,\ g_i(x^*) \leq 0,\ \lambda_i g_i(x^*) = 0, \quad i = 1, \ldots, r, \tag{9.6}$$

$$h_j(x^*) = 0, \quad j = 1, \ldots, m. \tag{9.7}$$

Proof. If $\lambda_0 = 0$, then

$$\sum_{i \in I(x^*)} \lambda_i \nabla g_i(x^*) + \sum_{j=1}^{m} \mu_j \nabla h_j(x^*) = 0.$$

The linear independence hypothesis of the vectors implies that the multipliers $\{\lambda_i\}_1^r$ and $\{\mu_j\}_1^m$ are all zero. But then the entire multiplier vector (λ, μ) is zero, contradicting the above theorem. □

The conditions (9.5)–(9.7) in Corollary 9.6 are the Karush–Kuhn–Tucker (KKT) conditions for the problem (P). The assumption of the linear independence of the gradient vectors in the statement of the corollary is an example of constraint qualification. There are other, less stringent, constraint qualifications that imply the KKT conditions. These will be discussed in Section 9.5 below.

9.2 Derivation of Fritz John Conditions Using Penalty Functions

The derivation of the FJ conditions using penalty functions originates from McShane's article [195], and uses little beyond elementary calculus. Nevertheless, we remark that the penalty function approach has close connections with the previous approach outlined above, since the implicit function theorem and Lyusternik's theorem can also be proved using this kind of approach; see Theorem 2.26 on page 45.

We now prove Theorem 9.4 using a penalty function approach.

Proof. Consider the penalty function

$$F_k(x) = f(x) + \frac{k}{2}\sum_{i=1}^{r} g_i^+(x)^2 + \frac{k}{2}\sum_{j=1}^{m} h_j^2(x) + \frac{1}{2}\|x - x^*\|^2,$$

where $g_i^+(x) = \max\{0, g_i(x)\}$ and $k > 0$ is a parameter. Pick an $\epsilon > 0$ small enough that $f(x^*) \le f(x)$ for all feasible $x \in \overline{B}_\epsilon(x^*)$.

Let x_k be a global minimizer of F_k over $\overline{B}_\epsilon(x^*)$, which exists by the Weierstrass theorem. Since $F_k(x^*) = f(x^*)$, we have

$$\begin{aligned} f(x_k) &\le f(x_k) + \frac{k}{2}\sum_{i=1}^{r} g_i^+(x_k)^2 + \frac{k}{2}\sum_{j=1}^{m} h_j^2(x_k) + \frac{1}{2}\|x_k - x^*\|^2 \\ &= F_k(x_k) \le F_k(x^*) = f(x^*). \end{aligned} \tag{9.8}$$

The functions g_i^+, h_j, and f are all bounded on $\overline{B}_\epsilon(x^*)$, and (9.8) shows that $kg_i^+(x_k)^2/2$, $kh_j^2(x_k)/2$ are also bounded. Thus, we have $g_i^+(x_k) \to 0$ and $h_j(x_k) \to 0$ as $k \to \infty$. Let $\overline{x} \in \overline{B}_\epsilon(x^*)$ be a limit point of the sequence $\{x_k\}_0^\infty$. We have $g_i^+(\overline{x}) = 0$ (that is, $g_i(\overline{x}) \le 0$) and $h_j(\overline{x}) = 0$. This proves that $\overline{x}$ is a feasible point of (P).

Taking limits in (9.8), we obtain

$$f(\overline{x}) \le f(\overline{x}) + \frac{1}{2}\|\overline{x} - x^*\|^2 \le f(x^*).$$

Since $f(x^*) \le f(x)$ for all feasible $x \in \overline{B}_\epsilon(x^*)$, we have also $f(x^*) \le f(\overline{x})$. This and the above inequalities immediately imply that $\|\overline{x} - x^*\|^2 = 0$, that is, $\overline{x} = x^*$.

Consequently, the minimization problem

$$\min\{F_k(x) : x \in \overline{B}_\epsilon(x^*)\}$$

becomes an unconstrained optimization problem for large enough k. Therefore, $\nabla F_k(x_k) = 0$, that is,

$$\begin{aligned} &\nabla f(x_k) + \sum_{i=1}^{r}(kg_i^+(x_k))\nabla g_i(x_k) + \sum_{j=1}^{m}(kh_j(x_k^*))\nabla h_j(x_k) \\ &\quad + (x_k - x^*) = 0. \end{aligned} \tag{9.9}$$

Define $\alpha_{i,k} = kg_i^+(x_k)$ and $\beta_{j,k} = kh_j(x_k^*)$, scale the vector

$$(1, \alpha_{1,k}, \ldots, \alpha_{r,k}, \beta_{1,k}, \ldots, \beta_{m,k}, 1)$$

so that the sum of the absolute values of the entries is 1 (that is, divide the entries by $\gamma_k := 2 + \sum_{i=1}^{r} \lambda_{i,k} + \sum_{j=1}^{m} |\beta_{j,k}|$), and denote the resulting vector by

$$\alpha_k := (\lambda_{0,k}, \lambda_{1,k}, \ldots, \lambda_{r,k}, \mu_{1,k}, \ldots, \mu_{m,k}, \lambda_{0,k}).$$

Dividing both sides of (9.9) by γ_k gives

$$\lambda_{0,k}\nabla f(x_k) + \sum_{i=1}^{r} \lambda_{i,k}\nabla g_i(x_k) + \sum_{j=1}^{m} \mu_{j,k}\nabla h_j(x_k) + \lambda_{0,k}(x_k - x^*) = 0.$$

Since the entries of α_k are bounded, we may assume that they converge as $k \to \infty$ (otherwise, we can take a convergent subsequence). Taking the limit as $k \to \infty$, we define $\lambda_{i,k} \to \lambda_i$ $(i = 0, \ldots, r)$, $\mu_{j,k} \to \mu_j$ $(j = 1, \ldots, m)$. We also have $\lambda_{0,k}(x_k - x^*) \to 0$, since $\lambda_{0,k}$ is bounded and $x_k \to x^*$. Thus,

$$\lambda_0\nabla f(x^*) + \sum_{i=1}^{r} \lambda_i\nabla g_i(x^*) + \sum_{j=1}^{m} \mu_j\nabla h_j(x^*) = 0.$$

Finally, we note that $\lambda_i = 0$ for an inactive constraint due to the fact that $x_k \to x^*$ and thus $kg_i(x_k)^+ = 0$ for large enough k. The theorem is proved.

□

9.3 Derivation of Fritz John Conditions Using Ekeland's ϵ-Variational Principle

In this section, we give a third, independent proof of the Fritz John conditions for the nonlinear program (9.1) using Ekeland's ϵ-variational principle. The only other tool used is Danskin's theorem; hence the proof below could have been given as early as in Chapter 3. Moreover, it has the added merit that it is valid in a Banach space, that is, if the functions f, g_i, h_j are defined in a Banach space; see [86] for the original proof. There are variants of the FJ optimality conditions for suitable nonsmooth functions that are proved using Ekeland's ϵ-variational principle; see [59, 60].

The proof below is an adaptation of the one in [60].

Proof. Let x^* be a local minimizer of problem (9.1), where $f(x^*) \le f(x)$ for all feasible x in a closed ball $C = \overline{B}_r(x^*) = \{x : \|x - x^*\| \le r\}$. Define the set

$$T := \{(\lambda_0, \lambda, \mu) \in \mathbb{R} \times \mathbb{R}^r \times \mathbb{R}^m \;:\; (\lambda_0, \lambda) \ge 0,\; \|(\lambda_0, \lambda, \mu)\| = 1\},$$

and for a given $\epsilon > 0$, where $\sqrt{\epsilon} < r$, define the function

$$F(x) := \max_T \Big\{\lambda_0\big(f(x) - f(x^*) + \epsilon\big) + \sum_{i=1}^{r} \lambda_i g_i(x) + \sum_{j=1}^{m} \mu_j h_j(x)\Big\}. \tag{9.10}$$

It is easy to see that $F(x^*) = \epsilon$. Moreover, the function F is positive on C, because if $x \in C$ and $F(x) \le 0$, then choosing $\lambda_0 = 1$, $\lambda_i = 1$, and $|\mu_j| = 1$, respectively, we obtain $f(x) \le f(x^*) - \epsilon$, $g_i(x) \le 0$, and $h_j(x) = 0$, that is,

x is feasible for problem (9.1). This is a contradiction, since x^* is a global minimizer of problem (9.1) on C. Therefore, we have

$$F(x^*) \le \inf_C F(x) + \epsilon.$$

It follows from Ekeland's ϵ-variational principle (Corollary 3.3) that there exists a point x_ϵ satisfying $\|x_\epsilon - x^*\| \le \sqrt{\epsilon}$ and

$$F(x_\epsilon) \le F(x) + \sqrt{\epsilon}\|x - x_\epsilon\| \quad \text{for all} \quad x \in C.$$

Thus, the point x_ϵ minimizes the function

$$G(x) := F(x) + \sqrt{\epsilon}\|x - x_\epsilon\|$$

on the ball $C = \overline{B}_r(x^*)$. Since $\|x_\epsilon - x^*\| \le \sqrt{\epsilon} < r$, we have $x_\epsilon \in \text{int}(C)$, and since $F(x_\epsilon) > 0$, the maximum in (9.10) is achieved at a unique point $\big(\lambda_0(\epsilon), \lambda(\epsilon), \mu(\epsilon)\big) \in T$. It follows from Danskin's theorem (Theorem 1.29) that if d is any unit vector in $\mathbb{R}^n$, then

$$\begin{aligned} G'(x_\epsilon; d) &= \Big\langle \lambda_0(\epsilon)\nabla f(x_\epsilon) + \sum_{i=1}^r \lambda_i(\epsilon)\nabla g_i(x_\epsilon) + \sum_{j=1}^m \mu_j(\epsilon)\nabla h_j(x_\epsilon),\ d\Big\rangle + \sqrt{\epsilon} \\ &= 0, \end{aligned}$$

where we used the fact that the function $N(x) = \|x - x_\epsilon\|$ has directional derivative $N'(x_\epsilon; d) = \|d\| = 1$. It follows that $\|\nabla F(x_\epsilon\| \le \sqrt{\epsilon}$, that is,

$$\Big\|\lambda_0(\epsilon)\nabla f(x_\epsilon) + \sum_{i=1}^r \lambda_i(\epsilon)\nabla g_i(x_\epsilon) + \sum_{j=1}^m \mu_j(\epsilon)\nabla h_j(x_\epsilon)\Big\| \le \sqrt{\epsilon}.$$

By the compactness of T, as $\epsilon \to 0$, there exists a convergent sequence $(\lambda_0(\epsilon), \lambda(\epsilon), \mu(\epsilon)) \to (\lambda_0, \lambda, \mu) \in T$. Since $x_\epsilon \to x^*$, we obtain

$$\lambda_0\nabla f(x^*) + \sum_{i=1}^r \lambda_i\nabla g_i(x^*) + \sum_{j=1}^m \mu_j\nabla h_j(x^*) = 0.$$

It is clear from (9.10) that if $g_i(x^*) < 0$, then $\lambda_i(\epsilon) = 0$ for small enough ϵ; hence the complementarity condition $\lambda_i g_i(x^*) = 0$ holds. □

9.4 First-Order Sufficient Optimality Conditions

In this short section, we give a sufficient condition for local minimizer in a nonlinear programming problem that seems to have been overlooked in the optimization literature but deserves to be better known. Perhaps it has been neglected because in a constrained optimization problem, one often imagines second-order sufficient optimality conditions and not first-order ones, analogous to the situation in unconstrained optimization. The original version of this optimality condition is due to Fritz John [148] and deals with a semi-infinite programming problem (see Chapter 12) in which he deals with only inequality constraints. It is interesting that the optimality condition below is valid even in abnormal cases in which $\lambda_0 = 0$.

Theorem 9.7. *Let x^* be a feasible solution to the optimization problem (P) in* (9.1), *satisfying the FJ conditions* (9.2) *and* (9.3), *where we write* (9.2) *in the form*

$$\lambda_0 \nabla f(x^*) + \sum_{i \in I(x^*)} \lambda_i \nabla g_i(x^*) + \sum_{j=1}^{m} \mu_j \nabla h_j(x^*) = 0.$$

If the totality of the vectors

$$\lambda_0 \nabla f(x^*), \quad \{\lambda_i \nabla g_i(x^*)\}_{i \in I(x^*)}, \quad \{\nabla h_j(x^*)\}_1^m,$$

span $\mathbb{R}^n$, *then x^* is a local minimizer of (P).*

Proof. Suppose that x^* is not a local minimizer of (P). Then there exists a feasible sequence of points $x_k \to x^*$ satisfying $f(x_k) < f(x^*)$. Writing $x_k = x^* + t_k d_k$ with $t_k > 0$, $\|d_k\| = 1$, we have

$$\begin{aligned}
&0 > f(x^* + t_k d_k) - f(x^*) = t_k \langle \nabla f(x^*), d_k \rangle + o(t_k), \\
&0 \geq g_i(x^* + t_k d_k) = t_k \langle \nabla g_i(x^*), d_k \rangle + o(t_k), \ i \in I(x^*), \\
&0 = h_j(x^* + t_k d_k) = t_k \langle \nabla h_j(x^*), d_k \rangle + o(t_k), \ j = 1, \ldots, m.
\end{aligned}$$

Since $\|d_k\| = 1$, we can assume, by taking a subsequence if necessary, that $d_k \to d$, $\|d\| = 1$. Dividing both sides of all equalities and inequalities above by t_k and letting $t_k \to 0$ gives $\langle \nabla f(x^*), d \rangle \leq 0$, $\langle \nabla g_i(x^*), d \rangle \leq 0$, and $\langle \nabla h_j(x^*), d \rangle = 0$. Since

$$\langle \lambda_0 \nabla f(x^*), d \rangle + \sum_{i \in I(x^*)} \lambda_i \langle \nabla g_i(x^*), d \rangle + \sum_{j=1}^{m} \mu_j \langle \nabla h_j(x^*), d \rangle = 0,$$

we have $\langle \lambda_0 \nabla f(x^*), d \rangle = 0$, $\langle \lambda_i \nabla g_i(x^*), d \rangle = 0$, and $\langle \nabla h_j(x^*), d \rangle = 0$. By virtue of our assumption on the gradient vectors, the vector d is orthogonal to every vector in $\mathbb{R}^n$. This implies $d = 0$, contradicting $\|d\| = 1$. □

9.5 Constraint Qualifications

As we discussed above, it is of interest whether λ_0 is zero or positive in the FJ conditions for optimality conditions. If $\lambda_0 = 0$, then $f(x)$ plays no role in determining the local optimizer, a very peculiar situation indeed, since we expect f, the function to be optimized, to be present in the FJ conditions.

However, there exist problems in which $\lambda_0 = 0$, as the following example shows.

Example 9.8. **(Failure of the KKT conditions)**
Consider the problem

$$\begin{aligned} \min\ & -x \\ \text{s.t.}\ & (x-1)^3 + y \le 0, \\ & x \ge 0,\ y \ge 0. \end{aligned}$$

We make the definitions $f(x,y) = -x$, $g_1(x,y) = (x-1)^3 + y \le 0$, $g_2(x,y) = -x \le 0$, and $g_3(x,y) = -y \le 0$. It is clear from a picture of the feasible region that the point $(1,0)$ is the (global) minimizer of the problem. Thus, the FJ conditions must hold there.

The gradients of the objection function and the two active constraints are $\nabla f(1,0) = (-1,0)$, $\nabla g_1(1,0) = (0,1)$, and $\nabla g_3(1,0) = (0,-1)$. The equation

$$\lambda_0 \nabla f(1,0) + \lambda_1 \nabla g_1(1,0) + \lambda_3 \nabla g_3(1,0) = (-\lambda_0, \lambda_1 - \lambda_3) = (0,0)$$

gives $\lambda_0 = 0$. Thus, *the KKT conditions fail at the optimal point* $(1,0)$.

Consequently, it is useful to identify additional conditions on the objective function f, and especially on the constraint functions g_i and h_j that guarantee that $\lambda_0 > 0$, that is, the KKT conditions hold. Any such condition is called a *constraint qualification*. Several such conditions are known in the literature. We already met one such condition in Corollary 9.6. We discuss some more conditions below.

First, however, we give a necessary and sufficient condition for the existence of the KKT multipliers.

Theorem 9.9. *Let x^* be an FJ point for problem (P) in (9.1). The KKT conditions*

$$\nabla f(x^*) + \sum_{i \in I(x^*)} \lambda_i \nabla g_i(x^*) + \sum_{j=1}^{m} \mu_0 \nabla h_j(x^*) = 0 \tag{9.11}$$

hold at x^* if and only if

$$\begin{aligned} &\{d : \langle \nabla f(x^*), d\rangle < 0\} \cap \{d : \langle \nabla g_i(x^*), d\rangle \le 0,\ i \in I(x^*)\} \\ &\quad \cap \{d : \langle \nabla h_j(x^*), d\rangle = 0,\ j = 1, \ldots, m\} = \emptyset. \end{aligned} \tag{9.12}$$

Proof. The equivalence of (9.12) and (9.11) follows immediately from the homogeneous version of Motzkin's transposition theorem. □

We remark that the difference between the FJ and KKT conditions is seemingly very small. In the FJ conditions, at a local minimizer x^* we must have

$$\begin{aligned} &\{d : \langle \nabla f(x^*), d\rangle < 0\} \cap \{d : \langle \nabla g_i(x^*), d\rangle < 0,\ i \in I(x^*)\} \\ &\quad \cap \{d : \langle \nabla h_j(x^*), d\rangle = 0,\ j = 1, \ldots, m\} = \emptyset, \end{aligned}$$

whereas in the KKT conditions we demand somewhat stronger conditions in that the strict inequalities $\langle \nabla g_i(x^*), d\rangle < 0$ for active constraints $g_i(x)$ are replaced by the weak inequalities $\langle \nabla g_i(x^*), d\rangle \le 0$.

Corollary 9.10. (Concave and linear constraints) *Let x^* be a local minimizer of problem (P) in (9.1). The KKT conditions hold at x^* if the active constraints $\{g_i\}_{i\in I(x^*)}$ are concave functions in a convex neighborhood of x^* and the equality constraints $\{h_j\}_1^m$ are affine functions on $\mathbb{R}^n$.*

In particular, the KKT conditions hold at every local minimizer if all the constraint functions g_i and h_j are affine, that is,

$$g_i(x) = \langle a_i, x\rangle + \alpha_i, \quad h_j(x) = \langle b_j, x\rangle + \beta_j.$$

Proof. Let d satisfy the conditions

$$\langle \nabla g_i(x^*), d\rangle \le 0,\ i \in I(x^*), \quad \langle \nabla h_j(x^*), d\rangle = 0,\ j = 1, \dots, m.$$

The point $x(t) = x^* + td$ is feasible for small $t > 0$, since

$$g_i(x^* + td) \le g_i(x^*) + t\langle \nabla g_i(x^*), d\rangle \le 0,$$

by Theorem 4.27; similarly

$$h_j(x^* + td) = h_j(x^*) + \langle \nabla h_j(x^*), d\rangle = 0,$$

because h_j is an affine function. Since x^* is a local minimizer of the problem (P), we have $f'(x; d) = \langle \nabla f(x^*), d\rangle \ge 0$ and (9.12) holds. It follows from Theorem 9.9 that the KKT conditions hold. □

We next prove the *Mangasarian–Fromovitz constraint qualification.*

Theorem 9.11. (Mangasarian–Fromovitz [193]) *Let x^* be an FJ point for problem (P) in (9.1). If the gradients $\{\nabla h_j(x^*)\}_1^m$ of the equality constraints are linearly independent and there exists a direction d satisfying the conditions*

$$\langle \nabla g_i(x^*), d\rangle < 0,\ i \in I(x^*), \quad \langle \nabla h_j(x^*), d\rangle = 0,\ j = 1, \dots, m, \tag{9.13}$$

then the KKT conditions are satisfied at x^*.

Proof. On the one hand, since (9.13) is consistent, the homogeneous version of Motzkin's transposition theorem implies that in any solution $0 \le \lambda := (\lambda_i : i \in I(x^*))$ and $\mu : (\mu_1, \dots, \mu_m)$ to the equation

$$\sum_{i\in I(x^*)} \lambda_i \nabla g_i(x^*) + \sum_{j=1}^{m} \mu_j \nabla h_j(x^*) = 0, \tag{9.14}$$

we must have $\lambda = 0$, and then $\mu = 0$ as well, since the gradients $\nabla h_j(x^*)$ are linearly independent.

On the other hand, Theorem 9.4 implies that if $\lambda_0 = 0$, then (9.14) has a solution with $(\lambda, \mu) \ne 0$. It follows that $\lambda_0 > 0$. □

One of the earliest and best known constraint qualifications is *Slater's constraint qualification*, which applies to nonlinear programs with convex constraints.

Corollary 9.12. (Slater [243]) *Let the functions $\{g_i\}_1^r$ in (9.1) be convex, and the functions $\{h_j\}_1^r$ affine. Let x^* be a local minimizer of problem (P). If there exists a feasible point x_0,* strictly feasible *for the active constraints g_i, that is,*

$$g_i(x_0) < 0, \quad i \in I(x^*),$$

then the KKT conditions are satisfied at x^*.

Proof. Let $h_j(x) = \langle a_j, x - x_0 \rangle$, $j = 1, \ldots, m$. If $\{a_j\}_1^m$ is linearly dependent, then we can choose a linearly independent subset of it, say $\{a_j\}_1^k$, such that $\text{span}\{a_1, \ldots, a_k\} = \text{span}\{a_1, \ldots, a_m\}$. Note that keeping only the constraints $\{h_i\}_1^k$ in the formulation of (P) does not change its feasible region.

Thus, we may assume that the gradient vectors $\{\nabla h_j(x^*)\}_1^m$ are linearly independent. We have

$$\begin{aligned} 0 > g_i(x_0) &\geq g_i(x^*) + \langle \nabla g_i(x^*), x_0 - x^* \rangle, \quad i \in I(x^*), \\ 0 = h_j(x_0) &= h_j(x^*) + \langle \nabla h_j(x^*), x_0 - x^* \rangle, \quad j = 1, \ldots, m, \end{aligned}$$

where the second inequality follows from Theorem 4.27. We see that the direction $d := x_0 - x^*$ satisfies the Mangasarian–Fromovitz constraint qualifications, and Theorem 9.11 implies that the KKT conditions hold. □

9.6 Examples of Nonlinear Programs

Example 9.13. This problem and its first-order necessary conditions were considered in the nineteenth century by the great American physicist J. W. Gibbs, one of the founders of thermodynamics:

$$\begin{aligned} \min \ & \sum_{i=1}^n f_i(x_i) \\ \text{s.t.} \ & \sum_{i=1}^n x_i = 1, \\ & x \geq 0. \end{aligned}$$

Since the constraint functions are all linear, it follows from Corollary 9.10 that the KKT conditions hold. We write the Lagrangian function

$$L(x, \lambda, \mu) = \sum_{i=1}^n f_i(x_i) - \sum_{i=1}^n \lambda_i x_i + \mu\Big(1 - \sum_{i=1}^n x_i\Big), \quad \lambda_i \geq 0,\ i = 1, \ldots, n,\ \mu \in \mathbb{R}.$$

The KKT conditions are

$$
\begin{aligned}
&(a) \quad \frac{\partial L}{\partial x_i} = f_i'(x_i) - \mu - \lambda_i = 0, \quad i = 1, \ldots, n, \\
&(b) \quad \sum_{i=1}^{n} x_i = 1, \\
&(c) \quad x_i \geq 0,\ \lambda_i \geq 0,\ x_i\lambda_i = 0, \quad i = 1, \ldots, n.
\end{aligned}
$$

If $x_i > 0$, then (c) implies that $\lambda_i = 0$, and then (a) implies $f_i'(x_i^*) = \mu$. If $x_i^* = 0$, then we have $f_i'(x_i^*) = \mu + \lambda_i \geq \mu$. Thus, we have the following optimality conditions that were known to Gibbs:

$$
\begin{aligned}
f_i'(x_i^*) &= \mu \ \text{ for all } \ i \text{ such that } x_i > 0, \\
f_i'(x_i^*) &\geq \mu \ \text{ for all } \ i \text{ such that } x_i = 0.
\end{aligned}
$$

Note that the problem of projecting a point $A \in \mathbb{R}^n$ onto the standard unit simplex (see Exercise 10, p. 190) is a particular case of this problem in which $f_i(x_i) = (x_i - a_i)^2/2$, $i = 1, \ldots, n$.

Example 9.14. We consider two related problems.

(i)

$$
\begin{aligned}
&\min \ x^2 + 4y^2 + 16z^2 \\
&\text{s.t.} \ \ xy = 1.
\end{aligned}
$$

Since the objective function is coercive, there exist global minimizer(s) to the problem. The constraint function $h(x, y, z) = xy - 1 = 0$ has the gradient $\nabla h(x, y, z) = (y, x, 0) \neq 0$ on the constraint set $xy = 1$. It follows from Corollary 9.6 that the KKT conditions hold. Thus, the Lagrangian function is

$$L = x^2 + 4y^2 + 16z^2 + \lambda(xy - 1),$$

and the KKT conditions are

$$
\begin{aligned}
&(a) \quad \tfrac{\partial L}{\partial x} = 2x + \lambda y = 0, \\
&(b) \quad \tfrac{\partial L}{\partial y} = 8y + \lambda x = 0, \\
&(c) \quad \tfrac{\partial L}{\partial z} = 32z = 0.
\end{aligned}
$$

Multiplying (a) and (b) by x and y, and using $xy = 1$, gives

$$2x^2 = -\lambda = 8y^2 = \frac{8}{x^2}.$$

This gives $x = \mp\sqrt{2}$. If $x = \mp\sqrt{2}$, then $y = 1/x = \mp 1/\sqrt{2}$. Thus, the KKT points are

$$\begin{pmatrix} \sqrt{2} \\ \frac{1}{\sqrt{2}} \\ 0 \end{pmatrix}, \begin{pmatrix} -\sqrt{2} \\ -\frac{1}{\sqrt{2}} \\ 0 \end{pmatrix}.$$

The objective value is the same at both points, so they are both global minimizers.

(ii)

$$\begin{aligned} \min\ & x^2 + 4y^2 + 16z^2 \\ \text{s.t.}\ & xyz = 1. \end{aligned}$$

As in part (i), the constraint function $h(x, y, z) = xyz - 1 = 0$ has a nonzero gradient $\nabla h(x, y, z) = (yz, xz, xy)$ on the constraint set $xyz = 1$, and we may assume that $\lambda_0 = 1$ by virtue of Corollary 9.6. The Lagrangian function can be written as

$$L = x^2 + 4y^2 + 16z^2 + \lambda(xyz - 1),$$

and the KKT conditions are

$$\begin{aligned} &(a) \quad \frac{\partial L}{\partial x} = 2x + \lambda yz = 0, \\ &(b) \quad \frac{\partial L}{\partial y} = 8y + \lambda xz = 0, \\ &(c) \quad \frac{\partial L}{\partial z} = 32z + \lambda xy = 0, \\ &(d) \quad xyz = 1. \end{aligned}$$

Multiplying (a)–(c) by x, y, and z, respectively, and using (d), we obtain

$$x^2 = 4y^2 = 16z^2 = -\frac{\lambda}{2} > 0.$$

These give $x^6 = 64x^2y^2z^2 = 64$, that is, $x = \mp 2$. Since there are four possible choices for the signs of x and y (the sign of z is then determined by (d)), the KKT points are

$$\begin{pmatrix} 2 \\ 1 \\ 1/2 \end{pmatrix}, \begin{pmatrix} 2 \\ -1 \\ -1/2 \end{pmatrix}, \begin{pmatrix} -2 \\ 1 \\ -1/2 \end{pmatrix}, \begin{pmatrix} -2 \\ -1 \\ 1/2 \end{pmatrix},$$

which are all global minimizers of the problem.

Example 9.15. We examine the problem

$$\begin{aligned} \max\ & (x + 1)^2 + (y + 1)^2 \\ \text{s.t.}\ & x^2 + y^2 \le 2, \\ & y \le 1. \end{aligned}$$

To avoid developing the FJ and KKT conditions for a maximization problem, we first convert our problem to a minimization problem:

$$\begin{aligned} \min \quad & \frac{-1}{2}(x+1)^2 - \frac{1}{2}(y+1)^2 \\ \text{s.t.} \quad & x^2 + y^2 - 2 \le 0, \\ & y - 1 \le 0. \end{aligned}$$

The latter problem has global minimizers, since the constraint region is compact. Since the constraints are convex functions, and there exists a strictly feasible solution (for example, the point $(x_0, y_0) = (0, 0)$), Slater's conditions hold. It follows from Corollary 9.12 that the KKT conditions must hold at any local minimizer. Thus, we have the Lagrangian function

$$L(x, y, \lambda) = -\frac{1}{2}(x+1)^2 - \frac{1}{2}(y+1)^2 + \frac{\lambda_1}{2}(x^2 + y^2 - 2) + \lambda_2(y - 1),$$

and the KKT conditions

$$\begin{aligned} (a) \quad & \frac{\partial L}{\partial x} = -(x+1) + \lambda_1 x = 0, \\ (b) \quad & \frac{\partial L}{\partial y} = -(y+1) + \lambda_1 y + \lambda_2 = 0, \\ (c) \quad & \lambda_1 \ge 0, \ x^2 + y^2 \le 2, \ \lambda_1(x^2 + y^2 - 2) = 0, \\ (d) \quad & \lambda_2 \ge 0, \ y \le 1, \ \lambda_2(y - 1) = 0. \end{aligned}$$

Now equations (a) and (b) simplify to the conditions

$$x = \frac{1}{\lambda_1 - 1}, \quad y = \frac{1 - \lambda_2}{\lambda_1 - 1}. \tag{9.15}$$

The complementarity conditions in (c) and (d), $\lambda_1(x^2 + y^2 - 2) = 0$ and $\lambda_2(y - 1) = 0$, are combinatorial conditions signifying that at least one of the multipliers in each equation must be zero. Thus, we are forced to examine the different possibilities. This can be done by examining all the possible choices of active constraints, or by examining all the possible choices for the signs of the multipliers. We choose the latter strategy in this problem; the reader is encouraged to try the former.

(i) *$\lambda_1 > 0$ and $\lambda_2 > 0$:* Note that (c) and (d) imply $x^2 + y^2 = 2$ and $y = 1$, which give $x = \mp 1$ and $y = 1$, that is, the two points $(x, y) = (1, 1)$ and $(x, y) = (-1, 1)$. In the first case, we have $x = y$, and (9.15) implies $\lambda_2 = 0$. This contradicts our assumption that $\lambda_2 > 0$. In the second case, $(x, y) = (-1, 1)$, which implies $-1 = \lambda_1 - 1$ or $\lambda_1 = 0$, which is again impossible. We see that it is impossible to have both multipliers λ_1 and λ_2 positive.

(ii) $\lambda_1 > 0$ *and* $\lambda_2 = 0$*:* The condition (c) implies $x^2 + y^2 = 2$, and the equations (9.15) give $x = y = 1/(\lambda_1 - 1)$. Thus, the points $(x, y) = (1, 1)$ and $(x, y) = (-1, -1)$ are the possible KKT points. At the point $(x, y) = (1, 1)$, we have $1 = 1/(\lambda_1 - 1)$, or $\lambda_1 = 2$. Therefore, the point $(x, y) = (1, 1)$ is a KKT point with the corresponding multipliers $(\lambda_1, \lambda_2) = (2, 0)$. We note that $(x, y) = (1, 1)$ is a feasible point. At the point $(x, y) = (-1, -1)$, we have $-1 = 1/\lambda_1 - 1$ or $\lambda_1 = 0$, which is impossible.

(iii) $\lambda_1 = 0$ *and* $\lambda_2 > 0$*:* The complementarity condition in (d) implies that $y = 1$, and the equations (9.15) imply $x = -1$ and $\lambda_2 = 2$. Thus, the point $(-1, 1)$ is a KKT point with the corresponding multipliers $(\lambda_1, \lambda_2) = (0, 2)$. We note that $(-1, 1)$ is a feasible point.

(iv) $\lambda_1 = 0$ *and* $\lambda_2 = 0$*:* The equations (9.15) give the point $(x, y) = (-1, -1)$, which is a KKT point. We note that this is a feasible point.

Therefore, the three KKT points and their corresponding multipliers are

$$\begin{pmatrix} 1 \\ 1 \end{pmatrix}, \ \lambda = \begin{pmatrix} 2 \\ 0 \end{pmatrix}; \quad \begin{pmatrix} -1 \\ 1 \end{pmatrix}, \ \lambda = \begin{pmatrix} 0 \\ 2 \end{pmatrix}; \quad \begin{pmatrix} -1 \\ -1 \end{pmatrix}, \ \lambda = \begin{pmatrix} 0 \\ 0 \end{pmatrix}.$$

It is obvious that the points $(1, 1)$ and $(-1, -1)$ are the global minimizer and maximizer, respectively, of the minimization problem, hence the global maximizer and minimizer, respectively, of the original, maximization problem. What about the point $(-1, 1)$? It is not possible to apply Theorem 9.7, since its conditions are not satisfied at any of the three KKT points, as the reader may verify. In Section 9.8, we will apply below second-order tests to obtain more information about this KKT point.

It should be instructive to draw the feasible region and illustrate the KKT conditions pictorially.

Example 9.16. Consider the problem

$$\begin{aligned} \max \ & (x+1)^2 + (y+1)^2 \\ \text{s. t.} \ & x^2 + y^2 \le 3, \\ & -x^2 + 2y \le 0. \end{aligned}$$

The problem has global maximizers, since the constraint region is compact.

The gradients of the constraints are $\nabla g_1(x, y) = 2(x, y)$ and $\nabla g_2(x, y) = 2(-x, 1)$. Note that the second gradient is always nonzero, and the first one is zero only at the origin, where g_1 is not active. Thus, the KKT conditions hold at any local minimizer (x^*, y^*) that has at most one active constraint. If both constraints are active, the equalities $x^2 + y^2 = 3$ and $x^2 = 2y \ge 0$ give $y = 1$, and the gradients $\{\nabla g_1(x, y), \nabla g_2(x, y)\}$ are linearly independent. Thus, the KKT conditions must hold at all possible KKT points.

We can also demonstrate that $\lambda_0 > 0$ by an independent algebraic argument in the following way. We form the weak Lagrangian function

$$L = -\frac{\lambda_0}{2}((x+1)^2 + (y+1)^2) + \frac{\lambda_1}{2}(x^2 + y^2 - 3) + \frac{\lambda_2}{2}(-x^2 + 2y),$$

and write down the FJ conditions

$$
\begin{array}{ll}
(a) & -\lambda_0(x+1)+\lambda_1 x-\lambda_2 x=0, \quad \lambda_0 \geq 0, \\
(b) & -\lambda_0(y+1)+\lambda_1 y+\lambda_2=0, \quad (\lambda_0,\lambda_1,\lambda_2) \neq 0, \\
(c) & \lambda_1 \geq 0,\ x^2+y^2 \leq 3,\ \lambda_1(x^2+y^2-3)=0, \\
(d) & \lambda_2 \geq 0,\ 2y \leq x^2,\ \lambda_2(-x^2+2y)=0.
\end{array}
$$

We claim that $\lambda_0 > 0$. If $\lambda_0 = 0$, then (a) and (b) become

$$(a') \quad (\lambda_1-\lambda_2)x=0, \qquad (b') \quad \lambda_1 y+\lambda_2=0.$$

These imply that

$$\lambda_1=\lambda_2>0,\ y<0 \quad \text{or} \quad x=0,\ \lambda_1>0,\ y\leq 0.$$

The first case is impossible, because we saw above that $y = 1$ when both constraints are active. In the second case, we have $x = 0$, and the equation $x^2 + y^2 = 3$ implies that $y = -\sqrt{3}$. But then g_2 is inactive, $\lambda_2 = 0$, and (b') implies that $\lambda_1 = 0$. This contradiction proves the claim, and we may assume that $\lambda_0 = 1$.

The conditions (a) and (b) give

$$x=\frac{1}{\lambda_1-\lambda_2-1}, \quad (\lambda_1-1)y=1-\lambda_2. \tag{9.16}$$

Let us consider all the possible cases for the signs of the multipliers:

(i) $\lambda_1 > 0$ *and* $\lambda_2 > 0$*:* The complementarity conditions in (c) and (d) give $x^2 + y^2 = 3$, $y = 1$ and $x^2 = 2y = 2$, that is, $x = \mp\sqrt{2}$. Thus, the points $(\sqrt{2}, 1)$ and $(-\sqrt{2}, 1)$ are the possible KKT points, but we need to verify that their multipliers are not negative.
At the point $(\sqrt{2}, 1)$, solving the linear equations in (9.16) gives $\lambda_1 = 3/2 + 1/(2\sqrt{2}) > 0$, $\lambda_2 = 2 - \lambda_1 = 1/2 - 1/(2\sqrt{2}) > 0$. At the second point $(-\sqrt{2}, 1)$, the same equations give $\lambda_1 = 3/2 - 1/(2\sqrt{2}) > 0$, $\lambda_2 = 1/2 + 1/(2\sqrt{2}) > 0$. Therefore, both points are KKT points.

(ii) $\lambda_1 > 0$ *and* $\lambda_2 = 0$*:* The condition (c) implies $x^2 + y^2 = 3$, and the equations (9.16) imply $x = \frac{1}{\lambda_1 - 1} = y$. Thus, the possible KKT points are $(\sqrt{3/2}, \sqrt{3/2})$ and $(-\sqrt{3/2}, -\sqrt{3/2})$. In the first case, we have $\lambda_1 = 1 + \sqrt{\frac{2}{3}} > 0$. However, this point is not feasible, since it violates the constraint $-x^2 + 2y \leq 0$. In the second case, $\lambda_1 = 1 - \sqrt{\frac{2}{3}} > 0$, and the point $(-\sqrt{3/2}, -\sqrt{3/2})$ is feasible, hence a KKT point.

(iii) $\lambda_1 = 0$ *and* $\lambda_2 > 0$*:* The complementarity condition in (d) implies $x^2 = 2y$, and the equations in (9.16) give $x = 1/(-\lambda_2 - 1)$ and $y = \lambda_2 - 1$. Thus, the equation $x^2 = 2y$ gives $1/(\lambda_2 + 1)^2 = 2(\lambda_2 - 1)$, that is, $\lambda_2^3 + \lambda_2^2 - \lambda_2 - \frac{3}{2} = 0$. Solving this cubic equation approximately (say with Newton's method), we obtain $\lambda_2 \approx 1.11208$. This gives the point $(x, y) = (-0.473465, 0.1120849)$. It is a feasible point, hence satisfies the KKT conditions.

(iv) $\lambda_1 = 0$ *and* $\lambda_2 = 0$*:* The equations in (9.16) give $(x, y) = (-1, -1)$, which is a KKT point.

Summarizing, we list the five KKT points with their multipliers:

$$\begin{pmatrix} \sqrt{2} \\ 1 \end{pmatrix}, \ \lambda = \begin{pmatrix} \frac{3}{2} + \frac{1}{2\sqrt{2}} \\ \frac{1}{2} - \frac{1}{2\sqrt{2}} \end{pmatrix}; \ \begin{pmatrix} -\sqrt{2} \\ 1 \end{pmatrix}, \ \lambda = \begin{pmatrix} \frac{3}{2} - \frac{1}{2\sqrt{2}} \\ \frac{1}{2} + \frac{1}{2\sqrt{2}} \end{pmatrix}; \ \begin{pmatrix} -1 \\ -1 \end{pmatrix}, \ \lambda = \begin{pmatrix} 0 \\ 0 \end{pmatrix};$$

$$\begin{pmatrix} -\sqrt{\frac{3}{2}} \\ -\sqrt{\frac{3}{2}} \end{pmatrix}, \ \lambda = \begin{pmatrix} 1 - \sqrt{\frac{2}{3}} \\ 0 \end{pmatrix}; \ \begin{pmatrix} -0.473465 \\ 0.1120849 \end{pmatrix}, \ \lambda = \begin{pmatrix} 0 \\ 1.1120849 \end{pmatrix}.$$

At these points, the objective function $f(x, y) := (x + 1)^2 + (y + 1)^2$ of the maximization problem has values 9.828, 4.1715, 0, 0.101020, and 1.51397, respectively. Hence, the point $(\sqrt{2}, 1)$ is the global maximizer of f over the feasible region, whereas $(-1, -1)$ is clearly the point where f is *minimized* over the same feasible set. We leave it to the reader to work out how Theorem 9.7 applies to the KKT points. It can be shown that only the first two KKT points, $(\mp\sqrt{2}, 1)$, satisfy the conditions of this theorem. This proves that the KKT point $(-\sqrt{2}, 1)$ is a local minimizer of the function $-f$ over the feasible region, hence a local maximizer of the original maximization problem. In Section 9.8, we will apply second-order tests to determine further the nature of all five KKT points.

Example 9.17. This is an interesting problem whose solution rests upon a novel observation:

$$\begin{aligned} \min \quad & \frac{1}{3} \sum_{i=1}^{n} x_i^3 \\ \text{s.t.} \quad & \sum_{i=1}^{n} x_i = 0, \\ & \sum_{i=1}^{n} x_i^2 = n. \end{aligned}$$

The feasible region is compact, so the Weierstrass theorem implies that the problem has a global minimizer and a global maximizer.

The gradients $\nabla g_1(x) = (1, \ldots, 1) = e$ and $\nabla g_2(x) = 2x$ of the constraint functions are linearly dependent only when x is a multiple of e, but there is no such feasible point. Thus, we may assume that $\lambda_0 = 1$.

We write the Lagrangian function

$$L(x,\lambda) = \frac{1}{3}\sum_{i=1}^{n} x_i^3 + \lambda_1 \sum_{i=1}^{n} x_i + \frac{\lambda_2}{2}\Big(\sum_{i=1}^{n} x_i^2 - n\Big),$$

and the KKT conditions

$$(a) \quad \frac{\partial L}{\partial x_i} = x_i^2 + \lambda_1 + \lambda_2 x_i = 0, \quad i = 1, \ldots, n,$$

$$(b) \quad \sum_{i=1}^{n} x_i = 0,$$

$$(c) \quad \sum_{i=1}^{n} x_i^2 = n.$$

Summing (a) over i gives

$$0 = \sum_{i=1}^{n} x_i^2 + n\lambda_1 + \lambda_2 \sum_{i=1}^{n} x_i = n + n\lambda_1,$$

that is, $\lambda_1 = -1$. Next, multiplying (a) by x_i and summing over i gives

$$\sum_{i=1}^{n} x_i^3 + \lambda_1 \sum_{i=1}^{n} x_i + \lambda_2 \sum_{i=1}^{n} x_i^2 = \sum_{i=1}^{n} x_i^3 + n\lambda_2 = 0,$$

that is, $\lambda_2 = -\sum_{i=1}^{n} x_i^3/n$. However, this does not seem to help much.

In this problem, the key observation for determining x_i is to perceive that when λ_1 and λ_2 are fixed, the equation (a) in the KKT conditions is a *quadratic* equation in x_i, so that each x_i can have at most *two* values,

$$x_i = \frac{-\lambda_2 \mp \sqrt{\lambda_2^2 + 4}}{2},$$

one positive x_+, one negative x_-. Since all the functions in the optimization problem are symmetric in the variables x_i, we may assume, without loss of generality, that

$$x := (x_1, \ldots, x_n) = (\underbrace{x_+, \ldots, x_+}_{k \text{ times}}, \underbrace{x_-, \ldots, x_-}_{n-k \text{ times}}).$$

Then the equation (b) in the KKT conditions becomes $kx_+ + (n-k)(x_-) = 0$, giving

$$x_- = \frac{k}{k-n} x_+,$$

and the equation (c) gives $n = kx_+^2 + (n-k)x_-^2$, which upon simplification becomes $(x_+)^2 = (n-k)/k$. Thus,

$$x_+ = \sqrt{\frac{n-k}{k}}, \quad x_- = -\frac{k}{n-k}\sqrt{\frac{n-k}{k}} = -\sqrt{\frac{k}{n-k}}.$$

Since x contains at least one positive and one negative entry, the possible values for k are $k = 1, \ldots, n-1$.

We have

$$\sum_{i=1}^{n} x_i^3 = kx_+^3 + (n-k)x_-^3 = k\left(\sqrt{\frac{n-k}{k}}\right)^3 + (n-k)\left(-\sqrt{\frac{k}{n-k}}\right)^3$$
$$= (n-k)\sqrt{\frac{n-k}{k}} - k\sqrt{\frac{k}{n-k}}.$$

Note that the first term on the right-hand side of the equation is minimized and the second term is maximized at $k = n-1$, and thus the objective function is minimized for $k = n-1$. (Similarly, it is maximized at $k = 1$.) Therefore, the global minimizer of the problem is the point

$$x^* = \left(\frac{1}{\sqrt{n-1}}, \ldots, \frac{1}{\sqrt{n-1}}, -\sqrt{n-1}\right),$$

and the global maximizer of the problem is the point

$$x^* = \left(\sqrt{n-1}, -\frac{1}{\sqrt{n-1}}, \ldots, -\frac{1}{\sqrt{n-1}}\right).$$

We will apply second-order tests in Section 9.8 to determine the nature of all the KKT points.

Example 9.18. Let Q be an $n \times n$ symmetric matrix, $c \in \mathbb{R}^n$, and $\Delta > 0$. Consider the problem

$$\begin{aligned} \min \quad & q(x) := \frac{1}{2}\langle Qx, x\rangle + \langle c, x\rangle \\ \text{s.t.} \quad & \|x\| \le \Delta. \end{aligned} \tag{9.17}$$

This problem appears in *trust region methods* for the numerical solution of an unconstrained minimization problem. In this context, suppose that we are trying to find numerically a local minimizer of a nonlinear function $f(x)$ of n variables. If x_k is a given approximate minimizer of f, then we can attempt to find a better approximate minimizer x_{k+1} by approximating f by its quadratic Taylor series at x_k and then minimizing this surrogate function in a suitable disk $\overline{B}(x_k, \Delta)$ around x_k. We end up with the optimization problem equivalent to (9.17). See [244, 201, 213, 62] for much more information on trust region methods and the numerical methods for solving (9.17).

Since the constraint function has a nonzero gradient at every point of the feasible region, we may assume that $\lambda_0 = 1$. We change the constraint to $\|x\|^2 \le \Delta^2$, and write the Lagrangian function

$$L = \frac{1}{2}\langle Qx, x\rangle + \langle c, x\rangle + \frac{\lambda}{2}(\|x\|^2 - \Delta^2).$$

At a global minimizer x^* of (9.17), we have the KKT conditions

$$\begin{aligned} &(a) \quad \nabla_x L = (Q + \lambda I)x^* + c = 0, \\ &(b) \quad \lambda \geq 0, \ \|x^*\| \leq \Delta, \ \lambda(\|x^*\| - \Delta) = 0. \end{aligned}$$

We claim that the conditions (a) and (b), together with the second-order condition

$$(c) \quad Q + \lambda I \text{ is positive semidefinite,}$$

characterize a global minimizer of (9.17).

First, assume that x^* is a global minimizer of (9.17). We need only to verify the condition (c). If $\|x^*\| < \Delta$, then $\lambda = 0$ and x^* is an unconstrained local minimizer of q, and hence Q is positive semidefinite by Theorem 2.12. If $\|x^*\| = \Delta$ and $\|x\| = \Delta$ is any feasible point, then $q(x) - q(x^*) \geq 0$, and

$$\begin{aligned} & q(x) - q(x^*) \\ = \ & \langle \nabla q(x^*), x - x^* \rangle + \frac{1}{2}\langle Q(x - x^*), x - x^* \rangle \\ = \ & -\lambda\langle x^*, x - x^* \rangle + \frac{1}{2}\langle Q(x - x^*), x - x^* \rangle \\ = \ & \frac{1}{2}\langle (Q + \lambda I)(x - x^*), x - x^* \rangle - \frac{\lambda}{2}\left(\|x - x^*\|^2 + 2\langle x^*, x - x^* \rangle\right) \\ = \ & \frac{1}{2}\langle (Q + \lambda I)(x - x^*), x - x^* \rangle - \frac{\lambda}{2}(\|x\|^2 - \|x^*\|^2), \end{aligned} \tag{9.18}$$

where the second equality follows from (a). This immediately implies that $\langle (Q + \lambda I)(x - x^*), x - x^* \rangle \geq 0$ for all $\|x\| = \Delta$. Since $\langle x^*, x - x^* \rangle \leq 0$, we have $\langle (Q + \lambda I)d, d \rangle \geq 0$ for all d satisfying $\langle x^*, d \rangle \leq 0$, hence for all $d \in \mathbb{R}^n$, proving that $Q + \lambda I$ is positive semidefinite.

Conversely, suppose that the conditions (a)–(c) are satisfied. If $\|x^*\| < \Delta$, then $\lambda = 0$, and (9.18) shows that x^* is a global minimizer of q on $\mathbb{R}^n$. If $\|x^*\| = \Delta$, the same equations (9.18) shows that x^* is a global minimizer of q on the disk $\overline{B}(0, \Delta)$.

9.7 Second-Order Conditions in Nonlinear Programming

We reconsider the nonlinear program (9.1),

$$\begin{aligned} \min \quad & f(x) \\ \text{s.t.} \quad & g_i(x) \leq 0, \ i = 1, \ldots, r, \qquad (P) \\ & h_j(x) = 0, \quad j = 1, \ldots, m \end{aligned}$$

where this time we assume that the functions f, g_i, h_j have continuous second-order partial derivatives in an open set containing the feasible region of (P).

The FJ and KKT conditions are first-order necessary conditions for a local minimizer. Thus, *any* local minimizer must satisfy these. However, the converse statement is false: a point satisfying these conditions is not necessarily a local minimizer.

The second-order conditions give us additional restrictions that help us narrow down the search for local minimizers of (P). As in the unconstrained optimization case, these are either *necessary* or *sufficient* conditions. Every local minimizer of (P) must satisfy the first-order conditions and the second-order *necessary* conditions, but not every point satisfying these is necessarily a local minimizer. By contrast, every point satisfying the first-order conditions and a second-order *sufficient* condition *must* be a local minimizer.

The second order optimality conditions given in this section have their origins in a paper by McCormick (see [96]), and contain some improvements. They are applicable, under some restrictions, at a KKT point x^* for (9.1) which has associated Lagrange multipliers λ^*, μ^*. It should be noted that the optimality conditions are stated in a form suitable for the minimization of the Lagrangian function $L(x, \lambda^*, \mu^*)$ and not the minimization of the objective function $f(x)$. Since the passage from f to L is not straightforward, this should explain the need for care (see, for example, the statements of Lemma 9.19 and Theorem 9.20) in formulating the correct form of the optimality conditions.

The research on second-order optimality conditions is active even today. More sophisticated second-order conditions, for a wide variety of optimization problems, are available; see for example [28] and [39].

9.7.1 Second-Order Necessary Conditions

Denote by $\nabla_x^2 L(x, \lambda, \mu)$ the Hessian of the Lagrangian function L with respect to the decision variables x, that is,

$$\nabla_x^2 L(x, \lambda, \mu) = \nabla^2 f(x) + \sum_{i=1}^{r} \lambda_i \nabla^2 g_i(x) + \sum_{j=1}^{m} \mu_j \nabla^2 h_j(x).$$

Lemma 9.19. *Let x^* be a local minimizer of (P) satisfying the KKT conditions with multipliers λ^*, μ^*. If $d \in \mathbb{R}^n$ is a feasible direction at x^* with the property that there exists a sequence of feasible points $x_k \to x^*$ satisfying the conditions $(x_k - x^*)/\|x_k - x^*\| \to d$, $g_i(x_k) = 0$, $i \in I(x^*)$, and $h_j(x_k) = 0$, then $\langle \nabla_x^2 L(x^*, \lambda^*, \mu^*)d, d\rangle \geq 0$.*

Proof. Let d and $\{x_k\}$ satisfy the assumptions of the lemma. Defining $d_k = x_k - x^*$, we have

$$\begin{aligned}
&0 \\
&\leq f(x_k) - f(x^*) \\
&= L(x_k, \lambda^*, \mu^*) - L(x^*, \lambda^*, \mu^*) \\
&= \langle \nabla_x L(x^*, \lambda^*, \mu^*), d_k \rangle + \frac{1}{2}\langle \nabla_x^2 L(x^*, \lambda^*, \mu^*) d_k, d_k \rangle + o(\|d_k\|^2) \\
&= \frac{1}{2}\langle \nabla_x^2 L(x^*, \lambda^*, \mu^*) d_k, d_k \rangle + o(\|d_k\|^2),
\end{aligned}$$

where the inequality follows because x^* is a local minimizer of (P), the second equality follows from Taylor's expansion, and the last equality follows because $\nabla_x L(x^*, \lambda^*, \mu^*) = 0$ from the KKT conditions. Dividing the above inequality by $\|d_k\|^2$ and letting $k \to \infty$, we conclude that $\langle \nabla_x^2 L(x^*, \lambda^*, \mu^*) d, d \rangle \geq 0$. □

Theorem 9.20. *Let x^* be a local minimizer of (P) satisfying the KKT conditions with multipliers λ^*, μ^*. If the active gradient vectors,*

$$\nabla g_i(x^*), \ i \in I(x^*), \ \ \nabla h_j(x^*), \ j = 1, \ldots, m$$

are linearly independent, then $\nabla_x^2 L(x^*, \lambda^*, \mu^*)$ must be positive semidefinite on the linear subspace

$$M = (\operatorname{span}\{\nabla g_i(x^*), i \in I(x^*), \nabla h_j(x^*), j = 1, \ldots, m\})^{\perp}.$$

That is, if a direction d satisfies

$$\langle d, \nabla g_i(x^*) \rangle = 0, \ i \in I(x^*), \quad \langle d, \nabla h_j(x^*) \rangle = 0, \ j = 1, \ldots, m,$$

then $\langle \nabla_x^2 L(x^*, \lambda^*, \mu^*) d, d \rangle \geq 0$.

We provide two proofs for this theorem, one based on feasible directions and Lyusternik's theorem, and the other using a penalty function approach.

Proof. Since the active gradients at x^* are linearly independent, it follows from Lyusternik's theorem that M coincides with the set of tangent directions to the set

$$\{x : g_i(x) = 0, \ i \in I(x^*), \ h_j(x) = 0, \ j = 1, \ldots, m\}$$

at the point x^*. If x is a point near x^* belonging to this set, then x is clearly a feasible point for (P). The theorem follows immediately from Lemma 9.19. □

Here is a second, independent proof of the same theorem based on the penalty function approach.

Proof. We use the twice continuously differentiable penalty function

$$F_k(x) = f(x) + \frac{k}{3}\sum_{i=1}^{r} g_i^+(x)^3 + \frac{k}{2}\sum_{j=1}^{m} h_j^2(x) + \frac{1}{4}\|x - x^*\|^4.$$

The same arguments used earlier in the penalty function approach to derive the FJ conditions applies here: $x_k \to x^*$, where x_k is global minimum of F_k over a small ball $\overline{B}_\epsilon(x^*)$. Thus, $\nabla F_k(x_k) = 0$ and $\nabla^2 F_k(x_k)$ is positive semidefinite for large enough k. Toward computing the expressions for $\nabla F_k(x_k)$ and $\nabla^2 F_k(x_k)$, we compute the Taylor expansions of the component functions in $F_k(x_k + td)$.

Define the function $\alpha(t) = (t^+)^3/3$. It is easy to verify that $\alpha'(t) = (t^+)^2$ and $\alpha''(t) = 2t^+$. Thus,

$$\begin{aligned}
\frac{1}{3} g_i^+(x+td)^3 &= \alpha(g_i(x+td)) \\
&= \alpha\left(g_i(x) + t\langle \nabla g_i(x), d\rangle + \frac{t^2}{2}\langle \nabla^2 g_i(x)d, d\rangle + o(t^2)\right) \\
&= \alpha(g_i(x)) + \alpha'(g_i(x))\left[t\langle \nabla g_i(x), d\rangle + \frac{t^2}{2}\langle \nabla^2 g_i(x)d, d\rangle\right] \\
&\quad + \frac{1}{2}\alpha''(g_i(x))[t\langle \nabla g_i(x), d\rangle]^2 + o(t^2) \\
&= \frac{1}{3} g_i^+(x)^3 + t\left[g_i^+(x)^2 \langle \nabla g_i(x), d\rangle\right] \\
&\quad + \frac{t^2}{2}\left[g_i^+(x)^2 \langle \nabla^2 g_i(x)d, d\rangle + 2g_i^+(x)\langle \nabla g_i(x), d\rangle^2\right] + o(t^2),
\end{aligned}$$

$$\begin{aligned}
h_j(x+td)^2 &= \left[h_j(x) + t\langle \nabla h_j(x), d\rangle + \frac{t^2}{2}\langle \nabla^2 h_j(x)d, d\rangle + o(t^2)\right]^2 \\
&= h_j(x)^2 + 2h_j(x)\left[t\langle \nabla h_j(x), d\rangle + \frac{t^2}{2}\langle \nabla^2 h_j(x)d, d\rangle\right] \\
&\quad + t^2\langle \nabla h_j(x), d\rangle^2 + o(t^2) \\
&= h_j(x)^2 + 2t[h_j(x)\langle \nabla h_j(x), d\rangle] \\
&\quad + t^2\left[h_j(x)\langle \nabla^2 h_j(x)d, d\rangle + \langle \nabla h_j(x), d\rangle^2\right] + o(t^2),
\end{aligned}$$

and

$$\begin{aligned}
\|x+td\|^4 = \langle x+td, x+td\rangle^2 &= \left[\|x\|^2 + 2t\langle x, d\rangle + t^2\|d\|^2\right]^2 \\
&= \|x\|^4 + 2\|x\|^2\left[2t\langle x, d\rangle + t^2\|d\|^2\right] + t^2[2\langle x, d\rangle + t\|d\|^2]^2 \\
&= \|x\|^4 + 4t\left[\|x\|^2\langle x, d\rangle\right] + 2t^2\left[\|x\|^2 \cdot \|d\|^2 + 2\langle x, d\rangle^2\right] + o(t^2).
\end{aligned}$$

Putting all these together, we obtain

$$\begin{aligned}
\nabla F_k(x_k) = \nabla f(x_k) &+ \sum_{i=1}^{r} k g_i^+(x_k)^2 \nabla g_i(x_k) + \sum_{j=1}^{m} k h_j(x_k)\nabla h_j(x_k) \\
&+ \|x_k - x^*\|^2 (x_k - x^*) = 0
\end{aligned} \tag{9.19}$$

and

$$\begin{aligned}
&\langle \nabla^2 F_k(x_k)d, d\rangle \\
&= \Big\langle \big[\nabla^2 f(x_k) + \sum_{i=1}^{r} k g_i^+(x_k)^2 \nabla^2 g_i(x_k) + \sum_{j=1}^{m} k h_j(x_k)\nabla^2 h_j(x_k)\big]d, d\Big\rangle \\
&\quad + \Big[\sum_{i=1}^{r} 2k g_i^+(x_k)\langle \nabla g_i(x_k), d\rangle^2 + \sum_{j=1}^{m} k\langle \nabla h_j(x_k), d\rangle^2\Big] \\
&\quad + \big[\|x_k - x^*\|^2 \cdot \|d\|^2 + 2\langle x_k - x^*, d\rangle^2\big]
\end{aligned} \tag{9.20}$$

Now we use the linear independence hypothesis: define the matrices

$$\begin{aligned}
A &:= [\nabla g_i(x^*),\ i \in I(x^*),\ \nabla h_j(x^*),\ j = 1, \dots, m], \\
A_k &:= [\nabla g_i(x_k),\ i \in I(x^*),\ \nabla h_j(x_k),\ j = 1, \dots, m].
\end{aligned}$$

We have $M = N(A^T)$ and

$$\pi_M = \pi_{N(A^T)} = I - \pi_{R(A)} = I - A(A^T A)^{-1} A^T.$$

($A^T A$ is nonsingular since the columns of A are linearly independent.) Since $x_k \to x^*$, we have $A_k^T A_k \to A^T A$, so that $A_k^T A_k$ is nonsingular for large k.

For a vector $d \in M$, define

$$d_k := \pi_{N(A_k^T)} d = (I - A_k(A_k^T A_k)^{-1} A_k^T)d.$$

Note that d_k is orthogonal to the vectors $\nabla g_i(x_k)$ ($i \in I(x^*)$) and $\nabla h_j(x_k)$ ($j = 1, \dots, m$), and that (this is the most crucial point) $d_k \to d$. Moreover, since the KKT conditions hold by Corollary 9.6, the scaling argument in the penalty function proof of Theorem 9.4 shows that the sequences $\{k g_i^+(x_k)^2\}_{k=1}^{\infty}$ and $\{k h_j(x_k)\}_{k=1}^{\infty}$ are bounded. Thus, there exist convergent subsequences $\lambda_i^* = \lim_{l\to\infty} k_l g_i^+(x_{k_l})^2$ and $\mu_j^* = \lim_{l\to\infty} k_l h_j(x_{k_l})$ such that $\lambda_i^* = 0$ for $i \notin I(x^*)$. Letting $l \to \infty$ in the expression $\langle \nabla^2 F_{k_l}(x_{k_l}) d_{k_l}, d_{k_l}\rangle \geq 0$ and using (9.19) and (9.20), we see that

$$\nabla_x L(x^*) = \nabla f(x^*) + \sum_{i=1}^{r} \lambda_i^* \nabla g_i(x^*) + \sum_{j=1}^{m} \mu_j^* \nabla h_j(x^*) = 0,$$

and for all $d \in M$,

$$\Big\langle \Big[\nabla^2 f(x^*) + \sum_{i=1}^{r} \lambda_i^* \nabla^2 g_i(x^*) + \sum_{j=1}^{m} \mu_j^* \nabla^2 h_j(x^*)\Big] d, d\Big\rangle \geq 0.$$

□

9.7.2 Second-Order Sufficient Conditions

Theorem 9.21. *Let x^* be a feasible point for (P) that satisfies the KKT conditions with multipliers λ^*, μ^*. If*

$$\langle \nabla_x^2 L(x^*, \lambda^*, \mu^*)d, d \rangle > 0 \tag{9.21}$$

for all $d \neq 0$ *satisfying the conditions*

$$\begin{aligned} &\langle d, \nabla g_i(x^*) \rangle \leq 0, \ i \in I(x^*), \\ &\langle d, \nabla g_i(x^*) \rangle = 0, \ i \in I(x^*) \ \textit{and} \ \lambda_i^* > 0, \\ &\langle d, \nabla h_j(x^*) \rangle = 0, \ j = 1, \ldots, m, \end{aligned} \tag{9.22}$$

then x^* *is a strict local minimizer of* (P)*, and there exist a constant* $c > 0$ *and a ball* $\overline{B}_\epsilon(x^*)$ *such that*

$$f(x) \geq f(x^*) + c\|x - x^*\|^2 \ \textit{for all feasible} \ x \in \overline{B}_\epsilon(x^*). \tag{9.23}$$

Proof. Suppose that (9.23) is not satisfied, and let ϵ_k be a sequence positive numbers converging to zero. Then there exists a sequence of feasible points $x_k \to x^*$ such that $f(x_k) < f(x^*) + \epsilon_k \|x_k - x^*\|^2$. Define $d_k = x_k - x^*$ and assume without any loss of generality that $d_k/\|d_k\| \to d$, $\|d\| = 1$. On the one hand, we have

$$\begin{aligned} &\epsilon_k \|d_k\|^2 \\ &> [f(x_k) - f(x^*)] + \sum_{i \in I(x^*)} \lambda_i^* g_i(x_k) \\ &= L(x_k, \lambda^*, \mu^*) - L(x^*, \lambda^*, \mu^*) \\ &= \langle \nabla_x L(x^*, \lambda^*, \mu^*), d_k \rangle + \frac{1}{2} \langle \nabla_x^2 L(x^*, \lambda^*, \mu^*) d_k, d_k \rangle + o(\|d_k\|^2) \\ &= \frac{1}{2} \langle \nabla_x^2 L(x^*, \lambda^*, \mu^*) d_k, d_k \rangle + o(\|d_k\|^2), \end{aligned}$$

where the last equality follows because of the KKT conditions $\nabla_x L(x^*, \lambda^*, \mu^*) = 0$. Dividing the above inequalities by $\|d_k\|^2/2$ and letting $k \to \infty$, we obtain

$$\langle \nabla_x^2 L(x^*, \lambda^*, \mu^*)d, d \rangle \leq 0,$$

that is, d does not satisfy (9.21).

On the other hand,

$$\begin{aligned} \epsilon_k \|d_k\|^2 &> f(x_k) - f(x^*) = \langle \nabla f(x^*), d_k \rangle + o(\|d_k\|), \\ 0 &\geq g_i(x_k) - g_i(x^*) = \langle \nabla g_i(x^*), d_k \rangle + o(\|d_k\|), \quad i \in I(x^*), \\ 0 &= h_j(x_k) - h_j(x^*) = \langle \nabla h_j(x^*), d_k \rangle + o(\|d_k\|), \end{aligned}$$

and dividing the inequalities by $\|d_k\|$ and letting $k \to \infty$ gives

$$\langle \nabla f(x^*), d \rangle \leq 0, \quad \langle \nabla g_i(x^*), d \rangle \leq 0, \quad \langle \nabla h_j(x^*), d \rangle = 0.$$

These imply that d satisfies (9.22), since multiplying the above inequalities by $1, \lambda_i^*, \mu_j^*$, respectively ($i \in I(x^*)$, $j = 1, \ldots, m$) gives

$$0 = \langle \nabla_x L(x^* \lambda^*, \mu^*), d \rangle = \left\langle \nabla f(x^*) + \sum_{i=1}^{r} \lambda_i^* \nabla g_i(x^*) + \sum_{j=1}^{m} \mu_j^* \nabla h_j(x^*), d \right\rangle$$

$$= \langle \nabla f(x^*), d \rangle + \sum_{i=1}^{r} \lambda_i^* \langle \nabla g_i(x^*), d \rangle,$$

which implies that $\lambda_i^* \langle \nabla g_i(x^*), d \rangle = 0$ for each $i \in I(x^*)$.

We have produced a vector d that satisfies (9.22) but not (9.21), contradicting the assumptions of the theorem. The theorem is proved. □

Corollary 9.22. *Let x^* be a feasible point satisfying the KKT conditions with multipliers λ^*, μ^*. If $\lambda_i^* > 0$ for all $i \in I(x^*)$ (this is called the strict complementarity condition) and the Hessian $\nabla_x^2 L(x^*, \lambda^*, \mu^*)$ is* positive definite *in the subspace*

$$\{d : \langle d, \nabla g_i(x^*) \rangle = 0,\ i \in I(x^*),\ \ \langle d, \nabla h_j(x^*) \rangle = 0,\ j = 1, \ldots, m\},$$

then x^* is a strict local minimizer of (P).

The corollary follows immediately from Theorem 9.21.

For some special classes of problems, it is possible to obtain stronger results.

Lemma 9.23. *Let x^* be a KKT point of the quadratic program*

$$\min\{f(x) := \frac{1}{2}\langle Qx, x \rangle + \langle c, x \rangle : Ax \leq b\}. \quad (P)$$

If $\langle Qd, d \rangle \geq 0$ for all directions d satisfying the condition that $\langle a_i, d \rangle \leq 0$ for all $i \in I(x^*)$, then x^* is a local minimizer of (P).

Proof. The individual linear constraints of the program are $\langle a_i, x \rangle \leq b_i$ where $\{a_i\}$ are the rows of A. Let x be a feasible point in a small enough neighborhood of x^* and define $d = x - x^*$. We have

$$f(x) - f(x^*) = \langle \nabla f(x^*), d \rangle + \frac{1}{2}\langle Qd, d \rangle = - \sum_{i \in I(x^*)} \lambda_i^* \langle a_i, d \rangle + \frac{1}{2}\langle Qd, d \rangle$$

$$\geq \frac{1}{2}\langle Qd, d \rangle \geq 0,$$

where the first equation follows since f is a quadratic function, the second equation is due to the KKT condition, and the first inequality follows since $\langle a_i, d \rangle = \langle a_i, x - x^* \rangle = \langle a_i, x \rangle - b_i \leq 0$ for $i \in I(x^*)$. □

Remark 9.24. If the quadratic program contains only equations, then it is clear from the above proof that a point x^* satisfying the conditions of Lemma 9.23 is actually a global minimizer of the quadratic program; see also Exercise 12.

9.8 Examples of Second-Order Conditions

Example 9.25. (Continuation of Example 9.15)
We recall that for this problem on page 222, we have

$$L = -\frac{1}{2}((x+1)^2 + (y+1))^2 + \frac{\lambda_1}{2}(x^2+y^2-2) + \lambda_2(y-1),$$

$$\nabla g_1(x,y) = 2\begin{pmatrix} x \\ y \end{pmatrix}, \ \nabla g_2(x,y) = \begin{pmatrix} 0 \\ 1 \end{pmatrix}, \ H := \nabla^2_{(x,y)} L = (\lambda_1 - 1)I,$$

and the KKT points are

$$\begin{pmatrix} 1 \\ 1 \end{pmatrix}, \ \lambda = \begin{pmatrix} 2 \\ 0 \end{pmatrix}; \quad \begin{pmatrix} -1 \\ 1 \end{pmatrix}, \ \lambda = \begin{pmatrix} 0 \\ 2 \end{pmatrix}; \quad \begin{pmatrix} -1 \\ -1 \end{pmatrix}, \ \lambda = \begin{pmatrix} 0 \\ 0 \end{pmatrix}.$$

At the KKT point $(1,1)$, we have $H = I$; hence the second-order necessary and sufficient conditions are trivially satisfied. Thus, the point $(1,1)$ *is* a strict local minimizer of the reformulated (minimization) problem, hence a local maximizer of the original (maximization) problem.

At the KKT point $x = (-1,1)$, both constraints are active, and the gradients of the active constraints are linearly independent. We also have $H = -I$. Since $M = 0$, the second-order necessary conditions in Theorem 9.20 trivially hold. A vector d in the cone $T(-1,1)$ must satisfy the conditions $\langle(-1,1),(d_1,d_2)\rangle = -d_1 + d_2 \geq 0$ and $\langle(0,1),(d_1,d_2)\rangle = d_2 \geq 0$. For example, the vector $d = (1,1)$ is in the cone and satisfies $\langle Hd, d\rangle < 0$. It follows from Theorem 9.20 that $(-1,1)$ is *not* a local minimizer of the reformulated problem, hence not a local maximizer of the original problem.

At the KKT point $x = (-1,-1)$, only the first constraint is active with the gradient $\nabla g_1(-1,1) = 2(-1,-1) \neq 0$, and $H = -I$. Since the subspace M in Theorem 9.20 is a line through the origin, the second-order necessary conditions fail. Thus, the point $(-1,-1)$ is *not* a local maximizer of the original problem.

Example 9.26. (Continuation of Example 9.16)
We recall (see page 224) that in this problem

$$\begin{aligned} L &= -\frac{1}{2}(x+1)^2 - \frac{1}{2}(y+1)^2 + \frac{\lambda_1}{2}(x^2+y^2-3) + \frac{\lambda_2}{2}(-x^2+2y), \\ \nabla g_1(x,y) &= 2\begin{pmatrix} x \\ y \end{pmatrix}, \ \nabla g_2(x,y) = 2\begin{pmatrix} -x \\ 1 \end{pmatrix}, \\ H &= \nabla^2_{(x,y)} L = \begin{bmatrix} -1+\lambda_1-\lambda_2 & 0 \\ 0 & -1+\lambda_1 \end{bmatrix}, \end{aligned}$$

and the KKT points are

$$\begin{pmatrix}\sqrt{2}\\1\end{pmatrix},\ \lambda=\begin{pmatrix}\frac{3}{2}+\frac{1}{2\sqrt{2}}\\\frac{1}{2}-\frac{1}{2\sqrt{2}}\end{pmatrix};\ \begin{pmatrix}-\sqrt{2}\\1\end{pmatrix},\ \lambda=\begin{pmatrix}\frac{3}{2}-\frac{1}{2\sqrt{2}}\\\frac{1}{2}+\frac{1}{2\sqrt{2}}\end{pmatrix};\ \begin{pmatrix}-1\\-1\end{pmatrix},\ \lambda=\begin{pmatrix}0\\0\end{pmatrix};$$

$$\begin{pmatrix}-\sqrt{\frac{3}{2}}\\-\sqrt{\frac{3}{2}}\end{pmatrix},\ \lambda=\begin{pmatrix}1-\sqrt{\frac{2}{3}}\\0\end{pmatrix};\ \begin{pmatrix}-0.473465\\0.1120849\end{pmatrix},\ \lambda=\begin{pmatrix}0\\1.1120849\end{pmatrix}.$$

At the KKT point $(\sqrt{2},1)$, both constraints are active, $\lambda>0$, the gradients $\nabla g_1(\sqrt{2},1)=(2\sqrt{2},2)$ and $\nabla g_2(\sqrt{2},1)=(-2\sqrt{2},2)$ are linearly independent, and the Hessian matrix $H=\operatorname{diag}(1/\sqrt{2},(2+\sqrt{2})/2)$ is positive definite. Thus, the assumptions of Corollary 9.22 are satisfied, and the point $(\sqrt{2},1)$ is a strict local minimizer of the reformulated (minimization) problem, that is, a strict local maximizer of the original (maximization) problem.

At the KKT point $(-\sqrt{2},1)$, both constraints are active, $\lambda>0$, the gradients $\nabla g_1(-\sqrt{2},1)=(-2\sqrt{2},2)$ and $\nabla g_2(-\sqrt{2},1)=(2\sqrt{2},2)$ are linearly independent, but the Hessian matrix $H=\operatorname{diag}(-1/\sqrt{2},(2-\sqrt{2})/4)$ is not positive semidefinite. However, since $M=0$ in Corollary 9.22, the second-order sufficient conditions are trivially satisfied, and the point $(-\sqrt{2},1)$ is a strict local maximizer of the original problem.

At the point $(-1,-1)$, both constraints are inactive and the Hessian matrix $H=\operatorname{diag}(-1,-1)$ is negative definite. Therefore, this point is *not* a local maximizer of the original problem. We remark that since there are no active constraints at $(-1,-1)$, the second-order necessary conditions for constrained and unconstrained optimization problems coincide.

At the point $(-\sqrt{3/2},-\sqrt{3/2})$, only the first constraint is active, and the gradient of g_1 is not zero. Thus, M is a line through the origin. Since the Hessian matrix $H=\operatorname{diag}(-\sqrt{2/3},-\sqrt{2/2})/2)$ is negative definite, Theorem 9.20 implies that the second-order necessary conditions fail. Thus, this KKT point is *not* a local maximizer of the original problem.

Finally, at the point $(-0.473465\ldots,0.1120849\ldots)$, only the second constraint is active, with a nonzero gradient, and the Hessian matrix $H=\operatorname{diag}(-2.1120849,-1)$ is negative definite. As in the preceding situation, this KKT point is *not* a local maximizer of the original problem.

In summary, we see that the only local maximizers of the original (maximization) problem are the two KKT points $(\mp\sqrt{2},1)$, which are actually strict local maximizers. The point $(\sqrt{2},1)$ is, of course, the global maximizer of the original problem.

Example 9.27. (Continuation of Example 9.17)

We reconsider this problem on page 226,

$$\min\left\{\frac{1}{3}\sum_{i=1}^{n}x_i^3:\sum_{i=1}^{n}x_i=0,\sum_{i=1}^{n}x_i^2=n\right\},$$

with the intention of applying to it the second-order necessary and sufficient tests in order to isolate its local minimizers (and maximizers). We recall that the KKT points of the problem consist of the vectors

$$x = (x_1, \ldots, x_n) = (\underbrace{x_+, \ldots, x_+}_{k \text{ times}}, \underbrace{x_-, \ldots, x_-}_{n-k \text{ times}}), \; k = 1, \ldots, n-1,$$

and all their permutations, where

$$x_+ = \sqrt{\frac{n-k}{k}}, \quad x_- = -\sqrt{\frac{k}{n-k}},$$

with the multipliers

$$\lambda_1 = -1, \;\; \lambda_2 = \frac{k}{n}\sqrt{\frac{k}{n-k}} - \frac{n-k}{n}\sqrt{\frac{n-k}{k}}.$$

The constraint functions have the gradients

$$\nabla h_1(x) = e = (1, \ldots, 1), \quad \nabla h_2(x) = 2x.$$

These gradients are linearly independent at each KKT point. We see that the assumptions of Theorem 9.20 are satisfied. In order to determine whether second-order necessary conditions are satisfied, we need to verify whether the Hessian matrix $\nabla_x^2 L(x, \lambda)$ is positive semidefinite on the subspace $\{e, x\}^{\perp}$.

We have

$$L(x, \lambda) = \frac{1}{3}\sum_{i=1}^{n} x_i^3 + \lambda_1 \sum_{i=1}^{n} x_i + \frac{\lambda_2}{2}\Big(\sum_{i=1}^{n} x_i^2 - n\Big),$$

so that $\nabla_x^2 L(x, \lambda) = 2\,\mathrm{diag}(x) + \lambda_2 I$, and

$$\langle \nabla_x^2 L(x, \lambda) d, d \rangle = \sum_{i=1}^{n} (2x_i + \lambda_2) d_i^2. \tag{9.24}$$

The conditions for a vector d to be orthogonal to the vectors $\{e, x\}$ are

$$0 = \langle d, e \rangle = \sum_{i=1}^{n} d_i = 0,$$

$$0 = \langle d, x \rangle = \sum_{i=1}^{k} d_i x_i + \sum_{i=k+1}^{n} d_i x_i = \sqrt{\frac{n-k}{k}} \sum_{i=1}^{k} d_i - \sqrt{\frac{k}{n-k}} \sum_{i=k+1}^{n} d_i.$$

Therefore, d is characterized by the equation $\sum_{i=1}^{k} d_i = 0 = \sum_{i=k+1}^{n} d_i$. Thus, (9.24) can be rewritten in the form

$$\begin{aligned}
\langle \nabla_x^2 L(x, \lambda) d, d \rangle &= \left(2\sqrt{\frac{n-k}{k}} + \lambda_2\right) \sum_{i=1}^{k} d_i^2 + \left(\lambda_2 - 2\sqrt{\frac{k}{n-k}}\right) \sum_{i=k+1}^{n} d_i^2 \\
&= \left(\frac{n+k}{n}\sqrt{\frac{n-k}{k}} + \frac{k}{n}\sqrt{\frac{k}{n-k}}\right) \sum_{i=1}^{k} d_i^2 \\
&\quad - \left(\frac{2n-k}{n}\sqrt{\frac{k}{n-k}} + \frac{n-k}{n}\sqrt{\frac{n-k}{k}}\right) \sum_{i=k+1}^{n} d_i^2.
\end{aligned}$$

Each term in the parentheses is positive, so that $\langle \nabla_x^2 L(x,\lambda)d,d\rangle \geq 0$ only when $\sum_{i=k+1}^n d_i^2 = 0$. Thus, the only local minimizer occurs exactly when $k = n-1$, in which case $\sum_{i=1}^{n-1} d_i = d_n = 0$. It follows from Theorem 9.20 that the KKT points corresponding to $k \neq n-1$ are *not* local minimizers. It is easily seen that the second-order sufficiency conditions in Corollary 9.22 are satisfied at the KKT point corresponding to $k = n-1$ (provided $\lambda_2 \neq 0$, that is, except for the trivial case $n = 2$) at the point

$$x = \left(\frac{1}{\sqrt{n-1}}, \ldots, -\frac{1}{\sqrt{n-1}}, -\sqrt{n-1}\right).$$

This is the only local (actually strict, global) minimizer of the problem.

In the same way, the quadratic form $\langle \nabla_x^2 L(x,\lambda)d,d\rangle$ is less than or equal to zero only when $k = 1$. Therefore, the vector

$$x = \left(\sqrt{n-1}, -\frac{1}{\sqrt{n-1}}, \ldots, -\frac{1}{\sqrt{n-1}}\right)$$

is the only local (global, strict) maximizer of the problem.

We already know from the previous analysis of the problem that the KKT point corresponding to $k = n-1$ ($k = 1$) is a global, hence local, minimizer (maximizer) of the problem. The new results we uncovered are the facts that these are strict optimizers, and that the remaining KKT points are *not* local optimizers.

9.9 Applications of Nonlinear Programming to Inequalities

Optimization often provides effective tools for proving inequalities. To illustrate the idea, we describe here one example. Many other possibilities exist and can be explored; see the exercises at the end of the chapter for some examples.

Consider the optimization problem

$$\min\{f(x) : g(x) = 1\}, \tag{9.25}$$

where f and g are positively homogeneous of degree α and β, respectively, that is,

$$f(tx) = t^\alpha f(x), \quad g(tx) = t^\beta g(x), \text{ for } t \geq 0.$$

Suppose that (9.25) has a *global* minimizer x^* with optimal objective value $z^* := f(x^*)$. Then $f(x) \geq z^*$ for all x satisfying $g(x) = 1$, which we can write in the form

$$\frac{f(x)^{1/\alpha}}{g(x)^{1/\beta}} \geq (z^*)^{1/\alpha} \text{ for all } x,\ g(x) = 1.$$

Note that above ratio is homogeneous of degree 0, that is, it is unchanged if x is replaced by tx, $t > 0$. This means that we have

$$f(x)^{1/\alpha} \geq (z^*)^{1/\alpha} g(x)^{1/\beta}, \text{ for all } x, g(x) > 0. \tag{9.26}$$

Frequently, it happens that g is a nonnegative function, and that (9.26) is also satisfied when $g(x) = 0$. Then, there is a one-to-one correspondence between the optimization problem (9.25) and the inequality (9.26). Furthermore, note that if (P) has a unique optimal solution x^*, then (9.26) becomes an equality only at x^*. This is useful, since often it is important to determine when an inequality becomes an equality.

As an example, we consider the well-known *Cauchy–Schwarz inequality*, sometimes called the *Cauchy–Schwarz–Buniakovsky inequality*, which states that

$$|\langle x, y\rangle| \leq \|x\| \cdot \|y\| \quad \text{for all} \quad x, y \in \mathbb{R}^n, \tag{9.27}$$

with equality holding if and only if the vectors x and y have the same direction.

We can prove this inequality by solving the optimization problem

$$\max\{\langle x, z\rangle : \|z\|^2 = 1\},$$

where $x \in \mathbb{R}^n$ is a fixed nonzero (the case $x = 0$ is trivial) vector, say $\|x\| = 1$.

Since the constraint set is compact, there exists a global maximizer. The gradient of the constraint function is never zero on the feasible set, so that the KKT conditions are satisfied at any local minimizer of the problem. We have

$$L(z, \mu) = -\langle x, z\rangle + \frac{\mu}{2}(\|z\|^2 - 1),$$

and the KKT conditions are

$$-x + \mu z = 0, \quad \|z\| = 1.$$

Thus, $\mu z = x$, and taking the norms of both sides of this equality, we obtain $\mu = \pm 1$, giving the KKT points $z = \pm x$. Since $\langle x, z\rangle = \mu\|x\|^2 = \mu \geq 0$ at a global maximizer, we see that the point $z = x$ is the unique global maximizer.

This proves the Cauchy–Schwarz inequality (9.27), and at the same time characterizes when it holds as an equality.

The well-known *Hölder's inequality* can be proved in the same way; see Exercise 29 on page 248.

9.10 Exercises

1. Find the minimal value of the function $f(x, y) = (x-2)^2 + (y-1)^2$ subject to the conditions that $y \geq x^2$ and $x + y \leq 2$.

2. The problem

$$\begin{aligned} \min\ & -xy \\ \text{s.t.}\ & x+y=8, \\ & x \geq 0, y \geq 0 \end{aligned}$$

codifies the problem of finding the rectangle of maximum area with perimeter 16.
   (a) Write down the FJ conditions, and show algebraically that $\lambda_0 \neq 0$, and thus the KKT conditions are satisfied at all points satisfying the FJ conditions.
   (b) Show that the point $(x, y) = (4, 4)$ satisfies the KKT conditions.
   (c) Determine all the KKT points of the problem.
   (d) Show that the point $(x, y) = (4, 4)$ satisfies an appropriate second-order sufficient condition, thus is a local (indeed global) minimizer of the problem.
3. Consider the problem $\max\{x^2 + (y+1)^2 : -x^2 + y \geq 0, x + y \leq 2\}$.
   (a) Write down the FJ conditions, and argue that $\lambda_0 \neq 0$.
   (b) Sketch the feasible region, and graphically determine the optimal solution(s).
   (c) Determine all the points satisfying the KKT conditions; then determine (global) maximizer(s) among these.
4. In the problem

$$\begin{aligned} \min\ & \sum_{j=1}^{n} \frac{c_j}{x_j} \\ \text{s.t.}\ & \sum_{j=1}^{n} a_j x_j = b, \\ & x_j \geq 0, \quad j = 1, \ldots, n, \end{aligned}$$

a_j, c_j, b are all positive constants. Write the FJ, and if applicable the KKT, conditions. Then solve for the optimal solution(s) $x^* = (x_1^*, \ldots, x_n^*)$.
5. Consider the problem

$$\begin{aligned} \min\ & x^2 + y^2 + z^2 \\ \text{s.t.}\ & xyz \geq 8, \\ & x \geq 0, y \geq 0, z \geq 0. \end{aligned}$$

   (a) Find all the points satisfying the KKT conditions.
   (b) Use appropriate second-order tests to locate all the local and global minimizers.

6. Consider the problem

$$\begin{aligned} \min\ & \ln x - y \\ \text{s.t.}\ & x^2 + y^2 \le 4 \\ & x \ge 1. \end{aligned}$$

   (a) Find all the points satisfying the FJ conditions.
   (b) Find all the points satisfying the KKT conditions.
   (c) Which of the KKT point(s) have the lowest objective value?
   (d) Verify whether an appropriate second-order sufficient condition is satisfied at each KKT point.

7. *(Absence of KKT points)* Consider the problem

$$\begin{aligned} \min\ & x^2 + y^2 \\ \text{s.t.}\ & x^2 - (y-1)^3 = 0. \end{aligned}$$

   (a) Solve the problem geometrically.
   (b) Show that there exist no points satisfying the KKT conditions.
   (c) Find all the points satisfying the FJ conditions.
   (d) One may be tempted to solve the optimization problem by substituting $x^2 = (y-1)^3$ in the objective, thereby reducing it to the unconstrained problem $\min y^2 + (y-1)^3$. But something is wrong with this approach. What is it, and how can it be corrected?

8. In the optimization problem

$$\begin{aligned} \min\ & (x+1)^2 - y^2 \\ \text{s.t.}\ & x + y \le 0, \\ & x^2 + y^2 = 1, \end{aligned}$$

   starting from different initial points, a numerical algorithm returns the following points (x^*, y^*) as candidates for a local minimizer:

$$(i)\ \left(\frac{-1}{\sqrt{2}}, \frac{1}{\sqrt{2}}\right), \quad (ii)\ (-1, 0), \quad (iii)\ (0, 0), \quad (iv)\ \left(\frac{-1}{2}, \frac{-\sqrt{3}}{2}\right).$$

   (a) Determine which of these points satisfy the KKT conditions.
   (b) Determine which KKT points satisfy a version of the second-order necessary conditions.
   (c) Determine which KKT points satisfy a version of the second-order sufficient conditions.

9. Consider the maximization problem

$$\begin{aligned} \max\ & x^2 + y \\ \text{s.t.}\ & x^2 + y^2 \le 9, \\ & x + y \le 1. \end{aligned}$$

(a) Sketch the feasible region and the level curves of the objective function. Based on this, guess the global maximizer of the problem.
(b) Justify why KKT must hold at local maximizers.
(c) Write down the KKT conditions, and use them to determine all the KKT points.
(d) Determine which KKT points satisfy the second-order (necessary and sufficient) conditions.

10. Consider the problem

$$\begin{aligned} \max \quad & x_1^3 + x_2^3 + \cdots + x_n^3 \\ \text{s.t.} \quad & x_1^2 + x_2^2 + \cdots + x_n^2 = 1. \end{aligned}$$

(a) Prove that the KKT conditions must be satisfied at each local maximizer.
(b) Determine all the KKT points.
(c) Determine the global maximizers of the problem.
(d) Use (c) to prove the inequality

$$\sum_{i=1}^{n} |x_i|^3 \le \Big(\sum_{i=1}^{n} x_i^2\Big)^{3/2} \quad \text{for all} \quad (x_1, \ldots, x_n) \in \mathbb{R}^n.$$

(e) If there are KKT points other than global maximizers, determine which ones are local maximizers.

11. Consider the following variant of Exercise 10:

$$\begin{aligned} \max \quad & x_1^3 + x_2^3 + \cdots + x_n^3 \\ \text{s.t.} \quad & x_1^4 + x_2^4 + \cdots + x_n^4 = 1. \end{aligned}$$

Answer the corresponding questions, and take care to formulate and prove the correct form of the inequality in part (c).

12. The equality constrained quadratic program

$$\begin{aligned} & \min q(x) = \tfrac{1}{2}\langle Qx, x\rangle + \langle c, x\rangle \\ & \text{s.t. } Ax = b \end{aligned}$$

has a symmetric $n \times n$ matrix Q and a vector $c \in \mathbb{R}^n$.
(a) Show that a local minimizer x^* must satisfy the KKT conditions $Qx^* + c \in R(A^T)$ and $Ax^* = b$.
(b) Show that a local minimizer x^* must satisfy the second-order necessary condition that Q is positive semidefinite on the subspace $N(A)$, the null space of A.
(c) Show that a KKT point satisfying the second-order necessary condition in (b) is in fact a *global* minimizer of the quadratic program.

13. Consider the nonlinear program (P) as in (9.1) but with only inequality constraints, and where the functions f, g_i are continuously differentiable

in a neighborhood of the feasible region. Let x^* be a local minimizer of (P).
The purpose of this problem is to show the validity of the FJ conditions at x^* via Danskin's theorem. Define the function

$$\varphi(x) := \max\{f(x) - f(x^*),\ g_1(x), \dots, g_r(x)\}.$$

(a) Show that x^* is a local minimizer of φ.
(b) Show that Theorem 1.29 implies that for any $d \in \mathbb{R}^n$,

$$0 \leq \varphi'(x^*; d) = \max\{\langle \nabla f(x^*), d\rangle,\ \langle \nabla g_i(x^*), d\rangle,\ i \in I\},$$

where I is the index set of active constraints at x^*.
(c) Show that (b) implies that the system

$$\langle \nabla f(x^*), d\rangle < 0,\ \langle \nabla g_i(x^*), d\rangle < 0,\ i \in I,$$

is inconsistent. Use this fact to prove that the FJ conditions hold at x^*.

14. *(Second-order conditions)* In the problem

$$\begin{aligned} &\min\ x^2 + (y-1)^2 \\ &\text{s.t.}\ -y + \tfrac{x^2}{k} \geq 0, \end{aligned}$$

show that $(x^*, y^*) = (0, 0)$ is a KKT point for all values of the parameter $k > 0$. However, the nature of the point depends on the value of k. Use the available second-order conditions to determine the values of k for which $(0, 0)$ is a local minimum. What is the status of the point for the remaining values of the parameter $k > 0$?

15. Consider the problem

$$\begin{aligned} \min\ & (x-2)^2 + y^2 \\ s.t.\ & x^2 \leq ky^2 + 1 \\ & x \geq 0, \end{aligned}$$

where $k \in \mathbb{R}$ is a parameter of the problem.
(a) Sketch the feasible region of the problem for $k > 0$, $k = 0$, and $k < 0$.
(b) Determine the status of the point $(1, 0)$ for each value of k. For what values of k is the point $(1, 0)$ is a KKT point, local minimizer, global minimizer?

16. Sketch the constraint set of the optimization problem

$$\begin{aligned} \min\ & x \\ \text{s.t.}\ & (x-3)^2 + (y-2)^2 \geq 13, \\ & (x-4)^2 + y^2 \leq 16. \end{aligned}$$

Then,

(a) Find all the points satisfying the KKT conditions.
(b) Verify which second-order sufficient conditions, if any, are satisfied at each KKT point.

17. *(Absence of strict complementarity)* In the problem

$$\begin{aligned}\min\ & xy\\ \text{s.t.}\ & x^2+y^2 \le 2,\\ & x+y \ge 0,\end{aligned}$$

show that the KKT conditions are satisfied at every local minimizer. Compute all the KKT points. Show that *strict* complementarity is satisfied at *no* KKT point. Finally, use the available second-order necessary/sufficient conditions to determine which KKT points are local minimizers.

18. Consider the problem of covering the triangle with vertices at the points $(0,0)$, $(0,1)$, and $(1,0)$ with a ball of smallest radius.
(a) By geometric considerations, show that the optimal ball has center at the point $(1/2,1/2)$ and radius $1/\sqrt{2}$.
(b) Show that the problem can be formulated as the nonlinear program

$$\begin{aligned}\min\ & r\\ \text{s.t.}\ & x^2+y^2 \le r\\ & (x-1)^2+y^2 \le r\\ & x^2+(y-1)^2 \le r.\end{aligned}$$

(c) Solve the above program.

19. Inscribe a triangle in the disk $\{(x,y) : x^2+y^2 \le 1\}$ with maximum area. (If you get stuck; see the general form of the problem on page 271.)

20. Inscribe a tetrahedron in the sphere $\{(x,y,z) : x^2+y^2+z^2 \le 1\}$ with maximum-volume. (If you get stuck; see the general form of the problem on page 271.)

21. Prove that if $-1 \le x_i \le 0$, $i=1,2,\ldots,n$, then

$$(1+x_1)(1+x_2)\cdots(1+x_n) \ge 1+x_1+x_2+\cdots+x_n,$$

and investigate when equality holds.
(a) Set up an appropriate optimization problem such that the optimal solutions x^* to this problem give equality in the above inequality.
(b) Show that each component x_i^* has only two possible values.
(c) Use second-order conditions to isolate the optimal solution(s) among the KKT points.

22. Let $x_1,x_2,\ldots,x_n$ be real numbers such that $\sum_1^n x_i = 0$. Show that the inequality

$$(x_n-x_1)^2+\sum_1^{n-1}(x_i-x_{i+1})^2 \ge 4\sin^2(\pi/n)\cdot\sum_1^n x_i^2$$

holds.

(a) Formulate the problem as an optimization problem and verify it for $n = 3, 4, 5$.
(b) Verify the inequality for all n.

Hint: The inequality can be viewed as an eigenvalue problem.

23. This problem has appeared in connection with Karmarkar's potential function for linear programming. For a fixed parameter α satisfying $0 < \alpha < 1$, consider the minimization problem

$$\begin{aligned} \min \quad & \prod_{i=1}^{n} x_i \\ \text{s.t.} \quad & \sum_{i=1}^{n} x_i = 1 \\ & \sum_{i=1}^{n} \left(x_i - \frac{1}{n} \right)^2 = \frac{\alpha^2}{n(n-1)}. \end{aligned}$$

(a) Show that constraints of the problem imply that each feasible vector x is positive, that is, $x_i > 0$ for all $i = 1, \ldots, n$. Thus, if we like, we may replace the objective function by its logarithm.
Hint: The fact that $x_i > 0$ may be verified by solving an optimization problem!
(b) Write down the FJ conditions for a local minimizer x^*, and show that $\lambda_0 > 0$ at each FJ point.
(c) Find all the KKT points. Determine which KKT point(s) are global minimizers by comparing their objective values.
(d) Determine whether any second-order conditions (necessary and sufficient) hold at the global minimizer(s) that you found in part (c).
(e) Consider the problem of maximizing the objective function over the constraint set. Repeat parts (b)–(d) for this problem.

24. Let $a, b \in \mathbb{R}^n$ be given nonzero vectors. It is known that if $x \in \mathbb{R}^n$ satisfies $\|x\| = 1$ and $\langle a, x \rangle = 0$, then

$$\langle b, x \rangle \leq \frac{\sum_{1 \leq i < j \leq n} (a_i b_j - a_j b_i)^2}{\|a\|^2}.$$

(You may assume that $\{a, b\}$ are linearly independent, since otherwise the inequality is trivial.) Reduce the inequality to an appropriate constrained optimization problem. Find the optimizer(s) of this problem; and finally show that the inequality follows.

25. This problem concerns inscribing into a given circle a triangle such that the sum of the squares of the sides of the triangle is maximized. This can be formulated as an optimization problem as follows:

$$\begin{aligned} \max \quad & \|x - y\|^2 + \|x - d\|^2 + \|y - d\|^2 \\ \text{s.t.} \quad & \|x\|^2 = 1, \\ & \|y\|^2 = 1. \end{aligned}$$

Here, we have assumed that the circle is the unit circle with center at the origin, one of the vertices of the inscribed triangle is at the point $d = (0, 1)$, and x and y are the remaining, unknown vertices of the triangle.

(a) Write down the FJ conditions, and show that all points satisfying the FJ conditions must be KKT points.

(b) Determine all the KKT points.

(c) Determine the optimal solutions among the KKT points. What can you say about the lengths of the sides of the optimal triangle? Are the second-order sufficient conditions satisfied at the optimal solution(s)?

26. **(Fagnano)** Find a point on each side of a given triangle A such that the triangle B formed by these three points has the smallest perimeter. Prove that if all three angles of A are acute, then the vertices of the triangle B are the base points of the perpendicular lines dropped from the vertices A onto the opposite sides of A.
Hint: One way to formulate the problem is to assume that one vertex of A is at the origin and the two sides adjacent to it form linearly independent vectors $a, b \in \mathbb{R}^2$. The vertices of B are then at the points sa, tb, and $a + u(b - a)$, say, where $0 \leq s, t, u \leq 1$.

27. The goal of this problem is to prove the inequality

$$\frac{\sum_{i=1}^n x_i}{n} \cdot \frac{\sum_{i=1}^n y_i}{n} \leq \frac{\sum_{i=1}^n x_i y_i}{n} \tag{9.28}$$

whenever x_i and y_i are nonincreasing,

$$x_1 \geq x_2 \geq \cdots \geq x_{n-1} \geq x_n, \quad y_1 \geq y_2 \geq \cdots \geq y_{n-1} \geq y_n, \tag{9.29}$$

and characterize its cases of equality.
Prove (9.28) by maximizing the function

$$\Big(\sum_{i=1}^n x_i\Big) \cdot \Big(\sum_{i=1}^n y_i\Big)$$

subject to the constraint $\sum_{i=1}^n x_i y_i = n$ and the constraints in (9.29). Note that it suffices to show that the objective value of the optimization problem is n^2.
Form the (weak) Lagrangian

$$L = -\lambda_0\Big(\sum_{i=1}^n x_i\Big)\Big(\sum_{i=1}^n y_i\Big) + \lambda\Big(\sum_{i=1}^n x_i y_i - n\Big) + \sum_{i=1}^{n-1} \delta_i (x_{i+1} - x_i)$$
$$+ \sum_{i=1}^{n-1} \xi_i (y_{i+1} - y_i), \quad (\lambda_0,\ \delta,\ \xi) \geq 0.$$

(a) Write the FJ conditions, and use them to prove that

$$n\lambda = \lambda_0\Big(\sum_{i=1}^{n} x_i^*\Big)\cdot\Big(\sum_{i=1}^{n} y_i^*\Big).$$

(This is the crucial result.)

(b) Use this to show that $\lambda_0 \neq 0$; thus the KKT conditions must hold. Put $\lambda_0 = 1$, and use the KKT conditions to show that

$$(\lambda - n)\Big(\sum_{i=1}^{n} y_i^*\Big) = (\lambda - n)\Big(\sum_{i=1}^{n} x_i^*\Big) = 0.$$

Argue that

$$\lambda = n.$$

(c) By considering the form of the optimization problem, argue that any (x^*, y^*) above cannot be a (local) minimizer, and then argue that it must be a global maximizer. Conclude that the inequality (9.28) must hold.

(d) Show that the optimal solution (x^*, y^*) must have $x_1^* = \cdots = x_n^*$ and $y_1^* = \cdots = y_n^*$. Conclude that the inequality (9.28) holds as an equality if and only if $x_1 = \cdots = x_n$ and $y_1 = \cdots = y_n$.

28. Let $x_1, \ldots, x_n$ $(n \geq 2)$ be real numbers subject to the conditions

$$\sum_{1}^{n} x_i = 0, \quad \max_{1\leq i\leq n} |x_i| = 1.$$

Set $x_{n+1} = x_1$, and define

$$\mu(x) = \max_{1\leq i\leq n} |x_i - x_{i+1}|.$$

(a) Show that the minimum of $\mu(x)$ on $\mathbb{R}^n$ equals $4/n$ if n is even, and $4n/(n^2-1)$ if n is odd.

(b) Determine the optimal solution(s) $x^* = (x_1^*, \ldots, x_n^*)$ in each case.

Hint: The symmetry of the problem allows one to set any $x_i^* = 1$; this simplifies the formulation of the optimization problem.

29. *(Hölder's inequality)* Let $p > 1$, $q > 1$ be such that $p^{-1} + q^{-1} = 1$. Then

$$|\langle x, y\rangle| \leq \|x\|_p \cdot \|y\|_q \text{ for all } x, y \in \mathbb{R}^n,$$

with equality holding if and only if x and y are parallel vectors. Here the p-norm, $\|x\|_p$, is defined by $\|x\|_p^p = |x_1|^p + \cdots + |x_n|^p$.
Prove Hölder's inequality by solving the optimization problem

$$\max\Big\{\langle z, y\rangle : \sum_{i=1}^{n} |z_i|^p = 1\Big\},$$

where $y \in \mathbb{R}^n$ is a fixed nonzero vector, say $\|y\|_q = 1$.

30. Solve the optimization problem

$$\max\Big\{\sum_{i=1}^{n} |x_i|^p : \sum_{i=1}^{n} |x_i|^q = 1\Big\},$$

where $q \geq p > 1$, and use it to prove the inequality

$$\left(\frac{1}{n}\sum_{i=1}^{n} |x_i|^p\right)^{1/p} \leq \left(\frac{1}{n}\sum_{i=1}^{n} |x_i|^q\right)^{1/q}, \quad q \geq p > 1,$$

and determine when equality holds.

31. **(Waterhouse [263])** Determine the optimal value of the problem

$$\begin{aligned} \min \ & (r-1)^2 + \left(\frac{s}{r} - 1\right)^2 + \left(\frac{t}{s} - 1\right)^2 + \left(\frac{4}{t} - 1\right)^2 \\ \text{s.t.} \ & 1 \leq r \leq s \leq t \leq 4. \end{aligned}$$

Hint: Introduce new variables r, s/r, t/s, and $4/t$.

32. **(Waterhouse [263])** Consider the following *symmetric* optimization problem:

$$\begin{aligned} \min / \max \ & f(x) \\ \text{s.t.} \ & g_i(x) = 0, \ 1 \leq i \leq m, \end{aligned}$$

where f and $\{g_i\}_1^m$ are symmetric, continuously differentiable functions defined on $\mathbb{R}^n$. The aim of this problem is to show that *symmetric problems generically admit symmetric solutions.*

Suppose that there exists a feasible point $x_0 = (\alpha, \alpha, \ldots, \alpha)$.

(a) Show that x_0 is a KKT point (except possibly when $\nabla g_i(x_0) = 0$ $(i = 1, \ldots, m)$ and $\nabla f(x_0) \neq 0$).

(b) Assuming that x_0 is a KKT point, investigate when x_0 satisfies the second-order sufficient conditions for a local optimizer (minimizer or maximizer) of the problem. Conclude that x_0 is generically a local optimizer. What can the second-order conditions tell us when the second-order sufficient conditions fail?

Hint: Let $h : \mathbb{R}^n \to \mathbb{R}$ be a symmetric function, σ a permutation of the set $\{1, 2, \ldots, n\}$, and $\overline{\sigma} : \mathbb{R}^n \to \mathbb{R}^n$ the linear map $\overline{\sigma}(x_1, \ldots, x_n) = (x_{\sigma(1)}, \ldots, x_{\sigma(n)})$. Define $\overline{x} := \overline{\sigma}x$ and $\overline{d} := \overline{\sigma}d$. Compare the Taylor expansions of $h(\overline{\sigma}(x + td))$ and $h(x + td)$ and relate $\langle \nabla h(\overline{x}), \overline{d} \rangle$ to $\langle \nabla h(x), d \rangle$ and $\nabla^2 h(\overline{x})[\overline{d}, \overline{d}]$ to $\nabla^2 h(x)[d, d]$. Conclude that $\nabla h(x_0)$ is a multiple of $e := (1, \ldots, 1)$ and that $\nabla^2 h(x_0)$ is a symmetric matrix with a diagonal a multiple of e and off-diagonal elements all equal.

33. **(Waterhouse [263])** Consider the symmetric optimization problem

$$\begin{aligned} \min / \max \quad & x^4 yz + xy^4 z + xyz^4 \\ \text{s.t.} \quad & x^3 y^3 + x^3 z^3 + y^3 z^3 - 3 = 0. \end{aligned}$$

Show that $x_0 = (1,1,1)$ is a KKT point, but not a local minimizer or maximizer. (This is a degenerate problem that forms an exception to part (b) of the above problem.)
Hint: Consider the values of $f(x(t))$ where $x(t) = (st, s, s)$, t is close to 1, and $s = s(t)$ is chosen to make $x(t)$ a feasible curve.

10

Structured Optimization Problems

In this chapter, we solve several important well-known problems using optimization techniques. These include the extensive theories of the eigenvalues of symmetric matrices and the singular values of a matrix, an optimization problem in Broyden's method for solving nonlinear systems of equations, an optimization problem appearing in quasi-Newton methods for unconstrained minimization of a nonlinear function, the inequalities of Kantorovich, Hadamard, and Hilbert, and the problem of inscribing a maximum-volume ellipsoid in a convex polytope in $\mathbb{R}^n$. The variational approach to the eigenvalues and singular values are especially important, both in finite and infinite dimensions, since they can be used to prove various inequalities among the eigenvalues (and the singular values), and to establish the spectral decomposition of compact operators in Hilbert spaces, for example.

Many other important problems can be treated by variational means.

10.1 Spectral Decomposition of a Symmetric Matrix

The eigenvalues and eigenvectors of a symmetric matrix can be obtained by variational means, by solving certain optimization problems. This approach avoids the use of determinants, and is particularly important in infinite-dimensional Hilbert spaces, where determinants are not always meaningful. The variational approach is due to Hilbert, who at the beginning of the twentieth century, developed the spectral theory of compact operators in Hilbert spaces using such an approach; see the book [227], Chapter 6, for a lucid presentation. Here we deal only with the finite-dimensional situation.

Let A be a symmetric $n \times n$ matrix. We will prove Theorem 2.19, the spectral decomposition of A, using nonlinear programming.

Consider the minimization of the quadratic form $\langle Ax, x\rangle$ over the unit sphere,

O. Güler, *Foundations of Optimization*, Graduate Texts in Mathematics 258,
DOI 10.1007/978-0-387-68407-9_10,

$$\begin{aligned} \min \quad & \langle Ax, x\rangle \\ \text{s.t.} \quad & \|x\|^2 = 1. \end{aligned}$$

Since the gradient of the constraint function is nonzero on the feasible region, the KKT conditions hold. Thus, we have the Lagrangian function

$$L(x, \lambda) = \langle Ax, x\rangle + \lambda_1(1 - \|x\|^2).$$

Since the constraint set is compact, there exists a global minimizer u_1 on the unit sphere that is the solution to the KKT conditions

$$\nabla_x L = 2Au_1 - 2\lambda_1 u_1 = 0, \quad \|u_1\| = 1.$$

Thus, $Au_1 = \lambda_1 u_1$ and $\|u_1\| = 1$, that is, λ_1 is the eigenvalue of A corresponding to the eigenvector u_1.

We have $\langle Au_1, u_1\rangle = \langle \lambda_1 u_1, u_1\rangle = \lambda_1$, and if a unit vector x is any eigenvector of A with the corresponding eigenvalue λ, then

$$\lambda = \langle \lambda x, x\rangle = \langle Ax, x\rangle \geq \langle Au_1, u_1\rangle = \lambda_1.$$

Thus, λ_1 is the *smallest* eigenvalue of A and u_1 is a corresponding eigenvector.

Now consider *sequentially* the following problems:

$$\begin{aligned} \min \quad & \langle Ax, x\rangle \\ \text{s.t.} \quad & \|x\|^2 = 1, \\ & \langle u_i, x\rangle = 0,\ i = 1, \ldots, k-1, \end{aligned} \tag{P_k}$$

for $k = 2, \ldots, n$, where the vector u_i in the last set of constraints is an optimal solution to problem (P_i), $i < k$.

Note that the vectors $\{u_i\}_1^{k-1}$ form an orthonormal set. Since the constraint set is compact, there exists a global minimizer x^* on the unit sphere; the gradients $\{-2x, u_1, \ldots, u_{k-1}\}$ of the constraints are orthogonal, hence linearly independent. It follows from Corollary 9.10 that the KKT conditions hold, and we have the Lagrangian function

$$L(x, \lambda, \delta_1, \ldots, \delta_{k-1}) = \frac{1}{2}\langle Ax, x\rangle + \frac{\lambda}{2}(1 - \|x\|^2) + \sum_{i=1}^{k-1} \delta_i^k \langle u_i, x\rangle.$$

The KKT conditions are

$$\begin{aligned} & Ax^* - \lambda x^* + \sum_{i=1}^{k-1} \delta_i^k u_i = 0, \quad \|x^*\| = 1, \\ & \langle x^*, u_i\rangle = 0,\ i = 1, \ldots, k-1. \end{aligned} \tag{10.1}$$

We claim that the multiplier vector $\delta^k := (\delta_1^k, \ldots, \delta_{k-1}^k)$ is zero. We prove this by induction on k. For $k = 1$, there is nothing to prove. Suppose that the induction hypothesis is true for all integers less than k. Taking the inner product of both sides of the first equation in (10.1) with u_j $(j < k)$, we obtain

$$\langle Ax^*, u_j\rangle - \delta_j^k = \langle x^*, Au_j\rangle - \delta_j^k = \langle x^*, \lambda_j u_j\rangle - \delta_j^k = 0,$$

where the first equality follows because A is a symmetric matrix, and the second equality follows from the induction hypothesis. (If the induction hypothesis holds, then $\delta^j = 0$ and (10.1) above shows that u_j is an eigenvalue of A.)

This proves the claim. Thus, $u_k := x^*$ is an eigenvector of A with the corresponding eigenvector $\lambda_k := \lambda$, which is the kth-smallest eigenvalue of A.

Define $U := [u_1, u_2, \ldots, u_n]$ and $\Lambda = \text{diag}(\lambda_1, \ldots, \lambda_n)$. We have $Au_i = \lambda_i u_i$, $i = 1, \ldots, n$, so that

$$\begin{aligned} AU &= A[u_1, \ldots, u_n] = [Au_1, \ldots, Au_n] = [\lambda_1 u_1, \ldots, \lambda_n u_n] \\ &= [u_1, \ldots, u_n] \,\text{diag}(\lambda_1, \ldots, \lambda_n) = U\Lambda. \end{aligned}$$

Since $\{u_i\}_1^n$ is an orthonormal set of vectors, the matrix U is orthogonal, that is, $UU^T = U^TU = I$.

In summary, we have proved Theorem 2.19.

Remark 10.1. It may appear quite remarkable that the multiplier vector δ vanishes in the optimization problem (P_k). This may be explained by the fact that A is *invariant* on the subspace

$$L_{k-1} := \{u_1, \ldots, u_{k-1}\}$$

and on its orthogonal complement $L_{k-1}^\perp$, that is, $A(L_{k-1}) \subseteq L_{k-1})$ and $A(L_{k-1}^\perp) \subseteq L_{k-1}^\perp)$. The invariance on L_{k-1} is easy to see, and if $x \in L_{k-1}^\perp$ and $j \leq k-1$, then

$$\langle Ax, u_j\rangle = \langle x, Au_j\rangle = \lambda_j\langle x, u_j\rangle = 0,$$

proving the invariance on $L_{k-1}^\perp$.

Thus, we may recast the problem (P_k) as the optimization problem

$$\min\{\langle Ax, x\rangle : \|x\|^2 = 1, \ x \in E_k\}$$

within the vector space $E_k = L_{k-1}^T$. Then problem (P_k) looks exactly like (P_1), which has no linear constraints, and it is seen as in (P_1) that the optimal solution u_k is an eigenvector of A. This is the original argument in [227].

The spectral decomposition above can be used to give further variational characterizations of the eigenvalues of a symmetric matrix.

Theorem 10.2. (Courant–Fischer) *Let A be an $n \times n$ symmetric real matrix with eigenvalues*

$$\lambda_1(A) \leq \lambda_2(A) \leq \cdots \leq \lambda_n(A)$$

arranged in ascending order. Then

$$\lambda_k(A) = \max_{L_{k-1}} \min_{x \in L_{k-1}^{\perp}, \|x\|=1} \langle Ax, x \rangle, \quad k = 1, \ldots, n,$$

where the maximization is over the set of all $(k-1)$-dimensional linear subspaces of $\mathbb{R}^n$. Furthermore,

$$\lambda_k(A) = \min_{L_{n-k}} \max_{x \in L_{n-k}^{\perp}, \|x\|=1} \langle Ax, x \rangle, \quad k = 1, \ldots, n.$$

Proof. We will prove only the first equality, since the second one follows from the first applied to the matrix $-A$. Let u_i be the eigenvector corresponding to λ_i obtained in problem (P_i). Denote by L_k^* the linear span of $\{u_i\}_1^k$, and note that

$$\lambda_{k+1} = \min_{x \in (L_k^*)^{\perp}, \|x\|=1} \langle Ax, x \rangle \le \max_{L_k} \min_{x \in L_k^{\perp}, \|x\|=1} \langle Ax, x \rangle.$$

It remains to prove the reverse inequality. Let L_k be an arbitrary k-dimensional linear subspace of $\mathbb{R}^n$. It follows from dimensional considerations that the subspace $L_{k+1}^* \cap L_k^{\perp}$ is nontrivial. Pick a unit vector $u = \sum_{i=1}^{k+1} \delta_i u_i$ in this subspace. We note that $1 = \|u\|^2 = \sum_{i=1}^{k+1} \delta_i^2$, and

$$\min_{x \in L_k^{\perp}, \|x\|=1} \langle Ax, x \rangle \le \langle Au, u \rangle = \left\langle \sum_{i=1}^{k+1} \lambda_i \delta_i u_i, \sum_{i=1}^{k+1} \delta_i u_i \right\rangle = \sum_{i=1}^{k+1} \lambda_i \delta_i^2 \le \lambda_{k+1}.$$

This proves the reverse inequality

$$\max_{L_k} \min_{x \in L_k^{\perp}, \|x\|=1} \langle Ax, x \rangle \le \lambda_{k+1},$$

and the theorem. □

Corollary 10.3. (Weyl) *Let A, B be $n \times n$ symmetric real matrix, and $\{\lambda_i(A)\}_1^n$, $\{\lambda_i(B)\}_1^n$, and $\{\lambda_i(A+B)\}_1^n$ the eigenvalues of A, B, and $A+B$, respectively. Then*

$$\lambda_i(A) + \lambda_1(B) \le \lambda_i(A+B) \le \lambda_i(A) + \lambda_n(B), \quad i = 1, \ldots, n.$$

Proof. We prove only the first inequality; the second one is proved similarly:

$$\begin{aligned}
\lambda_i(A+B) &= \max_{L_{i-1}} \min_{x \in L_{i-1}^{\perp}, \|x\|=1} \langle (A+B)x, x \rangle \\
&\ge \max_{L_{i-1}} \Big(\min_{x \in L_{i-1}^{\perp}, \|x\|=1} \langle Ax, x \rangle + \min_{x \in L_{i-1}^{\perp}, \|x\|=1} \langle Bx, x \rangle \Big) \\
&\ge \max_{L_{i-1}} \Big(\min_{x \in L_{i-1}^{\perp}, \|x\|=1} \langle Ax, x \rangle + \min_{x \in \mathbb{R}^n, \|x\|=1} \langle Bx, x \rangle \Big) \\
&= \lambda_i(A) + \lambda_1(B).
\end{aligned}$$

□

Remark 10.4. There are many other inequalities satisfied by the eigenvalues of A, B, and $A+B$. *Horn's conjecture* [142] states that a certain set of inequalities gives the complete set of inequalities between these three sets of eigenvalues. Horn's conjecture has only recently been settled, in the affirmative, using advanced algebraic techniques. See Fulton [100] for an exposition of Horn's conjecture and its proof.

The Courant–Fischer equalities can be used to prove other interesting results.

Corollary 10.5. *Let A be an $n \times n$ symmetric real matrix, and A_k a $k \times k$ principal submatrix of A that is obtained by deleting $n-k$ rows and the corresponding columns of A. If $\{\lambda_i(A)\}_1^n$ and $\{\lambda_i(A_k)\}_1^k$ are the eigenvalues of A and A_k respectively, arranged in ascending order, then*

$$\lambda_i(A) \le \lambda_i(A_k) \le \lambda_{n-k+i}(A), \quad i = 1, \ldots, k.$$

Proof. We have

$$\begin{aligned}\lambda_i(A_k) &= \max_{M_{i-1}} \min_{x \in M_{i-1}^{\perp}, \|x\|=1} \langle A_k x, x\rangle \\ &\le \max_{L_{n-k+i-1}} \min_{x \in L_{n-k+i-1}^{\perp}, \|x\|=1} \langle A_k x, x\rangle = \lambda_{n-k+i}(A),\end{aligned}$$

where M_{i-1} varies over the set of all $(i-1)$-dimensional subspaces of $\mathbb{R}^k$, and $L_{n-k+i-1}$ varies over the set of all $(n-k+i-1)$-dimensional subspaces of $\mathbb{R}^n$. The equalities above follow from Theorem 10.2, and the inequality follows from the fact that the set of subspaces M_{i-1} is a restricted set of $(n-k+i-1)$-dimensional subspaces of $\mathbb{R}^n$.

This proves the second inequality of the corollary; the first inequality is proved similarly. □

An immediate consequence of the corollary is the *interlacing property* between the eigenvalues of A and A_{n-1}, an $(n-1) \times (n-1)$ principal submatrix A_{n-1} of A:

$$\begin{aligned}\lambda_1(A) &\le \lambda_1(A_{n-1}) \le \lambda_2(A) \le \lambda_2(A_{n-1}) \le \cdots \\ &\le \lambda_{n-1}(A) \le \lambda_{n-1}(A_{n-1}) \le \lambda_n(A).\end{aligned}$$

10.2 Singular-Value Decomposition of a Matrix

Let A be an $m \times n$ matrix. The *singular-value decomposition (SVD)* expresses A in the form $A = U\Sigma V^T$, where $U \in \mathbb{R}^{m\times m}$ and $V \in \mathbb{R}^{n\times n}$ are orthogonal matrices and Σ is an $m \times n$ diagonal matrix with nonnegative entries. The singular-value decomposition is a very important tool in numerical linear algebra; see for example [108]. Here we give a derivation of the SVD using optimization techniques similar to the orthogonal decomposition of a symmetric matrix above.

Theorem 10.6. (Singular-value decomposition of a matrix) *Let A be an $m \times n$ real matrix. There exist an orthogonal $m \times m$ matrix U, an orthogonal $n \times n$ matrix V, and an $m \times n$ matrix Σ whose only nonzero elements are the diagonal entries $\Sigma_{ii} = \sigma_i$, $i = 1, \ldots, p := \min\{m, n\}$,*

$$\sigma_1 \geq \sigma_2 \geq \cdots \geq \sigma_p \geq 0$$

such that

$$A = U \Sigma V^T.$$

The scalars $\{\sigma_i\}_1^p$ are called the *singular values* of A. Note that $AA^T = U(\Sigma\Sigma^T)U^T$ is the orthogonal decomposition of the positive semidefinite matrix AA^T. Similarly, $A^T A = V(\Sigma^T \Sigma)V^T$ is the orthogonal decomposition of the positive semidefinite matrix $A^T A$. We see that the vectors $\{u_i\}_1^p$ and $\{v_i\}_1^p$ are the eigenvectors of AA^T and $A^T A$, respectively, both with the corresponding eigenvalues $\{\sigma_i^2\}_1^p$. The remaining vectors $\{u_i\}$ and $\{v_i\}$ (if any) are also the eigenvectors of AA^T and $A^T A$, respectively, with the corresponding eigenvalues equal to zero.

Proof. We consider the optimization problem

$$\begin{aligned} \min \quad & -\langle Ax, y\rangle \\ \text{s.t.} \quad & \|x\|^2 - 1 = 0, \\ & \|y\|^2 - 1 = 0. \end{aligned}$$

Since the feasible set is compact, there exists a global minimizer $(x^*, y^*) = (v_1, u_1)$. The gradients of the constraint functions, $\{(x^T, 0^T)^T, (0^T, y^T)^T\}$, are clearly linearly independent, and thus the KKT conditions hold. We have the Lagrangian function

$$L(x, y; \delta, \mu) = -\langle Ax, y\rangle + \frac{\delta}{2}(\|x\|^2 - 1) + \frac{\mu}{2}(\|y\|^2 - 1)$$

and the corresponding KKT conditions

$$A^T u_1 = \delta v_1, \quad A v_1 = \mu u_1, \quad \|u_1\| = 1, \quad \|v_1\| = 1.$$

First of all, we have

$$\delta = \langle \delta v_1, v_1\rangle = \langle A^T u_1, v_1\rangle = \langle u_1, A v_1\rangle = \langle u_1, \mu u_1\rangle = \mu \geq 0,$$

where the inequality follows since the optimal objective value $-\langle Av_1, u_1\rangle$ is clearly nonpositive. We set

$$\sigma_1 := \delta = \mu \geq 0.$$

Note that the KKT conditions give $AA^T u_1 = \sigma_1 A v_1 = \sigma_1^2 u_1$, and similarly $A^T A v_1 = \sigma_1^2 v_1$, that is, v_1 and u_1 are the eigenvalues of the positive semidefinite matrices AA^T and $A^T A$, respectively, corresponding to the same eigenvalue σ_1^2.

Next, we consider sequentially the following problems:

$$\begin{aligned} \min \quad & -\langle Ax, y\rangle \\ \text{s.t.} \quad & \|x\|^2 - 1 = 0, \quad \|y\|^2 - 1 = 0, \\ & \langle v_i, x\rangle = 0, \ \langle u_i, y\rangle = 0, \ i = 1, \ldots, k-1, \end{aligned} \tag{P_k}$$

for $k = 2, \ldots, p$, where (v_i, u_i) in the last two sets of constraints is an optimal solution to problem (P_i), $i < k$.

Again the constraint set is compact, and there exists a global minimizer (v_k, u_k). The gradient vectors of the constraints,

$$\begin{pmatrix} v_i \\ 0 \end{pmatrix}, \ \begin{pmatrix} 0 \\ u_i \end{pmatrix}, \quad i = 1, \ldots, k,$$

are linearly independent, since $\{v_i\}_1^k$ and $\{u_i\}_1^k$ are both sets of orthonormal vectors. Thus, the KKT conditions hold. We write the Lagrangian function

$$\begin{aligned} L(x, y; \delta, \mu, \alpha, \beta) = -\langle Ax, y\rangle + \frac{\delta}{2}(\|x\|^2 - 1) + \frac{\mu}{2}(\|y\|^2 - 1) \\ + \sum_1^{k-1} \alpha_i \langle v_i, x\rangle + \sum_1^{k-1} \beta_i \langle u_i, y\rangle. \end{aligned}$$

Setting the optimal solution $(x^*, y^*) = (v_k, u_k)$, the KKT conditions give the equations

$$A^T u_k = \delta v_k + \sum_1^{k-1} \alpha_i v_i, \ Av_k = \mu u_k + \sum_1^{k-1} \beta_i u_i, \ \|v_k\| = 1, \ \|u_k\| = 1.$$

We claim that the multiplier vectors α, β are zero. We prove this by induction on k. For $k = 1$, there is nothing to prove. Suppose that the induction hypothesis is true for all integers less than k. We have for a fixed $j < k$,

$$\alpha_j = \langle A^T u_k, v_j\rangle = \langle Av_j, u_k\rangle = \sigma_j \langle u_j, u_k\rangle = 0,$$

where the first equality follows from the first KKT condition above, the third equality follows from the induction hypothesis, and the last equality follows because of the constraints in problem (P_k). Thus $\alpha = 0$, and a similar proof shows that $\beta = 0$, proving the claim. Consequently, we have

$$Av_i = \sigma_i u_i, \ \ A^T u_i = \sigma_i v_i, \ \ i = 1, \ldots, p.$$

These can be written more compactly in matrix notation. First, assume that $n \le m$. Then $p = n$, and we have

$$\begin{aligned} AV &= A[v_1, v_2, \ldots, v_n] = [\sigma_1 u_1, \sigma_2 u_2, \ldots, \sigma_n u_n] \\ &= [u_1, u_2, \ldots, u_n] \operatorname{diag}(\sigma_1, \sigma_2, \ldots, \sigma_n) = U_n \Sigma_n, \end{aligned}$$

where we have defined $V := [v_1, v_2, \ldots, v_n]_{n\times n}$, $U_n := [u_1, u_2, \ldots, u_n]_{m\times n}$, and $\Sigma_n := \mathrm{diag}(\sigma_1, \sigma_2, \ldots, \sigma_n)_{n\times n}$. Extend the orthonormal vectors $\{u_i\}_1^n$ to a full set of mutually orthogonal vectors $\{u_i\}_1^m$ in $\mathbb{R}^m$, say by the Gram–Schmidt process, and define the matrices

$$U := [u_1, u_2, \ldots, u_n, u_{n+1}, \ldots, u_m] := [U_n|U']$$

and

$$\Sigma := \begin{bmatrix} \Sigma_n \\ 0 \end{bmatrix}_{m\times n}.$$

We thus have the singular-value decomposition of A,

$$AV = U_n\Sigma_n = [U_n|U'] \begin{bmatrix} \Sigma_n \\ 0 \end{bmatrix} = U\Sigma,$$

or

$$A = U\Sigma V^T, \tag{10.2}$$

where U and V are orthogonal matrices and Σ is an $m \times n$ diagonal matrix.

If $n > m$, we have the singular-value decomposition of A^T, say in the form $A^T = V\Sigma^T U^T$. Transposing both sides yields (10.2) once more. □

We note that the largest singular value σ_1 is the ℓ_2-norm of A, that is,

$$\|A\| := \max_{\|x\|=1} \|Ax\| = \max_{\|x\|=1} \max_{\|y\|=1} \langle Ax, y\rangle.$$

Remark 10.7. It may again appear remarkable that the multiplier vectors α, β vanish in the optimization problem (P_k). This can be explained as follows: A feasible vector x in (P_k) is orthogonal to the vectors $\{v_i\}_1^{k-1}$. Thus, if $j < k$, then

$$\langle Ax, u_j\rangle = \langle x, A^T u_j\rangle = \langle x, \sigma_j v_j\rangle = 0,$$

which means that the linear map $x \mapsto Ax$ maps the linear subspace $L_{k-1} := \{v_1, \ldots, v_{k-1}\}^\perp \subseteq \mathbb{R}^n$ into the subspace $M_{k-1} := \{u_1, \ldots, u_{k-1}\}^\perp \subseteq \mathbb{R}^m$, that is, $A(L_{k-1}) \subseteq M_{k-1}$. Consequently, the problem (P_k) can be rewritten as

$$\begin{aligned} \min \quad & -\frac{1}{2}\langle Ax, y\rangle \\ \text{s.t.} \quad & \|x\|^2 - 1 = 0, \; x \in L_{k-1}, \\ & \|y\|^2 - 1 = 0, \; y \in M_{k-1}, \end{aligned}$$

thereby avoiding the need for the multiplier vectors α and β altogether.

10.3 Variational Problems in Quasi-Newton Methods

Example 10.8. **(Broyden's method)**

This problem appears in Broyden's method for approximating a root of a nonlinear system of equations $G(x) = 0$, where $G : \mathbb{R}^n \to \mathbb{R}^n$ is a nonlinear map. It is the problem

$$\begin{aligned} \min \quad & \|X\|_F^2 \\ \text{s.t.} \quad & Xa = b, \end{aligned}$$

where $X = (x_{ij})$ is an $m \times n$ matrix, $a \in \mathbb{R}^n$, $0 \neq b \in \mathbb{R}^m$, and where the Frobenius (or Hilbert–Schmidt) norm $\|X\|_F$ is given by

$$\|X\|_F^2 = \sum_{i=1}^{m} \sum_{j=1}^{n} x_{ij}^2.$$

This norm is a Euclidean norm that comes from the trace inner product

$$\langle X, Y \rangle = \operatorname{tr}(X^T Y) = \sum_{i,j=1}^{n} x_{ij} y_{ij}$$

on $\mathbb{R}^{n \times n}$, the vector space of $n \times n$ matrices. Thus,

$$\|X\|_F^2 = \langle X, X \rangle.$$

Since the objective function is coercive, there exists a global minimizer. The constraint functions are linear, so it follows from Corollary 9.10 that the KKT conditions must hold at any local minimizer.

We form the Lagrangian

$$L = \frac{1}{2} \sum_{i=1}^{m} \sum_{j=1}^{n} x_{ij}^2 + \sum_{i=1}^{m} \lambda_i \Big(b_i - \sum_{j=1}^{n} x_{ij} a_j \Big).$$

The KKT conditions are given by

$$(a) \quad \frac{\partial L}{\partial x_{ij}} = x_{ij} - \lambda_i a_j = 0, \quad i = 1, \ldots, m, \quad j = 1, \ldots, n,$$

$$(b) \quad \sum_{j=1}^{n} x_{ij} a_j = b_i, \quad i = 1, \ldots, m.$$

Multiplying (a) by a_j and summing over j, we obtain

$$b_i = \sum_{1}^{n} x_{ij} a_j = \lambda_i \sum_{j=1}^{n} a_j^2 = \lambda_i \|a\|^2, \ i = 1, \ldots, m,$$

where the first equality follows from (b).

The condition (a) implies that $x_{ij} = \lambda_i a_j$, that is,

$$X = [x_{ij}] = [\lambda_i a_j] = \lambda a^T = \lambda \otimes a.$$

Since $b_i = \lambda_i \|a\|^2$, we have $\lambda_i = b_i/\|a\|^2$, that is, $\lambda = b/\|a\|^2$. This gives

$$X = \frac{ba^T}{\|a\|^2} = \frac{b \otimes a}{\|a\|^2}.$$

Example 10.9. **(Symmetric matrix updates in quasi-Newton methods)**

We now consider the symmetric version of the preceding problem,

$$\begin{aligned} \min \quad & \|X\|_F^2 \\ \text{s.t.} \quad & Xa = b, \\ & X^T = X, \end{aligned}$$

where X is an $n \times n$ symmetric matrix, $b, c \in \mathbb{R}^n$, and $b \neq 0$. Slight modifications of the above problem appear in the variational characterization of the matrix updates in quasi-Newton methods such as those of Davidon–Fletcher–Powell (DFP) and Broyden–Fletcher–Goldfarb–Shanno (BFGS). We refer the reader to [97, 73, 74, 213, 120] for more details.

The symmetry constraint $X^T = X$ on the matrix X can be enforced through the system of constraints $x_{ij} = x_{ji}$ for $1 \leq i < j \leq n$. We arrive at the Lagrangian

$$L = \frac{1}{2} \sum_{i,j=1}^{n} x_{ij}^2 + \sum_{i=1}^{n} \lambda_i \Big(b_i - \sum_{j=1}^{n} x_{ij} a_j \Big) + \sum_{i<j} \delta_{ij} (x_{ij} - x_{ji}).$$

We leave to the reader to solve the problem using this Lagrangian.

The problem may be solved more elegantly by setting it up *within* the vector space S^n of the space of $n \times n$ symmetric matrices equipped with the trace inner product inherited from $\mathbb{R}^{n \times n}$,

$$\langle X, Y \rangle = \operatorname{tr}(X^T Y) = \operatorname{tr}(XY) = \sum_{i,j=1}^{n} x_{ij} y_{ij}.$$

In this setup, the vector constraint equation $Xa = b$ will enter into the Lagrangian in the term $\langle \lambda, b - Xa \rangle = \lambda^T b - \langle \lambda, Xa \rangle$, where $\lambda \in \mathbb{R}^n$ is the Lagrange multiplier and the inner product is the usual one in $\mathbb{R}^n$. We can write the last as an inner product in S^n: note that

$$\langle \lambda, Xa \rangle = \lambda^T (Xa) = \operatorname{tr}(\lambda^T (Xa)) = \operatorname{tr}(X(a\lambda^T)) = \langle X, \lambda a^T \rangle,$$

where the last inner product is the trace inner product in $\mathbb{R}^{n \times n}$, the vector space of $n \times n$ matrices *not* in S^n, since the matrix λa^T is not symmetric. However,

$$\langle X, \lambda a^T \rangle = \mathrm{tr}(X(a\lambda^T)) = \mathrm{tr}((\lambda a^T)X) = \langle X, a\lambda^T \rangle,$$

and we have

$$\langle \lambda, Xa \rangle = \left\langle X, \frac{a\lambda^T + \lambda a^T}{2} \right\rangle,$$

which *is* an inner product in S^n, since both matrices X and $a\lambda^T + \lambda a^T$ are symmetric.

We thus obtain the Lagrangian function

$$L = \frac{1}{2}\langle X, X \rangle - \left\langle X, \frac{a\lambda^T + \lambda a^T}{2} \right\rangle + \langle b, \lambda \rangle.$$

The KKT conditions are $\nabla_X L = X - (a\lambda^T + \lambda a^T) = 0$, or

$$X = a\lambda^T + \lambda a^T.$$

Then

$$b = Xa = (a\lambda^T + \lambda a^T)a = (\lambda^T a)a + \|a\|^2 \lambda$$

and

$$b^T a = (\lambda^T a)\|a\|^2 + \|a\|^2(\lambda^T a) = 2\|a\|^2 (\lambda^T a).$$

The last equation gives $\lambda^T a = a^T b / 2\|a\|^2$, and substituting it into the first equation yields the equality

$$\lambda = \frac{b}{\|a\|^2} - \frac{a^T b}{2\|a\|^4} a.$$

Finally, substituting this value of λ into the equation $X = a\lambda^T + \lambda a^T$, we obtain

$$X = \frac{ab^T + ba^T}{\|a\|^2} - \frac{a^T b}{\|a\|^4} aa^T = \frac{1}{\|a\|^2}(a \otimes b + b \otimes a) - \frac{\langle a, b \rangle}{\|a\|^4} a \otimes a.$$

10.4 Kantorovich's Inequality

Let A be an $n \times n$ symmetric positive definite matrix with eigenvalues

$$\lambda_1 \geq \lambda_2 \geq \cdots \geq \lambda_n > 0.$$

Kantorovich's inequality states that

$$\max\{\langle Ax, x \rangle \cdot \langle A^{-1}x, x \rangle : \|x\| = 1\} \leq \frac{(\lambda_1 + \lambda_n)^2}{4\lambda_1 \lambda_n}.$$

A proof of this inequality is given in Lemma 14.8 on page 369. Here we give a perhaps more natural proof by setting it up as an optimization problem.

Consider the problem

$$\begin{array}{ll} \max & \langle Ax, x\rangle \cdot \langle A^{-1}x, x\rangle \\ \text{s.t.} & \|x\|^2 = 1. \end{array}$$

Kantorovich's inequality will follow if we can show that the optimal objective value of this problem is $(\lambda_1+\lambda_n)^2)/(4\lambda_1\lambda_n)$. Considering the spectral decomposition $A = U\Lambda U^T$, U orthogonal and $\Lambda = \operatorname{diag}\{\lambda_1, \ldots, \lambda_n\}$ and noting that $\|x\| = 1$ if and only if $\|Ux\| = 1$, the optimization problem reduces to maximizing $\left(\sum_{j=1}^n \lambda_j x_j^2\right) \cdot \left(\sum_{j=1}^n \lambda_j^{-1} x_j^2\right)$ subject to the constraint $\sum_{j=1}^n x_j^2 = 1$. A further substitution $y_i = x_i^2$ yields the optimization problem

$$\begin{array}{ll} \max & \Big(\sum_{j=1}^n \lambda_j y_j\Big) \cdot \Big(\sum_{j=1}^n \lambda_j^{-1} y_j\Big) \\ \text{s.t.} & \sum_{j=1}^n y_j = 1, \\ & y \geq 0. \end{array}$$

From the Lagrangian function (for the equivalent minimization problem)

$$L(y, \delta, \mu) = -\Big(\sum_{j=1}^n \lambda_j y_j\Big) \cdot \Big(\sum_{j=1}^n \lambda_j^{-1} y_j\Big) + \delta\Big(\sum_{j=1}^n y_j - 1\Big) - \langle \mu, y\rangle,$$

we deduce the KKT conditions (the KKT conditions hold since the constraints are linear)

$$\begin{aligned} &-\lambda_i\Big(\sum_{j=1}^n \lambda_j^{-1} y_j\Big) - \lambda_i^{-1}\Big(\sum_{j=1}^n \lambda_j y_j\Big) + \delta - \mu_i = 0, \quad i = 1, \ldots, n, \\ &\sum_1^n y_j = 1, \quad y \geq 0, \quad \mu \geq 0, \quad \langle \mu, y\rangle = 0. \end{aligned} \tag{10.3}$$

Multiplying the first equation above by y_i, summing up over i, and using the feasibility and complementarity conditions, we get

$$\delta = 2\Big(\sum_{j=1}^n \lambda_j^{-1} y_j\Big) \cdot \Big(\sum_{j=1}^n \lambda_j y_j\Big). \tag{10.4}$$

If $i \in I := \{i : y_i > 0\}$, then $\mu_i = 0$ by complementarity. Dividing both sides of the first equation in (10.3) by $\left(\sum_{j=1}^n \lambda_j^{-1} y_j\right) \cdot \left(\sum_{j=1}^n \lambda_j y_j\right)$ and using (10.4) gives

$$\frac{\lambda_i}{\sum_{j=1}^n \lambda_j y_j} + \frac{\lambda_i^{-1}}{\sum_{j=1}^n \lambda_j^{-1} y_j} = 2, \quad i \in I. \tag{10.5}$$

If $i, j \in I$, then

$$\frac{\lambda_i}{\sum_{k=1}^n \lambda_k y_k} + \frac{\lambda_i^{-1}}{\sum_{k=1}^n \lambda_k^{-1} y_k} = \frac{\lambda_j}{\sum_{k=1}^n \lambda_k y_k} + \frac{\lambda_j^{-1}}{\sum_{k=1}^n \lambda_k^{-1} y_k},$$

which upon simplification becomes

$$\lambda_i \lambda_j = \frac{\sum_{k=1}^n \lambda_k y_k}{\sum_{k=1}^n \lambda_k^{-1} y_k}, \quad i, j \in I,\ \lambda_i \neq \lambda_j. \tag{10.6}$$

Solving (10.5) and (10.6) for $\sum_{k=1}^n \lambda_k y_k$ and $\sum_{k=1}^n \lambda_k^{-1} y_k$ gives

$$2 = \frac{\lambda_i}{\sum_1^n \lambda_k y_k} + \frac{\lambda_i^{-1}}{\frac{\sum_1^n \lambda_k y_k}{\lambda_i \lambda_j}} = \frac{\lambda_i + \lambda_j}{\sum_1^n \lambda_k y_k},$$

that is,

$$\sum_1^n \lambda_k y_k = \frac{\lambda_i + \lambda_j}{2}$$

and

$$\sum_1^n \lambda_k^{-1} y_k = \frac{\sum_{k=1}^n \lambda_k y_k}{\lambda_i \lambda_j} = \frac{\lambda_i + \lambda_j}{2\lambda_i \lambda_j}.$$

Thus, the objective function value is

$$\Big(\sum_1^n \lambda_k y_k\Big) \cdot \Big(\sum_1^n \lambda_k^{-1} y_k\Big) = \frac{(\lambda_i + \lambda_j)^2}{4\lambda_i \lambda_j}.$$

Setting $t = \lambda_i/\lambda_j$, the right-hand side equals $(1/2) + (t + t^{-1})/4$, and is maximized, in the worst case, at the extreme eigenvalues λ_1 and λ_n.

10.5 Hadamard's Inequality

Let $X = (x_{ij}) = [x_1, \ldots, x_n]$ be an $n \times n$ matrix with columns $\{x_i\}_1^n$. We consider the problem

$$\begin{aligned} &\max \quad \det X \\ &\text{s.t.} \quad \|x_i\|^2 = 1, \quad i = 1, \ldots, n. \end{aligned}$$

Since the n-volume of a parallelepiped with side vectors $x_1, \ldots, x_n$ is known to be $|\det X|$, this problem is equivalent to the problem of finding the largest-volume parallelepiped with unit sides. Geometric intuition tells us that the solution must be a cube, that is, the sides of the parallelepiped must be mutually orthogonal.

The gradient vectors of the constraints are linearly independent, because each one contains different columns of the matrix X. It follows from Corollary 9.10 that the KKT conditions hold at each local optimizer.

We write the Lagrangian for the equivalent minimization problem,

$$L = -\det X + \sum_{j=1}^{n} \frac{\lambda_j}{2}\Big(\sum_{i=1}^{n} x_{ij}^2 - 1\Big).$$

Toward obtaining the KKT conditions, we first calculate the derivatives of the determinant function (see also Exercise 22 on page 29). Recall the Laplace expansion formula for determinants,

$$\det X = \sum_{j=1}^{n} (-1)^{i+j} x_{ij} X_{ij},$$

where X_{ij} is the ij-minor of X, that is, the determinant of the matrix obtained from X by striking out the ith row and jth column of X. Thus,

$$\frac{\partial \det X}{\partial x_{ij}} = b_{ij} := (-1)^{i+j} X_{ij},$$

where the expression b_{ij} is called the ij-cofactor of X, and the KKT conditions are

$$(a) \quad -b_{ij} = \lambda_j x_{ij}, \;\; i,j = 1,\dots,n,$$

$$(b) \quad \sum_{i=1}^{n} x_{ij}^2 = 1, \;\; j = 1,\dots,n.$$

Recall that $B = (b_{ji}) = [b_1,\dots,b_n]^T := \operatorname{Adj}(X)$ is called the *adjoint* matrix of X and is characterized by the equation

$$X \cdot \operatorname{Adj}(X) = (\det X)I.$$

Thus, in matrix notation, (a) can be written as

$$(a') \quad \operatorname{Adj}(X) = X\Lambda,$$

where $\Lambda = \operatorname{diag}(\lambda_1,\dots,\lambda_n)$. These give $-(\det X)I = -X^T \operatorname{Adj}(X) = X^T X \Lambda$, meaning that $X^T X$ is a diagonal matrix, that is, the matrix X has orthogonal columns. In fact, since the columns X_i have unit length, we have $X^T X = I$, so that X is an orthogonal matrix. This implies that $|\det X| = 1$.

Consequently, we obtain *Hadamard's inequality*

$$|\det X| \le \|x_1\| \cdots \|x_n\| \;\text{ for all }\; X = [x_1,\dots,x_n] \in \mathbb{R}^{n\times n},$$

with equality holding if and only if the columns $\{x_i\}_1^n$ are mutually orthogonal.

See also Exercise 24 on page 113 for a different treatment of Hadamard's inequality.

10.6 Maximum-Volume Inscribed Ellipsoid in a Symmetric Convex Polytope

Let

$$P = \{x \in \mathbb{R}^n : |\langle a_k, x\rangle| \leq 1, \ k = 1, \dots, p\}$$

be a symmetric polytope (in the sense that $P = -P$, that is, if x lies in P, then so does $-x$). Suppose

$$E = \{x \in \mathbb{R}^n : x^T X^{-1} x \leq 1\}$$

is the ellipsoid with maximal volume contained (inscribed) in P, where X is an $n \times n$ symmetric, positive semidefinite matrix. (The uniqueness of the ellipsoid is well known and assuming that the center of the optimal ellipsoid is the origin, follows from the fact that the optimization problem (10.8) below is a convex programming problem with a strictly convex objective function.)

We show here, using optimization techniques, that the ellipsoid $\sqrt{n}E$ contains P, that is,

$$E \subseteq P \subseteq \sqrt{n}E, \tag{10.7}$$

an important inequality essentially due to Fritz John [148].

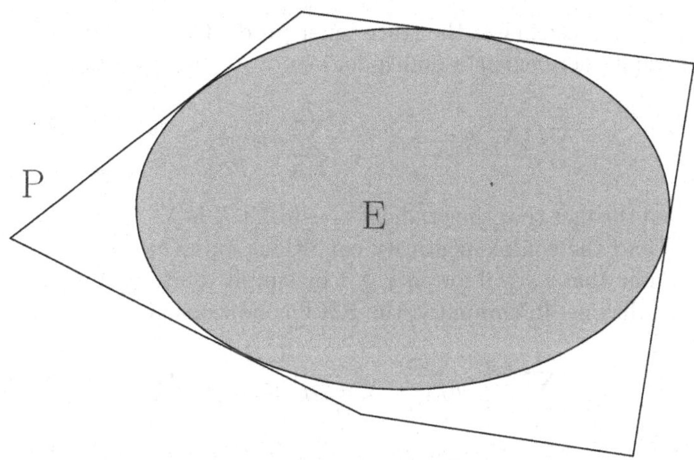

Fig. 10.1. Inscribed ellipsoid in a polytope.

The problem of finding the maximum-volume ellipsoid can be set up as an optimization problem as follows. Let $X^{1/2}$ be the symmetric square root of X. ($X^{1/2}$ is the unique symmetric matrix whose square is X. If $X = U\Lambda U^T$ is the spectral decomposition of X, then $X^{1/2} = U\Lambda^{1/2}U^T$, where $\Lambda^{1/2} = \operatorname{diag}(\sqrt{\lambda_1}, \dots, \sqrt{\lambda_n})$.) We have

$$E = \{x \in \mathbb{R}^n : \|X^{-1/2}x\| \leq 1\} = \{X^{1/2}y \in \mathbb{R}^n : \|y\| \leq 1\} = X^{1/2}(B_n),$$

where $B_n = \{x \in \mathbb{R}^n : \|x\| \leq 1\}$ is the unit ball in $\mathbb{R}^n$. Thus, $\text{vol}(E) = \det(X^{1/2})\,\text{vol}(B_n)$. The condition $E \subseteq P$ can be turned into a constraint as follows: if $x^T X^{-1} x = \|X^{-1/2}x\| \leq 1$, then we want to have $|\langle a_k, x\rangle| = |\langle X^{1/2}a_k, X^{-1/2}x\rangle| \leq 1$. This is equivalent to requiring that

$$\langle X^{1/2}a_k, y\rangle| \leq 1 \ \text{ for all } \ y, \ \|y\| \leq 1,$$

or to the constraint $\|X^{1/2}a_k\| \leq 1$.

Thus, we arrive at the optimization problem

$$\begin{aligned} \min \quad & -\ln\det X \\ \text{s.t.} \quad & \langle Xa_k, a_k\rangle \leq 1, \ \ k = 1, \ldots, p. \end{aligned} \tag{10.8}$$

Since the constraints are linear, it follows from Corollary 9.10 that the KKT conditions hold at each local optimizer.

The Lagrangian of the above problem is

$$\begin{aligned} L(X, \lambda) &= -\ln\det X + \sum_{k=1}^{p} \lambda_k(\langle Xa_k, a_k\rangle - 1) \\ &= -\ln\det X + \left\langle X, \sum_k \lambda_k a_k a_K^T \right\rangle - \sum_{k=1}^{p} \lambda_k, \end{aligned}$$

where $\langle\cdot,\cdot\rangle$ on the last line is the trace inner product in S^n. The KKT conditions imply that there exists a multiplier vector $\lambda \geq 0$ such that

$$\nabla L(X, \lambda) = -X^{-1} + \sum_{k=1}^{p} \lambda_k a_k a_k^T = 0$$

(we have used the fact that the gradient of $-\ln\det X$ is X^{-1}; see Example 1.27 on page 18) and the complementarity conditions $\lambda_k(\langle Xa_k, a_k\rangle - 1) = 0$ hold. We may assume that $\lambda_k > 0$ for all $k \geq 1$ by simply omitting any $\lambda_k = 0$ from the KKT conditions. In summary, the KKT conditions are

$$X^{-1} = \sum_{k=1}^{l} \lambda_k a_k a_k^T, \quad \langle Xa_k, a_k\rangle = 1, \ \ k = 1, \ldots, l.$$

The first equation gives $I = \sum_{k=1}^{l} \lambda_k X a_k a_k^T$, and taking traces of both sides, we obtain

$$n = \text{tr}(I) = \text{tr}\Big(\sum_{k=1}^{l} \lambda_k X a_k a_k^T\Big) = \sum_{k=1}^{l} \lambda_k \langle Xa_k, a_k\rangle = \sum_{k=1}^{l} \lambda_k.$$

We can now finish the proof. If $x \in P$, that is, $\langle x, a_k\rangle^2 \leq 1$, then

$$\langle X^{-1}x, x\rangle = \sum_{k=1}^{l} \lambda_k \langle a_k, x\rangle^2 \leq \sum_{k=1}^{l} \lambda_k = n,$$

which immediately implies that $x \in \sqrt{n}E$, proving (10.7).

The problem of inscribing a maximum-volume ellipsoid into an arbitrary convex body $K \subset \mathbb{R}^n$ is treated in Section 12.4 using semi-infinite programming. Chapter 12 also treats the problem of circumscribing a minimum-volume ellipsoid around K.

10.7 Hilbert's Inequality

Consider the Hilbert space of square summable sequences,

$$l^2 = \Big\{ x = \{x_i\}_0^\infty : \sum_{i=0}^{\infty} x_i^2 < \infty \Big\}.$$

The inner product and the norm in l^2 are given by

$$\langle x, y\rangle = \sum_0^\infty x_i y_i, \quad \|x\|_2 = \Big(\sum_0^\infty x_i^2\Big)^{1/2}.$$

Let $\{x_i\}_0^\infty$ and $\{y_j\}_0^\infty$ be nonnegative sequences in l^2. *Hilbert's inequality* states that

$$\sum_{i,j=0}^{\infty} \frac{x_i y_j}{i+j+1} < \pi \cdot \Big(\sum_{i=0}^{\infty} x_i^2\Big)^{1/2} \cdot \Big(\sum_{j=0}^{\infty} y_j^2\Big)^{1/2} = \pi \cdot \|x\|_2 \cdot \|y\|_2,$$

unless $\{x_i\}$ or $\{y_j\}$ is identically zero. This is an important inequality in analysis, treated extensively in the book *Inequalities* by Hardy, Littlewood, and Polya [127].

Hilbert's inequality can be extended to l^p spaces. Let $p > 1$ and $q > 1$ be *conjugate exponents*, that is,

$$\frac{1}{p} + \frac{1}{q} = 1.$$

The space l^p is the p-summable sequences

$$l^p = \Big\{ x = \{x_i\}_0^\infty : \sum_{i=0}^{\infty} |x_i|^p < \infty \Big\}.$$

The norm of a sequence $x \in l^p$ is given by

$$\|x\|_p = \Big(\sum_0^\infty |x_i|^p\Big)^{1/p}.$$

Recall that Hölder's inequality states

$$\langle x, y\rangle \le \|x\|_p \cdot \|y\|_q \quad \text{for all } x \in l^p,\ y \in l^q.$$

Hilbert's inequality is a weighted version of Hölder's inequality.

Theorem 10.10. (Hilbert's inequality) *If $\{x_i\}_0^\infty$ and $\{x_j\}_0^\infty$ are nonnegative sequences in conjugate spaces l^p and l^q, respectively, then*

$$\sum_{i,j=0}^{\infty} \frac{x_i y_j}{i+j+1} < \frac{\pi}{\sin(\pi/p)} \cdot \|x\|_p \cdot \|y\|_q,$$

unless $\{x_i\}$ or $\{y_j\}$ is identically zero.

Proof. Consider the finite-dimensional optimization problem

$$\begin{aligned} \max \quad & \langle Ax, y \rangle \\ \text{s. t.} \quad & \sum_{i=0}^{k} x_i^p = 1, \ \sum_{j=0}^{k} y_j^q = 1, \\ & x \geq 0, \ y \geq 0, \end{aligned}$$

where $x \in \mathbb{R}^{k+1}$, $y \in \mathbb{R}^{k+1}$, and

$$\langle Ax, y \rangle := \sum_{i,j=0}^{k} \frac{x_i y_j}{i+j+1}.$$

Note that A is a symmetric matrix with ij entry $1/(i+j+1)$, a well-known matrix in numerical linear algebra. We form the (weak) Lagrangian (for the equivalent minimization problem):

$$L = -\lambda_0 \langle Ax, y \rangle + \frac{\delta}{p}\Big(\sum_{i=0}^{k} x_i^p - 1\Big) + \frac{\gamma}{q}\Big(\sum_{j=0}^{k} y_j^q - 1\Big) - \langle \lambda, x \rangle - \langle \mu, y \rangle.$$

The FJ conditions are

$$\begin{aligned} & -\lambda_0 Ay + \delta x^{p-1} - \lambda = 0, \\ & -\lambda_0 Ax + \gamma y^{q-1} - \gamma = 0, \\ & x \geq 0, \ y \geq 0, \ \langle \lambda, x \rangle = 0 = \langle \mu, y \rangle, \\ & (\lambda_0, \delta, \gamma, \lambda, \mu) \neq 0, (\lambda_0, \lambda, \mu) \geq 0, \end{aligned}$$

where $x^{p-1} = (x_1^{p-1}, \ldots, x_k^{p-1})$, and where y^{q-1} is defined similarly. Taking the inner product of the first equation by x and using the above conditions, we obtain $\lambda_0 \langle Ax, y \rangle = \delta$. Similarly, $\lambda_0 \langle Ax, y \rangle = \gamma$, and thus $\delta = \gamma$.

Now if $\lambda_0 = 0$, then we immediately obtain $\delta = 0 = \gamma$, and also $\lambda = 0 = \mu$ by the FJ conditions, so that all the multipliers are zero, contradicting the FJ conditions. Therefore, the KKT conditions hold, and we may take $\lambda_0 = 1$. Thus, we have

$$\delta = \langle Ax, y \rangle = \gamma.$$

We now estimate δ. If $x_i > 0$, the KKT conditions give

$$\delta x_i^{p-1} = \langle e_i, Ay \rangle = \sum_{j=0}^{k} \frac{y_j}{i+j+1},$$

and similarly if $y_j > 0$,

$$\delta y_j^{q-1} = \sum_{i=0}^{k} \frac{x_i}{i+j+1}.$$

Let i^* and j^* be indices maximizing $\{(i+1/2)^{1/p} x_i : 0 \le i \le k\}$ and $\{(j+1/2)^{1/q} y_j : 0 \le j \le k\}$, respectively. We have

$$\begin{aligned}\delta x_{i^*}^{p-1} &= \sum_{j=0}^{k} \frac{y_j}{i^*+j+1} = \sum_{j=0}^{k} \frac{y_j (j+1/2)^{1/q}}{(i^*+j+1)(j+1/2)^{1/q}} \\ &\le \left(y_{j^*} (j^*+1/2)^{1/q} \right) \cdot \sum_{j=0}^{k} \frac{1}{(i^*+j+1)(j+1/2)^{1/q}}.\end{aligned}$$

Now

$$\begin{aligned}\sum_{j=0}^{k} \frac{1}{(i^*+j+1)(j+1/2)^{1/q}} &\le \int_0^{k+1} \frac{dx}{(i^*+1/2+x)x^{1/q}} \\ &\le (i^*+1/2)^{-1/q} \int_0^{2k+2} \frac{dy}{y^{1/q}(y+1)},\end{aligned}$$

where the second inequality follows by substituting $y = x/(i^*+1/2)$ and using $y \le (k+1)/(i^*+1/2) \le 2(k+1)$. These give us

$$\delta\, x_{i^*}^{p-1} \le y_{j^*} (j^*+1/2)^{1/q} \cdot (i^*+1/2)^{-1/q} \int_0^{2k+2} \frac{dy}{y^{1/q}(y+1)},$$

and taking qth powers of both sides and using $p+q = pq$,

$$\delta^q\, x_{i^*}^p \le y_{j^*}^q (j^*+1/2) \cdot (i^*+1/2)^{-1} \cdot \left(\int_0^{2k+2} \frac{dy}{y^{1/q}(y+1)} \right)^q.$$

Similarly, we also have

$$\delta^p\, y_{j^*}^q \le x_{i^*}^p (i^*+1/2) \cdot (j^*+1/2)^{-1} \cdot \left(\int_0^{2k+2} \frac{dy}{y^{1/p}(y+1)} \right)^p.$$

Multiplying the last two inequalities and simplifying gives

$$\delta^{pq} = \delta^{p+q} \le \left(\int_0^{2k+2} \frac{dy}{y^{1/p}(y+1)} \right)^p \cdot \left(\int_0^{2k+2} \frac{dy}{y^{1/q}(y+1)} \right)^q,$$

or equivalently,

$$\delta \leq \left(\int_0^{2k+2} \frac{dy}{y^{1/p}(y+1)} \right)^{1/q} \cdot \left(\int_0^{2k+2} \frac{dy}{y^{1/q}(y+1)} \right)^{1/p}.$$

Now let $k \to \infty$. It can be shown, by contour integration for example, that

$$\int_0^{\infty} \frac{dy}{y^{1/p}(y+1)} = \frac{\pi}{\sin(\pi/p)}.$$

Thus, we obtain $\delta \leq \pi/\sin(\pi/p)$.

It is known that the constant $\pi/\sin(\pi/p)$ is optimal in Hilbert's inequality; see [127]. □

The proof of Hilbert's inequality given above is due to Cassels [56]. It should be noticed that even though optimization techniques help, the proof of Hilbert's inequality requires considerable ingenuity.

10.8 Exercises

1. Example 10.8 may be solved more efficiently by expressing its Lagrangian in terms of the trace inner product on $\mathbb{R}^{n \times n}$.
   (a) Show that the Lagrangian has the form $L = \|X\|^2/2 + \langle \lambda, b - Xa, \rangle$.
   (b) Show that the term $\langle \lambda, Xa \rangle$ in L can be written as $\lambda^T Xa = \operatorname{tr}(a\lambda^T X) = \langle X, \lambda \otimes a \rangle$, giving
   $$L = \frac{1}{2}\|X\|^2 - \langle X, \lambda \otimes a \rangle + \langle b, \lambda \rangle.$$
   (c) Proceed as in Example 10.9 and solve Example 10.8 in a coordinate-free manner.
2. *(Simultaneous diagonalization of two symmetric matrices)* Let A, B be two symmetric $n \times n$ matrices, where B is also positive definite. Imitate the procedure in Section 10.1 to obtain a spectral decomposition of A with respect to B, by replacing the constraint $\|x\|^2 = 1$ with $\langle Bx, x \rangle = 1$, and the constraints $\langle x, u_j \rangle = 0$ with $\langle Bx, u_j \rangle = 0$. Show that this leads to the following results:
   (i) The $n \times n$ matrix $U := [u_1, \ldots, u_n]$ satisfies the conditions $\langle Bu_i, u_i \rangle = 1$, $\langle Bu_i, u_j \rangle = 0$, $Au_i = \lambda_i Bu_i$ for $i, j = 1, \ldots, n$, $i \neq j$.
   (ii) Define $\Lambda := \operatorname{diag}(\lambda_1, \ldots, \lambda_n)$. Prove that (i) implies that we have the simultaneous diagonalization of A and B,
   $$U^T AU = \Lambda, \quad \text{and} \quad U^T BU = I.$$

3. (***Abu Ali al-Hasan ibn al-Haytham, 965–1039***) There are two balls on a circular billiard table with center O at the origin and radius r; see Figure 10.2. The balls are at the points $p = (a, b)$ and $q = (c, d)$. How should one strike the ball p so that it hits the ball q after rebounding from the cushion? The problem can be posed as the problem of minimizing the distance traveled by the ball p, that is,

$$\begin{aligned} \min \quad & \|z - p\| + \|z - q\| \\ \text{s.t.} \quad & \|z\| = r, \end{aligned}$$

where $z = (x, y)$ is the point where p hits the cushion. It is well known (Snell's law) that the point z is characterized by the condition that the angles α and β are equal. The object of this problem is to prove this fact, and then characterize the four possible points $z = (x, y)$ based on this condition.

(a) Show that the KKT conditions hold at an optimal point z, and give the equation

$$\frac{z - p}{\|z - p\|} + \frac{z - q}{\|z - q\|} = \lambda \frac{z}{\|z\|}.$$

(b) Show that the equation in (a) implies that $\alpha = \beta$.
Hint: Obtain two equations, first by taking the inner product of both sides with z, and then square of the norms of both sides. Use the fact $\|z\| = r$ and simplify.

(c) Show that the four possible (in general) optimal points z lie in the intersection of the circle $x^2 + y^2 = r^2$ and the hyperbola

$$A(x^2 - y^2) + Bxy + r^2(Cx + Dy) = 0,$$

where $A = ad + bc$, $B = 2(bd - ac)$, $C = -(b + d)$, and $D = a + c$.
Hint: Let γ, θ, and δ be the angles the lines pz, Oz, and qz make with the positive x-axis. Show that $\alpha = \theta - \gamma$, $\beta = \delta - \theta$. Thus, $\tan(\theta - \alpha) = \tan(\delta - \theta)$. Use the formula $\tan(\phi_1 - \phi_2) = (\tan\phi_1 - \tan\phi_2)/(1 + \tan\phi_1 \tan\phi_2)$, and substitute the expressions for $\tan\gamma$, $\tan\theta$, and $\tan\delta$.

4. *(Inscribe a maximum-volume simplex in an n-dimensional sphere)* This can be set up as an optimization problem. We take as our sphere the unit sphere with center at $e_n = (0, \ldots, 0, 1) \in \mathbb{R}^n$, that is, $B = \{x \in R^n : \|x - e_n\| \le 1\}$. Let $\{x_1, \ldots, x_n, x_{n+1}\}$ be the $n + 1$ vertices of the desired simplex Δ, where we assume that $x_{n+1} = 0$. Then

$$\begin{aligned} \Delta &= \left\{ \sum_1^{n+1} \lambda_i x_i : \sum_1^{n+1} \lambda_i = 1,\ \lambda_i \ge 0,\ i = 1, \ldots, n + 1 \right\} \\ &= \left\{ \sum_1^{n} \lambda_i x_i : \sum_1^{n} \lambda_i \le 1,\ \lambda_i \ge 0,\ i = 1, \ldots, n \right\}. \end{aligned}$$

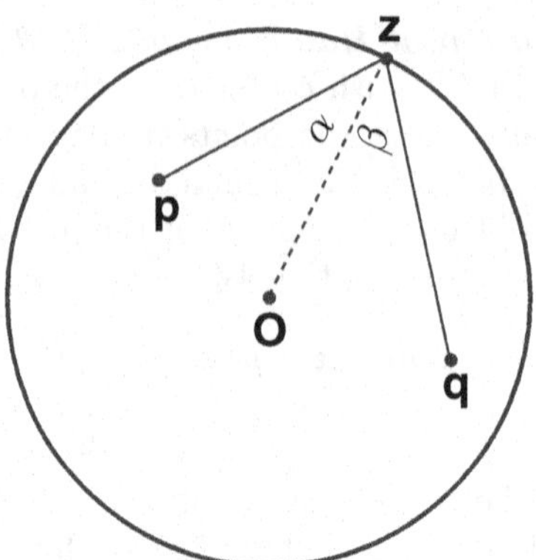

Fig. 10.2. Al-Haytham's problem

(a) Show that $\Delta = T(S_1)$, where $T(\lambda) = X\lambda$, $X = [x_1, \ldots, x_n]$, and $S_1 := \{\lambda \in \mathbb{R}^n : \lambda \geq 0, \sum_1^n \lambda_i \leq 1\}$ is the solid unit simplex in $\mathbb{R}^n$.

(b) Show that $\{x_i\}_1^n$ are linearly independent and T in (a) is a one-to-one linear map. Conclude that $\text{vol}(\Delta) = |\det X| \text{ vol}(S_1)$ by the change of variables formula.
Consequently, determining the largest-volume simplex Δ can be set up as the constrained optimization problem

$$\max\{\det X : \|x_i - e_n\| \leq 1, \ i = 1, \ldots, n\}.$$

(c) Write the Lagrangian (for the equivalent minimization problem) as

$$L(X, \lambda) = -\det X + \sum_{j=1}^{n} \frac{\lambda_j}{2}(\|x_j - e_n\|^2 - 1).$$

Give a theoretical justification as to why we have taken $\lambda_0 = 1$ in L.

(d) Show that the KKT conditions give

$$\langle x_i, x_j - e_n \rangle = 0, \ \ 1 \leq i, j \leq n, \ i \neq j.$$

Hint: Expand $L(X + tH, \lambda)$ using Exercises 22 and 23 in Chapter 1.

(e) Use the results of (d) above to prove that for some α,

$$\|x_i - x_j\|^2 = 2\alpha, \quad 1 \leq i, j \leq n+1, i \neq j,$$

and

$$\langle x_i - x_j, x_i - x_k \rangle = \alpha, \ \ 1 \leq i, j, k \leq n+1, \ \ i, j, k \text{ distinct}.$$

(f) Conclude from (e) that the optimal simplex has edges of equal lengths, and the angle between any two adjacent edges is 60 degrees.

5. This problem is inspired by electrostatics. Loosely speaking, it is the problem of how to place n point charges on a given set so that the charges are as "far away" from each other as possible. Its generalizations, when the number of points charges is large, lead to important problems that are still unsolved. One form of the general problem is

$$\begin{aligned} \max \quad & \prod_{1 \le i < j \le n} \|x_i - x_j\| \\ \text{s.t.} \quad & x_i \in D, \ i = 1, \ldots, n, \end{aligned} \qquad (P)$$

where $D \subset \mathbb{R}^n$ is a given bounded set. In this generality, the problem is unsolved, that is, no explicit solution is known at present.
In this problem, we take $D = [-1, 1] \subseteq \mathbb{R}$; the solution is known in this case. We reformulate the problem as the maximization of the function $\prod_{1 \le i < j \le n}(x_i - x_j)$ subject to the conditions $-1 \le x_n \le x_{n-1} \le \cdots \le x_1 \le 1$.
   (a) Show that the problem can be solved as an unconstrained minimization problem.
   Hint: Use logarithms.
   (b) Show that we must have $x_1 = 1$ and $x_n = -1$.
   (c) Solve the problem for $n = 4$.
   (c) Solve the problem for $n = 5$.
6. Solve the problem (P) for the square $D = \{(x, y) : 0 \le x, y \le 1\} \subset \mathbb{R}^2$ for $n = 3$ charges.

11

Duality Theory and Convex Programming

The theory of duality is an important tool in optimization, both in theory and in computation. Although duality is not universally applicable in optimization, it holds under mild regularity assumptions in some important classes of problems such as convex programs and some of its close relatives.

The beginnings of duality theory go back a long time. Already in 1921, E. Helly [130] gave necessary and sufficient conditions for the solvability of a linear system in an infinite-dimensional normed space subject to some norm conditions. This was a preliminary step in the evolution to the Hahn–Banach theorem in functional analysis.

The famous minimax theorem of von Neumann [211] is also a kind of duality theorem. This theorem from 1928 was the starting point of *game theory.*

Theorem 11.1. *Let $L : C_1 \times C_2 \to \mathbb{R}$ be a continuous function where $C_1 \subset \mathbb{R}^n$ and $C_2 \subset \mathbb{R}^m$ are compact convex sets. If L is a convex–concave function, that is, $x \mapsto L(x, y)$ is convex for fixed y, and $y \mapsto L(x, y)$ is concave for fixed x, then*

$$\min_{x \in C_1} \max_{y \in C_2} L(x, y) = \max_{y \in C_2} \min_{x \in C_1} L(x, y).$$

Shortly after the simplex method for linear programming was invented in 1947, the existence of a duality theory was hinted at by von Neumann; see Dantzig [68]. In 1951, this duality theorem was extended to convex programming by Kuhn and Tucker [182], who also brought to light the connection between duality and saddle points.

11.1 Perspectives on Duality

In this chapter, we treat duality through the Lagrangian function and the minimax theorems for such functions. For this approach to succeed in a given optimization (say minimization) problem (P), one needs somehow to devise

O. Güler, *Foundations of Optimization*, Graduate Texts in Mathematics 258,
DOI 10.1007/978-0-387-68407-9_11, © Springer Science+Business Media, LLC 2010

a Lagrangian function $L(x, y)$ such that (P) is equivalent to the minimax problem $\min_x \max_y L(x, y)$, and then show that the minimax theorem holds,

$$\inf_x \sup_y L(x, y) = \sup_y \inf_x L(x, y),$$

that is, the optimal values of the minimax and maximin problems are equal. One can then denote the maximin problem $\sup_y \inf_x L(x, y)$ by (D), call it the dual of (P), and declare that the strong duality theorem holds for the pair (P) and (D). We treat several classes of problems in this chapter using this approach, and additional examples are given in the exercises at the end of the chapter.

It is possible to reverse this process: if $L : A \times B \to \mathbb{R}$ is any function, then we can define a primal problem (P) by $\inf_{x \in A} \sup_{y \in B} L(x, y)$ (say) and a dual problem (D) by $\sup_{y \in B} \inf_{x \in A} L(x, y)$, and state that the strong duality holds for (P) and (D) if the minimax theorem is valid for L.

Within convex programming, there is another approach to duality, called the *Fenchel duality*. This approach depends on the *Fenchel transform* within the class of proper, lower semicontinuous functions. Such a function $f : \mathbb{R}^n \to \mathbb{R} \cup \{\infty\}$ achieves a finite value somewhere, and $\text{epi}(f) = \{(x, \alpha) \in \mathbb{R}^n \times \mathbb{R} : f(x) \le \alpha\}$ is a closed convex set in $\mathbb{R}^{n+1}$. Denote the class of proper, lower semicontinuous convex functions on $\mathbb{R}^n$ by $\Gamma_0(\mathbb{R}^n)$. If $f \in \Gamma_0(\mathbb{R}^n)$, its *Fenchel dual function* (or Fenchel conjugate function) [93, 94] is the function

$$f^*(x^*) := \sup\{\langle x^*, x - f(x) : x \in \mathbb{R}^n\},$$

which is also in $\Gamma_0(\mathbb{R}^n)$. The most important property of the Fenchel transform is that it is an *involution* on $\Gamma_0(\mathbb{R}^n)$, that is, $f^{**} := (f^*)^* = f$. (In this respect, it is similar to the Fourier transform. The recent paper [9] characterizes the Fenchel transform in terms of involution and order-reversing properties.)

To successfully apply the Fenchel duality approach, one first writes a convex programming problem (P) as the "unconstrained" minimization of a function $f \in \Gamma_0(\mathbb{R}^n)$. If f_0 is the objective function of (P), then the function f defined simply by setting $f(x) = f_0(x)$ if x is a feasible point in (P), and $f(x) = \infty$ otherwise. The next step is somehow to *embed* f into a function $\Phi(x, p) \in \Gamma_0(\mathbb{R}^{n+m})$ such that $\Phi(x, 0) = f(x)$. Here the variable p is regarded as a perturbation. Finally, one defines the dual problem (D) as the minimization problem

$$\max_{p^*} -\Phi^*(0, p^*). \qquad (D)$$

The Lagrangian and Fenchel duality approaches are theoretically equivalent within the class of convex programming. In fact, one can switch between the Lagrangian function L and the Fenchel transform Φ^* by the partial Fenchel conjugation. Both approaches and the connection between them are developed in Rockafellar's book [228]. A short, elegant, but complete treatment of the basic theory can be found in the book Ekeland and Temam [89], Chapters 1–3,

6, in a general setting. The short book [107] treats the Lagrangian approach and some associated minimax theorems.

While they are theoretically equivalent, the Lagrangian and Fenchel duality approaches may offer different advantages in practice. On the one hand, the Fenchel duality approach offers valuable sensitivity analysis information through the use of subgradients and subdifferentials, and also explains why we should expect the strong duality theorem to hold generically. On the other hand, a Lagrangian function may be easier to obtain than a perturbation function for a given convex program, and the Lagrangian approach offers a connection between optimization and game theory. Finally, the Fenchel duality theory is valid only within convex programming, but the Lagrangian approach may extend beyond it, since there exist several version of minimax theorems that hold beyond convex–concave Lagrangian functions.

11.2 Saddle Points and Their Properties

Consider a function

$$L : \mathcal{A} \times \mathcal{B} \to \mathbb{R},$$

where $\mathcal{A}$ and $\mathcal{B}$ are arbitrary sets and L is an arbitrary function. We associate with L the primal and dual problems

$$\inf_{x\in\mathcal{A}} \sup_{y\in\mathcal{B}} L(x,y), \qquad (P)$$

$$\sup_{y\in\mathcal{B}} \inf_{x\in\mathcal{A}} L(x,y). \qquad (D)$$

We start with the following elementary result.

Theorem 11.2. (***Weak duality theorem***) *If (P) and (D) are the associated primal and dual programs with L, then*

$$\sup(D) \le \inf(P),$$

that is,

$$\sup_{y\in\mathcal{B}} \inf_{x\in\mathcal{A}} L(x,y) \le \inf_{x\in\mathcal{A}} \sup_{y\in\mathcal{B}} L(x,y).$$

Proof. It is easy to see that

$$\inf_{x\in\mathcal{A}} L(x,y) \le L(u,y) \quad \text{for all} \quad u \in \mathcal{A},\ y \in \mathcal{B}.$$

This implies that

$$\sup_{y\in\mathcal{B}} \inf_{x\in\mathcal{A}} L(x,y) \le \sup_{y\in\mathcal{B}} L(u,y) \quad \text{for all} \quad u \in \mathcal{A},$$

and

$$\sup_{y\in\mathcal{B}} \inf_{x\in\mathcal{A}} L(x,y) \le \inf_{u\in\mathcal{A}} \sup_{y\in\mathcal{B}} L(u,y) = \inf_{x\in\mathcal{A}} \sup_{y\in\mathcal{B}} L(x,y).$$

This proves the theorem. □

The weak duality theorem already gives us some useful information concerning the problems (P) and (D). In particular, we have the following.

Corollary 11.3. *Let L be a Lagrangian function, and let (P) and (D) be the associated primal and dual programs with L. If the primal minimization problem (P) is* unbounded, *then the dual maximization problem (D) is* infeasible.

Proof. (P) is unbounded if and only if $\inf(P) = -\infty$. Then $\sup(D) = -\infty$ as well, which means that (D) is has an empty feasible region. □

The difference in the objective values,

$$\inf(P) - \sup(D) \geq 0,$$

is called the *duality gap*. It is of great importance to know when the duality gap is zero, that is, when the optimal objective values of the primal and dual problems coincide.

Definition 11.4. *A point $(x^*, y^*) \in \mathcal{A} \times \mathcal{B}$ is a saddle point of L if*

$$L(x^*, y) \leq L(x^*, y^*) \leq L(x, y^*) \quad \text{for all} \quad x \in \mathcal{A},\ y \in \mathcal{B}. \tag{11.1}$$

The following fundamental theorem gives a characterization of saddle points.

Theorem 11.5. (*Saddle point theorem*) *Let $L : \mathcal{A} \times \mathcal{B} \to \mathbb{R}$ and $(x^*, y^*) \in \mathcal{A} \times \mathcal{B}$. The following conditions are equivalent:*

(a) (x^*, y^*) is a saddle point of $L(x, y)$.
(b) x^* is a solution of (P), $y*$ is a solution of (D), and $\min(P) = \max(D)$, that is,

$$\min_{x \in \mathcal{A}} \sup_{y \in \mathcal{B}} L(x, y) = \max_{y \in \mathcal{B}} \inf_{x \in \mathcal{A}} L(x, y). \tag{11.2}$$

The minimization on the left-hand side is achieved at x^*, and the maximization on the right-hand side is achieved at y^*.

Furthermore, if either (a) or (b) is satisfied, then the common optimal value of (P) and (D) must be $L(x^*, y^*)$.

Proof. Suppose that (a) holds. We have

$$\sup_{y \in \mathcal{B}} L(x^*, y) = \max_{y \in \mathcal{B}} L(x^*, y) = L(x^*, y^*) = \min_{x \in \mathcal{A}} L(x, y^*) = \inf_{x \in \mathcal{A}} L(x, y^*),$$

where the middle equalities follow directly from (11.1), and the first and last equalities are trivial. Then,

$$\begin{aligned} \inf_{x \in \mathcal{A}} \sup_{y \in \mathcal{B}} L(x, y) &\leq \sup_{y \in \mathcal{B}} L(x^*, y) = L(x^*, y^*) = \inf_{x \in \mathcal{A}} L(x, y^*) \\ &\leq \sup_{y \in \mathcal{B}} \inf_{x \in \mathcal{A}} L(x, y) \leq \inf_{x \in \mathcal{A}} \sup_{y \in \mathcal{B}} L(x, y), \end{aligned}$$

where the last inequality follows from the weak duality theorem, Theorem 11.2. Since the first and the last terms in the above inequalities are the same, we must have equalities throughout. This proves (b) and the fact that the common optimal value of (P) and (D) equals $L(x^*, y^*)$.

Conversely, suppose that (b) holds. Then (11.2) gives

$$\inf_{x \in \mathcal{A}} L(x, y^*) = \sup_{y \in \mathcal{B}} L(x^*, y),$$

which in turn implies

$$L(x^*, y) \leq L(x^*, y^*) \leq L(x, y^*) \quad \text{for all} \quad x \in \mathcal{A}, \; y \in \mathcal{B},$$

proving (a). □

The theorem reveals the intimate connection between saddle points and a pair of primal–dual optimization problems. This connection was first brought to light by Kuhn and Tucker [182], inspired by the linear programming duality and the game theory results of von Neumann [211].

Corollary 11.6. *The set of saddle points of $L : \mathcal{A} \times \mathcal{B} \to \mathbb{R}$ is a direct product $\mathcal{A}_0 \times \mathcal{B}_0$, where $\mathcal{A}_0 \subseteq \mathcal{A}$ and $\mathcal{B}_0 \subseteq \mathcal{B}$.*

Proof. This result follows immediately from Theorem 11.5. An independent proof runs as follows: suppose (x_1^*, y_1^*) and (x_2^*, y_2^*) are saddle points of L. We need to show that the points (x_1^*, y_2^*) and (x_2^*, y_1^*) are also saddle points of L. We have

$$L(x_1^*, y_2^*) \leq L(x_1^*, y_1^*) \leq L(x_2^*, y_1^*) \leq L(x_2^*, y_2^*) \leq L(x_1^*, y_2^*),$$

where the first two inequalities follow from (11.1) applied to the saddle point (x_1^*, y_1^*), and the remaining ones from applying (11.1) to (x_2^*, y_2^*). Since the end terms are equal, we must have equalities throughout, $L(x_1^*, y_2^*) = L(x_1^*, y_1^*) = L(x_2^*, y_1^*) = L(x_2^*, y_2^*)$. These imply

$$L(x_1^*, y) \leq L(x_1^*, y_1^*) = L(x_1^*, y_2^*) = L(x_2^*, y_2^*) \leq L(x, y_2^*) \quad \text{for all} \quad x \in \mathcal{A}, \; y \in \mathcal{B},$$

where the inequalities follow from (11.1) applied to the saddle points (x_i^*, y_i^*), $i = 1, 2$. In particular, we have

$$L(x_1^*, y) \leq L(x_1^*, y_2^*) \leq L(x, y_2^*) \quad \text{for all} \quad x \in \mathcal{A}, \; y \in \mathcal{B},$$

which proves that (x_1^*, y_2^*) is a saddle point. A similar proof shows that (x_2^*, y_1^*) is a saddle point. □

We also have the following result, useful in convex programming.

Lemma 11.7. *Let $\mathcal{A} \subseteq \mathbb{R}^n$ and $\mathcal{B} \subseteq \mathbb{R}^m$ be convex sets, and let $L(x, y)$ be a convex–concave function.*

The set of saddle points of L is a set of the form $\mathcal{A}_0 \times \mathcal{B}_0$, where $\mathcal{A}_0$ and $\mathcal{B}_0$ are convex sets.

Moreover, if $\mathcal{A}_0 \times \mathcal{B}_0 \neq \emptyset$, and L is strictly convex in x (strictly concave in y), then $\mathcal{A}_0$ $(\mathcal{B}_0)$ is a singleton.

Proof. We prove only the statements concerning $\mathcal{A}_0$; the corresponding proofs for $\mathcal{B}_0$ are the same. Let $y^* \in \mathcal{B}_0$, and $x_1^*, x_2^* \in \mathcal{A}_0$, and $0 < \lambda < 1$. Define the point $x_\lambda^* := (1-\lambda)x_1^* + \lambda x_2^*$. Since (x_i^*, y^*) $(i = 1, 2)$ is a saddle point by Corollary 11.6, it follows that

$$L(x_i^*, y) \leq L(x_i^*, y^*) \leq L(x, y^*),\ i = 1, 2, \quad \text{for all } \ x \in \mathcal{A},\ y \in \mathcal{B}.$$

We multiply the above inequality for $i = 1$ by $1 - \lambda$, the one for $i = 2$ by λ, and add them, and use the convexity of L in x to obtain

$$\begin{aligned} L(x_\lambda^*, y) &\leq (1-\lambda)L(x_1^*, y) + \lambda L(x_2^*, y) \leq (1-\lambda)L(x_1^*, y^*) + \lambda L(x_2^*, y^*) \\ &= L(x_i^*, y^*) \leq L(x_\lambda^*, y^*) \quad (i = 1, 2), \quad \text{for all } \ y \in \mathcal{B}. \end{aligned}$$

Here the equality follows since $L(x_1^*, y^*) = L(x_2^*, y^*)$, a consequence of the fact that the points (x_i^*, y^*) $(i = 1, 2)$ are saddle points, and the last inequality follows, again since (x_i^*, y^*) is a saddle point.

Next, we have

$$\begin{aligned} L(x_\lambda^*, y^*) &\leq (1-\lambda)L(x_1^*, y^*) + \lambda L(x_2^*, y^*) \leq (1-\lambda)L(x, y^*) + \lambda L(x, y^*) \\ &= L(x, y^*) \ \text{ for all } \ x \in \mathcal{A}, \end{aligned}$$

where the first inequality follows from the convexity of L in x, the second one since the points (x_i^*, y^*) $(i = 1, 2)$ are saddle points. The two facts above give

$$L(x_\lambda^*, y) \leq L(x_\lambda^*, y^*) \leq L(x, y^*) \qquad \text{for all } \ x \in \mathcal{A},\ y \in \mathcal{B},$$

meaning that (x_λ^*, y^*) is a saddle point. Consequently, $\mathcal{A}_0$ is a convex set.

Suppose that L be strictly convex in x and $\mathcal{A}_0 \times \mathcal{B}_0 \neq \emptyset$. Let $x_1^*, x_2^* \in \mathcal{A}_0$ and $y^* \in \mathcal{B}_0$. If $x_1^* \neq x_2^*$, then

$$L\Big(\frac{x_1^* + x_2^*}{2}, y^*\Big) < \frac{L(x_1^*, y^*)}{2} + \frac{L(x_2^*, y^*)}{2} = L(x_1^*, y^*) = L(x_2^*, y^*).$$

This is a contradiction, since $\mathcal{A}_0$ is convex, so that the point $(x_1^* + x_2^*)/2$ belongs to $\mathcal{A}_0$, and it follows from Corollary 11.6 that $((x_1^* + x_2^*)/2, y^*)$ is a saddle point. Thus $\mathcal{A}_0$ must be a singleton, and the lemma is proved. □

11.3 Nonlinear Programming Duality

Consider the nonlinear program

$$\begin{aligned} \min \quad & f(x) \\ \text{s.t.} \quad & g_i(x) \leq 0,\ i = 1, \ldots, r, \\ & h_j(x) = 0,\ j = 1, \ldots, m, \\ & x \in C. \end{aligned} \qquad (P)$$

We associate with (P) the Lagrangian function

$$L(x,\lambda,\mu) = f(x) + \sum_{i=1}^{r} \lambda_i g_i(x) + \sum_{j=1}^{m} \mu_j h_j(x), \quad x \in C, \ \lambda \geq 0, \ \mu \in \mathbb{R}^m. \quad (11.3)$$

Note that

$$\sup_{\lambda_i \geq 0} \lambda_i g_i(x) = \begin{cases} 0, & \text{if } g_i(x) \leq 0, \\ \infty, & \text{otherwise.} \end{cases}$$

Similarly,

$$\sup_{\mu_j \in \mathbb{R}} \mu_j h_j(x) = \begin{cases} 0, & \text{if } h_j(x) = 0, \\ \infty, & \text{otherwise.} \end{cases}$$

Thus, we have

$$\sup_{0 \leq \lambda \in \mathbb{R}^r, \ \mu \in \mathbb{R}^m} L(x,\lambda,\mu) = \begin{cases} f(x), & \text{if } g_i(x) \leq 0, \ i = 1,\ldots,r, \\ & \quad h_j(x) = 0, \ j = 1,\ldots,m, \\ +\infty, & \text{otherwise.} \end{cases}$$

Consequently, we can write (P) in the form

$$\inf_{x \in C} \sup_{0 \leq \lambda \in \mathbb{R}^r, \ \mu \in \mathbb{R}^m} L(x,\lambda,\mu). \qquad (P)$$

We formulate the *dual program* to (P) by switching the orders of inf and sup above,

$$\sup_{0 \leq \lambda \in \mathbb{R}^r, \ \mu \in \mathbb{R}^m} \inf_{x \in C} L(x,\lambda,\mu). \qquad (D)$$

We call (P) a *convex program* or a *convex programming problem* if the functions f, $\{g_i\}_1^r$ are convex, $\{h_j\}_1^m$ affine, and $C \subseteq \mathbb{R}^n$ is a nonempty convex set. In Section 11.4, we describe the important and deep duality relationships between a convex program (P) and its dual. In the rest of the chapter, we will work out the duality relationships in some special classes of convex programs and closely related problems.

We now undertake a more detailed study of the Lagrangian function $L(x,\lambda,\mu)$ in (11.3).

Theorem 11.8. *A point (x^*,λ^*,μ^*) is a saddle point of the Lagrangian L, that is,*

$$L(x^*,\lambda,\mu) \leq L(x^*,\lambda^*,\mu^*) \leq L(x,\lambda^*,\mu^*) \ \textit{for all} \ x \in C, \ 0 \leq \lambda \in \mathbb{R}^r, \ \mu \in \mathbb{R}^m,$$

if and only if

(i) x^* *is a (global) minimizer of* (P),

(ii) (λ^*,μ^*) *is a (global) maximizer of* (D),

(iii) $\min(P) = \max(D)$.

Furthermore, if (x^*, λ^*, μ^*) is a saddle point, then

$$(iv)\quad \min(P) = L(x^*, \lambda^*, \mu^*) = \max(D),$$

or equivalently

$$(v)\quad \lambda_i^* g_i(x^*) = 0, \quad i = 1, \ldots, r.$$

Consequently, if (x^*, λ^*) is a saddle point, then the complementarity conditions hold,

$$\lambda_i^* \geq 0, \quad g_i(x^*) \leq 0, \quad \lambda_i^* g_i(x^*) = 0, \quad i = 1, \ldots, r.$$

Proof. Theorem 11.5 implies immediately that (x^*, λ^*, μ^*) is a saddle point if and only if (i)–(iii) hold, and if (x^*, λ^*, μ^*) is a saddle point, then (iv) holds, that is,

$$f(x^*) = L(x^*, \lambda^*, \mu^*) = f(x^*) + \sum_{i=1}^{r} \lambda_i^* g_i(x^*) + \sum_{j=1}^{m} \mu_j^* h_j(x^*).$$

Since $h_j(x^*) = 0$ $(j = 1, \ldots, m)$, we get

$$\sum_{i=1}^{m} \lambda_i^* g_i(x^*) = 0.$$

Since $\lambda_i^* \geq 0$ and $g_i(x^*) \leq 0$, we conclude that $\lambda_i^* g_i(x^*) = 0$. This proves (v). □

Corollary 11.9. *If the functions f, $\{g_i\}_1^r$, and $\{h_j\}_1^m$ are differentiable, and $C = \mathbb{R}^n$, then the KKT conditions hold,*

$$0 = \nabla_x L(x^*, \lambda^*, \mu^*) = \nabla f(x^*) + \sum_{i=1}^{r} \lambda_i^* \nabla g_i(x^*) + \sum_{j=1}^{m} \mu_j^* \nabla h_j(x^*).$$

Proof. The inequality $L(x^*, \lambda^*, \mu^*) \leq L(x, \lambda^*, \mu^*)$ shows that x^* minimizes $L(x, \lambda^*, \mu^*)$ over $\mathbb{R}^n$. This immediately implies the corollary. □

Remark 11.10. Since the Lagrangian function for (P) is an affine function of the multipliers (λ, μ), it follows that the objective function of the dual program (D),

$$q(\lambda, \mu) := \min_{x \in C} L(x, \lambda, \mu),$$

is a concave function, being the pointwise infimum of the family of affine functions $\{q_x\}_{x \in C}$, where $q_x(\lambda, \mu) = L(x, \lambda, \mu)$. Thus, (D) is equivalent to the minimization of the convex function $-q$ over its domain, which is a convex subset of $\{(\lambda, \mu) : \lambda \geq 0\}$. Thus, (D) is a convex programming problem, regardless of whether (P) is. Thus, in order to obtain symmetric duality results between (P) and (D), such as the results in Section 11.4, one needs to impose some convexity conditions on the primal program (P).

The following result gives an interesting method to construct a constrained optimization problem out of an unconstrained one.

Lemma 11.11. *Let $\lambda^* \in \mathbb{R}^r_+$ and $\mu^* \in \mathbb{R}^m$ be arbitrary. If $x^* \in C$ is a global minimizer of the function*

$$L(x, \lambda^*, \mu^*) = f(x) + \sum_{i=1}^{r} \lambda_i^* g_i(x) + \sum_{j=1}^{m} \mu_j^* h_j(x)$$

on C, then x^* is a global minimizer of the nonlinear program

$$\begin{aligned} \min \quad & f(x) \\ \text{s.t.} \quad & g_i(x) \le g_i(x^*), \quad i = 1, \ldots, r, \\ & h_j(x) = h_j(x^*), \quad j = 1, \ldots, m, \\ & x \in C \subseteq \mathbb{R}^n. \end{aligned} \qquad (P)$$

Proof. If x satisfies the constraints of (P), then

$$\begin{aligned} f(x^*) &= L(x^*, \lambda^*, \mu^*) - \sum_{i=1}^{r} \lambda_i^* g_i(x^*) - \sum_{j=1}^{m} \mu_j^* h_j(x^*) \\ &\le L(x, \lambda^*, \mu^*) - \sum_{i=1}^{r} \lambda_i^* g_i(x^*) - \sum_{j=1}^{m} \mu_j^* h_j(x^*) \\ &= f(x) + \sum_{i=1}^{r} \lambda_i^* (g_i(x) - g_i(x^*)) + \sum_{j=1}^{m} \mu_j^* (h_j(x) - h_j(x^*)) \\ &\le f(x), \end{aligned}$$

proving that x^* is a global minimizer of (P). □

Remark 11.12. The above theorem leads to a simple method to construct a constrained minimization problem whose Lagrangian has a saddle point, and thus to construct an optimization problem for which no duality gap exists. It was discovered by Everett [90] in connection with finding approximate solutions to some integer programming problems.

11.4 Strong Duality in Convex Programming

In this section, we will be concerned with the duality theory of the convex programming problem

$$\begin{aligned} \min \quad & f(x) \\ \text{s.t.} \quad & g_i(x) \le 0, \ i = 1, \ldots, r, \\ & h_j(x) \le 0, \ j = 1, \ldots, m, \\ & x \in C, \end{aligned} \qquad (P) \qquad (11.4)$$

where $C \subseteq \mathbb{R}^n$ is a nonempty convex set, f and $\{g_i\}_1^r$ are convex functions on C, and $\{h_j\}_1^m$ are affine functions, say

$$h_j(x) = \langle a_j, x \rangle + \beta_j, \quad j = 1, \ldots, m,$$

with $a_j \in \mathbb{R}^n$ and $\beta_j \in \mathbb{R}$. Associated with (P), we have the Lagrangian function

$$L(x, \lambda, \mu) = f(x) + \sum_{i=1}^{r} \lambda_i g_i(x) + \sum_{j=1}^{m} \mu_j h_j(x), \quad x \in C, \ \lambda \in \mathbb{R}_+^r, \ \mu \in \mathbb{R}_+^m.$$

Recall that the dual program is given by

$$\begin{aligned} \max \quad & g(\lambda, \mu) \\ \text{s.t.} \quad & \lambda \in \mathbb{R}_+^r, \ \mu \in \mathbb{R}_+^m, \end{aligned} \qquad (D)$$

where

$$g(\lambda, \mu) := \inf_{x \in C} L(x, \lambda, \mu).$$

The main result of this section, indeed of the whole chapter, is the deep result that under minor restrictions, the optimal multipliers (λ, μ) *exist, and there is no duality gap between the primal–dual programming pair* (P) *and* (D). The proof of this fundamental result will be based on a nonlinear analogue of Motzkin's transposition theorem for a system of nonlinear convex inequalities and linear inequalities. It is necessary to single out linear constraints, since it will be seen that the nonlinear constraints need a Slater-type constraint qualification, but the linear ones do not.

We first prove a technical lemma.

Lemma 11.13. *Let* l *be a nonnegative affine function on a convex set* $C \subseteq \mathbb{R}^n$. *If* $l(x_0) = 0$ *at some point in* $x_0 \in \operatorname{ri}(C)$, *then* l *is identically zero on* C.

Proof. Suppose that $l(x) > 0$ at some point $x \in C$. Since $x_0 \in \operatorname{ri}(C)$, there exists $t > 1$ such that the point $x_1 = x + t(x_0 - x) = (1-t)x + tx_0$ is in C. This gives the contradiction $0 \leq l(x_1) = (1-t)l(x) + tl(x_0) = (1-t)l(x) < 0$. □

The theorem below is a nonlinear generalization of the homogeneous version of Motzkin's transposition theorem; see Theorem 3.15, Theorem 7.17, or Theorem A.3.

Theorem 11.14. (***Convex transposition theorem***) *Let* $C \subseteq \mathbb{R}^n$ *be a nonempty convex set,* $\{g_i\}_1^r$ *convex functions,* $\{h_j\}_1^m$ *affine functions on* $\mathbb{R}^n$, *such that* $\operatorname{dom}(g_i)$ *contain* C, $i = 1, \ldots, r$. *Assume that there exists a point* $x^* \in \operatorname{ri}(C)$ *satisfying the linear inequalities, that is,*

$$x^* \in \operatorname{ri}(C), \quad h_j(x^*) \leq 0, \quad j = 1, \ldots, m. \tag{11.5}$$

Then exactly one of the following alternatives holds:

(a) $\exists\ \overline{x} \in C,\ \ g_i(\overline{x}) < 0,\ i = 1, \ldots, r,\ \ h_j(\overline{x}) \leq 0,\ j = 1, \ldots, m,$

(b) $\exists\ (\lambda, \mu) \geq 0,\ \lambda \neq 0,\ \sum_{i=1}^{r} \lambda_i g_i(x) + \sum_{j=1}^{m} \mu_j h_j(x) \geq 0$ *for all* $x \in C.$

Proof. The statements (a) and (b) cannot both be true: if $\overline{x} \in C$ satisfies (a) and (λ, μ) satisfies (b), then we have a contradiction

$$0 \leq \sum_{i=1}^{r} \lambda_i g_i(\overline{x}) + \sum_{j=1}^{m} \mu_j h_j(\overline{x}) < 0,$$

where the first inequality follows from (b) and the second one follows from (a), the nonnegativity of the multipliers, and the fact that $\lambda \neq 0$.

Suppose that (a) is false. We will show that (b) must hold, using a separation argument. We define a set of "right-hand-side vectors" of the constraint functions on C,

$$R := \{(y, z) \in \mathbb{R}^r \times \mathbb{R}^m\ :\ \exists x \in C,\ \ g_i(x) < y_i,\ i = 1, \ldots, r,\\ h_j(x) = z_j,\ j = 1, \ldots, m.\}.$$

Since $C \subseteq \operatorname{dom}(g_i)$ for all $i = 1, \ldots, r$, R is a nonempty convex set. Since (a) is false, we clearly have that the set R is disjoint from the nonpositive orthant

$$N := \{(u, v) \in \mathbb{R}^r \times \mathbb{R}^m : u \leq 0, v \leq 0\}.$$

Since N is a polyhedral set, it follows from Theorem 6.21 that there exists a hyperplane $H_{(\lambda,\mu,\alpha)}$ separating R and N and not containing R, say

$$\langle \lambda, y \rangle + \langle \mu, z \rangle \geq \alpha \geq \langle \lambda, u \rangle + \langle \mu, v \rangle \ \text{ for all } (y, z) \in R,\ (u, v) \in N,$$

such that the first inequality is strict for some $(\overline{y}, \overline{z}) \in R$. Setting $(u, v) = (0, 0)$ gives $\alpha \geq 0$. Moreover, if $\lambda_i < 0$, then picking $(u, v) = (te_i, 0)$, $t \to -\infty$ contradicts the second inequality above; thus, $\lambda \geq 0$, and similarly, $\mu \geq 0$. Since $C \subseteq \operatorname{dom}(g_i)$ for each $i = 1, \ldots, m$, this means that

$$\sum_{i=1}^{r} \lambda_i (g_i(x) + s_i) + \sum_{j=1}^{m} \mu_j h_j(x) \geq \alpha \geq 0, \ \text{ for all } x \in C,\ s > 0, \tag{11.6}$$

and the first inequality is strict for some $\overline{x} \in C$, $\overline{s} > 0$. Letting $s_i \to 0$ gives

$$l(x) := \sum_{i=1}^{r} \lambda_i g_i(x) + \sum_{j=1}^{m} \mu_j h_j(x) \geq 0 \ \text{ for all } x \in C.$$

The proof will be complete once we establish the claim that $\lambda \neq 0$. Suppose that $\lambda = 0$. Then the affine function $l(x)$ is nonnegative on C, and it follows from assumption (11.5) that $l(x^*) = 0$. Lemma 11.13 implies that $l(x)$ is identically zero on C (and that $\alpha = 0$). However, this contradicts the fact in (11.6) that $l(\overline{x}) > 0$ for some $\overline{x} \in C$. This proves the claim. □

It is now easy to prove the following fundamental duality theorem of convex programming.

Theorem 11.15. (***Strong duality theorem of convex programming***) *Suppose that the convex program (P) in (11.4) has a finite infimum, that is,* $-\infty < \inf(P) < \infty$, *and the conditions* $C \subseteq \mathrm{dom}(f)$, $C \subseteq \mathrm{dom}(g_i)$, $i = 1, \ldots, m$ *hold.*

If Slater's conditions *are satisfied, that is,*

$$\exists \overline{x} \in \mathrm{ri}(C), \;\; g_i(\overline{x}) < 0, \; i = 1, \ldots, r, \;\; h_j(\overline{x}) \le 0, \; j = 1, \ldots, m, \tag{11.7}$$

then there exist multipliers $(\lambda^*, \mu^*) \in \mathbb{R}^r_+ \times \mathbb{R}^m_+$ *such that*

$$\inf(P) = \inf_{x \in C} \left\{ f(x) + \sum_{i=1}^r \lambda_i^* g_i(x) + \sum_{j=1}^m \mu_j^* h_j(x) \right\}.$$

The multiplier vector (λ^*, μ^*) *is an optimal solution to the dual program (D), and*

$$\inf(P) = \max(D).$$

Furthermore, if (P) has an optimal solution x^*, *then* (x^*, λ^*, μ^*) *is a saddle point of the Lagrangian function* $L(x, \lambda, \mu)$.

We emphasize that Slater's conditions (11.7) put no restrictions on the linear constraints.

Proof. Define $f^* := \inf(P)$, and consider the system of constraints

$$\begin{aligned} f(x) - f^* &< 0, \\ g_i(x) &< 0, \quad i = 1, \ldots, r, \\ h_j(x) &\le 0, \quad j = 1, \ldots, m, \\ x &\in C. \end{aligned}$$

Theorem 11.14 applies to the above system, since $C \subseteq \mathrm{dom}(f)$, $C \subseteq \mathrm{dom}(g_i)$, $= 1, \ldots, r$, and Slater's conditions (11.7) hold. Since $f^* \le f(x)$ for any feasible solution of (P), the system is inconsistent; thus there exists a vector $(\lambda_0^*, \lambda_1^*, \ldots, \lambda_r^*, \mu_1^*, \ldots, \mu_m^*) \ge 0$, $(\lambda_0^*, \lambda_1^*, \ldots, \lambda_r^*) \ne 0$, such that

$$\lambda_0^*(f(x) - f^*) + \sum_{i=1}^r \lambda_i^* g_i(x) + \sum_{j=1}^m \mu_j^* h_j(x) \ge 0 \;\text{ for all }\; x \in C. \tag{11.8}$$

We claim that $\lambda_0^* \ne 0$. Otherwise, $(\lambda_1^*, \ldots, \lambda_r^*) \ne 0$,

$$\sum_{i=1}^r \lambda_i^* g_i(x) + \sum_{j=1}^m \mu_j^* h_j(x) \ge 0 \;\text{ for all }\; x \in C,$$

and therefore by Theorem 11.14, the system

$$x \in C, \quad g_i(x) < 0, \ (1 \leq i \leq r), \quad h_j(x) \leq 0 \ (1 \leq j \leq m)$$

cannot have a solution, contradicting Slater's condition (11.7). This proves that $\lambda_0^* > 0$, and we set $\lambda_0^* = 1$.

The inequality (11.8) gives that for all $x \in C$,

$$\inf(P) =: f^* \leq f(x) + \sum_{i=1}^{r} \lambda_i^* g_i(x) + \sum_{j=1}^{m} \mu_j^* h_j(x) = L(x, \lambda^*, \mu^*).$$

Thus, we have

$$\inf(P) \leq \inf_{x \in C} L(x, \lambda^*, \mu^*) \leq \sup_{(\lambda,\mu) \geq 0} \inf_{x \in C} L(x, \lambda, \mu) = \sup(D) \leq \inf(P),$$

where the last inequality follows the weak duality theorem, Theorem 11.2. This proves that all inequalities above are equalities, that (λ^*, μ^*) is an optimal solution to (D), and that $\inf(P) = \max(D)$.

If x^* is an optimal solution to (P), it follows from Theorem 11.5 that (x^*, λ^*, μ^*) is a saddle point of L. □

Remark 11.16. A special version of the convex transposition theorem, Theorem 11.14, in which the affine functions do not appear, was first proved by Fan, Glicksberg, and Hoffman [91]. Theorems 11.14 and 11.15 in the generality stated above seem to have first been proved by Rockafellar [228] as Theorem 21.2 and Theorem 28.2, respectively, under slightly more general assumptions than $C \subseteq \operatorname{dom}(f), \operatorname{dom}(g_i)$. His proof of Theorem 11.14 uses the special separation result, Theorem 6.21, in which one of the convex sets is a polyhedral set. Later, several direct proofs of Theorem 11.15 were published avoiding the use of this special separation theorem, but utilizing other tricks [247, 221, 214, 147]. Our proof of Theorem 11.14 is essentially the same as Rockafellar's, but we have simplified the proof of its main ingredient, Theorem 6.21, using a device in [147].

Corollary 11.17. *Let the convex program (P) have only linear constraints;*

$$\begin{array}{lll} \min & f(x) & \\ \text{s.t.} & h_j(x) \leq 0, \quad j = 1, \ldots, m, & \qquad (P) \\ & x \in \mathbb{R}^n, & \end{array}$$

and an objective function f with $\operatorname{dom}(f) = \mathbb{R}^n$.

If (P) has a finite infimum, then there exists a multiplier vector $\mu^* \geq 0$ in $\mathbb{R}^m$ that is an optimal solution to the dual program (D), and strong duality holds, that is, $\inf(P) = \max(D)$.

Furthermore, if (P) has an optimal solution x^*, then (x^*, μ^*) is a saddle point of the Lagrangian function

$$L(x, \mu) = f(x) + \sum_{j=1}^{m} \mu_j h_j(x).$$

11.4.1 Failure of Strong Duality in Convex Optimization

We now give an example of a convex programming problem for which there exist no optimal multipliers and that has a positive duality gap.

Consider the following convex program in which the convex set C is the nonnegative quadrant $C = \{(x_1, x_2) : x_1 \geq 0, x_2 \geq 0\}$,

$$\begin{aligned} \min \quad & f(x) = e^{-\sqrt{x_1 x_2}} \\ \text{s.t.} \quad & x_2 = 0, \qquad\qquad (P) \\ & x = (x_1, x_2) \in C. \end{aligned}$$

We note that any point $x^* = (x_1^*, 0)$, $x_1^* \geq 0$, is an optimal solution for (P) and that $\min(P) := f^* = f(x^*) = 1$. The dual problem is

$$\max_{\lambda \in \mathbb{R}} g(\lambda), \quad (D)$$

where

$$g(\lambda) := \inf_{x \geq 0} \; e^{-\sqrt{x_1 x_2}} + \lambda x_2.$$

Note that if $\lambda \geq 0$, then $e^{-\sqrt{x_1 x_2}} + \lambda x_2 \to 0$ as $x_2 \to 0$ and $x_1 x_2 \to \infty$. Also, if $\lambda < 0$, then $e^{-\sqrt{x_1 x_2}} + \lambda x_2 \to -\infty$ as $x_2 \to \infty$. Thus, we have

$$g(\lambda) = \begin{cases} 0, & \text{if } \lambda \geq 0, \\ -\infty, & \text{if } \lambda < 0. \end{cases}$$

Consequently, the dual program is

$$\begin{aligned} \max \quad & g(\lambda) \\ \text{s.t.} \quad & \lambda \geq 0, \end{aligned} \qquad (D)$$

where $g(\lambda) \equiv 0$ is the zero function. Any $\lambda \geq 0$ is an optimal solution to (D), and we have $\sup(D) = 0$. Therefore, we have a positive duality gap $\inf(P) - \sup(D) = 1 - 0 = 1$.

Of course, Slater's conditions are not satisfied, since $\operatorname{int}(C) = \{(x_1, x_2) : x_1 > 0, x_2 > 0\}$, and an optimal point $(x_1, 0)$ to (P) does not belong to it.

11.5 Examples of Dual Problems

In this section, we consider the duality theory of several special common classes of problems in detail.

11.5.1 Linear Programming

The duality theory of linear programming is developed in Chapter 8 using Farkas's lemma and the theory of convex polyhedra. The development below offers a different perspective, and is largely independent of the previous treatment, although in the final analysis, both approaches depend on separation of convex sets.

Consider the linear program

$$\begin{aligned} \min \quad & \langle c, x\rangle \\ \text{s. t.} \quad & Ax = b, \qquad (P) \\ & x \geq 0, \end{aligned}$$

where A is an $m \times n$ matrix, $c \in \mathbb{R}^n$, and $b \in \mathbb{R}^m$.

Define $C = \mathbb{R}^n_+$. (The reader is encouraged to work out the formulation of the dual problem by choosing $C = \mathbb{R}^n$.) We have the Lagrangian function

$$L(x, \mu) = \langle c, x\rangle + \langle \mu, b - Ax\rangle = \langle b, \mu\rangle + \langle c - A^T\mu, x\rangle.$$

The dual problem is, by definition,

$$\sup_{\mu \in \mathbb{R}^m} \inf_{x \geq 0} L(x, \mu). \quad (D)$$

Let us give an explicit description of (D). We first deal with the inner minimization problem in (D). We have

$$\inf_{x \geq 0} L(x, \mu) = \langle b, \mu\rangle + \inf_{x \geq 0} \langle c - A^T\mu, x\rangle = \begin{cases} \langle b, \mu\rangle, & \text{if } c - A^T\mu \geq 0, \\ -\infty, & \text{otherwise.} \end{cases}$$

Therefore, (D) can be written as

$$\sup_{\mu \in \mathbb{R}^m} \{\langle b, \mu\rangle : c - A^T\mu \geq 0\}.$$

In other words, (D) is the linear programming problem (replacing μ with y)

$$\begin{aligned} \max \quad & \langle b, y\rangle \\ \text{s. t.} \quad & A^T y \leq c. \end{aligned} \qquad (D)$$

If (P) is unbounded, then Corollary 11.3 (weak duality theorem) implies that (D) is infeasible. If (P) is feasible and bounded from below, then Corollary 11.17 implies that (D) has an optimal solution, and the duality gap is zero. Furthermore, Lemma 8.1 on page 195 implies that (P) has an optimal solution. The same result can be obtained more easily by observing that (P) is the dual of (D) and applying Corollary 11.17.

Finally, if (P) is infeasible, then the above considerations show that (D) is either unbounded or infeasible. There are examples showing that each of these two cases is possible.

We leave to the reader as an exercise to calculate the dual to the linear program

$$\begin{aligned} \min \quad & \langle c, x\rangle \\ \text{s.t.} \quad & Ax \geq b, \\ & x \geq 0. \end{aligned}$$

11.5.2 Quadratic Programming

We now consider the convex quadratic program

$$\begin{aligned} \min \quad & \frac{1}{2}\langle Qx, x\rangle + \langle c, x\rangle \\ \text{s.t.} \quad & Ax = b, \qquad\qquad (P) \\ & x \geq 0, \end{aligned}$$

where Q is an $n \times n$ symmetric positive semidefinite matrix, A an $m \times n$ matrix, $c \in \mathbb{R}^n$, and $b \in \mathbb{R}^m$.

Defining $C = \mathbb{R}^n$ (the reader is invited to work out the dual problem by choosing $C = \mathbb{R}^n_+$), we have the Lagrangian function

$$\begin{aligned} L(x, \lambda, \mu) &= \frac{1}{2}\langle Qx, x\rangle + \langle c, x\rangle + \langle \mu, b - Ax\rangle - \langle \lambda, x\rangle \\ &= \langle b, \mu\rangle + \frac{1}{2}\langle Qx, x\rangle - \langle A^T\mu + \lambda - c, x\rangle. \end{aligned}$$

The dual problem (D) is, by definition,

$$\sup_{\mu\in\mathbb{R}^m, 0\leq\lambda\in\mathbb{R}^n} \inf_{x\in\mathbb{R}^n} L(x, \lambda, \mu).$$

The inner optimization problem is the unconstrained minimization of the convex quadratic function

$$q(x) := \frac{1}{2}\langle Qx, x\rangle - \langle A^T\mu + \lambda - c, x\rangle$$

on $\mathbb{R}^n$. It follows from Corollary 4.30 on page 99 that q is bounded from below if and only if it has a minimizer, and this happens if and only if the equation

$$Qx = A^T\mu + \lambda - c \tag{11.9}$$

is solvable for x. The set X of solutions to this linear equation is then the set of minimizers of q. Note the important fact that

$$\langle Qx^*, x^* \rangle = \langle A^T \mu + \lambda - c, x^* \rangle = -2 \min_{\mathbb{R}^n} q \quad \text{for all } x^* \in X.$$

Consequently, the dual program (D) is given by

$$\begin{aligned} \max \quad & \langle b, \mu \rangle - \frac{1}{2} \langle A^T \mu + \lambda - c, x^* \rangle \\ \text{s.t.} \quad & A^T \mu + \lambda - c = Qx^*, \\ & \mu \in \mathbb{R}^m, \quad \mathbb{R}^n \ni \lambda \geq 0, \end{aligned}$$

where x^* is *any* point satisfying the equation (11.9). We can choose x^* as a linear function of λ, μ; for example, we can choose

$$x^* = Q^\dagger (A^T \mu + \lambda - c),$$

where $Q^\dagger$ is the *pseudoinverse* or *Moore–Penrose inverse* of Q. We recall that

$$Q^\dagger = U \Lambda^\dagger U^T, \qquad \Lambda^\dagger = \operatorname{diag}\{\lambda_1^\dagger, \ldots, \lambda_n^\dagger\},$$

where $Q = U \Lambda U^T$ is the spectral decomposition of Q, $\lambda_i^\dagger = (\lambda_i)^{-1}$ if $\lambda_i \neq 0$, and $\lambda_i^\dagger = 0$ when $\lambda_i = 0$; see Golub and van Loan [108] and [218]. Since Q is positive semidefinite, we see that $Q^\dagger$ is also positive semidefinite. Thus, the dual program is

$$\begin{aligned} \max \quad & \langle b, \mu \rangle - \frac{1}{2} \left\langle Q^\dagger (A^T \mu + \lambda - c), A^T \mu + \lambda - c \right\rangle \\ \text{s.t.} \quad & A^T \mu + \lambda - c \in R(Q), \ \lambda \geq 0. \end{aligned} \qquad (D)$$

This is a maximization quadratic program whose objective function is a concave function; thus (D) is equivalent to the convex quadratic program

$$\begin{aligned} \min \quad & \frac{1}{2} \left\langle Q^\dagger (A^T \mu + \lambda - c), A^T \mu + \lambda - c \right\rangle - \langle b, \mu \rangle \\ \text{s.t.} \quad & A^T \mu + \lambda - c \in R(Q), \ \lambda \geq 0. \end{aligned}$$

If one wishes, one can further eliminate μ from this formulation by minimizing over $\mu \in \mathbb{R}^m$.

In the case that Q is positive definite, then $Q^\dagger = Q^{-1}$, and (D) becomes

$$\begin{aligned} \max \quad & \langle b, \mu \rangle - \frac{1}{2} \langle Q^{-1} (A^T \mu + \lambda - c), A^T \mu + \lambda - c \rangle \\ \text{s.t.} \quad & \lambda \geq 0, \ \mu. \end{aligned}$$

We summarize the strong duality properties of the primal–dual pair (P) and (D):

Theorem 11.18. *Let (P) be the convex quadratic program above. The dual program is also a convex quadratic program.*

If (P) is unbounded, then (D) is infeasible.

If (P) is feasible and bounded from below, then both (P) and (D) have optimal solutions that are then the saddle points of the Lagrangian function $L(x, \lambda, \mu)$.

Finally, if (P) is infeasible, then (D) is either unbounded or infeasible.

Proof. If (P) is unbounded, then Corollary 11.3 implies that (D) is infeasible. If (P) is feasible and bounded from below, then Corollary 11.17 on page 287 implies that the dual program (D) has an optimal solution (λ^*, μ^*). Since the dual of (D) is (P), applying the same theorem, this time to (D), shows that (P) also has an optimal solution x^*. Then Theorem 11.5 shows that (x^*, λ^*, μ^*) is a saddle point of $L(x, \lambda, \mu)$, and conversely.

Finally, if (P) is infeasible, then the above considerations show that (D) is either unbounded or infeasible. Furthermore, since linear programming is a quadratic program with $Q = 0$, there are examples showing that each of these two cases is possible. □

Remark 11.19. Part of the above theorem asserts that if a convex quadratic program is bounded from below, then it has a (global) minimizer. The same result is true for any quadratic programming problem regardless of whether the quadratic objective function is convex. This was first proved by Frank and Wolfe [98] in 1956 and has been re-proved numerous times since then. A particularly interesting solution to this problem is given in [99].

We leave to the reader as an exercise (see Exercise 7 on page 303) to compute the dual of the quadratic program

$$\begin{aligned} \min \quad & \frac{1}{2}\langle Qx, x\rangle + \langle c, x\rangle \\ \text{s.t.} \quad & Ax \geq b. \end{aligned}$$

11.5.3 A Minimax Problem

Consider the problem

$$\min_{x \in \mathbb{R}^n} \max\{f_1(x), \ldots, f_m(x)\}, \qquad (P)$$

where $f_i : \mathbb{R}^n \to \mathbb{R}$ are convex functions. We will reformulate (P) in order to first define a Lagrangian function for it. We begin with a preliminary result.

Lemma 11.20. *Let $a_1, \ldots, a_m$ be real numbers. We have*

$$\max\{a_1, \ldots, a_m\} = \max_{\lambda \in \Delta_{m-1}} \sum_{i=1}^{m} \lambda_i a_i,$$

where Δ_{m-1} is the standard unit simplex in $\mathbb{R}^m$.

Proof. We may assume that $a_1 = \max\{a_1, \ldots, a_m\}$ without any loss of generality. If $\lambda \in \Sigma_m$, we clearly have

$$\sum_{i=1}^{m} \lambda_i a_i \le \sum_{i=1}^{m} \lambda_i a_1 = a_1.$$

This proves that

$$\max_{\lambda \in \Sigma_m} \sum_{i=1}^{m} \lambda_i a_i \le \max\{a_1, \ldots, a_m\}.$$

We achieve equality in the above inequality by taking $\lambda = (1, 0, \ldots, 0) \in \Sigma_m$. The lemma is proved. □

Using this result, we have

$$\max\{f_1(x), \ldots, f_m(x)\} = \max_{\lambda \in \Sigma_m} \sum_{i=1}^{m} \lambda_i f_i(x),$$

so that (P) can be rewritten in the form

$$\min_{x \in \mathbb{R}^n} \max_{\lambda \in \Sigma_m} \sum_{i=1}^{m} \lambda_i f_i(x).$$

A natural Lagrangian function for (P) is then

$$L(x, \lambda) = \sum_{i=1}^{m} \lambda_i f_i(x), \quad x \in \mathbb{R}^n, \lambda \in \Sigma_m.$$

We note that the Lagrangian L is linear, thus concave in λ; since the functions $\{f_i(x)\}_1^m$ are convex, L is convex in x.

The dual of the problem (P) is

$$\max_{\lambda \in \Sigma_m} \min_{x \in \mathbb{R}^n} \sum_{i=1}^{m} \lambda_i f_i(x). \qquad (D)$$

Let (x^*, λ^*) be a saddle point of L. We have

$$L(x^*, \lambda) \le L(x^*, \lambda^*) \le L(x, \lambda^*) \text{ for all } x \in \mathbb{R}^n, \ \lambda \in \Sigma_m.$$

Note that if all the functions f_i are differentiable, then the second inequality above gives

$$\sum_{i=1}^{m} \lambda_i^* \nabla f_i(x^*) = 0.$$

Also, it follows immediately from Theorem 11.5 that x^* is a minimizer of (P).

Define $I(x^*) := \{i : f_i(x^*) = \max\{f_1(x^*), \ldots, f_m(x^*)\}\}$, that is, $I(x^*)$ is the indices of f_i that attain the maximum value $\max_{i \le j \le m}\{f_j(x^*)\}$.

Lemma 11.21. *If $i \notin I(x^*)$, then $\lambda_i^* = 0$.*

Proof. We have

$$\begin{aligned} f^* &:= \max_{i \in I(x^*)} f_i(x^*) = \max_{1 \le i \le m} f_i(x^*) = \sum_{i=1}^{m} \lambda_i^* f_i(x^*) \\ &= \Big(\sum_{i \in I(x^*)} \lambda_i^* \Big) f^* + \sum_{i \notin I(x^*)} \lambda_i^* f_i(x^*), \end{aligned}$$

where the third equality follows because (x^*, λ^*) is a saddle point of L. If $\lambda_k^* > 0$ for an index $k \notin I(x^*)$, then we have $\lambda_k^* f_k(x^*) < \lambda_k^* f^*$, and the above equalities imply

$$f^* < (\sum_{i \in I(x^*)} \lambda_i^*) f^* + \sum_{i \notin I(x^*)} \lambda_i^* f^* = f^*,$$

a contradiction. □

Remark 11.22. An alternative approach to dualizing (P) is to reformulate it as the constrained optimization problem

$$\begin{aligned} \min \quad & z \\ \text{s.t.} \quad & f_i(x) - z \le 0, \quad i = 1, \ldots, m, \end{aligned}$$

and then apply the standard Lagrangian approach to obtain a dual for it. This approach leads to the same dual (D) obtained above; we leave the details to the reader. In addition, we can now apply the strong duality theorem, Theorem 11.15: since the constraints satisfy Slater's condition (pick any point x_0 and z_0 large enough that $f_i(x_0) - z_0 < 0$), we see that if (P) is feasible and $\inf(P) > -\infty$, then (D) has an optimal solution and $\inf(P) = \max(D)$.

11.6 Conic Programming Duality

Let E be a *finite Euclidean space*. Thus, E is a finite-dimensional vector space over $\mathbb{R}$ equipped with an inner product $\langle \cdot, \cdot \rangle$ and the associated Euclidean norm $\| \cdot \|$ given by $\|x\|^2 = \langle x, x \rangle$. A closed, convex cone in E with a nonempty interior and containing no whole lines is called a *regular convex cone*.

In recent years, *interior-point methods* have made it popular to consider conic programs and their duality theory. A conic-form programming problem is a programming problem on the vector space E that generalizes a linear program in that the linear constraints $x \ge 0$ in the linear program are replaced by a constraint of the form $x \in K$, where K is a regular convex cone in E. Linear programming corresponds to the choices $E = \mathbb{R}^n$ and $K = \mathbb{R}^n_+$.

Thus, a conic programming problem has the form

$$\begin{aligned} \min\quad & \langle c, x\rangle \\ \text{s. t.}\quad & \langle a_i, x\rangle = b_i,\ i = 1, \ldots, m, \quad (P) \\ & x \in K, \end{aligned} \tag{11.10}$$

where c and $\{a_i\}_1^m$ are vectors in E, $\{b_i\}_1^m$ are real scalars, and K is a regular convex cone in E. The program (P) is a convex program in which the only nonlinearities appear in the cone constraint $x \in K$.

Important examples of conic programming beyond linear programming include *semidefinite programming*, in which E is the space of $n \times n$ symmetric matrices equipped with the trace inner product $\langle X, Y\rangle = \mathrm{tr}(XY)$ and K is the cone of symmetric positive semidefinite matrices, and *second-order cone programming*, in which $E = \mathbb{R}^{n+1}$ equipped with the usual inner product and K is the second-order or Lorentz cone $K = \{x \in \mathbb{R}^{n+1} : \|(x_1, \ldots, x_n)\| \le x_{n+1}\}$. These are important examples of *symmetric cone programming* in which K is a *symmetric (homogeneous self-dual) cone* [118, 210, 226]. Still more general examples in interior-point methods include *homogeneous cone programming* [118], hyperbolic cone programming [119], and several others.

In the theorem below,

$$K^* = \{z \in E : \langle x, z\rangle \ge 0 \text{ for all } x \in K\}$$

is the (modified) dual cone, which is the reflection through the origin of the usual dual cone. We use the notation K^* (in this section only) for the modified cone out of respect for the established terminology in conic programming. The primal problem (P) is represented in the form below in order to bring out the similarity between (P) and (D).

Theorem 11.23. *Let E and F be two finite-dimensional Euclidean spaces. Let $A : E \to F$ be a linear operator, $A^* : F \to E$ its adjoint, $c \in E$, $b \in F$, and $K \subset E$ a regular convex cone in E. Consider the conic programming pairs*

$$\begin{aligned} \min\quad & \langle c, x\rangle & \qquad \max\quad & \langle b, y\rangle \\ \text{s. t.}\quad & Ax = b \quad (P) & \qquad \text{s. t.}\quad & A^*y + s = c \quad (D) \\ & x \in K, & & s \in K^*, \end{aligned}$$

The conic programs (P) and (D) form a primal–dual pair. If (P) has an interior feasible point $x \in \mathrm{int}(K)$ and $\inf(P) > -\infty$, then (D) has an optimal solution, and the strong duality theorem holds, that is, $\inf(P) = \max(D)$.

Similarly, If (D) has an interior feasible point $s \in \mathrm{int}(K^*)$ and $\sup(D) < \infty$, then (P) has an optimal solution, and the strong duality theorem holds, that is, $\min(P) = \sup(D)$.

Consequently, if both programs (P) and (D) have interior feasible solutions, then both programs have optimal solutions, and $\min(P) = \max(D)$.

Proof. Consider the Lagrangian function $L : K \times F \to \mathbb{R}$,

$$L(x, y) := \langle c, x\rangle + \langle b - Ax, y\rangle = \langle b, y\rangle + \langle c - A^*y, x\rangle,$$

and observe that (P) can be written as the minimax problem

$$\min_{x \in K} \max_{y \in F} L(x, y).$$

The Lagrangian dual of (P) with respect to L is given by

$$\begin{aligned} \max_{y \in F} \min_{x \in K} L(x, y) &= \max_{y \in F} \{ \langle b, y \rangle + \min_{x \in K} \langle c - A^* y, x \rangle \} \\ &= \max \{ \langle b, y \rangle : c - A^* y \in K^* \}, \end{aligned}$$

where the second equality follows by setting $s := c - A^* y$ and noting that the term $\min_{x \in K} \langle s, x \rangle$ has value zero if $s \in K^*$, and $-\infty$ otherwise. This proves that (P) and (D) form a primal–dual pair.

Now suppose that (P) has a feasible point $\overline{x} \in \operatorname{int}(K)$ and that $\inf(P) > -\infty$. Theorem 11.15 implies that (D) has an optimal solution and strong duality holds, that is, $\inf(P) = \max(D)$. Since (P) is the dual of (D), the same argument shows that if (D) has a feasible point $\overline{s} \in \operatorname{int}(K^*)$ (an easy argument show that K^* is a regular convex cone) and $\sup(D) < \infty$, then (P) has an optimal solution and strong duality holds, that is, $\min(P) = \sup(D)$. □

Corollary 11.24. *Consider the conic programming pairs*

$$\begin{array}{lll} \min & \langle c, x \rangle & \\ \text{s. t.} & \langle a_i, x \rangle = b_i, \ i = 1, \ldots, m & (P) \\ & x \in K, & \end{array} \qquad \begin{array}{lll} \max & \sum_{i=1}^m b_i y_i & \\ \text{s. t.} & \sum_{i=1}^m y_i a_i + s = c, & (D) \\ & s \in K^*. & \end{array}$$

The conic programs (P) *and* (D) *form a primal–dual pair. If* (P) *has an interior feasible point* $x \in \operatorname{int}(K)$ *and* $\inf(P) > -\infty$*, then* (D) *has an optimal solution, and the strong duality theorem holds, that is,* $\inf(P) = \max(D)$.

Similarly, If (D) *has an interior feasible point* $s \in \operatorname{int}(K^*)$ *and* $\sup(D) < \infty$*, then* (P) *has an optimal solution, and the strong duality theorem holds, that is,* $\min(P) = \sup(D)$.

Consequently, if both programs (P) *and* (D) *have interior feasible solutions, then both programs have optimal solutions, and* $\min(P) = \max(D)$.

The corollary follows immediately from Theorem 11.23 by defining the linear operator $A : E \to \mathbb{R}^m$ by

$$Ax := (\langle a_1, x \rangle, \ldots, \langle a_1, x \rangle)^T,$$

and noting that

$$\langle Ax, y \rangle = \sum_{i=1}^m y_i \langle a_i, x \rangle = \left\langle x, \sum_{i=1}^m y_i a_i \right\rangle = \langle x, A^* y \rangle,$$

so that $A^* y = \sum_{i=1}^m y_i a_i$.

11.7 The Fermat–Torricelli–Steiner Problem

Here we consider how to formulate a dual problem to the minimization problem

$$\min_{x \in R^n} \|x - a_1\| + \cdots + \|x - a_k\|, \qquad (P) \tag{11.11}$$

where $\{a_i\}_{i=1}^k$ are given vectors in $\mathbb{R}^n$ and $\|\cdot\|$ is the Euclidean norm.

The case $n = 3$ is called the *Fermat–Torricelli–Steiner* problem and has a distinguished history.

This type of problem occurs in *plant location problems*, for example. The vectors a_i could then be the locations of existing plants, and we may want to locate a new plant at x such that the sum of the distances from the new plant to the existing ones is as short as possible.

It is a fact of life that most problems cannot be dualized easily in the sense that the dual problem cannot be given explicitly (or computed easily). Here, too, there does not seem to be a straightforward approach to dualize the problem.

The following trick permits us to obtain a workable dual problem. Observe that $\|u\| = \max_{\|y\| \le 1} \langle u, y \rangle$ for any vector $u \in \mathbb{R}^n$. Thus, we can write (P) as the minimax problem

$$\min_{x \in \mathbb{R}^n} \max_{\|y_i\| \le 1} \sum_{i=1}^{k} \langle a_i - x, y_i \rangle, \tag{11.12}$$

with the Lagrangian function

$$L(x, y) = L(x, y_1, \ldots, y_k) := \sum_{i=1}^{k} \langle a_i - x, y_i \rangle.$$

Clearly, the objective function $f(x) := \|x - a_1\| + \cdots + \|x - a_k\|$ of the primal problem is coercive, and we may assume that x lies in a compact set, say in $\{x : f(x) \le f(0) = \|a_1\| + \cdots + \|a_k\|\}$. By von Neumann's minimax theorem, Theorem 11.1, there exists a saddle point $(x^*, y_1^*, \ldots, y_k^*)$. Therefore, we have

$$\min_{x \in \mathbb{R}^n} \max_{\|y_i\| \le 1} L(x, y) = \max_{\|y_i\| \le 1} \min_{x \in \mathbb{R}^n} L(x, y),$$

and we can regard the maximin problem on the right-hand side as the dual problem to (P). When simplified, the dual problem becomes

$$\max\Big\{\sum_{i=1}^{k} \langle a_i, y_i \rangle : \|y_i\| \le 1,\ 1 \le i \le k,\ \sum_{i=1}^{k} y_i = 0\Big\}. \qquad (D) \tag{11.13}$$

Observe that the knowledge of a dual optimal solution $y^* = (y_1^*, \ldots, y_k^*)$ does not help us determine the primal one x^*, since the problem $\min_x L(x, y^*)$

becomes vacuous. However, if we have a primal optimal solution x^*, then we can explicitly compute y^* by solving the simple optimization problem

$$\max_{\|y_i\|\le 1} \sum_{i=1}^{k} \langle a_i - x^*, y_i \rangle.$$

The dual problem (D) has a nice form that can be exploited in numerical algorithms. Indeed, if it did not have the coupling constraint $\sum_{i=1}^{k} y_i = 0$, then (D) would be separated into k simple subproblems. There exist algorithms that exploit this fact and try to solve (D) in parallel. There are also algorithms that try to solve (P) and (D) together, using proximal-point methods.

Interior-point methods can be used to solve (P). We reformulate (P):

$$\begin{aligned} \min \quad & \sum_{i=1}^{k} t_i \\ \text{s.t.} \quad & \|x - a_i\| \le t_i, \; i = 1, \ldots, k. \end{aligned}$$

Each constraint may be written as $x - a_i \in K$, where

$$K := \{(x,t) \in \mathbb{R}^n \times \mathbb{R} : \|x\| \le t\}$$

is the second-order, or Lorentz, cone. Thus, (P) reduces to a second-order cone problem

$$\begin{aligned} \min \quad & \sum_{i=1}^{k} t_i \\ \text{s.t.} \quad & (x - a_i, t_i) \in K, \; i = 1, \ldots, k. \end{aligned}$$

Its standard dual is also a second-order cone program. The primal, dual, or primal–dual pair can be solved in polynomial time. See [269] for more details.

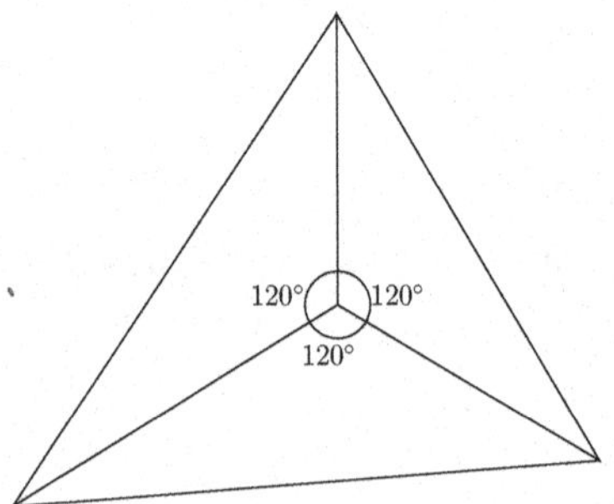

Fig. 11.1. Torricelli point of a triangle.

Remark 11.25. When $n = 2$ and $k = 3$, and $\{a_i\}$ form the vertices of a triangle T whose angles at the vertices are smaller than 120°, then the optimal solution x^* is the Torricelli point, which is the point inside T such that the angle between the two lines connecting x^* to any two vertices of T is 120°; see Figure 11.1. Also, the dual problem is equivalent to determining the *largest* equilateral triangle circumscribing the given triangle; see [71], pp. 325–326. This dual problem was apparently known to Torricelli as early as 1810–1811.

11.8 Hoffman's Lemma

Let P be a nonempty polyhedron in $\mathbb{R}^n$ given in the form

$$P := \{z \in \mathbb{R}^n : Az \leq b\},$$

where $A \in \mathbb{R}^{m \times n}$. Suppose that we have a point x not lying in P but that *almost* satisfies the inequalities $Az \leq b$ defining P. What can we say about x? For example, is it close to the polyhedron P?

In this section, we give an estimate of the Euclidean distance from such a point x to P, that is, we estimate the optimal value of the minimization problem

$$d(x, P) := \min\{\|z - x\| : z \in P\}$$

in terms of the size of the *residual vector* $(Ax - b)^+ = \Pi_{\mathbb{R}^n_+}(Ax - b)$, the nonnegative part of the vector $Ax - b$, obtained from $Ax - b$ by replacing its negative entries by zero.

Theorem 11.26. (***Hoffman's lemma***) *There exists a constant $c(A) > 0$, which depends only on the matrix A defining the polyhedron $P = \{z : Az \leq b\}$, such that*

$$d(x, P) \leq c(A)\|(Ax - b)^+\| \quad \textit{for all} \quad x \in \mathbb{R}^n. \tag{11.14}$$

We emphasize that this is a nontrivial result. For example, in the ill-conditioned polyhedron in Figure 11.2, the point x almost satisfies both of the inequalities defining the shaded polyhedron, but is far away from the polyhedron. Thus, the constant $c(A)$ can get arbitrarily large as the angle between the two lines gets smaller.

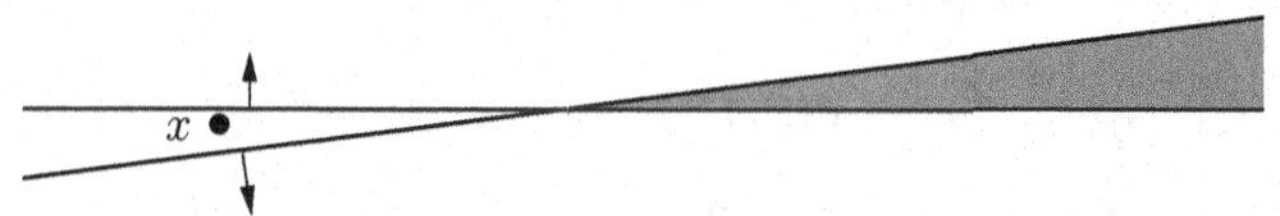

Fig. 11.2. An ill-conditioned polyhedron.

Hoffman's lemma shows that the size of the vector $(Ax - b)^+$ can be used as a stopping criterion in algorithms: if $\|(Ax - b)^+\|$ is small, then x is almost feasible, optimal, etc., depending on the nature of the polyhedron P.

Theorem 11.26 was first proved in Hoffman [136], and has led to many subsequent results in the optimization literature; see for example [121]. Hoffman's original proof is not very transparent, but later studies have shown that such estimates are deeply connected with the duality theory of convex analysis.

Proof. In its broad outlines, the idea of the proof is simple. The most important idea is the second equation below, which follows from the fact that $\|v\| = \max_{\|z\|\le 1}\langle v, z\rangle$ for a vector in $\mathbb{R}^n$. Assuming that equality holds in the minimax problem (the third equation below), we would have

$$\begin{aligned}
d(x,P) &= \min_{Az\le b}\|x-z\| \\
&= \min_{Az\le b}\max_{\|u\|\le 1}\langle x-z, u\rangle \\
&= \max_{\|u\|\le 1}\min_{Az\le b}\langle x-z, u\rangle \\
&= \max_{\|u\|\le 1}\min_{Aw\ge \bar b}\langle w, u\rangle \qquad [w := x-z,\ \bar b := Ax-b] \\
&= \max_{\|u\|\le 1}\max\{\langle \bar b, \lambda\rangle : A^T\lambda = u,\ \lambda\ge 0\} \\
&= \max\{\langle Ax-b, \lambda\rangle : \|A^T\lambda\|\le 1,\ \lambda\ge 0\},
\end{aligned} \tag{11.15}$$

where the fifth equality follows from linear programming duality. These would then give us the formula

$$d(x,P) = \max\{\langle Ax-b, \lambda\rangle : \|A^T\lambda\|\le 1, \lambda\ge 0\}. \tag{11.16}$$

The estimate (11.14) then follows from an appropriate estimation of the right-hand side above.

There exist minimax theorems, such as the *lop-sided minimax theorem* in [10], p. 319, that show that in fact equality holds in the third equation in (11.15); however, we prefer to give a simple, self-contained proof of this fact.

Note that the weak duality theorem (Theorem 11.2) applied to the minimax problem in (11.15) already gives

$$d(x,P) \ge \max\{\langle Ax-b, \lambda\rangle : \|A^T\lambda\|\le 1, \lambda\ge 0\}.$$

To prove the reverse inequality, define the convex sets

$$C_1 := \{w : \|w\|\le d(x,P)\}, \qquad C_2 := \{w : Aw\ge \bar b\},$$

and notice that $C_1^0\cap C_2 = \emptyset$. By Theorem 6.15, C_1 and C_2 can be properly separated. Thus, there exists a vector $\mu\in\mathbb{R}^n$, $\|\mu\| = 1$, such that

$$\langle \mu, w_1\rangle \le \langle \mu, w_2\rangle \ \text{ for all } \ w_1\in C_1,\ w_2\in C_2. \tag{11.17}$$

This gives

$$d(x,P) = \max_{w \in C_1} \langle \mu, w\rangle \le \min_{Aw \ge \bar{b}} \langle \mu, w\rangle = \max\{\langle \lambda, \bar{b}\rangle : \lambda \ge 0, A^T\lambda = \mu\}$$
$$\le \max\{\langle Ax - b, \lambda\rangle : \|A^T\lambda\| \le 1, \lambda \ge 0\},$$

where the first inequality follows from (11.17), the last equality from linear programming duality, and last inequality from the fact $\|\mu\| = 1$. This proves (11.16). Thus, all the inequalities above are actually equalities, and we have

$$d(x,P) = \max\{\langle Ax - b, \lambda\rangle : A^T\lambda = \mu, \lambda \ge 0\}.$$

The linear program above has an optimal solution λ^* that is an extreme point (vertex) of the polyhedron

$$P_* := \{\lambda : A^T\lambda = \mu, \lambda \ge 0\}.$$

An extreme point $\lambda^* \in P_*$ is given by

$$\lambda^* = \begin{pmatrix} \lambda_B^* \\ 0 \end{pmatrix}, \quad B^T\lambda_B^* = \mu,$$

where $A^T = (B^T, N^T)$ and B^T a nonsingular matrix. Thus, $A = \left[\begin{smallmatrix} B \\ N \end{smallmatrix}\right]$, that is, B is a nonsingular submatrix of A whose rows are selected from the rows of A. (It is assumed here, for notational convenience, that the rows of A^T are ordered so that $A^T = (B^T, N^T)$.) We then have

$$\begin{aligned} d(x,P) &= \langle Ax - b, \lambda^*\rangle \le \langle (Ax-b)^+, \lambda^*\rangle \\ &= \langle (Ax-b)_B^+, \lambda_B^*\rangle \le \|\lambda_B^*\| \cdot \|(Ax-b)_B^+\| \\ &\le \|\lambda_B^*\| \cdot \|(Ax-b)^+\|. \end{aligned}$$

The norm of the vector λ_B^* is dependent only on the submatrix B, since $B^T\lambda_B^* = \mu$, B is nonsingular, and $\|\mu\| = 1$. This completes the proof of the theorem. □

Remark 11.27. Some estimates in the spirit of Hoffman's lemma, but distinct from it, appear in Lemma 4 in Rosenbloom's paper [231]. His results, established almost at the same time as Hoffman's lemma, and which deserve to be better known, use essentially duality techniques similar to the ones we have used above. This is noteworthy, since the machinery of convex duality theory in its full generality was developed a decade later.

Hoffman's lemma can also be proved using Ekeland's variational principle and metric regularity; see [144, 145, 15].

11.9 Exercises

1. Consider the convex programming problem

$$\begin{aligned} \min \quad & -\ln x - \ln y \\ \text{s.t.} \quad & (x-1)^2 + y^2 \le 1. \end{aligned}$$

(a) Give a theoretical reason as to why we can assert that $\lambda_0 \neq 0$ in the FJ conditions.
(b) Write the KKT conditions for the program, and find all solutions to the problem by solving the KKT conditions.
(c) Use the Lagrangian function with $C = \mathbb{R}^2$ to explicitly determine the dual problem (D), which should not contain any of the primal variables x, y in its final form.

2. Consider the optimization problem

$$\begin{aligned} \min \quad & \frac{1}{2}x_1^2 + \frac{1}{2}(x_2 - 3)^2 \\ \text{s.t.} \quad & x_1^2 - x_2 \leq 0 \\ & -x_1 + x_2 \leq 2. \end{aligned}$$

(a) Is this a convex programming problem? Justify your answer.
(b) Solve the problem geometrically.
(c) Give a theoretical reason why there should exist KKT points. Give a theoretical reason for the uniqueness of the KKT point.
(d) Write down the KKT conditions and solve them to determine the KKT point.
(e) Write the Lagrangian for the problem using $C = \mathbb{R}^2$, and use it to determine explicitly the dual problem. The primal variables x_1, x_2 should not appear in the dual program.
(f) Determine an optimal solution to the dual problem on the basis of the information obtained in parts (a)–(e).

3. Use Lagrangian duality to determine the dual to the linear program

$$\begin{aligned} \min \quad & x_1 - 3x_2 - x_3 \\ \text{s.t.} \quad & 3x_1 - x_2 + 2x_3 \geq 1 \\ & -2x_1 + 4x_2 \leq 12 \\ & -4x_1 + 3x_2 + 3x_3 = 14. \end{aligned}$$

Do not transform the above linear program into any other format (say into the standard form LP) before applying duality.

4. Consider the optimization problem

$$\begin{aligned} \min \quad & \frac{1}{2}\langle Qx, x\rangle \\ \text{s.t.} \quad & \langle a, x\rangle \leq b, \end{aligned} \qquad (P)$$

where Q is an $n \times n$ symmetric, positive definite matrix, $0 \neq a \in \mathbb{R}^n$, and $b < 0$.

(a) Formulate explicitly the dual program (D) corresponding to (P) in such a way that the primal variable x does not appear in (D).
(b) Determine whether the strong duality theorem holds between (P) and (D), justifying carefully your reasoning.

(c) Solve the dual program (D), that is, determine its optimal solutions λ^*.
(d) Use (c) to determine the primal optimal solution x^*.

5. Consider the convex quadratic program

$$\min\Big\{\frac{1}{2}\langle Qx, x\rangle + \langle c, x\rangle : x \geq 0\Big\}, \qquad (P)$$

where Q is an $n \times n$ symmetric positive semidefinite matrix.
(a) Let (P) have an optimal solution x^*. Argue that there exists $\lambda^* \geq 0$ in $\mathbb{R}^n$ such that (x^*, λ^*) is a saddle point of the Lagrangian function corresponding to (P).
(b) Show that the variational inequality holds at x^*:

$$\langle Qx^* + c, x - x^*\rangle \geq 0 \text{ for all } x \geq 0,$$

and deduce from it the equality $\langle Qx^* + c, x^*\rangle = 0$.
(c) Alternatively, show that the KKT conditions for (P) result in the linear complementarity problem

$$x^* \geq 0, \quad Qx^* + c \geq 0, \quad \text{and} \quad \langle Qx^* + c, x^*\rangle = 0.$$

(d) Using the Lagrangian function, construct the dual program to (P).

6. Consider the quadratic program pair in Section 11.5.2. If both programs have optimal solutions, then show that

$$Qx^* = A^T\mu^* + \lambda^* - c = \alpha$$

is a *constant* for all optimal solutions x^* to (P) and for all optimal solutions (λ^*, μ^*) to (D).
Hint: Use Corollary 11.6.

7. Consider the convex quadratic program

$$\begin{aligned} \min \quad & \frac{1}{2}\langle Qx, x\rangle + \langle c, x\rangle \\ \text{s.t.} \quad & Ax \geq b, \end{aligned} \qquad (P)$$

where A is an $m \times n$ matrix, and Q an $n \times n$ positive semidefinite matrix. Work out the dual program (D), following the example in Section 11.5.2. In particular, answer the following questions:
(a) Write the Lagrangian and formulate its dual program (D).
(b) Assume that $-\infty < \inf(P) < +\infty$. State the strong duality for the pair (P) and (D). Does this theorem hold for (P) and (D)? Explain and justify your answer.
(c) Assume that Q is positive definite. Show that (P) has an optimal solution.
(d) Assuming again that Q is positive definite, formulate the dual problem explicitly, that is, write its objective function in terms of only the dual variables.

(e) Repeat part (d), assuming only that Q is positive semidefinite.

8. The problem

$$\begin{aligned} \min \quad & \frac{1}{2}\|x\|^2 \\ \text{s.t.} \quad & Ax = b,\ x \geq 0 \end{aligned} \qquad (P)$$

seeks the point in the polyhedron $\{x : Ax = b, x \geq 0\}$ closest to the origin.
   (a) Write down the KKT conditions for (P).
   (b) Calculate (determine explicitly) the dual program (D).
   (c) Use the variational inequality to show that a feasible point x^* solves (P) if and only if x^* is a solution to the linear program

$$\begin{aligned} \min \quad & (x^*)^T x \\ \text{s.t.} \quad & Ax = b,\ x \geq 0. \end{aligned}$$

9. Consider the constrained optimization problem

$$\min\{f(x) : g_i(x) \leq 0,\ i = 1, \ldots, m\}, \tag{11.18}$$

where $f, g_i : \mathbb{R}^n \to \mathbb{R}$ are continuous functions. Denote by X^* the set of (global) optimal solutions to (11.18), and assume that $X^* \neq \emptyset$.
   (a) Suppose that $x^* \in X^*$ is and there exists a multiplier $\lambda^* \geq 0 \in \mathbb{R}^m$ such that (x^*, λ^*) form a saddle point of the Lagrangian function

$$L(x, \lambda) = f(x) + \sum_{i=1}^{m} \lambda_i g_i(x).$$

   Show that if $K > \|\lambda^*\|$, then X^* coincides with the optimal solution set X^{**} of the unconstrained optimization problem

$$\min_{x \in \mathbb{R}^n} \{f(x) + K\|g^+(x)\|\}, \tag{11.19}$$

   where $g^+(x) = (g_1^+(x), \ldots, g_m^+(x))$ and $g_i^+(x) = \max\{0, g_i(x)\}$.
   Hint: Use the definition of saddle points. It is easier to show that $X^* \subseteq X^{**}$.
   (b) Suppose that f, g_i are convex functions and that there exists x_0 such that $g_i(x_0) < 0$, $i = 1, \ldots, m$. Show that there exist saddle points of the Lagrangian function L. Show that if K is large enough, then X^* coincides with X^{**}.
   (c) Suppose f, g_i are linear functions (so that (11.18) is a linear program). Show that there exist saddle points of the Lagrangian function L, and that if K is large enough, then X^* coincides with X^{**}.
   Discuss the advantages and challenges associated with trying to solve the linear program (11.18) by solving an unconstrained minimization problem (11.19).

10. Compute the dual of the problem in Section 11.5.3 using the nonlinear programming approach outlined at the end of that section.
11. Consider the primal–dual linear programming pair

$$\begin{array}{llll} \min & \langle c, x\rangle & & \\ \text{s. t.} & Ax = b & (P) \\ & x \geq 0, & \end{array} \qquad \begin{array}{llll} \max & \langle b, y\rangle & \\ \text{s. t.} & A^T y + s = c & (D) \\ & s \geq 0, & \end{array}$$

where A is an $m \times n$ matrix. Assume that both programs have *interior* feasible points, that is, there exist a feasible point $x > 0$ for (P) and a feasible point (y, s) for (D) such that $s > 0$.
A class of *interior-point methods*, called *path-following methods*, attempt to follow the *primal–dual central path* $\{(x(t), y(t), s(t)) : t > 0\}$ by Newton's method, where $x(t)$ and $(y(t), s(t))$ are the solutions to the pair of primal–dual convex programming pair (P_t) and (D_t), respectively, where

$$\min\Big\{\langle c, x\rangle - t\sum_{j=1}^{n} \ln x_j : Ax = b\Big\}, \qquad (P_t)$$

and

$$\max\Big\{\langle b, y\rangle + t\sum_{j=1}^{n} \ln s_j : A^T y + s = c\Big\}. \qquad (D_t)$$

It is well known that the primal–dual central path exists for (P) and (D) under the interior-point assumptions above, and converges to specific optimal solutions of (P) and (D) as $t \downarrow 0$; see for example Ye [270].
   (a) Show that (P_t) and (D_t) form a primal–dual convex programming pair.
   (b) Show that there exist a unique solution $x(t)$ to (P_t) and a solution to (D_t) where $s(t)$ is unique.
   (c) Determine the KKT conditions that characterize $(x(t), y(t), s(t))$ on the central path.
12. Consider the minimax problem

$$\min_{x\in\mathbb{R}^n} \max_{1\leq i\leq m} \{|\langle a^i, x\rangle| : \langle c, x\rangle = 1\},$$

where A is an $m \times n$ matrix, a^i, $i = 1, \ldots, m$ are the rows of A, and $c \in \mathbb{R}^n$, $c \neq 0$.
   (a) Formulate the problem as a minimization problem using the suggestion in Remark 11.22.
   (b) Argue that the KKT conditions must hold for a minimizer x^*, and write down the KKT conditions.
   (c) Let z^* be the optimal objective value of the minimax problem. Prove that the multiplier δ corresponding to the constraint $\langle c, x\rangle = 1$ cannot be zero. Prove that, in fact, $\delta = z^*$.

13. **(von Neumann)** In this problem, the famous *von Neumann minimax theorem* for bimatrix games will be proved using linear programming duality. This result, which von Neumann [211] proved in 1928 by advanced methods (Brouwer's fixed point theorem), was the starting point of game theory.
Let A be an $m \times n$ matrix and denote by Δ_{m-1} and Δ_{n-1} the unit simplices, where

$$\Delta_{n-1} = \Big\{x \in \mathbb{R}^n : x \geq 0, \ \sum_{j=1}^{n} x_j = 1\Big\},$$

and where Δ_{m-1} is defined similarly. The von Neumann minimax theorem states that

$$\min_{x \in \Delta_{n-1}} \max_{y \in \Delta_{m-1}} \langle Ax, y\rangle = \max_{y \in \Delta_{m-1}} \min_{x \in \Delta_{n-1}} \langle Ax, y\rangle. \tag{11.20}$$

Denote by a^i and a_j the ith row and jth column of A, respectively.
(a) Show that

$$\max_{y \in \Delta_{m-1}} \langle Ax, y\rangle = \max_{1 \leq i \leq m} \{\langle a^i, x\rangle\},$$

so that

$$\min_{x \in \Delta_{n-1}} \max_{y \in \Delta_{m-1}} \langle Ax, y\rangle = \min_{x \in \Delta_{n-1}} \max_{1 \leq i \leq m} \{\langle a^i, x\rangle\}.$$

(b) Show that the right-hand side of the last equation above can be written as a linear program

$$\min\Big\{z : \langle a^i, x\rangle \leq z, \ i = 1, \ldots, m, \ \sum_{j=1}^{n} x_j = 1, \ x \geq 0\Big\},$$

in the variables (x, z).
(c) Show that the dual of the above linear program is

$$\max\Big\{\delta : \sum_{i=1}^{m} y_i a^i - \delta e \geq 0, \ \sum_{i=1}^{m} y_i = 1, \ y \geq 0\Big\},$$

in the variables (y, δ). Here $e \in \mathbb{R}^n$ is the vector with all components 1.
Hint: One way to proceed is to write the Lagrangian function

$$L(x, z; y, \delta) = z + \sum_{i=1}^{m} y_i(\langle a^i, x\rangle - z) + \delta(1 - \langle e, x\rangle)$$

on the set $X \times \Lambda$, where $X = \{x : x \geq 0\}$ and $\Lambda = \{(y, \delta) : y \geq 0\}$, and to formulate the dual using L.

(d) Using the already encountered techniques in parts (a) and (b), show that the dual linear program above is equivalent to the right-hand side of (11.20), thereby proving von Neumann's minimax theorem.

14. (**M. Riesz**; see Rosenbloom [231]) Let $K \subset \mathbb{R}^n$ be a closed convex cone.
   (a) For $u \in \mathbb{R}^n$, consider the problem

$$\begin{aligned} \max \quad & \langle u, y \rangle \\ \text{s.t.} \quad & \|y\| \le 1, \qquad (P) \\ & y \in K, \end{aligned}$$

   and the equalities

$$\begin{aligned} \max_{\|y\|\le 1, y \in K} \langle u, y\rangle &= \max_{\|y\|\le 1} \{\langle u, y\rangle - \max_{w\in K^*} \langle y, w\rangle\} \\ &= \max_{\|y\|\le 1} \min_{w \in K^*} \langle u - w, y\rangle \\ &= \min_{w\in K^*} \max_{\|y\|\le 1} \langle u - w, y\rangle \\ &= \min_{w \in K^*} \|u - w\|. \qquad (D) \end{aligned}$$

   Show that a saddle point (y^*, w^*), $\|y^*\| \le 1$, $w^* \in K^*$, exists for this minimax problem and justify all the equations above. Show that $u = (u - w^*) + w^*$ is a decomposition of u into two elements $u - w^* \in K$ and $w^* \in K^*$ that are orthogonal, $\langle w^*, u - w^*\rangle = 0$.
   (b) Show that if K is the polyhedral cone

$$K = \{y \in \mathbb{R}^n : \langle a_i, y\rangle = 0,\ i = 1, \dots, r,\ \langle a_i, y\rangle \ge 0,\ i = r+1, \dots, s\},$$

   then the dual problem is given by

$$\min\Big\{\Big\|u + \sum_1^s c_i a_i\Big\| : c_i \ge 0,\ i = r+1, \dots, s\Big\}. \quad (D)$$

15. (**Teboulle [252]**) Consider the minimization problem

$$\begin{aligned} \min \quad & -\sum_{i=1}^n \ln x_i \\ & \qquad\qquad\qquad\qquad (P) \\ \text{s.t.} \quad & \frac{1}{2}\langle Qx, x\rangle + \langle b, x\rangle + c \le 0, \end{aligned}$$

where Q is an $n \times n$ symmetric positive definite matrix. Assume that the constraint set contains a vector $\overline{x} > 0$ such that $\frac{1}{2}\langle Q\overline{x}, \overline{x}\rangle + \langle b, \overline{x}\rangle + c < 0$.
   (a) Show that (P) has a minimizer.
   (b) The direct approach to formulating the dual will lead to a problem that is hard to write down explicitly. The following trick will lead to a manageable dual problem. Since Q is positive semidefinite, it has a

square root, that is, an $n \times n$ positive semidefinite matrix A such that $A^2 = Q$. Introducing the variable $u = Ax$, we can rewrite (P) in the form

$$\begin{aligned} \min \quad & -\sum_{i=1}^{n} \ln x_i \\ \text{s.t.} \quad & \frac{1}{2}\|u\|^2 + \langle b, x\rangle + c \leq 0, \\ & Ax - u = 0. \end{aligned} \qquad (P')$$

Using the standard Lagrangian approach, formulate the dual problem explicitly, that is, as a maximization problem in which the objective function is given explicitly.

(c) Suppose an optimal solution to the dual problem is somehow found. Use this knowledge to determine the optimal solution $x^* \in \mathbb{R}^n$ to the original problem (P).
Hint: Use Theorem 11.5.

(d) Using the above approach, formulate the dual problem when (P) has more than one $(k > 1)$ convex quadratic inequality.

16. This problem is about different ways to dualize the problem of projecting a point $a \in \mathbb{R}^n$ onto a linear subspace L in $\mathbb{R}^n$.

(a) One way to formulate the projection problem is

$$\begin{aligned} \min \quad & \frac{1}{2}\|x - a\|^2 \\ \text{s.t.} \quad & Ax = 0, \end{aligned} \qquad (P_1)$$

where A is a matrix such that $L = \{x : Ax = 0\}$.

(i) Formulate the Lagrange dual (D_1) of (P_1), and prove whether strong duality between (P_1) and (D_1) holds.

(ii) The dual problem (D_1) has a geometric interpretation similar to that of (P_1); describe it.

(b) A second way to formulate the projection problem is

$$\begin{aligned} \min \quad & \|x - a\| \\ \text{s.t.} \quad & x \in L. \end{aligned} \qquad (P_2)$$

(i) Using the fact that $\|u\| = \max_{\|y\| \leq 1} \langle u, y \rangle$ for any $u \in \mathbb{R}^n$, write (P_2) as a minimax problem.

(ii) Write the dual (D_2) of the minimax problem above, and show that it can be written as

$$\begin{aligned} \max \quad & \langle a, y \rangle \\ \text{s.t.} \quad & \|y\| \le 1,\ y \in M, \end{aligned}$$

where M is a certain subset of $\mathbb{R}^n$. Give a description of M. Prove that the strong duality theorem holds between (P_2) and (D_2).

17. *(The von Neumann economic growth problem)* This is an early duality result formulated by von Neumann for an important economic growth problem. Consider the problem

$$\begin{aligned} \max \quad & \gamma \\ \text{s.t.} \quad & B - \gamma A)x \ge 0 \\ & \sum_1^n x_i = 1,\ x \ge 0, \end{aligned}$$

where A, B are nonnegative $m \times n$ matrices such that $Ax > 0$, $Bx > 0$ for all x in the standard unit simplex Δ_{n-1} in $\mathbb{R}^n$. Denote by $a_i, b_i \in \mathbb{R}^n$ the rows of A, B, respectively.

(a) Show that the vector constraints can be written as

$$\gamma \le \frac{\langle b_i, x \rangle}{\langle a_i, x \rangle}.$$

Use this to transform the minimization problem into the form

$$\max_{x \in \Delta_{n-1}} \min_{1 \le i \le m} \frac{\langle b_i, x \rangle}{\langle a_i, x \rangle}.$$

(b) Show that the auxiliary problem

$$\min_{y \in \Delta_{m-1}} \frac{\sum_1^m \beta_i y_i}{\sum_1^m \alpha_i y_i},$$

where the denominator $\sum_1^m \alpha_i y_i$ is positive on Δ_{m-1}, has the optimal value

$$\min_{1 \le i \le m} \frac{\beta_i}{\alpha_i}.$$

Use this to prove that the optimization problem in (a) can be written in the form

$$\max_{x \in \Delta_{n-1}} \min_{y \in \Delta_{m-1}} \frac{\sum_1^m y_i \langle b_i, x \rangle}{\sum_1^m y_i \langle a_i, x \rangle}. \qquad (P)$$

Von Neumann proved that the minimax theorem holds, that is, the optimal value of the maximin problem (P) equals the optimal value of the minimax problem

$$\min_{y \in \Delta_{m-1}} \max_{x \in \Delta_{n-1}} \frac{\sum_1^m y_i \langle b_i, x \rangle}{\sum_1^m y_i \langle a_i, x \rangle}, \qquad (D)$$

so that (D) can be regarded as the dual problem to (P) and thus to the original economic growth problem. Also, the function

$$L(x,y) = \frac{\sum_1^m y_i\langle b_i, x\rangle}{\sum_1^m y_i\langle a_i, x\rangle}, \qquad x \in \Delta_{n-1},\ y \in \Delta_{m-1},$$

can be considered a Lagrangian function (albeit an unusual one).

(c) Retrace the arguments in (a) and (b) and prove that the dual problem can be written in the form

$$\begin{aligned} \min \quad & \nu \\ \text{s.t.} \quad & (\nu A - B)^T y \geq 0 \\ & \sum_1^m y_j = 1,\ y \geq 0. \end{aligned}$$

The novelty of this problem and the duality result is that the Lagrangian function L is not a concave–convex function, that is, $x \mapsto L(x,y)$ is not a concave function for a fixed y, and $y \mapsto L(x,y)$ is not a convex function for a fixed x. However, L is a quasi-concave–convex function, that is, the level sets $\{x : L(x,y) \geq \alpha\}$ and $\{y : L(x,y) \leq \alpha\}$ are convex sets for a fixed y and x, respectively. Minimax theorems generally apply to such functions on compact spaces; see [242], for example.

18. Consider Fermat's problem

$$\min_{x\in R^2} \|x - a_1\| + \|x - a_2 + \|x - a_3\|,$$

where a_1, a_2, and a_3 are the vertices of a triangle T in $\mathbb{R}^2$, and $\|\cdot\|$ is the Euclidean norm.

(a) Suppose that the optimal solution x^* is not at a vertex. Show that x^* must be the *Torricelli point* of the triangle; see Figure 11.1. *Hint:* Use the constraints of (11.13) and argue that $\|y_i^*\| = 1$.

(b) If the angle at a vertex of T is at least 120°, then show that this vertex is the optimal solution.

(c) Show that the dual of Fermat's problem is equivalent to determining the largest equilateral triangle circumscribing T.

19. **(Courant–Hilbert [64])** Consider the Euclidean distance problem

$$\begin{aligned} \min \quad & \frac{1}{2}\|p - p_0\|^2 \\ \text{s.t.} \quad & p \in q_0 + L, \end{aligned} \tag{11.21}$$

where $L \subset \mathbb{R}^n$ is a linear subspace.

(a) Show that the dual of (11.21) is equivalent to the Euclidean distance problem

$$\begin{aligned} \min \quad & \frac{1}{2}\|q - q_0\|^2 \\ \text{s.t.} \quad & q \in p_0 + L^{\perp}, \end{aligned} \tag{11.22}$$

where

$$L^{\perp} = \{z \in \mathbb{R}^n : \langle x, z\rangle = 0 \ \text{ for all } \ x \in L\}$$

is the orthogonal complement of L.
Hint: Write $L = N(A) = \{x : Ax = 0\}$ for some linear map $A : \mathbb{R}^n \to \mathbb{R}^m$, and construct the Lagrangian dual of (11.21).

(b) Show that the (unique) solutions to (11.21) and (11.22) are *identical*, that is, if p^* is the solution to (11.21) and q^* is the solution to (11.22), then $p^* = q^*$; see Figure 11.3.
Hint: Use the optimality conditions of the two optimization problems to obtain $p^* - q^* \in L$ and $p^* - q^* \in L^{\perp}$.

(c) Show that if p primal feasible ($p \in q_0 + L$) and q is dual feasible ($q \in p_0 + L^{\perp}$), then

$$\left\| p^* - \frac{p+q}{2} \right\| = \frac{1}{2}\|p - q\|.$$

(d) Generalize the above results to the problem

$$\begin{aligned} \min \quad & \frac{1}{2}\left\|Q(p - p_0)\right\|^2 \\ \text{s.t.} \quad & p \in q_0 + L, \end{aligned}$$

where Q is a symmetric positive definite matrix, and $L \subset \mathbb{R}^n$ is a linear subspace.
Hint: Define the inner product $\langle u, v\rangle_Q := \langle Qu, v\rangle$ and the corresponding norm $\|u\|_Q^2 = \langle Qu, u\rangle$.

20. *(Projection onto the standard unit simplex)* Consider once more the problem of projecting onto a simplex previously treated on page 190. Write the problem as the optimization problem

$$\min\{\|x - a\|^2/2 : \langle e, x\rangle = 1, x \geq 0\}.$$

Formulate the dual problem, and simplify it by eliminating the multiplier vector λ^* corresponding to the constraints $x \geq 0$. Solve the resulting search problem for the remaining multiplier μ^* as before.

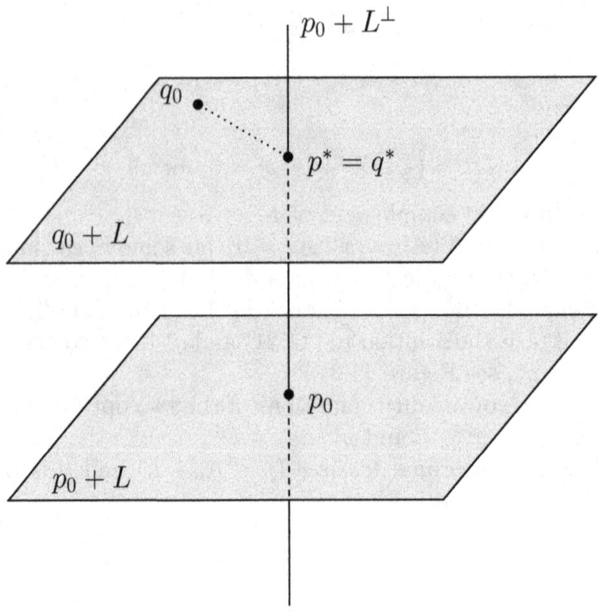

Fig. 11.3. Primal–dual least squares problem.

12

Semi-infinite Programming

Semi-infinite programs are constrained optimization problems in which the number of decision variables is finite, but the number of constraints is infinite. In this chapter, we treat a class semi-infinite programming problems in which the constraints are indexed by a compact set. We will demonstrate the usefulness of such problems by casting several important optimization problems in this form and then using semi-infinite programming techniques to solve them. Historically, Fritz John [148] initiated semi-infinite programming precisely to deduce important results about two such geometric problems: the problems of covering a compact body in $\mathbb{R}^n$ by the minimum-volume disk and the minimum-volume ellipsoid. In the same landmark paper, he derived what are now called *Fritz John optimality conditions* for this class of semi-infinite programs.

12.1 Fritz John Conditions for Semi-infinite Programming

We need the following technical result.

Lemma 12.1. *Let $K \subset \mathbb{R}^n$ be a compact set. Then $0 \in \operatorname{co}(K)$ if and only if*

$$\{h : \langle x, h\rangle < 0 \;\; \text{for all} \;\; x \in K\} = \emptyset. \tag{12.1}$$

Proof. If $0 \in \operatorname{co}(K)$, then $\sum_i \lambda_i x_i = 0$ for some $x_i \in K$ and $0 \leq \lambda \neq 0$. Then $\sum_i \lambda_i \langle x_i, h\rangle = 0$ and we cannot have $\langle x_i, h\rangle < 0$ for all i. Thus (12.1) holds.

Conversely, suppose that $0 \notin \operatorname{co}(K)$. It follows from Corollary 4.15 that $co(K)$ is compact, and from Theorem 6.10 that there exists $h \in \mathbb{R}^n$ such that $\langle h, x\rangle < 0$ for all $x \in \operatorname{co}(K)$, hence for all $x \in K$. □

Theorem 12.2. *Consider the semi-infinite program*

O. Güler, *Foundations of Optimization*, Graduate Texts in Mathematics 258,
DOI 10.1007/978-0-387-68407-9_12, © Springer Science+Business Media, LLC 2010

$$\begin{aligned} \min_{x \in X} \max_{y \in Y} \quad & f(x, y) \\ \text{s.t.} \quad & g(x, z) \le 0, \quad z \in Z, \end{aligned}$$

where $f(x,y)$ and $\nabla_x f(x,y)$ are continuous functions defined on the set $X \times Y$, where $X \subseteq \mathbb{R}^n$ is open and Y is a compact set in some topological space E, and $g(x,z)$ and $\nabla_x g(x,z)$ are continuous functions defined on the set $X \times Z$, where Z is a compact set in a topological space F.

If x^* is a local minimizer, then there exist vectors $\{y_i\}_{i \in I}$, $\{z_j\}_{j \in J}$, at most $n+1$ in number, satisfying

$$f(x^*, y_i) = \max_{y \in Y} f(x^*, y), \quad g(x^*, z_j) = 0,$$

and a nonzero, nonnegative vector (λ, μ) such that

$$\sum_{i \in I} \lambda_i \nabla_x f(x^*, y_i) + \sum_{j \in J} \mu_j \nabla_x g(x^*, z_j) = 0.$$

Proof. Define the functions

$$\begin{aligned} \varphi(x) &:= \max_{y \in Y} f(x, y), \\ \gamma(x) &:= \max_{z \in Z} g(x, z), \\ \vartheta(x) &:= \max\{\varphi(x) - \varphi(x^*), \gamma(x)\}. \end{aligned}$$

Note that $\vartheta(x^*) = 0$. If $\vartheta(x) < 0$ for some $x \in X$, then we must have $\varphi(x) < \varphi(x^*)$ and $g(x,z) < 0$ for all $z \in Z$. Since x^* is a local minimizer, this cannot happen if x is sufficiently close to x^*. Thus, we see that x^* is a local minimizer of ϑ. It follows from Theorem 1.29 (Danskin's theorem) that

$$\begin{aligned} 0 &\le \vartheta'(x^*; h) \\ &= \max\Big\{ \max_{y \in Y(x^*)} \langle \nabla_x f(x^*, y), h \rangle, \max_{z \in Z(x^*)} \langle \nabla_x g(x^*, y), h \rangle \Big\} \quad \text{for all} \quad h \in \mathbb{R}^n, \end{aligned}$$

where

$$Y(x^*) = \arg\max_{y \in Y} f(x^*, y), \quad Z(x^*) = \arg\max_{z \in Z} g(x^*, z).$$

This implies that there exists no direction $h \in \mathbb{R}^n$ satisfying the conditions

$$\langle \nabla_x f(x^*, y), h \rangle < 0, \ y \in Y(x^*), \ \ \langle \nabla_x g(x^*, z), h \rangle < 0, \ z \in Z(x^*).$$

Since $Y(x^*)$ and $Z(x^*)$ are compact, so is the set

$$K := \{\nabla_x f(x^*, y), \ y \in Y(x^*), \ \nabla_x g(x^*, z), z \in Z(x^*)\}.$$

It follows from Lemma 12.1 that $0 \in \text{co}(K)$. Theorem 4.13 implies that there exist at most $n+1$ vectors from K such that zero is in the convex hull of these vectors. □

The following two theorems are special cases of the above theorem.

Theorem 12.3. *Consider the minimax problem*

$$\min_{x \in X} \max_{y \in Y} f(x, y),$$

where $X \subseteq \mathbb{R}^n$ *is an open set,* Y *is a compact set in some topological space, and* $f(x, y)$, $\nabla_x f(x, y)$ *are continuous functions.*

If x^* *is a local minimizer, then there exist at most* $n+1$ *points* $\{y_i\}_1^k$ *satisfying*

$$f(x^*, y_i) = \max_{y \in Y} f(x^*, y),$$

and a nontrivial, nonnegative multiplier vector

$$0 \neq \lambda = (\lambda_1, \ldots, \lambda_k) \geq 0$$

such that

$$\sum_{i=1}^{k} \lambda_i \nabla_x f(x^*, y_i) = 0.$$

Theorem 12.4. *Consider the optimization problem*

$$\begin{aligned} \min \quad & f(x) \\ \text{s.t.} \quad & g(x, y) \leq 0, \quad y \in Y, \end{aligned} \tag{12.2}$$

where $f(x)$ *is a continuously differentiable function defined on an open set* $X \subseteq \mathbb{R}^n$, *and* $g(x, y)$ *and* $\nabla_x g(x, y)$ *are continuous functions defined on* $X \times Y$, *where* Y *is a compact set in some topological space.*

If x^* *is a local minimizer of* (12.2), *then there exist* k *active constraints* $\{g(x, y_i)\}_1^k$ *(that is,* $g(x^*, y_i) = 0$*), and a nontrivial, nonnegative multiplier vector*

$$\lambda^* := (\lambda_0^*, \lambda_1^*, \ldots, \lambda_k) \quad \textit{with at most } n+1 \textit{ positive entries}$$

such that

$$\lambda_0^* \nabla f(x^*) + \sum_{i=1}^{k} \lambda_i^* \nabla_x g(x^*, y_i) = 0.$$

In the following sections, we treat several important problems from geometry and analysis using semi-infinite programming techniques. These should serve as convincing examples of the power and importance of semi-infinite programming.

12.2 Jung's Inequality

We start with Jung's theorem on the relationship between the diameter and the inradius of a compact set $S \subset \mathbb{R}^n$. This is one of the problems treated in John's paper [148]. It is solved there by converting it to the semi-infinite program

$$\min_{x,z}\{z : \|x-y\|^2 - z \leq 0 \text{ for all } y \in S\},$$

which is in a form suitable for applying Theorem 12.4. The details of the solution below are very similar to John's, but we formulate the problem in a somewhat more natural form and apply Theorem 12.3 instead.

Define the *diameter* of S as

$$D(S) := \max\{\|x-y\| : x, y \in S\},$$

and the *inradius* the radius of the smallest ball containing S,

$$R(S) := \min_{x\in\mathbb{R}^n} \max_{y\in S} \|x-y\|.$$

Theorem 12.5. **(*Jung's inequality*)** *Let S be a compact set in R^n. The inequality*

$$D(S) \geq \sqrt{\frac{2(n+1)}{n}}\,R(S) \tag{12.3}$$

holds between the diameter $D(S)$ and the inradius $R(S)$.

Proof. Notice that $R(S)^2 = z^*$, where z^* is the optimal objective value of the minimax problem

$$\min_{x\in\mathbb{R}^n} \max_{y\in S} \|x-y\|^2.$$

Define

$$\varphi(x) = \max_{y\in S} \|x-y\|^2.$$

Evidently, $\varphi(x)$ is coercive, and so has a global minimizer x^* by the Weierstrass theorem. It follows from Theorem 12.3 that there exist k $(k \leq n+1)$ points $\{y_i\}_1^k$ in S satisfying $\varphi(x^*) = \|x^* - y_i\|^2$, and multipliers $0 \neq \lambda = (\lambda_1, \ldots, \lambda_k) \geq 0$ such that

$$\sum_{i=1}^{k} \lambda_i (x^* - y_i) = 0.$$

Without loss of generality, we assume that $\lambda_i > 0$ for all i and $\sum_{i=1}^{k} \lambda_i = 1$. Thus,

$$x^* = \sum_{i=1}^{k} \lambda_i y_i^*,$$

that is, x^*, the center of the enclosing optimal ball, is in the convex hull of the points $\{y_i\}_1^k$.

We compute

$$\begin{aligned}
\sum_{1\le i\neq j\le k} \lambda_i\lambda_j\|y_i-y_j\|^2 &= \sum_{1\le i,j\le k} \lambda_i\lambda_j\|y_i-y_j\|^2 \\
&= \sum_{1\le i,j\le k} \lambda_i\lambda_j\|(y_i-x^*)-(y_j-x^*)\|^2 \\
&= \sum_{j=1}^k \lambda_j\Big(\sum_{i=1}^k \lambda_i\|y_i-x^*\|^2\Big)+\sum_{i=1}^k \lambda_i\Big(\sum_{j=1}^k \lambda_j\|y_j-x^*\|^2\Big) \\
&\quad -2\Big\langle \sum_{i=1}^k \lambda_i(y_i-x^*), \sum_{j=1}^k \lambda_j(y_j-x^*)\Big\rangle \\
&= 2z^* = 2R(S)^2.
\end{aligned}$$

By the Cauchy–Schwarz inequality,

$$1=\Big(\sum_{i=1}^k \lambda_i\Big)^2 \le k\sum_{i=1}^k \lambda_i^2,$$

so that

$$\sum_{1\le i\neq j\le k} \lambda_i\lambda_j = 1-\sum_1^k \lambda_i^2 \le 1-\frac{1}{k}=\frac{k-1}{k}.$$

Consequently, we have

$$\begin{aligned}
2R(S)^2 &= \sum_{1\le i\neq j\le k} \lambda_i\lambda_j\|y_i-y_j\|^2 \le \Big(\sum_{1\le i\neq j\le k} \lambda_i\lambda_j\Big)\cdot \max_{i,j}\|y_i-y_j\|^2 \\
&\le \frac{k-1}{k}D(S)^2.
\end{aligned}$$

The theorem is proved. □

The inequality (12.3) is sharp and is attained for the standard unit simplex. In this case, we have $D(\Delta_n)=\|(1,0,\dots,0)-(0,1,0,\dots,0)\|=\sqrt{2}$, $R(\Delta_n)=\|1/(n+1)(1,1,\dots,1)-(1,0,\dots,0)\|=\sqrt{n/(n+1)}$, and $D(\Delta_n)=\sqrt{2(n+1)/n}R(\Delta_n)$.

12.3 The Minimum-Volume Circumscribed Ellipsoid Problem

The *circumscribed ellipsoid problem* is the problem of finding a minimum-volume ellipsoid circumscribing a convex body K in $\mathbb{R}^n$. This is the main problem treated in Fritz John [148]. In that paper, John shows that such an ellipsoid exists and is unique; let us denote it by E^K. John introduces semi-infinite programming and develops his optimality conditions to prove the following deep result about the ellipsoid E^K: the ellipsoid with the same center as E^K but shrunk by a factor n is *contained* in K, and if K is symmetric ($K = -K$), then E^K needs to be shrunk by the smaller factor $\sqrt{n}$ to be contained K. This fact is very important in the geometric theory of Banach spaces. In that theory, a symmetric convex body K is considered the unit ball of a Banach space, and if K is an ellipsoid, then the Banach space is a Hilbert space. Consequently, the shrinkage factor indicates how close the Banach space is to being a Hilbert space. In this context, it is not important to compute the exact ellipsoid E^K.

However, in some convex programming algorithms including the ellipsoid method and its variants, the exact or nearly exact ellipsoid E^K needs to be computed. If K is sufficiently simple, E^K can computed analytically. In more general cases, interior-point-type algorithms can be developed to approximately compute E^K. This, however, is a relatively challenging task.

In this section, we deal with the E^K problem more or less following John's approach. However, in the interest of brevity and clarity, we use more modern notation and give new and simpler proofs for some of the technical results, including for John's containment results mentioned above.

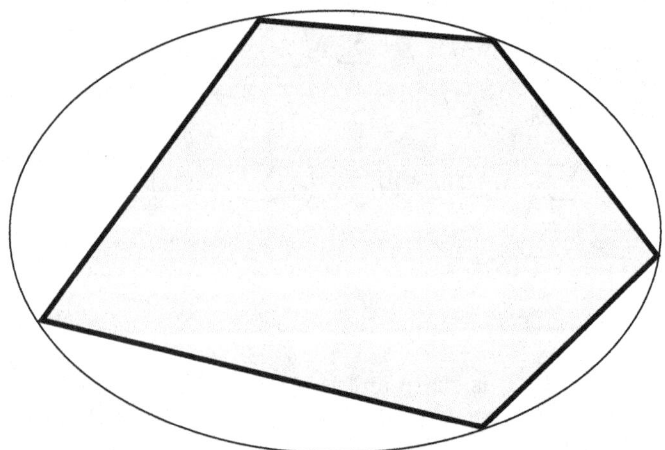

Fig. 12.1. Circumscribed ellipsoid around a convex body.

An ellipsoid E with center c is an affine image of the unit ball

$$B := \{u \in \mathbb{R}^n : \|u\| \leq 1\},$$

that is, $E = c + A(B)$, where A is a nonsingular $n \times n$ matrix. We may assume that A is a symmetric positive definite matrix: if $A = U\Lambda V^T$ is the singular-value decomposition of A, we can write

$$A = (U\Lambda U^T)(UV^T) = XO,$$

where $X := U\Lambda U^T$ is a symmetric positive definite matrix and $O = UV^T$ is an orthogonal matrix. Since $O(B) = B$, we have $E = c + X(O(B)) = c + X(B)$.

For technical reasons, we will represent E in the form $E = c + X^{-1/2}(B)$, where X a symmetric positive definite matrix. Setting $v = c + X^{-1/2}u$, that is, $u = X^{1/2}(v - c)$, we have $E = \{v : \|X^{1/2}(v - c)\| \leq 1\}$. We denote this ellipsoid by $E(X, c)$,

$$E(X, c) := \{v : \langle X(v - c), v - c\rangle \leq 1\}. \tag{12.4}$$

Since $E(X, c) = c + X^{-1/2}(B)$,

$$\mathrm{vol}(E(X, c)) = (\det X)^{-1/2}\omega_n, \tag{12.5}$$

where $\omega_n = \mathrm{vol}(B)$. Consequently, we can set up the circumscribed ellipsoid problem as a semi-infinite program

$$\begin{aligned} \min \quad & -\ln\det X \\ \text{s.t.} \quad & \langle X(y - c), y - c\rangle \leq 1, \quad y \in K, \end{aligned} \tag{12.6}$$

in which the decision variables are $(X, c) \in S^n \times \mathbb{R}^n$, where S^n is the vector space of $n \times n$ symmetric matrices.

Lemma 12.6. *If $K \subset \mathbb{R}^n$ is a convex body, then there exists an ellipsoid of minimum volume circumscribing K.*

Proof. We claim that the feasible sublevel sets in the problem (12.6) are compact; then the Weierstrass theorem implies that (12.6) has an optimal solution. Being a convex body, K contains a ball of radius r, which is then contained in every ellipsoid $E(X, c) \supseteq K$ containing K. Note that every such ellipsoid $E(X, c)$ contains the ball of radius r centered at c. This means that every vector $u \in \mathbb{R}^n$, $\|u\| = r$, must satisfy the inequality $\langle Xu, u\rangle \leq 1$. Picking $u = re_i$ gives $0 < X_{ii} \leq 1/r^2$, for all $i = 1, \ldots, n$. Since X is symmetric positive definite, a 2×2 submatrix $\begin{bmatrix} X_{ii} & X_{ij} \\ X_{ij} & X_{jj} \end{bmatrix}$ is positive semidefinite. This implies that $X_{ij}^2 \leq X_{ii}X_{jj} \leq 1/r^4$, meaning that all entries of X are bounded above by $1/r^2$ in absolute value. Thus, the feasible matrices X in problem (12.6) form a compact set. As the norm of the center c of an ellipsoid $E(X, c)$ circumscribing K goes to infinity, the volume of the ellipsoid must obviously go to infinity as well. This proves the claim and the lemma. □

We now derive the optimality conditions for (12.6).

Theorem 12.7. *Let $K \subset \mathbb{R}^n$ be a convex body. There exists an ellipsoid of minimum volume circumscribing K. If $E(X, c)$ is such an ellipsoid, then there* *exist a multiplier vector* $\lambda = (\lambda_1, \ldots, \lambda_k) > 0$, $0 \le k \le n(n+3)/2$, *and contact points* $\{u_i\}_1^k$ *in* K *such that*

$$\begin{aligned} X^{-1} &= \sum_{i=1}^{k} \lambda_i (u_i - c)(u_i - c)^T, \\ 0 &= \sum_{i=1}^{k} \lambda_i (u_i - c), \\ u_i &\in \partial K \cap \partial E(X, c), \quad i = 1, \ldots, k, \\ K &\subseteq E(X, c). \end{aligned} \tag{12.7}$$

Proof. The existence of a minimum-volume ellipsoid has already been proved in Lemma 12.6. Let $E(X, c)$ be such an ellipsoid. Since the constraints in (12.6) are indexed by $y \in K$, a compact set, Theorem 12.4 applies. Therefore, there exist a nonzero multiplier vector $(\lambda_0, \lambda_1, \ldots, \lambda_k) \ge 0$, where $k \le n(n+1)/2 + n = n(n+3)/2$, $\lambda_i > 0$ for $i > 0$, and points $\{u_i\}_1^k$ in K such that the Lagrangian function

$$\begin{aligned} L(X, c, \lambda) &:= -\lambda_0 \ln \det X + \sum_{i=1}^{k} \lambda_i \langle X(u_i - c), u_i - c \rangle \\ &= -\lambda_0 \ln \det X + \Big\langle X, \sum_{i=1}^{k} \lambda_i (u_i - c)(u_i - c)^T \Big\rangle, \end{aligned}$$

where the inner product on the last line is the trace inner product on S^n, satisfies the optimality conditions

$$0 = \nabla_c L(X, c, \lambda) = X \sum_{i=1}^{k} \lambda_i (u_i - c),$$

$$0 = \nabla_X L(X, c, \lambda) = -\lambda_0 X^{-1} + \sum_{i=1}^{k} \lambda_i (u_i - c)(u_i - c)^T.$$

Here we used the first formula for $L(X, c, \lambda)$ to differentiate L with respect to c, and the second formula for $L(X, c, \lambda)$ together with (1.13) to differentiate it with respect to X.

If $\lambda_0 = 0$, then $0 = \mathrm{tr}(\sum_{i=1}^{k} \lambda_i (u_i - c)(u_i - c)^T) = \sum_{i=1}^{k} \lambda_i \|u_i - c\|^2$. This implies that $\lambda_i = 0$ for all i, contradicting $\lambda \ne 0$. We let $\lambda_0 = 1$ without loss of generality, and arrive at the Fritz John conditions (12.7). □

For most theoretical purposes, we may assume that the optimal ellipsoid is the unit ball $E(I, 0)$. This can be accomplished by an affine change of coordinates, if necessary. This results in the more transparent optimality conditions

$$\begin{aligned} &I = \sum_{i=1}^{k} \lambda_i u_i u_i^T, \quad \sum_{i=1}^{k} \lambda_i u_i = 0, \\ &u_i \in \partial K \cap \partial B, \quad i = 1, \ldots, k, \quad K \subseteq B. \end{aligned} \tag{12.8}$$

Taking traces of both sides in the first equation above gives

$$n = \operatorname{tr}(I) = \operatorname{tr}\Big(\sum_{i=1}^{k} \lambda_i u_i u_i^T\Big) = \sum_{i=1}^{k} \lambda_i u_i^T u_i = \sum_{i=1}^{k} \lambda_i.$$

In this section and the next, convex duality will play an important role. If $C \subset \mathbb{R}^n$ is a convex body, the Minkowski *support function* is defined by

$$s_C(d) := \max_{u \in C} \langle d, u \rangle.$$

It is obviously defined on $\mathbb{R}^n$ and is a convex function, since it is a maximum of linear functions indexed by u. By duality results of convex programming, two convex bodies satisfy $C \subseteq D$ if and only if $s_C \le s_D$. (In fact, $s_C = \delta_C^*$, where δ_C is the indicator function of C, and $*$ denotes the Fenchel dual. By the fundamental theorem on the Fenchel dual functions, we have $s_C^* = \delta^{**} = \delta_C$; see [228] or [89] for further details. This implies our assertions.) We compute

$$\begin{aligned} s_{E(X,c)}(d) &= \max\left\{\langle d, u\rangle : \langle X(u-c), u-c\rangle \le 1\right\} \\ &= \max\left\{\langle d, c + X^{-1/2} v\rangle : \|v\| \le 1\rangle\right\} \\ &= \langle c, d\rangle + \max_{\|v\| \le 1} \langle X^{-1/2} d, v\rangle = \langle c, d\rangle + \|X^{-1/2} d\| \\ &= \langle c, d\rangle + \langle X^{-1} d, d\rangle^{1/2}, \end{aligned} \tag{12.9}$$

where we have defined $v := X^{1/2}(u-c)$, or $u = c + X^{-1/2} v$.

The polar the set C is defined by

$$C^* := \{d : s_C(d) \le 1\} = \{x : \langle x, u \le 1 \ \text{ for all } \ u \in C\}.$$

An easy calculation shows that

$$\left(\operatorname{co}\left(\{u_i\}_1^k\right)\right)^* = \{x : \langle x, u_i\rangle \le 1, \ i = 1, \ldots, k\}.$$

The following *key* result shows that the optimality conditions (12.8) are powerful enough to prove the uniqueness of the minimum-volume circumscribing ellipsoid in this section and the uniqueness of the maximum-volume inscribed ellipsoid in the next section.

Lemma 12.8. *Let $\{u_i\}_1^k$ be set of unit vectors in $\mathbb{R}^n$ satisfying the conditions $\sum_{i=1}^k \lambda_i u_i u_i^T = I$ and $\sum_{i=1}^k \lambda_i u_i = 0$. Define the polytope $P = \operatorname{co}\left(\{u_i\}_1^k\right)$ and its polar $P^* = \{x : \langle u_i, x\rangle \le 1, i = 1, \ldots, k\}$. The unit ball is both the unique minimum-volume ellipsoid circumscribing P and the unique maximum-volume ellipsoid inscribed in P^*.*

Proof. Let $E(X,c)$ be any ellipsoid covering the points $\{u_i\}_1^k$. We have $\langle X(u_i - c), u_i - c\rangle \le 1$ and

$$
\begin{aligned}
n &= \sum_{i=1}^{k} \lambda_i \geq \sum_{i=1}^{k} \lambda_i \langle X(u_i - c), u_i - c \rangle = \Big\langle X, \sum_{i=1}^{k} \lambda_i (u_i - c)(u_i - c)^T \Big\rangle \\
&= \Big\langle X, \sum_{i=1}^{k} \lambda_i u_i u_i^T \Big\rangle - \Big\langle X, \sum_{i=1}^{k} \lambda_i c u_i^T \Big\rangle - \Big\langle X, \sum_{i=1}^{k} \lambda_i u_i c^T \Big\rangle + \Big\langle X, (\sum_{i=1}^{k} \lambda_i) c c^T \Big\rangle \\
&= \langle X, I \rangle + n \langle X, cc^T \rangle = \operatorname{tr}(X) + n \langle Xc, c \rangle \\
&\geq n \left(\det(X)^{1/n} + \langle Xc, c \rangle \right).
\end{aligned}
$$

Here the fourth equality follows from (12.8), and the last inequality follows from the fact that $\det(X)^{1/n} \leq \operatorname{tr}(X)/n$, which is precisely the arithmetic–geometric mean inequality applied to the eigenvalues of X. Thus

$$
\det(X)^{1/n} + \langle Xc, c \rangle \leq 1,
$$

and the equality $\det(X) = 1$ holds if and only if $c = 0$, $\langle X(u_i - c), u_i - c \rangle = 1$ for all $i = 1, \ldots, k$, and the arithmetic–geometric mean inequality holds as an equality. The last condition holds if and only if X is a positive multiple of the identity matrix (and then $\det(X) = 1$ implies $X = I$). Thus, the minimum-volume ellipsoid covering the points $\{u_i\}_1^k$ must be the unit ball.

Next, let $E(X, c)$ be any ellipsoid inscribed in P^*. By (12.5), $\operatorname{vol}(E(X, c)) = \det(X^{-1}) \omega_n$. By virtue of (12.9), the inclusion $E(X, c) \subseteq P^*$ implies

$$
s_{E(X,c)}(u_i) = \langle c, u_i \rangle + \|X^{-1/2} u_i\| \leq s_{P^*}(u_i) = \max_j \langle u_i, u_j \rangle \leq 1, \quad i = 1, \ldots, k.
$$

By the Cauchy–Schwarz inequality, we have $\langle X^{-1/2} u_i, u_i \rangle \leq \|X^{-1/2} u_i\| \cdot \|u_i\| = \|X^{-1/2} u_i\|$; therefore,

$$
\begin{aligned}
n &= \sum_{i=1}^{k} \lambda_i \geq \sum_{i=1}^{k} \lambda_i \left(\langle c, u_i \rangle + \langle X^{-1/2} u_i, u_i \rangle \right) \\
&= \Big\langle X^{-1/2}, \sum_{i=1}^{k} \lambda_i u_i u_i^T \Big\rangle = \operatorname{tr}(X^{-1/2}) \geq n \det(X)^{-1/2n},
\end{aligned}
$$

where the last inequality follows from the arithmetic–geometric mean inequality applied to the eigenvalues of $X^{-1/2}$. Thus $\det X \geq 1$, and the equality $\det X = 1$ holds if and only if (i) X is a positive multiple of the identity matrix (and then $\det X = 1$ implies $X = I$), and (ii) $1 = \langle c, u_i \rangle + \langle X^{-1/2} u_i, u_i \rangle = \langle c, u_i \rangle + 1$, that is, $\langle c, u_i \rangle = 0$ for all $i = 1, \ldots, k$. Then the equation $\sum_{i=1}^{k} \lambda_i u_i u_i^T = I$ implies that $\|c\|^2 = \sum_{i=1}^{k} \lambda_i \langle c, u_i \rangle^2 = 0$. The lemma is proved. □

Theorem 12.9. *Let K be a convex body in $\mathbb{R}^n$. The minimum-volume ellipsoid circumscribing K is unique. Moreover, the optimality conditions (12.7) are necessary and sufficient for an ellipsoid $E(X, c)$ to be the minimum-volume ellipsoid circumscribing K.*

Proof. The necessity of the conditions (12.7) has already been proved in Theorem 12.7. To prove the sufficiency, we assume, without loss of generality, that $E(I,0)=B$ satisfies the optimality conditions (12.8) for some set of multipliers $\{\lambda_i\}$. Let $E \supseteq K$ be a minimum-volume ellipsoid circumscribing K. Since $P \subseteq K \subseteq E$, $P \subseteq K \subseteq B$, and E is the optimal covering ellipsoid for K, we have $\mathrm{vol}(E) \le \mathrm{vol}(B)$. Similarly, since B is the optimal covering ellipsoid for P by Lemma 12.8, $\mathrm{vol}(B) \le \mathrm{vol}(E)$. These give $\mathrm{vol}(E) = \mathrm{vol}(B)$, and the uniqueness of the covering ellipsoid for P implies $E = B$. □

Remark 12.10. The uniqueness of the circumscribed ellipsoid E^K can be seen by recasting problem (12.6) by setting $X = Y^2$ and $d := Y^{-1}c$. Then we have the semi-infinite program

$$\begin{aligned} \min \quad & -\ln\det Y \\ \text{s.t.} \quad & \|Yy - d\|^2 \le 1, \quad y \in K, \end{aligned} \tag{12.10}$$

with decision variables (Y,d). This is a convex semi-infinite program, since the objective function is strictly convex in Y by (1.13), and the constraint for each parameter $y \in K$ is convex. This proves the sufficiency of the conditions in (12.7) and the uniqueness of the matrix X in $E(X,c)$. The uniqueness of the center c can then be completed by a special argument.

We end this section by proving Fritz John's results mentioned at the beginning of the chapter. Our proof is simpler, and uses ideas from Ball [19] and Juhnke [150].

Theorem 12.11. *Let K be a convex body in $\mathbb{R}^n$ and $E(X,c) = E^K$ its optimal circumscribing ellipsoid. The ellipsoid with the same center c but shrunk by a factor n is contained in K. If K is symmetric, that is, $K = -K$, then the ellipsoid with the same center c but shrunk by a factor of only $\sqrt{n}$ is contained in K.*

Proof. Without loss of generality, we assume that $E^K = E(I,0) = B$. The theorem states that $n^{-1}B \subseteq K$. Let

$$P = \mathrm{co}\left(\{u_i\}_1^k\right)$$

be the convex hull of the contact points. We claim that the stronger statement $n^{-1}B \subseteq P$ holds. Since $P \subseteq K$, we will then have $n^{-1}B \subseteq K$. By duality, the claim is equivalent to showing that the polar sets satisfy $P^* \subseteq (n^{-1}B)^* = nB$. Let $x \in P^*$. Since $-\|x\| = -\|x\| \cdot \|u_i\| \le \langle x, u_i\rangle \le 1$, we have

$$\begin{aligned} 0 &\le \sum_{i=1}^k \lambda_i (1 - \langle x, u_i\rangle)(\|x\| + \langle x, u_i\rangle) \\ &= \Big(\sum_{i=1}^k \lambda_i\Big)\|x\| - \sum_{i=1}^k \lambda_i (\langle x, u_i\rangle)^2 = n\|x\| - \|x\|^2, \end{aligned}$$

where the second equality follows from $\sum_i \lambda_i = n$ and (12.8). This implies $\|x\| \leq n$, and proves that $P^* \subseteq nB$.

If K is symmetric, we define $Q = \operatorname{co}(\{\pm u_i\}_1^k) \subseteq K$, and claim that $n^{-1/2}B \subseteq Q$, or equivalently, that $Q^* \subseteq (n^{-1/2}B)^* = \sqrt{n}B$. It is easily shown that $Q^* = \{x : |\langle x, u_i\rangle| \leq 1, i = 1, \ldots, k\}$. Let $x \in Q^*$. Since $-1 \leq \langle x, u_i\rangle \leq 1$, we have

$$0 \leq \sum_{i=1}^{k} \lambda_i(1 - \langle x, u_i\rangle)(1 + \langle x, u_i\rangle) = n - \|x\|^2.$$

This gives $\|x\| \leq \sqrt{n}$ and proves the claim. □

12.4 The Maximum-Volume Inscribed Ellipsoid Problem

The *inscribed ellipsoid problem* is the problem of finding a maximum-volume ellipsoid inscribed in a convex body K in $\mathbb{R}^n$. It will be seen that this ellipsoid is unique as well, and we denote it by E_K. This ellipsoid is referred to as the *John ellipsoid* in the Banach space literature, and a reference is given to Fritz John [148], although the inscribed ellipsoid problem is not treated in John's paper.

In this section, we again use semi-infinite programming to treat this problem. The inscribed ellipsoid has properties similar to those of the circumscribed ellipsoid. For example, the ellipsoid with the same center but blown up n times contains K, and in the case $K = -K$ is symmetric, the ellipsoid needs to be blown up by a smaller factor $\sqrt{n}$. The ellipsoid E_K is very useful in the geometric theory of Banach spaces. It is also useful in some convex programming algorithms, such as the *inscribed ellipsoid method* of Tarasov, Erlikh, and Khachiyan [251].

As a first step, using (12.5), we can formulate the inscribed ellipsoid problem as a semi-infinite program

$$\max\{\det X : E(X, c) \subseteq K\}.$$

However, this is hard to work with, due to the inconvenient form of the constraints $E(X, c) \subseteq K$. Using support functions, we convert them into inequalities

$$\langle c, d\rangle + \langle X^{-1}d, d\rangle^{1/2} = s_{E(X,c)}(d) \leq s_K(d), \quad \|d\| = 1,$$

where we restrict d to the unit sphere, since support functions are homogeneous (of degree 1).

Defining $Y = X^{-1}$, we can therefore rewrite our semi-infinite program in the form

$$\begin{aligned} \min \quad & -\ln\det Y \\ \text{s.t.} \quad & \langle c, d\rangle + \langle Yd, d\rangle^{1/2} \leq s_K(d), \quad \|d\| = 1, \end{aligned} \tag{12.11}$$

in which the decision variables are $(Y, c) \in S^n \times \mathbb{R}^n$ and we have infinitely many constraints indexed by the unit vector $\|d\| = 1$.

Since s_K is a convex function on $\mathbb{R}^n$, it is continuous. Therefore, there exists a positive constant $M > 0$ such that if (Y, c) is a feasible decision variable, then $|\langle c, d\rangle| \leq M$, and $\langle Yd, d\rangle \leq M$ for all $\|d\| = 1$. This proves that the feasible set of (Y, c) for (12.11) is compact, and implies that there exists a maximum-volume ellipsoid inscribed in K.

We derive the optimality conditions for the maximum-volume inscribed ellipsoid.

Theorem 12.12. *Let $K \subset \mathbb{R}^n$ be a convex body. There exists an ellipsoid of maximum-volume inscribed in K. If $E(X, c)$ is such an ellipsoid, then there exist a multiplier vector $\lambda = (\lambda_1, \ldots, \lambda_k) > 0$, $0 \leq k \leq n(n+3)/2$, and contact points $\{u_i\}_1^k$ such that*

$$\begin{aligned} X^{-1} &= \sum_{i=1}^{k} \lambda_i (u_i - c)(u_i - c)^T, \\ 0 &= \sum_{i=1}^{k} \lambda_i (u_i - c), \\ u_i &\in \partial K \cap \partial E(X, c), \quad i = 1, \ldots, k, \\ E(X, c) &\subseteq K. \end{aligned} \tag{12.12}$$

Proof. The existence of a maximum-volume ellipsoid inscribed in K has already been proved above. Let $E(X, c)$ denote this ellipsoid, and define $Y = X^{-1}$. Since the constraints in (12.11) are indexed by $\|d\| = 1$, Theorem 12.4 applies. Therefore, there exist a nonzero multiplier vector $(\delta_0, \delta_1, \ldots, \delta_k) \geq 0$, where $k \leq n(n+3)/2$, $\delta_i > 0$ for $i > 0$, and directions $\{d_i\}_1^k$, $\|d_i\| = 1$, satisfying the conditions

$$\langle c, d_i\rangle + \langle Yd_i, d_i\rangle^{1/2} = s_K(d_i)$$

such that the Lagrangian function

$$L(Y, c, \delta) := -\delta_0 \ln \det Y + 2 \sum_{i=1}^{k} \delta_i \left[\langle c, d_i\rangle + \langle Yd_i, d_i\rangle^{1/2} - s_K(d_i) \right]$$

satisfies the optimality the conditions

$$0 = \frac{1}{2} \nabla_c L(Y, c, \delta) = \sum_{i=1}^{k} \delta_i d_i,$$

$$0 = \nabla_Y L(Y, c, \delta) = -\delta_0 Y^{-1} + \sum_{i=1}^{k} \frac{\delta_i}{\langle Yd_i, d_i\rangle^{1/2}} d_i d_i^T.$$

Recalling that $\|d_i\| = 1$ and taking the trace of the right-hand side of the last equation above gives $\delta_0 \operatorname{tr}(Y^{-1}) = \sum_{i=1}^k \delta_i \langle Yd_i, d_i\rangle^{-1/2}$. If δ_0, then all $\delta_i = 0$, which contradicts $\delta \neq 0$. Therefore, $\delta_0 \neq 0$, and we let $\delta_0 = 1$. Define

$$u_i := c + \frac{Yd_i}{\langle Yd_i, d_i\rangle^{1/2}}, \quad \lambda_i := \langle Yd_i, d_i\rangle^{1/2}\delta_i, \quad i = 1, \ldots, k.$$

Note that

$$s_{E(X,c)}(d_i) = s_K(d_i) = \langle c, d_i\rangle + \langle Yd_i, d_i\rangle^{1/2} = \langle d_i, u_i\rangle,$$

which means that $u_i \in \partial K \cap \partial E(X, c)$, that is, u_i is a contact point. Rewriting the above optimality conditions in terms of $\{u_i\}$ and $\{\lambda_i\}$ and simplifying, we arrive at the conditions (12.12). □

As in the circumscribed ellipsoid case, we can simplify these conditions by assuming that the optimal ellipsoid is the unit ball $E(I, 0)$. Then the Fritz John conditions become

$$\begin{aligned} I = \sum_{i=1}^k \lambda_i u_i u_i^T, \quad 0 = \sum_{i=1}^k \lambda_i u_i, \\ u_i \in \partial K \cap \partial B, \quad i = 1, \ldots, k, \quad K \subseteq B. \end{aligned} \tag{12.13}$$

We note that the optimality conditions (12.13) are exactly the *same* as the corresponding optimality conditions (12.8) in the circumscribed ellipsoid case, except for the feasibility constraint $K \subseteq B$.

Theorem 12.13. *Let K be a convex body in $\mathbb{R}^n$. The maximum-volume ellipsoid inscribed in K is unique. Furthermore, the optimality conditions* (12.12) *are necessary and sufficient for an ellipsoid $E(X, c)$ to be the maximum-volume ellipsoid inscribed in K.*

The proof uses Lemma 12.8. It is omitted, since it is very similar to the proof of Theorem 12.9.

We end this section by proving an analogue of Fritz John's containment results concerning E^K.

Theorem 12.14. *Let K be a convex body in $\mathbb{R}^n$ and let $E(X, c) = E_K$ be its optimal inscribed ellipsoid. The ellipsoid with the same center c but blown up by a factor n contains K. If $K = -K$ is symmetric, then the ellipsoid with the same center c but blown up by a factor $\sqrt{n}$ contains K.*

Proof. The proof here is similar to the proof of Theorem 12.11. Without loss of generality, we assume that $E_K = E(I, 0) = B$. The first part of the theorem follows if we can prove the claim that

$$K \subseteq P^* \subseteq nB.$$

Since $1 = s_B(u_i) = s_K(u_i) = \max_{x \in K}\langle u_i, x\rangle$, the first inclusion holds. If $x \in P^*$, then $-\|x\| = -\|x\| \cdot \|u_i\| \le \langle x, u_i\rangle \le 1$, and

$$0 \le \sum_{i=1}^{k} \lambda_i (1 - \langle x, u_i\rangle)(\|x\| + \langle x, u_i\rangle) = n\|x\| - \|x\|^2,$$

where the equality follows from $\sum_i \lambda_i = n$ and (12.13). This implies $\|x\| \le n$, and proves the second inclusion in the claim.

If K is symmetric, we define $Q = \mathrm{co}\left(\{\pm u_i\}_1^k\right) \subseteq K$ and claim that $K \subseteq Q^* \subseteq \sqrt{n}B$. Since $1 = s_K(\pm u_i)$, we have $|\langle u_i, x\rangle| \le 1$, and the first inclusion follows. To prove the second inclusion in the claim, let $x \in Q^*$. We have $|\langle x, u_i\rangle| \le 1$, and the rest of the proof follows as in the proof of Theorem 12.11. □

12.5 Chebyshev's Approximation Problem

In this section, we solve *Chebyshev's approximation problem* using semi-infinite programming techniques. Another interesting approach to tackling this and similar problems is given in Section 13.2.1 using Helly's theorem in convex analysis.

Let $I := [\alpha, \beta]$ be an interval. The supremum norm of a continuous function $f : I \to \mathbb{R}$ is

$$\|f\| := \max_{t \in [\alpha,\beta]} |f(t)|.$$

Chebyshev's approximation problem is one of the central problems in *approximation theory*. One version of it is formulated as follows. Let γ be a point not in I. We are looking for the nth-degree polynomial $q(t)$ whose supremum norm is minimal among all nth-degree polynomials $p(t)$ that satisfy the condition $p(\gamma) = 1$.

Mathematically, this is the minimax problem

$$\min_{p \in \mathcal{P}_n} \max_{t \in [\alpha,\beta]} |p(t)|, \tag{12.14}$$

where

$$\mathcal{P}_n := \{p(t) = a_n t^n + \cdots + a_1 t + a_0 : p(\gamma) = 1\}.$$

The solution is given by *Chebyshev's polynomials* as described below.

Theorem 12.15. *Let $I = [\alpha, \beta]$ be an interval and γ a point not in I. The optimal solution of the problem (12.14) is given by the polynomial*

$$q(t) = \frac{T_n\left(1 + 2\frac{t-\beta}{\beta-\alpha}\right)}{T_n\left(1 + 2\frac{\gamma-\beta}{\beta-\alpha}\right)}, \tag{12.15}$$

where

$$T_n(t) := \cos(n\cos^{-1} t)$$

is Chebyshev's polynomial of the first kind *of degree* n.

Moreover, the optimal objective value of the problem (12.14) *is*

$$\|q\| = \frac{1}{\left|T_n\left(1+2\frac{\gamma-\beta}{\beta-\alpha}\right)\right|} = \frac{1}{\left|T_n\left(2\frac{\gamma-\mu}{\beta-\alpha}\right)\right|},$$

where $\mu = \frac{\alpha+\beta}{2}$ *is the midpoint of the interval* $[\alpha, \beta]$.

Proof. Note that the linear transformation $l(t) = 1+2\frac{t-\beta}{\beta-\alpha}$ maps the interval $[\alpha, \beta]$ into the interval $[-1, 1]$ and sends the point γ into the point $\bar{\gamma} = 1 + 2\frac{\gamma-\beta}{\beta-\alpha}$. If $q(t)$ and $\bar{q}(t)$ are the optimal solutions of (12.15) with parameters $([\alpha, \beta], \gamma)$ and $([-1, 1], \bar{\gamma})$, respectively, then $\bar{q}(t) = q(l(t))$. Thus, it suffices to consider the case $([-1, 1], \gamma)$ from the start, and show that the optimal solution is $q(t) = T_n(t)/T_n(\gamma)$ and $\|q\| = \frac{1}{|T_n(\gamma)|}$ in that case.

We first demonstrate the existence of a solution. This will follow if we can show that a sublevel set of polynomials $L_\eta := \{p \in \mathcal{P}_k : \|p\| \leq \eta\}$ is compact. Let $\{p^i\}_1^\infty$ be a sequence of polynomials in L_η. We claim that the sequence of the coefficient vectors $a^i = (a^i_n, \ldots, a^i_1, a^i_0)$ is bounded. If the claim is true, then a subsequence $\{a^{i_j}\}$ can be extracted from $\{a^i\}$ converging to a vector $\bar{a} = (\bar{a}_n, \ldots, \bar{a}_1, \bar{a}_0)$. Then the polynomials p^{i_j} converge to $p(t) = \sum_{k=0}^n \bar{a}_k t^k$ in L_η, and this will show that L_η is compact. Suppose that the claim is not true. Then we there exists a subsequence $\{i_j\}$ such that $\|a^{i_j}\| \to \infty$ and $a^{i_j}/\|a^{i_j}\| \to \bar{a} \neq 0$. We have

$$\frac{p^{i_j}(t)}{\|a^{i_j}\|} = \sum_{k=0}^{n} \frac{a_k^{i_j}}{\|a^{i_j}\|} t^k.$$

As $j \to \infty$, the left-hand side of the equality above approaches zero on $[-1, 1]$, while the right-hand side approaches the nonzero polynomial $p(x) = \sum_{k=0}^n \bar{a}_k t^k$, a contradiction. This proves the claim.

Next, we reformulate the minimax problem as an optimization problem. We introduce an extra variable z, and write it as a semi-infinite program in the variables (z, a),

$$\begin{aligned} \min \quad & z \\ \text{s.t.} \quad & -z \leq (a_n t^n + a_{n-1} t^{n-1} + \cdots + a_1 t + a_0) \leq z, \quad t \in [-1, 1] \\ & a_n \gamma^n + a_{n-1}\gamma^{n-1} + \cdots + a_1\gamma + a_0 = 1. \end{aligned} \tag{12.16}$$

At the optimal solution $(z^*, a) \in \mathbb{R}^{n+2}$, it follows from a slight extension of Theorem 12.4 (to take into account the last, equality, constraint above) that there exist $s \leq n+2$ values of t ($s \leq n+1$ if $\lambda_0 > 0$ below)

$$-1 \le t_1 < t_2 < \cdots < t_s \le 1$$

for which the inequality constraints in the problem are active. We form the (weak) Lagrangian

$$\begin{aligned} L(z, a; \lambda, \delta) &= \lambda_0 z + \sum_{i=1}^{s} \lambda_i \left(\sum_{k=0}^{n} a_k t_i^k - \operatorname{sgn}(\lambda_i) z \right) + \delta \left(\sum_{k=0}^{n} a_k \gamma^k - 1 \right) \\ &= \left(\lambda_0 - \sum_{i=1}^{n} |\lambda_i| \right) z + \sum_{k=0}^{n} a_k \left(\sum_{i=1}^{s} \lambda_i t_i^k + \delta \gamma^k \right) - \delta, \end{aligned}$$

where $\operatorname{sgn}(t) = 1$ if $t > 0$ and $\operatorname{sgn}(t) = -1$ if $t < 0$. The multipliers satisfy the conditions $(\lambda, \delta) := (\lambda_0, \lambda_1, \ldots, \lambda_s, \delta) \ne 0$, $\lambda_0 \ge 0$, and λ_i is nonnegative or nonpositive depending on whether the corresponding constraint is of the form $t_i^n + \sum_{k=0}^{n-1} a_k t_i^k \le z$ or $-(t_i^n + \sum_{k=0}^{n-1} a_k t_i^k) \le z$, respectively. The FJ conditions consist of these conditions on the multipliers and the equations

$$\sum_{i=1}^{s} |\lambda_i| = \lambda_0, \qquad \sum_{i=1}^{s} \lambda_i t_i^k + \delta \gamma^k = 0, \quad k = 0, \ldots, n. \tag{12.17}$$

If $\lambda_0 = 0$, then all the multipliers $\{\lambda_i\}_1^s$ and δ are zero, contradicting the FJ conditions. We set $\lambda_0 = 1$ and obtain $\sum_{i=1}^{s} |\lambda_i| = 1$. The second set of equalities in (12.17) becomes

$$\begin{bmatrix} 1 & 1 & \cdot & 1 & 1 \\ t_1 & t_2 & \cdot & t_s & \gamma \\ \vdots & \vdots & \vdots & \vdots & \vdots \\ t_1^n & t_2^n & \cdot & t_s^n & \gamma^n \end{bmatrix} \begin{pmatrix} \lambda_1 \\ \lambda_2 \\ \vdots \\ \lambda_s \\ \delta \end{pmatrix} = 0, \quad (\lambda_1, \ldots, \lambda_s, \delta) \ne 0.$$

We must have $s = n + 1$, since otherwise $s \le n$ and the Vandermonde matrix above has full rank $s + 1$ (for example, its submatrix consisting of the first $s + 1$ rows is a nonsingular Vandermonde matrix), implying that $\lambda = 0$ and $\delta = 0$.

At the points $\{t_i\}_1^{n+1}$, the optimal polynomial $q(t) := \sum_{k=0}^{n} a_k t^k$ achieves its supremum

$$\nu := z^* = \max_{t \in [-1,1]} |q(t)|.$$

Since the derivative q' is a polynomial of degree at most $n - 1$, the number of t_i in the open interval $(-1, 1)$ can be at most $n - 1$. This means that the number of such t_i is exactly $n-1$ and that endpoints $-1, 1$ of I are also among the t_i's. Thus, we have

$$-1 = t_1 < t_2 < \cdots < t_n < t_{n+1} = 1.$$

Next, we see that the polynomial $(1 - t^2)(q')^2$ has the $2n$ roots

$$-1, t_2, t_2, \ldots, t_n, t_n, 1.$$

The polynomial $\nu^2 - q(t)^2$ is of degree $2n$, also with roots $\{t_i\}_1^{n+1}$. Since $(\nu^2 - q(t)^2)' = -2q(t)q'(t) = 0$ at an interior root t_i, each such t_i is a double root of $\nu^2 - q(t)^2$. Thus, the two polynomials are multiples of each other. By comparing the coefficients of their highest terms, we obtain

$$n^2(\nu^2 - q(t)^2) = (1 - t^2)q'(t)^2.$$

The problem of determining $q(t)$ is thus reduced to solving this equation.

Let $J = (t_i, t_{i+1})$ be an interval where t_i, t_{i+1} are in $(-1, 1)$ such that $q'(t) > 0$ on J. The above equation can be written as

$$\frac{q'(t)}{\sqrt{\nu^2 - q(t)^2}} = \frac{n}{\sqrt{1 - t^2}}.$$

Integration both sides, and making a substitution $u = q(t)/\nu$ on the left-hand side, we obtain

$$\cos^{-1}\left(\frac{q(t)}{\nu}\right) = \cos^{-1}(u) = \int \frac{du}{\sqrt{1 - u^2}} = n \int \frac{dt}{\sqrt{1 - t^2}} = c + n\cos^{-1}(t),$$

giving

$$q(t) = \nu\cos(c + n\cos^{-1}(t)) = \nu\cos(c)\cos(n\cos^{-1} t) - \nu\sin(c)\sin(n\cos^{-1} t).$$

Since the function $\sin(n\cos^{-1} t)$ is not a polynomial, the right-hand side above can be a polynomial only when $\sin c = 0$, implying that $\cos c = \mp 1$. We set $\cos c = 1$ without losing any generality. Thus, $q(t) = \nu T_n(t)$ and $q(\gamma) = 1$ implies $\nu = 1/T_n(\gamma)$. This proves that $q(t) = \frac{T_n(t)}{T_n(\gamma)}$. Since $\|T_n\| = 1$ on $[-1, 1]$, we also have $\|q\| = \frac{1}{|T_n(\gamma)|}$. □

12.6 Kirszbraun's Theorem and Extension of Lipschitz Continuous Functions

Kirszbraun's theorem [165] states that if $S \subset \mathbb{R}^m$ and $f : S \to \mathbb{R}^n$ is a Lipschitz continuous function, then f can be extended to a Lipschitz continuous function $\tilde{f} : \mathbb{R}^m \to \mathbb{R}^n$ without changing its Lipschitz constant. Kirszbraun's theorem is valid in infinite-dimensional Hilbert spaces as well, but may fail in spaces with non-Euclidean norms.

Kirszbraun's theorem has important applications, for example in the theory of *maximal monotone operators*, where it is used to establish Minty's theorem, which forms the cornerstone of that field; see for example [10, 48].

The key ingredients in the proof of Kirszbraun's theorem (Theorem 12.18 below) are the following two related results due to Schoenberg [237]. The first is an appealing, and geometrically plausible, result.

Lemma 12.16. *Let $\{D_i\}_1^l$ and $\{D_i'\}_1^l$ be disks in $\mathbb{R}^n$ and $\mathbb{R}^m$, respectively, where*

$$D_i := B_{r_i}(x_i), \quad D_i' := B_{r_i}(y_i) \quad (i = 1, \ldots, l).$$

If the disks $\{D_i\}_1^l$ have nonempty intersection, and the centers of the disks D_i' are closer together than the centers of D_i, that is,

$$\|y_i - y_j\| \leq \|x_i - x_j\| \quad (1 \leq i, j \leq l), \tag{12.18}$$

then the disks $\{D_i'\}_1^l$ also have nonempty intersection.

This lemma is equivalent to the following result, which states that a nonexpansive function defined on a finite set S can be extended to a nonexpansive function on a set containing S and one more point:

Lemma 12.17. *Let $\{x_i\}_1^l \subset \mathbb{R}^n$ and $\{y_i\}_1^l \subset \mathbb{R}^m$ be points satisfying* (12.18). *Let $x \in \mathbb{R}^n$ be given. There exists a point $y \in \mathbb{R}^m$ such that*

$$\|y - y_i\| \leq \|x - x_i\| \quad (i = 1, \ldots, l). \tag{12.19}$$

Lemma 12.16 implies Lemma 12.17: consider $D_i = B_{\|x-x_i\|}(x_i)$. Each D_i contains the point x; hence Lemma 12.16 implies that there exists a point $y \in \cap B_{\|x-x_i\|}(y_i)$. Clearly, y satisfies (12.19).

Conversely, Lemma 12.17 implies Lemma 12.16: if $x \in \cap B_{r_i}(x_i)$, then (12.19) states that there exists a point y satisfying $\|y - y_i\| \leq \|x - x_i\| \leq r_i$. Thus, $y \in \cap B_{r_i}(y_i) \neq \emptyset$.

We proceed with the proof of Lemma 12.17.

Proof. If $x = x_j$ for some j, then the lemma holds with the choice of $y = y_j$; thus we consider the case that x is distinct from $\{x_i\}_1^l$. Consider the minimax problem

$$\min_{v \in \mathbb{R}^m} \max_{1 \leq i \leq l} \frac{\|v - y_i\|^2}{\|x - x_i\|^2}. \tag{12.20}$$

The function

$$\varphi(v) := \max_i \frac{\|v - y_i\|^2}{\|x - x_i\|^2}$$

is coercive, since we clearly have $\varphi(v) \to \infty$ as $\|v\| \to \infty$. Thus, (12.20) has an optimal solution y.

Let $\{1, \ldots, k\}$ be the set of "active constraints," that is, $\varphi(y) = \|y - y_i\|^2/\|x - x_i\|^2$ if and only if $i \leq k$. It follows from Theorem 12.3 that there exists a nonzero vector $0 \leq \delta \in \mathbb{R}^k$ such that

$$\sum_{i=1}^{k} \lambda_i (y - y_i) = 0, \tag{12.21}$$

where $\lambda_i := \delta_i / \|x - x_i\|^2$. Thus, $y \in \text{co}(\{y_i\}_1^k)$.

The lemma is proved if we can show that $\varphi(y) \le 1$. Suppose this is not true. Then, defining

$$u_i := x_i - x, \quad v_i := y_i - y \quad (1 \le i \le k),$$

we have $\|u_i\|^2 < \|v_i\|^2$. Note that (12.18) implies $\|v_i - v_j\|^2 \le \|u_i - u_j\|^2$. Upon expanding this inequality and using $\|u_i\|^2 < \|v_i\|^2$, we get

$$\begin{aligned}\|u_i\|^2 + \|u_j\|^2 - 2\langle v_i, v_j\rangle &< \|v_i\|^2 + \|v_j\|^2 - 2\langle v_i, v_j\rangle \\ &\le \|u_i\|^2 + \|u_j\|^2 - 2\langle u_i, u_j\rangle,\end{aligned}$$

or simply $\langle v_i, v_j\rangle > \langle u_i, u_j\rangle$. Then we have

$$0 = \Big\|\sum_{i=1}^{k} \lambda_j v_i\Big\|^2 = \sum_{i,j=1}^{k} \lambda_i\lambda_j\langle v_i, v_j\rangle > \sum_{i,j=1}^{k} \lambda_i\lambda_j\langle u_i, u_j\rangle = \Big\|\sum_{i=1}^{k} \lambda_i u_i\Big\|^2,$$

where the first equality follows from (12.21). This is a contradiction, which proves that $\varphi(y) \le 1$. □

Theorem 12.18. *If $S \subset \mathbb{R}^m$ and $f : S \to \mathbb{R}^n$ is Lipschitz continuous, then f has an extension to a function $\tilde{f} : \mathbb{R}^m \to \mathbb{R}^n$ having the same Lipschitz constant as f.*

Proof. Without loss of generality, we may assume that f is nonexpansive, that is, $\|f(x_1) - f(x_2)\| \le \|x_1 - x_2\|$ for all $x_1, x_2 \in S$. By Zorn's lemma, f has a maximal extension to a nonexpansive function $\tilde{f} : T \to \mathbb{R}^n$.

We will show that if $T \neq \mathbb{R}^n$ and $p \notin T$, then there exists a point $q \in \mathbb{R}^m$ satisfying

$$\|q - \tilde{f}(x)\| \le \|p - x\| \quad \text{for all} \quad x \in T. \tag{12.22}$$

If this is true, then the domain of $\tilde{f}$ can be extended to include p. This contradicts the maximality of $\tilde{f}$, and proves that $T = \mathbb{R}^n$.

Now, (12.22) is equivalent to the statement that

$$\bigcap_{x \in T} B_{\|p-x\|}(\tilde{f}(x)) \neq \emptyset.$$

Since the disks $B_{\|p-x\|}(\tilde{f}(x))$ are compact, it is sufficient to show that every *finite* intersection of such disks is nonempty. But every such finite intersection is nonempty by Lemma 12.17. The theorem is proved. □

Remark 12.19. Theorem 12.18 also holds for Lipschitz functions of order α, $0 < \alpha < 1$, that is, for functions satisfying the condition

$$\|f(x) - f(y)\| \le L\|x - y\|^{\alpha} \quad \text{for all} \quad x, y \in S.$$

The proof is similar, but is a bit more involved.

12.7 Exercises

1. *(First-order sufficient conditions for a local minimizer)* Consider the semi-infinite program

$$\min\{f(x) : g(x,y) \le 0, y \in Y,\ h(x,z) = 0, z \in Z\}, \quad (P)$$

where $f(x)$, $g(x,y)$, and $h(x,z)$ are functions continuously differentiable with respect to x, and defined on X, $X \times Y$, and $X \times Z$, respectively, where $X \subseteq \mathbb{R}^n$ is an open set, and Y, Z are compact sets in some topological spaces. Let x^* be a feasible point of (P) such that

$$\lambda_0 \nabla f(x^*) + \sum_{i=1}^{k} \lambda_i \nabla g(x, y_i) + \sum_{j=1}^{l} \mu_j \nabla h(x^*, z_j) = 0,$$

for some $\lambda_0 \ge 0$, $\lambda_i > 0$, $y_i \in Y$, $g(x^*, y_i) = 0$, $i = 1, \ldots, k$, and $\mu_j \ne 0$, $z_j \in Z$, $j = 1, \ldots, l$. Show that if the gradients

$$\lambda_0 \nabla f(x^*), \nabla g(x, y_i), i = 1, \ldots, k, \nabla h(x^*, z_j) = 0, j = 1, \ldots, l,$$

span $\mathbb{R}^n$, then x^* is a local minimizer of (P), thereby generalizing Theorem 9.7.

2. Let $K \subset \mathbb{R}^n$ be a convex body. Show that the equation $\sum_{i=1}^{k} \lambda_i (u_i - c) = 0$ implies that the contact points of E^K (E_K) cannot lie on one side of any hyperplane passing through the center of E^K (E_K).
3. Use the equations $\sum_{i=1}^{k} \lambda_i u_i u_i^T = I$ and $\sum_{i=1}^{k} \lambda_i u_i = 0$ in (12.8) or (12.13) to prove that the vectors $\{u_i\}_1^k$ span $\mathbb{R}^n$. Show that this implies $k \ge n+1$. Conclude that the number of contact points must be at least $n+1$ in both the circumscribed and inscribed ellipsoid problems.
4. Let E be an ellipsoid and $K \subset \mathbb{R}^n$. The inclusion $K \subseteq E$ is equivalent to the inequalities

$$s_E(d) \ge s_K(d) \ \text{ for all } \ d, \ \|d\| = 1.$$

Use the method in Section 12.4 to develop an alternative semi-infinite programming method to derive the optimality conditions (12.7) for the minimum-volume ellipsoid circumscribing K.

5. Let C be a compact, convex set in $\mathbb{R}^n$ with nonempty interior. Let $b(C)$ be the breadth of C, that is, the minimum among the distances of two parallel support hyperplanes of C. Define also $r(C)$ to be the radius of the largest ball contained in C. Prove that

$$r(C) \ge \frac{\sqrt{n+2}}{2n+2} b(C), \qquad n = 2k,$$

and

$$r(C) \ge \frac{1}{2\sqrt{n}} b(C), \quad n = 2k+1,$$

by setting up an appropriate semi-infinite programming problem.

6. Let $C \subset \mathbb{R}^n$ be a closed set, and $d_C(x) := \min_{z \in \mathbb{R}^n} \|x - z\|$ the distance function to C, where $\|x\| = \sqrt{x^T x}$ is the Euclidean norm. Show that:
   (a) The function $f(x) = d_C(x)^2$ is directionally differentiable, with
   $$f'(x; h) = \min\{\langle x - z, h\rangle : \|x - z\| = d_C(x)\}.$$
   (b) If C is convex, then $f(x)$ is continuously differentiable. Compute $\nabla f(x)$.
   (c) If C is convex, then $d_C(x)$ is also continuously differentiable on $\mathbb{R}^n \backslash C$. Compute $\nabla d_C(x)$.
7. ***(Pólya)*** Let $0 < a < b$ be given. Solve the following problems using semi-infinite programming:
   (a) $\min_x \max_{y \in [a,b]} (x - y)^2$,
   (b) $\min_x \max_{y \in [a,b]} \frac{(x-y)^2}{y^2}$.
8. Determine the nth-degree polynomial $p(t) = t^n + a_{n-1}t^{n-1} + \cdots + a_i t + a_0$ whose supremum norm $\|p\| := \max_{|t| \le 1} |p(t)|$ on the interval $I = [-1, 1]$ is minimum. This is the problem of best approximation of the monomial t^n by an $(n-1)$th-degree polynomial on the interval $[-1, 1]$.
   Show that the solution of this problem is $\frac{T_n(t)}{2^{n-1}}$, where T_n is Chebyshev's polynomial.
   Hint: Mimic the proof of Theorem 12.15.
9. Show that Theorem 12.15 can be solved by ordinary nonlinear programming using Helly's theorem in Chapter 13 or one of its consequences.
10. Let $f : A \subset X \to \mathbb{R}$ be a Lipschitz continuous function on a subset A of a metric space (X, d), that is,
    $$|f(x) - f(y)| \le Ld(x, y) \quad \text{for all} \quad x, y \in A.$$
    Show that f can be extended to a Lipschitz continuous function $g : X \to \mathbb{R}$ with the same constant L.
    Hint: Show that $g(x) = \inf\{f(y) + Ld(x, y) : y \in A\}$ satisfies the desired properties.

13

Topics in Convexity

In this chapter, we probe several topics that use significant ideas from convexity theory and that have significant applications in various fields. In particular, we prove theorems of Radon, Helly, Kirchberger, Bárány, and Tverberg on the combinatorial structure of convex sets, application of Helly's theorem to semi-infinite programming, in particular to Chebyshev's approximation problem, homogeneous convex functions, and their applications to inequalities, attainment of optima in maximization of convex functions, decompositions of convex cones, and finally the relationship between the norms of a homogeneous polynomial and its associated symmetric form. The last result has an immediate application to self-concordant functions in interior-point algorithms.

Many interesting applications of the topics can be found in the exercises at the end of the chapter.

13.1 Combinatorial Theory of Convex Sets

Intersections of convex sets have interesting combinatorial properties. We start with three classical results by Radon, Helly, and Kirchberger dating back to the early part of the twentieth century.

Theorem 13.1. (***Radon***) *If $A \subseteq \mathbb{R}^n$ is an affinely dependent set, then A can be partitioned into two sets B, C such that* $\operatorname{co}(B) \cap \operatorname{co}(C) \neq \emptyset$.

Proof. Pick an affinely dependent set $\{x_i\}_1^k$ in A; then there exists $\lambda := (\lambda_1, \ldots, \lambda_k) \neq 0$ such that

$$\sum_{i=1}^{k} \lambda_i x_i = 0 \quad \text{and} \quad \sum_{i=1}^{k} \lambda_i = 0.$$

Suppose that $\lambda_i \geq 0$ for $1 \leq i \leq j$ and $\lambda_i < 0$ for $i = j+1, \ldots, k$. Defining

O. Güler, *Foundations of Optimization*, Graduate Texts in Mathematics 258,
DOI 10.1007/978-0-387-68407-9_13, © Springer Science+Business Media, LLC 2010

$$\lambda := \sum_{i=1}^{j} \lambda_i = - \sum_{i=j+1}^{k} \lambda_i \neq 0,$$

we see that the vector

$$x := \sum_{i=1}^{j} \frac{\lambda_i}{\lambda} x_i = \sum_{i=j+1}^{k} \frac{-\lambda_i}{\lambda} x_i$$

belongs to $\mathrm{co}(\{x_1, \ldots, x_k\}) \cap \mathrm{co}(\{x_{j+1}, \ldots, x_k\})$. Clearly, $B := \{x_1, \ldots, x_k\}$ and $C := A \setminus B$ satisfy $\mathrm{co}(B) \cap \mathrm{co}(C) \neq \emptyset$. □

Theorem 13.2. (*Helly*) *Let $\{A_i\}_{i=1}^k$ be a finite collection of convex sets in $\mathbb{R}^n$. If the intersection of any $n+1$ sets from this collection is nonempty, then $\cap_{i=1}^k A_i \neq \emptyset$.*

Proof. The theorem is trivially true for $k \leq n+1$, so we consider the case $k > n+1$. The proof is by induction on k. Suppose that the theorem has been proved for $k-1$, and let $\{A_i\}_1^k$ be a collection of convex sets satisfying the hypothesis of the theorem. By the induction hypothesis any $k-1$ of the sets have nonempty intersection; pick

$$x_i \in \cap_{j \neq i} A_j \neq \emptyset, \quad i = 1, \ldots, k.$$

The set $A := \{x_i\}_1^k$ is affinely dependent, so Theorem 13.1 implies that $A = B \cup C$, where $B = \{x_1, \ldots, x_j\}$ and $C = \{x_{j+1}, \ldots, x_k\}$, say, such that

$$x \in \mathrm{co}(\{x_1, \ldots, x_j\}) \cap \mathrm{co}(\{x_{j+1}, \ldots, x_k\}).$$

If $i \leq j$, then $x_i \in \cap_{j+1}^k A_l$, so that $x \in \mathrm{co}(\{x_1, \ldots, x_j\}) \subseteq \cap_{j+1}^k A_l$; similarly, $x \in \cap_1^j A_l$. We conclude that $x \in \cap_1^k A_l \neq \emptyset$. □

Here are some quick applications of Helly's theorem; more substantial examples, some important in optimization, will be given in Section 13.2 and in the exercises at the end of the chapter.

Example 13.3. Let K and $\{C_i\}_{i=1}^k$, $k > n+1$, be convex sets in $\mathbb{R}^n$.

(a) If the intersection of every $n+1$ of the sets C_i *contains* a translated copy of K, then $\cap_1^k C_i$ must also *contain* a translated copy of K.
(b) If the intersection of every $n+1$ of the sets C_i is *contained in* a translated copy of K, then $\cap_1^k C_i$ must also be *contained in* a translated copy of K.
(c) If the intersection of every $n+1$ of the sets C_i *intersects* a translated copy of K, then $\cap_1^k C_i$ must also *intersect* a translated copy of K.

To prove (a), define the sets $D_i := \{x \in \mathbb{R}^n : K \subseteq x + C_i\}$; it is easy to verify that the set D_i is convex. Since $x \in D_i$ means that $-x + K \subseteq C_i$, the statement $x \in \cap_{j=1}^{n+1} D_{i_j} \neq \emptyset$ is equivalent to $-x + K \subseteq \cap_{j=1}^{n+1} C_{i_j}$, which holds for some x by our assumption. It follows by Theorem 13.2 that $\cap_1^k D_i \neq \emptyset$, so there exists a point $a \in \mathbb{R}^n$ such that $a + K \subseteq \cap_i^k C_i$. This proves our claim.

The proofs of (b) and (c) are similar; define the sets $D_i := \{x \in \mathbb{R}^n : x + C_i \subseteq K\}$ and $D_i := \{x \in \mathbb{R}^n : (x + C_i) \cap K \neq \emptyset\}$ for (b) and (c), respectively. The verification of the details is left to the reader.

Corollary 13.4. *Let $\{A_\alpha\}_{\alpha\in\mathcal{A}}$ be any collection of closed convex sets in $\mathbb{R}^n$ such that some finite intersection of sets from this collection is bounded and nonempty. If the intersection of any $n+1$ sets from this collection is nonempty, then $\cap_{\alpha\in\mathcal{A}} A_\alpha \neq \emptyset$.*

Proof. If $\mathcal{F}$ is any finite subset of $\mathcal{A}$, it follows from Theorem 13.2 that $\cap_{\alpha\in\mathcal{F}} A_\alpha \neq \emptyset$. Let $A_0 := \cap_{i\in\mathcal{F}_0} A_i \neq \emptyset$ be a bounded set, and define $\hat{A}_\alpha := A_\alpha \cap A_0$. Each set $\hat{A}_\alpha$ is compact, and $\{\hat{A}_\alpha\}_{\alpha\in\mathcal{A}}$ has the property that any finite intersection of sets from this collection is nonempty. It follows from the finite intersection property of compact sets that $\cap_{\alpha\in\mathcal{A}} \hat{A}_\alpha \neq \emptyset$. Thus,

$$\cap_{\alpha\in\mathcal{A}} A_\alpha = \cap_{\alpha\in\mathcal{A}} \hat{A}_\alpha \neq \emptyset.$$

□

Theorem 13.5. (***Kirchberger***) *Let S and T be two finite subsets of $\mathbb{R}^n$. The sets S and T can be strictly separated if and only if every subset of S and T, consisting of at most $n+2$ points can be strictly separated.*

Here is an amusing application of Kirchberger's theorem [114]. Suppose that in a flock consisting of black sheep and white sheep, any four sheep may be separated by a straight fence, that is, black and white sheep lie on different sides of the fence. Then the whole flock can be separated by a straight fence.

Proof. We may assume that $|S \cup T| \geq n+2$. For each $s \in S$ and each $t \in T$, define the open half-spaces in $\mathbb{R}^{n+1}$

$$I_s := \{(\lambda_0, \lambda) \in \mathbb{R} \times R^n : \langle s, \lambda\rangle > \lambda_0\},$$
$$J_t := \{(\lambda_0, \lambda) \in \mathbb{R} \times R^n : \langle t, \lambda\rangle < \lambda_0\}.$$

By assumption, each $n+2$ members of the family $\{I_s : s \in S\} \cup \{J_t : t \in T\}$ have a nonempty intersection. Theorem 13.2 implies that there exists a point $(\lambda_0, \lambda) \in \cap_{s\in S} I_s \cap \cap_{t\in T} J_t$; this means that the sets S and T are strictly separated by the hyperplane $\{x : \langle \lambda, x\rangle = \lambda_0\}$ in $\mathbb{R}^n$. □

Helly's theorem and its relatives, Radon's, Carathéodory's, and Kircherberger's theorems among others, are the beginnings of an extensive literature on the combinatorial properties of convex sets. The reader is directed to the survey articles [69, 83, 156] for more information on this subject.

13.2 Applications of Helly's Theorem to Semi-infinite Programming

Helly's theorem has important applications to optimization, especially when the number of constraints is infinite (semi-infinite programming); see [224], [187], [228].

The theorem below is the main result of this section.

Theorem 13.6. *Consider the problem*

$$\rho := \inf\Big\{f(x) : g(x,y) \le 0, \ \ y \in Y\Big\}, \qquad (P)$$

where x belongs to a convex set in $\mathbb{R}^n$, $f(x)$ and $g(x,y)$ are lower semicontinuous convex functions in x, Y is a (possibly infinite) index set, and the feasible set

$$F := \{x : g(x,y) \le 0, \ \ \textit{for all } y \in Y\}$$

is not empty. Assume that

$$\begin{array}{c}\textit{there exists a finite subset } Z \subset Y \textit{ such that} \\ \{x : g(x,y) \le 0, \ y \in Z\} \quad \textit{is a bounded set.}\end{array} \tag{13.1}$$

Let Ω_n be the collection of all sets $\omega \subseteq Y$ of cardinality at most n, and for each $\omega \in \Omega_n$, define the subproblem

$$\rho(\omega) := \inf\Big\{f(x) : g(x,y) \le 0, \ y \in \omega\Big\}. \qquad (P_\omega)$$

Then

(a) $\rho = \sup_{\omega \in \Omega_n} \rho(\omega)$, that is,

$$\inf\Big\{f(x) : g(x,y) \le 0, \ y \in Y\Big\} = \sup_{\omega \in \Omega_n} \inf\Big\{f(x) : g(x,y) \le 0, \ y \in \omega\Big\}.$$

Moreover, if ρ is finite, then:

(b) If each subproblem (P_ω) has an optimal solution, then problem (P) also has an optimal solution.

(c) If each subproblem (P_ω) has an optimal solution $x^*(\omega)$ and the maximum of $\rho(\omega)$ is achieved at a set $\omega^* \in \Omega_n$, then the point $x^* := x^*(\omega^*)$ is an optimal solution of (P) provided it is a feasible solution of (P).

(d) If each subproblem (P_ω) has an optimal solution $x^*(\omega)$, the maximum of $\rho(\omega)$ is achieved at a (possibly nonunique) set $\omega^* \in \Omega_n$, and the optimal solution $x_0^* := x^*(\omega_0^*)$ to at least one $(P_{\omega_0^*})$ is unique, *then the point x_0^* is the* unique *optimal solution of (P).*

Proof. We assume that $|Y| > n$; otherwise the theorem is trivial. For each subset $Z \subset Y$, define the corresponding feasible set $F_Z := \{x : g(x,y) \leq 0, \text{ for all } y \in Z\}$; for convenience, set $F_y := F_{\{y\}}$. Define

$$\bar{\rho} := \sup_{\omega \in \Omega_n} \rho(\omega).$$

If $\omega \in \Omega_n$, then $F \subseteq F_\omega$ and $\rho(\omega) \leq \rho$; consequently $\bar{\rho} \leq \rho$. We claim that the reverse inequality $\bar{\rho} \geq \rho$ also holds. This is trivially true when $\rho = -\infty$, so assume that $\rho \in \mathbb{R} \cup \{\infty\}$. If $\rho \in \mathbb{R}$, for $\epsilon > 0$, define the set

$$C_\epsilon := \{x : f(x) \leq \rho - \epsilon\}.$$

Note that $C_\epsilon \cap F = C_\epsilon \cap \cap_{y \in Y} F_y = \emptyset$, but $\cap_{y \in Y} F_y \neq \emptyset$. It follows from Corollary 13.4 that there exists a subset $\omega \in \Omega_n$ such that $C_\epsilon \cap F_\omega = \emptyset$; this gives $\bar{\rho} \geq \rho(\omega) \geq \rho - \epsilon$ for all $\epsilon > 0$, proving the claim. If $\rho = \infty$ (that is, (P) is infeasible), we want to prove that $\bar{\rho} = \infty$. If $\bar{\rho} < M < \infty$, then define $D := \{x : f(x) \leq M\}$ and note that $D \cap F_\omega \neq \emptyset$ for every $\omega \in \Omega_n$; this implies that $D \cap F \neq \emptyset$ and gives the contradiction $\rho \leq M < \infty$.

To prove part (b), note that if $\omega \in \Omega_n$, then

$$\{x : f(x) \leq \rho(\omega)\} \cap F_\omega \subseteq \{x : f(x) \leq \rho\} \cap F_\omega = C_0 \cap F_\omega \neq \emptyset,$$

since the first intersection is nonempty by hypothesis, and the inclusion follows because $\rho(\omega) \leq \rho$. By Corollary 13.4, $C_0 \cap F \neq \emptyset$, and any point x in this intersection is clearly a minimizer of (P).

Part (c) is trivial: $f(x^*) = \bar{\rho} = \rho$ and x^* is feasible for (P) by assumption.

It remains to prove part (d). We know by part (b) that (P) has an optimal solution $\bar{x}$. Clearly, $\bar{x}$ is a feasible solution to $(P_{\omega_0^*})$, and we have

$$f(\bar{x}) = \rho = \bar{\rho} = \rho(\omega_0^*) = f(x_0^*),$$

which proves that $\bar{x} = x_0^*$ is the unique optimal solution to (P). □

We remark that it is possible to obtain results similar to Theorem 13.6 by imposing compactness assumptions on the index set Y instead of the assumption (13.1) on the feasible set; see Levin [187] (Theorem 1) for more details.

13.2.1 Chebyshev's Approximation Problem

Let $f(x)$ be a continuous function on the interval $[a,b]$, and n a given positive integer. *Chebyshev's approximation problem* we consider here is concerned with approximating the function $f(x)$ on $[a,b]$ by a polynomial p of degree at most $n-1$ such that the uniform norm

$$\|f - p\|_\infty := \max_{x \in [a,b]} |f(x) - p(x)|$$

is minimized. This is a central problem in *approximation theory*. A variant of the problem is solved explicitly in Section 12.5 by semi-infinite programming techniques. Following [224], we use Theorem 13.6 to solve Chebyshev's problem. Many other important problems in approximation theory, and in general in semi-infinite programming, can be solved similarly using Helly's theorem as a main tool.

Our problem is a minimax problem

$$\min_{p \in \mathcal{P}_{n-1}} \max_{x \in [a,b]} |f(x) - p(x)|, \tag{13.2}$$

where $\mathcal{P}_{n-1}$ is the vector space of polynomials of degree at most $n-1$. We can write it as the semi-infinite (linear) program

$$\min_{(z,a) \in \mathbb{R}^{n+1}} \Big\{ z : -z \le f(x) - \sum_{i=0}^{n-1} a_i x^i \le z,\ x \in [a,b] \Big\}. \tag{13.3}$$

We give a characterization of the solution in Theorem 13.10. As dictated by Theorem 13.6, we first need to solve a finite-constraint version of Chebyshev's approximation problem. This is done in the following two lemmas.

For $x \in \mathbb{R}^k$, we define

$$\|x\|_\infty = \max\{|x_i| : i = 1, \ldots, k\}.$$

Lemma 13.7. *Consider the problem*

$$\min\{\|x\|_\infty : \langle b, x \rangle = \beta\},$$

where $b = (b_0, b_1, \ldots, b_n)$, *and* $b_i \neq 0$ *for all* i.

The optimal solution x^* *is unique, and* x^* *and the optimal objective value* ρ *are given by*

$$\rho = \frac{|\beta|}{\sum_{i=0}^n |b_i|}, \quad x_i^* = \rho \operatorname{sgn}(\beta b_i), \quad i = 0, \ldots, n.$$

Proof. Clearly, we can assume that $\beta \ge 0$. We have

$$\beta = \sum_{i=0}^n b_i x_i \le \sum_{i=0}^n |b_i| \cdot |x_i| \le \|x\|_\infty \Big(\sum_{i=0}^n |b_i| \Big),$$

and $\|x\|_\infty$ is minimized only when the inequalities above are equalities, which happens if and only if $\rho = \beta / (\sum_{i=0}^n |b_i|)$ and $b_i x_i = |b_i| \cdot |x_i| = |b_i| \rho$ for all i, that is, $x_i = \operatorname{sgn}(b_i)\rho$ for all i. □

The next result gives the best approximating polynomial on a finite set of points.

Lemma 13.8. *Let $\{(x_i, y_i)\}_{i=0}^n$ be $n+1$ given points in the plane such that*

$$x_0 < x_1 < \cdots < x_n,$$

and consider the minimax problem

$$\min_{p \in \mathcal{P}_{n-1}} \max_i |y_i - p(x_i)|, \tag{13.4}$$

which seeks a polynomial $p(x) = a_{n-1}x^{n-1} + \cdots + a_1 x + a_0$ of degree at most $n-1$ that best fits the data points $\{(x_i, y_i)\}$ in the sense that it minimizes the quantity

$$\rho(x_0, x_1, \ldots, x_n) := \max_i |y_i - p(x_i)|.$$

The optimal objective value $\rho(x_0, x_1, \ldots, x_n)$ is the absolute value of the quantity

$$\begin{vmatrix} y_0 & 1 & x_0 & \ldots & x_0^{n-1} \\ y_1 & 1 & x_1 & \ldots & x_1^{n-1} \\ & & \vdots & & \\ y_n & 1 & x_n & \ldots & x_n^{n-1} \end{vmatrix} \cdot \begin{vmatrix} 1 & 1 & x_0 & \ldots & x_0^{n-1} \\ -1 & 1 & x_1 & \ldots & x_1^{n-1} \\ & & \vdots & & \\ (-1)^n & 1 & x_n & \ldots & x_n^{n-1} \end{vmatrix}^{-1},$$

and there exists a unique polynomial p^* achieving $\rho(x_0, x_1, \ldots, x_n)$, which is characterized by the condition that the discrepancies $\{y_i - p^*(x_i)\}_{i=0}^n$ are equal in absolute value to $\rho(x_0, x_1, \ldots, x_n)$ and alternate in sign.

In light of Theorem 13.6, it is perhaps not surprising that the characteristic *alternation* property of the optimal solution in Chebyshev's approximation problem (see Theorem 13.10 and Lemma 13.11) is already present in this finite-dimensional approximation subproblem.

Proof. Writing $u_i = y_i - p(x_i)$ for the discrepancies, problem (13.4) becomes the optimization problem

$$\begin{aligned} \min \quad & \|u\|_\infty \\ \text{s.t.} \quad & Xa = y - u \end{aligned} \tag{13.5}$$

in the decision variables a and u, where

$$X = \begin{bmatrix} 1 & x_0 & \ldots & x_0^{n-1} \\ 1 & x_1 & \ldots & x_1^{n-1} \\ & & \vdots & \\ 1 & x_n & \ldots & x_n^{n-1} \end{bmatrix}, \quad a = \begin{pmatrix} a_0 \\ a_1 \\ \vdots \\ a_{n-1} \end{pmatrix}, \quad y = \begin{pmatrix} y_0 \\ y_1 \\ \vdots \\ y_n \end{pmatrix}, \quad \text{and} \quad u = \begin{pmatrix} u_0 \\ u_1 \\ \vdots \\ u_n \end{pmatrix}.$$

Since the vector a can be chosen at will, the vectors u in the constraint $Xa = y - u$ fill the affine space $y + R(X)$. The matrix X has linearly independent columns, since augmenting it with the column $(x_0^n, \ldots, x_n^n)^T$ yields a Vandermonde matrix with determinant $\prod_{i>j}(x_i - x_j) > 0$. This implies that

$y + R(X)$ is an n-dimensional affine space in $\mathbb{R}^{n+1}$, hence a hyperplane, and can thus be described by a single linear equation in u. This equation is

$$\begin{vmatrix} u_0 - y_0 \; 1 \; x_0 \; \dots \; x_0^{n-1} \\ u_1 - y_1 \; 1 \; x_1 \; \dots \; x_1^{n-1} \\ \vdots \\ u_n - y_n \; 1 \; x_n \; \dots \; x_n^{n-1} \end{vmatrix} = 0,$$

which simply expresses the linear dependence of the vectors consisting of $y - u$ and the columns of X. Therefore, the constraint of (13.5) is given by

$$\begin{vmatrix} u_0 \; 1 \; x_0 \; \dots \; x_0^{n-1} \\ u_1 \; 1 \; x_1 \; \dots \; x_1^{n-1} \\ \vdots \\ u_n \; 1 \; x_n \; \dots \; x_n^{n-1} \end{vmatrix} = \begin{vmatrix} y_0 \; 1 \; x_0 \; \dots \; x_0^{n-1} \\ y_1 \; 1 \; x_1 \; \dots \; x_1^{n-1} \\ \vdots \\ y_n \; 1 \; x_n \; \dots \; x_n^{n-1} \end{vmatrix}.$$

Denoting the right-hand by β and expressing the left-hand side by expanding the determinant using the first column, we can write this equation in the form $\sum_{i=0}^{n} b_i u_i = \beta$, where

$$|b_i| = (-1)^i b_i = \det[(x_k)^l]_{k \neq i, 0 \leq l \leq n} = \prod_{k > l; k, l \neq i} (x_k - x_l) > 0$$

and

$$\sum_{i=0}^{n} |b_i| = \begin{vmatrix} 1 & 1 \; x_0 \; \dots \; x_0^{n-1} \\ -1 & 1 \; x_1 \; \dots \; x_1^{n-1} \\ & \vdots \\ (-1)^n & 1 \; x_n \; \dots \; x_n^{n-1} \end{vmatrix}.$$

The lemma is completed by invoking Lemma 13.7. The optimal polynomial p^* is unique, since a can be determined uniquely from $y - u$. □

Theorem 13.9. *Let $f(x)$ be a bounded (not necessarily continuous) function on the interval $[a, b]$, and n a given positive integer.*

The problem

$$\min\{\|f - p\|_\infty : p \in \mathcal{P}_{n-1}\}$$

has an optimal solution $p^* \in \mathcal{P}_{n-1}$, and

$$\rho^* := \|f - p^*\|_\infty = \sup\{\rho(x_0, x_1, \dots, x_n) : a \leq x_0 < x_1 < \dots < x_n \leq b\}.$$

The supremum is achieved at some $\{x_i\}_0^n$, and the optimal value ρ^* is the absolute value of

$$\begin{vmatrix} f(x_0) \; 1 \; x_0 \; \dots \; x_0^{n-1} \\ f(x_1) \; 1 \; x_1 \; \dots \; x_1^{n-1} \\ \vdots \\ f(x_n) \; 1 \; x_n \; \dots \; x_n^{n-1} \end{vmatrix} \div \begin{vmatrix} 1 & 1 \; x_0 \; \dots \; x_0^{n-1} \\ -1 & 1 \; x_1 \; \dots \; x_1^{n-1} \\ & \vdots \\ (-1)^n & 1 \; x_n \; \dots \; x_n^{n-1} \end{vmatrix}.$$

Proof. We write our problem as the semi-infinite linear program (13.3) in the decision variables $(z, a) = (z, a_0, \ldots, a_{n-1}) \in \mathbb{R}^{n+1}$. We can clearly assume that $|z| \le \sup_{x\in[a,b]} |f(x)|$ (pick $p = 0$), and if we choose $a \le x_0 < x_1 < \cdots < x_n \le b$, then it is easy to show, using the invertibility of the Vandermonde matrix, that the linear constraints

$$-z \le f(x_k) - \sum_{i=0}^{n-1} a_i x_k^i \le z, \quad i = 0, \ldots, n,$$

have a bounded set of feasible a. The theorem follows immediately from part (b) of Theorem 13.6 and Lemma 13.8. □

Theorem 13.10. (Chebyshev's theorem) *Let $f(x)$ be a continuous function on the interval $[a, b]$, and n a given positive integer.*

There exists a unique *polynomial p^* of degree at most $n - 1$ minimizing the norm $\|f - p\|_\infty$ on $[a, b]$.*

Moreover, there exist $n + 1$ distinct points $x_0 < x_1 < \cdots < x_n$ in $[a, b]$ such that the minimum norm $\rho^* := \|f - p^*\|_\infty$ satisfies

$$\rho^* = \rho(x_0, x_1, \ldots, x_n),$$

and the discrepancies $f(x_i) - p^*(x_i)$ are equal to ρ^* in absolute value and alternate in sign.

Proof. Theorem 13.9 guarantees everything except the uniqueness of p^*. Extend the function $\rho(x_0, x_1, \ldots, x_n)$ to all of $[a, b]^{n+1}$ by defining it to be zero when $\{x_i\}_0^n$ are not all distinct. Note that the uniqueness will follow from part (d) of Theorem 13.6, provided we can show that $\rho(x_0, x_1, \ldots, x_n)$ is continuous on $[a, b]^{n+1}$.

We may assume $\rho^* > 0$, because otherwise f itself is a polynomial of degree at most $n - 1$, $p^* = f$, and the function ρ is identically zero. The formula for $\rho(x_0, x_1, \ldots, x_n)$ in Lemma 13.8 shows that it is a continuous function if all x_i are distinct. To prove that it is everywhere continuous, let

$$x^{(k)} := \left(x_0^{(k)}, x_1^{(k)}, \ldots, x_n^{(k)}\right) \to x := (x_0, x_1, \ldots, x_n),$$

where the components of x are not all distinct. By the Lagrange interpolation formula, there exists $p \in \mathcal{P}_{n-1}$ such that $p(x_i) = f(x_i)$ for $i = 0, \ldots, n$. Since f is continuous, we have for $i = 0, \ldots, n$,

$$\rho(x^{(k)}) \le |f(x_i^{(k)}) - p(x_i^{(k)})| \to |f(x_i) - p(x_i| = 0,$$

proving $\rho(x^{(k)}) \to \rho(x)$. □

The alternation property in Theorem 13.10 actually characterizes the optimal solution to Chebyshev's approximation theorem. The following result goes back to de la Vallée-Poussin.

Lemma 13.11. *Let $f(x)$ be a continuous function on the interval $[a,b]$, and n a given positive integer.*

Let $p^* \in \mathcal{P}_{n-1}$ be a polynomial of degree at most $n-1$ such that there exist $n+1$ distinct points $x_0 < x_1 < \cdots < x_n$ in $[a,b]$ with the property that the discrepancies $f(x_i) - p^*(x_i)$ alternate in sign and their absolute value is equal to a constant $\mu > 0$.

Then p is an optimal solution to Chebyshev's approximation problem (13.2), *and μ is the optimal deviation $\mu = \rho^* := \min_{p \in \mathcal{P}_{n-1}} \|f - p\|_\infty$.*

Proof. Suppose that there exists $p \in \mathcal{P}_{n-1}$ satisfying $\|f - p\|_\infty < \mu$. Then the polynomial $r := p - p^* \in \mathcal{P}_{n-1}$ has the property that

$$r(x_i) = (f(x_i) - p^*(x_i)) - (f(x_i) - p(x_i)) \in \mp\mu + (-\mu, \mu), \quad i = 0, \ldots, n,$$

so that two consecutive values $r(x_i)$ and $r(x_{i+1})$ lie in the disjoint intervals $(-2\mu, 0)$ and $(0, 2\mu)$, thus are nonzero and alternate in sign. By the intermediate value theorem, r must have n distinct roots in $[a,b]$. This means that $r(x)$ is identically zero, a contradiction. We have proved that $\rho^* \geq \mu$, and since $|f(x_i) - p^*(x_i)| = \mu$, we must have $\rho^* = \mu$. □

13.3 Bárány's and Tverberg's Theorems

We describe two results here, one by Bárány [22] generalizing Theorem 4.13 (Carathéodory's theorem) and the other by Tverberg [257] generalizing Theorem 13.1 (Radon's theorem). The proofs below are taken from [63]; the elegant proof of Tverberg's theorem uses an idea from [235].

Theorem 13.12. (*Bárány*) *Let $\{A_i\}_{i=1}^{n+1}$ be $n+1$ nonempty sets in $\mathbb{R}^n$. If $x \in \mathbb{R}^n$ belongs to the convex hull of each set A_i, then there exists a point $x_i \in A_i$ such that x belongs to the convex hull of $\{x_1, \ldots, x_{n+1}\}$.*

Note that the theorem reduces to Carathéodory's theorem if $A_1 = A_2 = \cdots = A_{n+1}$. Bárány's theorem is referred to as the *colorful Carathéodory's theorem* for the following reason: assume that all points of A_i are given a certain color, say red, all points of A_i are given a certain color, say blue, etc. The theorem states that if the convex hulls of all monochromatic sets A_i have a common point x, then x must be in the convex hull of a colorful set $A = \{a_1, \ldots, a_{n+1}\}$ with $a_i \in A_i$.

Proof. We may assume, without loss of generality, that $x = 0$, and by virtue of Theorem 4.13 that each A_i contains at most $n+1$ affinely independent points. Consider the collection $\mathcal{A}$ of all colorful sets $A = \{x_1, \ldots, x_{n+1}\}$ with $x_i \in A_i$; $\mathcal{A}$ is a finite collection of sets. For a set $A \in \mathcal{A}$, define

$$d(A) = \min\{\|x\| : x \in \mathrm{co}(A)\}.$$

Since $\mathrm{co}(A)$ is compact, the minimum is achieved, say at point $a \in \mathrm{co}(A)$.

Suppose that $d(A) > 0$ for all $A \in \mathcal{A}$; we will obtain a contradiction by showing that if $A \in \mathcal{A}$ is the set with the smallest $d(A) > 0$, we can find $\overline{A} \in \mathcal{A}$ such that $d(\overline{A}) < d(A)$. Since $a \in A$ is the closest point to zero, a lies in a proper face of A, so that $a \in \mathrm{co}(A \setminus \{a_j\})$ for some $a_j \in A_j$. The set $\{x \in \mathbb{R}^n : \langle x - a, a \rangle < 0\}$ contains the point zero. Some point $\hat{a}_j \in A_j$ must also lie in this set, since otherwise we have $\langle y - a, a \rangle \geq 0$ for all $y \in A_j$, and since $0 \in \mathrm{co}(A_j)$, this leads to the contradiction $\langle -a, a \rangle = -\|a\| \geq 0$. Now define the set $\overline{A}$ that is obtained from A by replacing a_j with $\overline{a}_j$. We have $[a, \overline{a}_j] \in \overline{A}$, because $a, \overline{a}_j \in \overline{A}$; if $t > 0$ is small enough, we obtain

$$\|(1-t)a + t\overline{a}_j\|^2 = \|a\|^2 + 2t\langle a, \overline{a}_j - a \rangle + t^2\|\overline{a}_j - a\|^2 < \|a\|^2.$$

This proves the claim and the theorem. □

Theorem 13.13. ***(Tverberg)*** *Let $r > 1$ be an integer, and $A \subset \mathbb{R}^n$ a set with $(r-1)(n+1)+1$ distinct elements. Then A can be partitioned into r sets whose convex hulls have a common point.*

Proof. The proof utilizes Bárány's theorem. Define $k := (r-1)(n+1)$. Let $A = \{a_0, a_1, \ldots, a_k\}$ and define

$$\overline{A} = \{\overline{a}_0, \overline{a}_1, \ldots, \overline{a}_k\}, \quad \overline{a}_i := \begin{pmatrix} a_i \\ 1 \end{pmatrix} \in \mathbb{R}^{n+1}, \quad i = 0, 1, \ldots, k.$$

For each $\overline{a}_i$, associate the set

$$\begin{aligned} \hat{A}_i &:= \{M_{i1}, M_{i2}, \ldots, M_{ir-1}, M_{ir}\} \\ &= \Big\{[\overline{a}_i, 0, \ldots, 0], [0, \overline{a}_i, 0, \ldots, 0], \ldots, [0, \ldots, 0, \overline{a}_i], [-\overline{a}_i, -\overline{a}_i, \ldots, -\overline{a}_i]\Big\} \end{aligned}$$

of $(n+1) \times (r-1)$ matrices; note that we have $M_{i1} + \cdots + M_{ir} = 0$, so that $0 \in \mathrm{co}(\hat{A}_i)$, $i = 1, \ldots, r$. Identifying $\mathbb{R}^k$ with the vector space of $(n+1) \times (r-1)$ matrices, Theorem 13.12 implies that there exist matrices $M_{ij_i} \in \hat{A}_i$ and $\lambda_i \geq 0$, $\sum_{i=0}^k \lambda_i = 1$, such that $\sum_{i=0}^k \lambda_i M_{ij_i} = 0$. Define

$$I_j := \{i : j_i = j, 0 \leq i \leq k\}, \quad j = 1, \ldots, r.$$

Then $\{I_j\}_1^r$ is a partition of the index set $\{0, 1, \ldots, k\}$, and we have

$$\begin{aligned} 0 &= \sum_{j=1}^{r} \sum_{i \in I_j} \lambda_i M_{ij} \\ &= \sum_{i \in I_1} \lambda_i [\overline{a}_i, 0, \ldots, 0] + \sum_{i \in I_2} \lambda_i [0, \overline{a}_i, 0, \ldots, 0] + \cdots + \sum_{i \in I_{r-1}} \lambda_i [0, \ldots, \overline{a}_i] \\ &\quad + \sum_{i \in I_r} \lambda_i [-\overline{a}_i, \ldots, -\overline{a}_i]. \end{aligned}$$

This implies that

$$a := \sum_{i\in I_1} \lambda_i a_i = \sum_{i\in I_2} \lambda_i a_i = \cdots = \sum_{i\in I_r} \lambda_i a_i,$$

and recalling that the last component of each $\overline{a}_i$ is 1, we also have

$$\sum_{i\in I_1} \lambda_i = \sum_{i\in I_2} \lambda_i = \cdots = \sum_{i\in I_r} \lambda_i = \frac{1}{r}.$$

Define

$$A_j := \{a_i : i \in I_j\}, \quad j = 1, \ldots, r.$$

We see that $\{A_j\}_1^r$ is a partition of A into r sets such that $ra \in \cap_{j=1}^r \operatorname{co}(A_j)$. The theorem is proved. □

Note that the theorem reduces to Radon's theorem when $r = 2$. It can be given a colorful interpretation: if $A \subset \mathbb{R}^n$ has $(r-1)(n+1)+1$ distinct elements, then A can be partitioned into r monochromatic sets whose convex hulls have a common point. It is known that $(r-1)(n+1)+1$ is the best constant, so that the theorem is false for any smaller integer.

The computational complexity of these colorful theorems and related problems are discussed in [23].

13.4 Homogeneous Convex Functions

Let $K \subseteq \mathbb{R}^n$ be a convex cone containing the origin, and $p > 0$. A function $f : K \to \mathbb{R}$ is *homogeneous of degree* p if

$$f(tx) = t^p f(x) \ \text{ for all } \ t \geq 0,\ x \in K.$$

A convex homogeneous function of degree one is called a *sublinear function.* The usual definition of strict convexity is not meaningful for a sublinear function, because if x and y are positively collinear, say $y = \alpha x$ for some $\alpha > 0$, then

$$f(x+y) = f(x+\alpha x) = f((1+\alpha)x) = (1+\alpha)f(x) = f(x)+\alpha f(x) = f(x)+f(y);$$

thus it is more natural to call a sublinear function f *strictly convex* if

$$f(x+y) < f(x) + f(y) \ \text{ for all } x, y \text{ not positively collinear.}$$

Recall that if f is a convex function, then the directional derivatives always exist (but may be $\mp\infty$); in fact,

$$f'(x;d) := \lim_{t\searrow 0} \frac{f(x+td)-f(x)}{t} = \inf_{t>0} \frac{f(x+td)-f(x)}{t},$$

see Exercise 9 on page 108.

Lemma 13.14. *The directional derivative of a convex function f is sublinear, that is,*

$$f'(x; d_1 + d_2) \le f'(x; d_1) + f'(x; d_2).$$

Proof. We have

$$\begin{aligned} f\Big(x + \frac{t}{2}(d_1 + d_2)\Big) &= f\Big(\frac{1}{2}(x + td_1) + \frac{1}{2}(x + td_2)\Big) \\ &\le \frac{1}{2} f(x + td_1) + \frac{1}{2} f(x + td_2), \end{aligned}$$

which implies

$$\frac{f\left(x + \frac{t}{2}(d_1 + d_2)\right) - f(x)}{t/2} \le \frac{f(x + td_1) - f(x)}{t} + \frac{f(x + td_2) - f(x)}{t};$$

we obtain the lemma by taking the limits of both sides as $t \searrow 0$. □

Lemma 13.15. *A sublinear function $f : K \to \mathbb{R}$ is strictly convex if and only if*

$$f(y) > f'(x; y) \quad \text{for all } x, y \text{ not positively collinear.}$$

Proof. Let x, y be not positively collinear. If f is sublinear, then $f(x + ty) < f(x) + f(ty) = f(x) + tf(y)$, and this implies for any $t_0 > 0$,

$$f'(x; y) = \inf_{t>0} \frac{f(x + ty) - f(x)}{t} \le \frac{f(x + t_0 y) - f(x)}{t_0} < f(y).$$

Conversely, we have $f(x) > f'(x + y; x)$ and $f(y) > f'(x + y; y)$. Since $f'(x; \cdot)$ is sublinear by the previous lemma, we obtain

$$f(x) + f(y) > f'(x + y; x + y) = f(x + y),$$

where the last equality follows because

$$f'(z; z) = \inf_{t>0} \frac{f(z + tz) - f(z)}{t} = \inf_{t>0} \frac{(1 + t) f(z) - f(z)}{t} = f(z).$$

□

Lemma 13.16. *Let $f : K \to [0, \infty)$ be a convex function on a convex cone $K \subseteq \mathbb{R}^n$, $0 \in K$, such that $f(x) > 0$ for $x \ne 0$.*

If f is homogeneous of degree p, then $g(x) = f(x)^{1/p}$ is a sublinear function on K. If f is strictly convex in the sense that

$$f((1 - t)x + ty) < (1 - t) f(x) + t f(y),$$

where $0 < t < 1$ and x, y are not positively collinear, then g is also strictly convex.

Proof. Since g is homogeneous of degree one, it suffices to prove the inequality

$$g(x+y) \le g(x) + g(y), \tag{13.6}$$

because we then have

$$g((1-t)x + ty) \le g((1-t)x) + g(ty) = (1-t)g(x) + tg(y).$$

The inequality (13.6) is clearly true if $x = 0$ or $y = 0$. If x and y are both nonzero, then $g(x) > 0$ and $g(y) > 0$; letting $t = g(y)/(g(x)+g(y))$ (and $1 - t = g(x)/(g(x)+g(y))$, and using the convexity of f, we obtain

$$\begin{aligned} g\left(\frac{x+y}{g(x)+g(y)}\right) &= g\left((1-t)\frac{x}{g(x)} + t\frac{y}{g(y)}\right) \\ &= f\left((1-t)\frac{x}{g(x)} + t\frac{y}{g(y)}\right)^{1/p} \\ &\le \left((1-t)f(\frac{x}{g(x)}) + tf(\frac{y}{g(y)})\right)^{1/p}. \end{aligned}$$

Note that

$$f\left(\frac{x}{g(x)}\right) = \frac{1}{g(x)^p} f(x) = \frac{f(x)}{f(x)} = 1;$$

similarly $f(y/g(y)) = 1$. Thus, we have

$$g\left(\frac{x+y}{g(x)+g(y)}\right) \le ((1-t)+t)^{1/p} = 1,$$

proving (13.6). If f is strictly convex, x and y are not positively collinear, and $0 < t < 1$, then we have strict inequalities above, and g is strictly convex. □

We use the above lemma to give a quick proof of Minkowski's inequality, which is an important result in analysis. Let $p > 0$. If $x = (x_1, x_2, \ldots, x_n)$, we define

$$\|x\|_p = \left(\sum_{i=1}^n |x_i|^p\right)^{1/p}.$$

Minkowski's inequality states that $\|x\|_p$ is a sublinear function, so that $\|x\|_p$ is a norm in $\mathbb{R}^n$.

Corollary 13.17. (***Minkowski's inequality***) *Let $p > 1$. Then*

$$\|x+y\|_p \le \|x\|_p + \|y\|_p \ \textit{for all}\ \ x, y \in \mathbb{R}^n,$$

with equality holding if and only if the vectors x and y are proportional, $y = \alpha x$ or $x = \alpha y$, with $\alpha \ge 0$.

Proof. The function $t \mapsto t^p$ is strictly convex for $t \geq 0$, since its derivative pt^{p-1} is strictly increasing. It follows that the function $f(x) = \sum_{i=1}^{n} x_i^p$ is convex on $\mathbb{R}_+^n$ (it is a sum of convex functions), and it is easy to verify that f is strictly convex. Thus, the function $g(x) = f(x)^{1/p} = \|x\|_p$ is strictly convex on $\mathbb{R}_+^n$, proving Minkowski's inequality in $\mathbb{R}_+^n$,

$$\|x+y\|_p \leq \|x\|_p + \|y\|_p \quad \text{for all} \quad x, y \in \mathbb{R}_+^n,$$

with equality holding if and only if x and y are proportional.

If $x, y \in \mathbb{R}^n$ are arbitrary, denote by u and v the vectors with components $|x_i|$ and $|y_i|$, respectively. Applying Minkowski's inequality u, v gives

$$\|x+y\|_p \leq \|u+v\|_p \leq \|u\|_p + \|v\|_p = \|x\|_p + \|y\|_p,$$

where the first inequality follows from $|x_i + y_i| \leq |x_i| + |y_i|$. Now the second inequality holds with equality if and only if the vectors u and v are proportional, and the first one if and only if $|x_i + y_i| = |x_i| + |y_i|$ for all $i = 1, \ldots, n$. It is easy to see these two conditions are equivalent to the condition that $y = \alpha x$ or $x = \alpha y$ with $\alpha \geq 0$. □

13.5 Attainment of Optima in Mathematical Programming

The following result [135] gives a nice illustration of the use of the decomposition theorem, Theorem 5.37, in the investigation of attainment of an optimizer.

Theorem 13.18. *Let $f : C \to \mathbb{R}$ be a convex function that is bounded from above on a closed convex set $C \subseteq \mathbb{R}^n$. Then,*

(a) The function f is constant along any direction on the lineality space L_C of C.
(b) The function f is nonincreasing along a recession direction of C.

Consequently,

(c) The function f attains its maximum on C if and only if it attains its maximum on the set $\operatorname{ext}(\hat{C})$, where $C = \operatorname{co}(\operatorname{ext}(\hat{C})) + \operatorname{rec}(C)$ is the decomposition of C given in Theorem 5.37.

Proof. Let $M = \sup_{x \in C} f(x) < \infty$. Pick a point $x \in C$, and consider the convex function $g(l) := f(l + x)$ on the lineality subspace L_C of C. Since $l = (1/t)(tl) + (1 - 1/t)0$, we have for $t \geq 1$,

$$g(l) \leq \frac{1}{t} g(tl) + \Big(1 - \frac{1}{t}\Big) g(0) \leq \frac{1}{t} M + \Big(1 - \frac{1}{t}\Big) g(0).$$

Letting $t \to \infty$ gives $g(l) \leq g(0)$. Since $-l \in L_C$, we also have

$$g(0) \le \frac{1}{2}g(l) + \frac{1}{2}g(-l) \le \frac{1}{2}g(l) + \frac{1}{2}g(0),$$

or $g(0) \le g(l)$. Therefore, we have $g(l) = g(0)$, that is, $f(x+l) = f(x)$ for all $x \in C$ and $l \in L_C$, proving (a).

Next, pick $x \in C$ and $d \in \operatorname{rec}(C)$. Since $x + d = (1 - \frac{1}{t})x + \frac{1}{t}(x + td)$, the convexity of f gives, for $t \ge 1$,

$$f(x+d) \le \Big(1 - \frac{1}{t}\Big)f(x) + \frac{1}{t}f(x+td) \le \Big(1 - \frac{1}{t}\Big)f(x) + \frac{1}{t}M,$$

and letting $t \to \infty$ gives $f(x+d) \le f(x)$, proving (b)

Finally, if $x_1, x_2 \in \operatorname{ext}(\hat{C})$ are two extreme points and $0 < t < 1$, then

$$f((1-t)x_1 + tx_2) \le (1-t)f(x_1) + tf(x_2) \le \max\{f(x_1), f(x_2)\}.$$

Since $C = \operatorname{co}(\operatorname{ext}(\hat{C})) + \operatorname{rec}(C)$ by Theorem 5.37, the proof is complete. □

Corollary 13.19. *Let $f : C \to \mathbb{R}$ be a convex function bounded from above on a closed convex set $C \subseteq \mathbb{R}^n$.*

If C is a convex polyhedron, then f attains a maximum on C. The same result is true if $\operatorname{ext}(C)$ is compact and f is upper semicontinuous on C.

Proof. If C is a convex polyhedron, then Theorem 7.13 implies that f attains its maximum on $\{v_i\}_1^k$. The rest of the corollary follows immediately from Theorems 13.18 and 2.3. □

13.6 Decomposition of Convex Cones

In this section, we show that any pointed (line-free) convex cone decomposes into a direct sum of indecomposable or *irreducible* components in a unique fashion. This is a special case of a result in [113]; the more general result there is in an affine setting, which renders the proof more technically involved. The proof here is more accessible.

Recall that the Minkowski sum of a collection $\{A_i\}_{i=1}^m$ of sets of E is defined as

$$A_1 + \cdots + A_m := \Big\{\sum_{i=1}^m x_i : x_i \in A_i\Big\}.$$

If all of the $A_i = E_i$ are linear subspaces $\{0\} \ne E_i \subseteq E$ that satisfy $E = E_1 + \cdots + E_m$ and $E_i \cap (\sum_{j \ne i} E_j) = \{0\}$, then we say that E is a *direct sum* of $\{E_i\}_1^m$ and write

$$E = E_1 \oplus E_2 \oplus \cdots \oplus E_m.$$

In this section we assume that every cone contains the origin.

Definition 13.20. *Let $K \subseteq E$ be a pointed convex cone. K is called* decomposable *if there exist cones $\{K_i\}_{i=1}^m$, $m \geq 2$, such that $K = K_1 + \cdots + K_m$, where each K_i lies in a linear subspace $E_i \subset E$, and where the spaces $\{E_i\}_{i=1}^m$ decompose E into a direct sum $E = E_1 \oplus E_2 \oplus \cdots \oplus E_m$. Each K_i is called a* direct summand *of K, and K is called the* direct sum *of the $\{K_i\}$. We write*

$$K = K_1 \oplus K_2 \oplus \cdots \oplus K_m \tag{13.7}$$

to denote this relationship between K and $\{K_i\}_{i=1}^m$. K is called indecomposable *or* irreducible *if it cannot be decomposed into a nontrivial direct sum.*

Let us define $\hat{E}_i := \oplus_{j \neq i} E_j$ and $\hat{K}_i := \oplus_{j \neq i} K_j$. If K is the direct sum (13.7), then every $x \in K$ has a unique representation $x = x_1 + \cdots + x_m$ with $x_i \in K_i \subseteq E_i$. Thus, $x_i = \Pi_{E_i} x$, where Π_{E_i} is the projection of E onto E_i along $\hat{E}_i$. Also, since $0 \in K_i$, we have $K_i = K_i + \sum_{j \neq i}\{0\} \subseteq \sum_{j=1}^m K_j = K$. Therefore,

$$\Pi_{E_i} K = K_i \subseteq K.$$

This implies that $K_i = \Pi_{E_i} K$ is a convex cone. Similarly, we have

$$(I - \Pi_{E_i})K = \Pi_{\hat{E}_i} K = \hat{K}_i \subseteq K.$$

We first prove a useful technical result.

Lemma 13.21. *Let K be a pointed convex cone that decomposes into the direct sum (13.7). If $x \in K_i$ is a sum $x = x_1 + \cdots + x_k$ of elements $x_j \in K$, then each $x_j \in K_i$.*

Proof. We have $0 = \Pi_{\hat{E}_i} x = \Pi_{\hat{E}_i} x_1 + \cdots + \Pi_{\hat{E}_i} x_k$. Each term $\hat{x}_j := \Pi_{\hat{E}_i} x_j$ belongs to $\hat{K}_i \subseteq K$, so that $\hat{x}_j \in K$ and $-\hat{x}_j = \sum_{l \neq j} \hat{x}_l \in K$. Since K contains no lines, we have $\hat{x}_j = 0$, that is, $x_j = \Pi_{E_i} x_j \in K_i$, $j = 1, \ldots, k$.

Theorem 13.22. *Let $K \subseteq E$ be a decomposable pointed convex cone. The irreducible decompositions of K are identical modulo indexing, that is, the set of cones $\{K_i\}_{i=1}^m$ is unique. Moreover, the subspaces E_i corresponding to the nonzero cones K_i are also unique.*

If K is a solid cone, then all the cones K_i are nonzero and the subspaces $\{E_i\}_1^m$ are unique.

Proof. Suppose that K admits two irreducible decompositions

$$K = \bigoplus_{i=1}^m K_i \subseteq \bigoplus_{i=1}^m E_i \quad \text{and} \quad K = \bigoplus_{j=1}^q C_j \subseteq \bigoplus_{j=1}^q F_j.$$

Note that each nonzero summand in either decomposition of K must lie in $\operatorname{span}(K)$ and that the subspace corresponding to each zero summand must be one-dimensional, for otherwise the summand would be decomposable.

This implies that the number of zero summands in both decompositions is codim(span(K)).

We may thus concentrate our efforts on span(K), that is, we can assume that K is solid and all the summands of both decompositions of K are nonzero. By (13.7), each $x \in C_j \subseteq K$ has a unique representation $x = x_1 + \cdots + x_m$, where $x_i = \Pi_{E_i} x \in K_i \subseteq K$. Also, Lemma 13.21 implies that $x_i \in C_j$, and hence $x_i \in K_i \cap C_j$. Consequently, every $x \in C_j$ lies in the set $(K_1 \cap C_j) + \cdots + (K_m \cap C_j)$. Conversely, we have $K_i \cap C_j \subseteq C_j$, implying that $(K_1 \cap C_j) + \cdots + (K_m \cap C_j) \subseteq C_j$; therefore,

$$C_j = (K_1 \cap C_j) + \cdots + (K_m \cap C_j).$$

Note that $K_i \cap C_j \subseteq E_i \cap F_j$, $F_j = (E_1 \cap F_j) + \cdots + (E_m \cap F_j)$, and that the intersection of any two distinct summands in the last sum is the trivial subspace $\{0\}$. The above decompositions of F_j and C_j are therefore direct sums. Since C_j is indecomposable, exactly one of the summands in the decomposition of C_j is nontrivial. Thus, $C_j = K_i \cap C_j$, and hence $C_j \subseteq K_i$ for some i. Arguing symmetrically, we also have $K_i \subseteq C_l$ for some l, implying that $C_j \subseteq C_l$. Therefore, $j = l$, for otherwise $C_j \subseteq F_j \cap F_l = \{0\}$, contradicting our assumption above. This shows that $C_j = K_i$. The theorem is proved by repeating the above arguments for the cone $\hat{K}_i = \oplus_{k \neq i} K_k = \oplus_{l \neq j} C_l$. □

Theorem 13.22 is reminiscent of the Krull–Remak–Schmidt theorem in algebra; see [184].

13.7 Norms of Polynomials and Multilinear Maps

Let E, F be two vector spaces over $\mathbb{R}$ or $\mathbb{C}$ endowed with some norms. A mapping $p : E \to F$ is called a *polynomial* if for fixed $x, y \in E$, the map $t \mapsto p(x + ty)$ is a polynomial in t. A homogeneous polynomial of degree k induces a k-multilinear symmetric mapping $\tilde{p} : E^k \to F$ such that

$$p(x) = \tilde{p}(x, x, \ldots, x).$$

In fact, it is a well-known result of Mazur and Orlicz [194] that

$$\tilde{p}(x_1, \ldots, x_k) = \frac{1}{k!} \sum_{\varepsilon \in \{0,1\}^k} (-1)^{k + \sum_1^k \varepsilon_j} p\Big(\sum_1^k \varepsilon_j x_j\Big); \tag{13.8}$$

see [35] and [141], p. 393. One may associate two norms with such a mapping,

$$\|p\| := \sup\{\|p(x)\| : \|x\| = 1\} = \sup\{\|\tilde{p}(x, \ldots, x)\| \ : \ \|x\| = 1\},$$
$$\|\tilde{p}\| = \sup\{\|\tilde{p}(x_1, \ldots, x_k)\| \ : \ \|x_1\| \le 1, \ldots, \|x_k\| \le 1\}.$$

Of course, $\|p\| \le \|\tilde{p}\|$; conversely, the formula (13.8) implies that if $\|x_i\| = 1$, $i = 1, \ldots, k$, then

$$\|\tilde{p}(x_1, \ldots, x_k)\| \leq \frac{1}{k!} \sum_{\varepsilon \in \{0,1\}^k} \|p\| \cdot \left\| \sum_1^k \varepsilon_j x_j \right\|^k \leq \frac{(2k)^k}{k!} \|p\|,$$

that is,

$$\|\tilde{p}\| \leq \frac{(2k)^k}{k!} \|p\|.$$

If E, F are finite vector spaces over $\mathbb{R}$ and E is a Euclidean space, then the above norms are in fact equal. This result plays an important role in deriving some properties of self-concordant barrier functions in the book by Nesterov and Nemirovski [209] and is proved there in Appendix 1. It has an interesting history and seems to have been rediscovered many times. The first proof seems to have been given by Kellogg [162]. Subsequently, independent proofs have been given in [258, 21, 139, 35, 264, 209], and possibly others.

The following simple and elegant proof of the result is in Bochnak and Siciak [35] and is attributed to Lojasiewicz.

Theorem 13.23. *Let E be a finite-dimensional real Euclidean space, and F a real normed space. Then*

$$\|p\| = \|\tilde{p}\|.$$

Proof. It suffices to show that $\|\tilde{p}\| \leq \|p\|$. Let $S = \{x \in E : \|x\| = 1\}$ be the unit ball in E.

First, consider the case $k = 2$. If $x, y \in S$ such that $\|\tilde{p}(x, y)\| = \|\tilde{p}\|$, we claim that

$$\|\tilde{p}(x + y, x + y)\| = \|\tilde{p}\| \cdot \|x + y\|^2.$$

Otherwise, $\|\tilde{p}(x+y, x+y)\| < \|\tilde{p}\| \cdot \|x+y\|^2$; since $\|\tilde{p}(x-y, x-y)\| \leq \|\tilde{p}\| \cdot \|x-y\|^2$ and

$$\tilde{p}(x, y) = \frac{\tilde{p}(x + y, x + y) - \tilde{p}(x - y, x - y)}{4},$$

we have

$$\begin{aligned}
\|\tilde{p}\| = \|\tilde{p}(x, y)\| &< \frac{\|\tilde{p}\|}{4} (\|x + y\|^2 + \|x - y\|^2) \\
&= \frac{\|\tilde{p}\|}{4} (2\|x\|^2 + 2\|y\|^2) = \|\tilde{p}\|,
\end{aligned}$$

a contradiction.

Next, we assume $k > 2$. There exist $\tilde{x}_1, \ldots, \tilde{x}_k \in S$ such that $\|\tilde{p}\| = \|\tilde{p}(\tilde{x}_1, \ldots, \tilde{x}_k)\|$. We can find $a \in S$ such that the inner products $\langle a, \tilde{x}_i \rangle$ are nonzero for all $i = 1, \ldots, k$. Consequently, by replacing x_i by $-x_i$ if necessary, we see that there exists $\varepsilon > 0$ such that the set

$$A_\epsilon := \Big\{ (x_1, \ldots, x_k) \in S^k : \langle a, x_i \rangle \geq \epsilon,\ \forall i,\ \|\tilde{p}(x_1, \ldots, x_k)\| = \|\tilde{p}\| \Big\}$$

is a nonempty compact set. There exists a point $(x_1^*, \ldots, x_k^*) \in A_\epsilon$ that maximizes the linear functional $\sum_{i=1}^k \langle a, x_i \rangle$ over A_ϵ; we claim that

$$x_1^* = \cdots = x_k^*,$$

which will prove the theorem.

Suppose that $x_i^* \neq x_j^*$ for a pair of indices i, j. Note that $x_i^* \neq -x_j^*$, otherwise, we would have the contradiction

$$\epsilon < \langle a, x_j^* \rangle = -\langle a, x_i^* \rangle \leq -\epsilon < 0.$$

The parallelogram law

$$\|x_i^* + x_j^*\|^2 + \|x_i^* - x_j^*\|^2 = 2\|x_i^*\|^2 + 2\|x_j^*\|^2 = 4$$

implies that $0 < \|x_i^* + x_j^*\| < 2$. Consider the point $(x_1', \ldots, x_k') \in S^k$, where

$$x_i' = x_j' = \frac{x_i^* + x_j^*}{\|x_i^* + x_j^*\|}, \quad x_l' = x_l^*, \ l \neq i, j,$$

which lies in A_ϵ, because $\|x_i^* + x_j^*\| \leq 2$. In fact, $\|x_i^* + x_j^*\| < 2$, and we have

$$\sum_{l=1}^{k} \langle a, x_l' \rangle > \sum_{l=1}^{k} \langle a, x_l^* \rangle,$$

which contradicts the maximality of $(x_1^*, \ldots, x_k^*)$. □

We remark that the lemma also holds in the case that E is a real or complex Hilbert space.

13.8 Exercises

1. Suppose that a convex set $C \subseteq \mathbb{R}^n$ is covered by a finite family of open (or closed) half-planes $\{H_\alpha\}_{\alpha \in \mathcal{A}}$, that is, $C \subseteq \cup_{\alpha \in \mathcal{A}} H_\alpha$. Show that C can be covered by at most $n + 1$ of the half-planes.
   Hint: Show that the sets $\tilde{H}_\alpha := C \setminus H_\alpha$, $\alpha \in \mathcal{A}$, have empty intersection.
2. Let $\{S_i\}_{i=1}^m$ be vertical line segments in the plane, $S_i = \{(x_i, y) : \alpha_i \leq y \leq \beta_i\} = \{x_i\} \times [\alpha_i, \beta_i]$.
   (a) If any three of the segments S_i can be cut by a line, then there is a line that cuts all of the line segments $\{S_i\}_1^m$.
   Hint: Let $L_i = \{(a, b) : \alpha_i \leq a x_i + b \leq \beta_i\} \subseteq \mathbb{R}^2$ be the set of lines that cut the segment S_i.
   (b) If any $k+2$ of the line segments can be cut by a kth-degree polynomial curve $\{(x, y) : y = a_0 x^k + a_1 x^{k-1} + \cdots + a_k\}$, then there is a kth-degree polynomial curve that cuts all of the line segments $\{S_i\}_1^m$.
3. A slab of width d in $\mathbb{R}^n$ is the region between two parallel hyperplanes in $\mathbb{R}^n$ that are d-distance apart. The width of a set $A \subset \mathbb{R}^n$ is the smallest width of a slab containing A. Let $\{C_i\}_1^k$ be convex sets in $\mathbb{R}^n$ such that the intersection of every $n + 2$ of them has width at least d. Show that the width of the intersection $\cap_1^k C_i$ is at least d.

4. An interesting consequence of Helly's theorem is the following result [69]. Let $C \subset \mathbb{R}^n$ be a convex body. Show that there exists a point $\bar{x} \in C$ such that each chord $[u, v]$ in C passing through $\bar{x}$ is divided by $\bar{x}$ into two parts satisfying the condition
$$\frac{\|\bar{x} - v\|}{\|\bar{x} - u\|} \geq \frac{1}{n}.$$
Thus, $\bar{x}$ is a sufficiently "central" point of C.
Hint: For each $x \in C$, define the set
$$C_x = x + \frac{n}{n+1}(C - x).$$
If $\{x_i\}_0^n$ are arbitrary points in C, prove that the point $y = (x_0 + \cdots + x_n)/(n+1)$ lies in the intersection $\cap_{i=0}^n C_{x_i}$. Then show that the set $\cap_{x \in C} C_x$ is nonempty, and that any point $\bar{x}$ in this set satisfies the required conditions.
5. State and prove a version of Theorem 13.6 for a finite set Y without the lower semicontinuity assumptions on the functions f and g, and without assuming that $\{x : g(x, y) \leq 0, y \in Z\}$ is a bounded set for a finite subset $Z \subset Y$.
6. State and prove an analogue of Theorem 13.6 for the minimax problem
$$\inf_{x \in C} \sup_{\alpha \in A} \varphi_\alpha(x),$$
where all φ_α are lower semicontinuous convex functions on a convex set C in $\mathbb{R}^n$.
Hint: Convert the minimax problem into the minimization problem $\inf\{z : \varphi_\alpha(x) - z \leq 0, \alpha \in A\}$ as in Section 13.2.1.
7. This problem outlines a straightforward proof of a result of Bohnenblust, Karlin, and Shapley (see [157], Theorem 1), which the last two authors use to prove a number of interesting results in convex sets (including Helly's theorem) and in approximation theory [157].
Let $\mathcal{F} := \{\varphi_\alpha\}_{\alpha \in A}$ be a family of continuous convex functions on a compact convex set $K \subset \mathbb{R}^n$. Assume that
$$\text{for each } x \in K,\ \varphi(x) \text{ is positive for some function } \varphi \in \mathcal{F}. \qquad (13.9)$$
Prove that there exist $k \leq n+1$ positive numbers $\{\lambda_i\}_1^k$ satisfying $\sum_1^k \lambda_i = 1$, and functions $\{\varphi_i\}_1^k \subset \mathcal{F}$ such that the function
$$\varphi(x) := \sum_{i=1}^k \lambda_i \varphi_i(x) \text{ is positive on } K. \qquad (13.10)$$
Prove this result by the following steps:

(a) Show that the assumption (13.9) is equivalent to

$$\inf_{x\in K} \sup_{\alpha\in A} \varphi_\alpha(x) > 0. \tag{13.11}$$

(b) Convert the minimax problem into the minimization problem

$$\inf\{z \,:\, \varphi_\alpha(x) - z \leq 0,\ \alpha \in A\},$$

and use Theorem 13.6 to prove that there exist $\varphi_i(x) \in \mathcal{F}, i = 1, \ldots, k$, where $k \leq n+1$ such that the function

$$\varphi(x) := \max_{1\leq i\leq k} \varphi_i(x)$$

is positive on K.

(c) Let Δ_{k-1} be the standard unit simplex in $\mathbb{R}^k$. Note that $\varphi(x) = \max_{\lambda\in\Delta_{k-1}} \sum_{i=1}^k \lambda_i\varphi_i(x)$. The proof of (13.10) then follows from

$$\min_{x\in K} \varphi(x) = \min_{x\in K} \max_{\lambda\in\Delta_{k-1}} \sum_{i=1}^k \lambda_i\varphi_i(x) = \max_{\lambda\in\Delta_{k-1}} \min_{x\in K} \sum_{i=1}^k \lambda_i\varphi_i(x),$$

where the last equality follows from an appropriate minimax theorem; see Chapter 6 in Aubin and Ekeland [10].

8. The result of Bohnenblust, Karlin, and Shapley described in the previous problem is equivalent to Helly's theorem. In particular, show that it implies the version of the Helly's theorem in Corollary 13.4 in which each set in the family is compact. (In fact, part (c) of the previous problem is not needed.) Since the proof of Corollary 13.4 is done in two stages, first for a family compact sets, then in the general case, we obtain its full proof in this way.
Hint: Define the function $\varphi_\alpha(x) := d(x, A_\alpha) = \min_{z\in A_\alpha} \|x - z\|$, the distance function to the set A_α.
We remark that Rademacher and Schoenberg [224] give a geometric proof of Helly's theorem in a similar manner, by characterizing (in a different way) the optimal solution to a minimax problem $\min_{x\in\mathbb{R}^n} \max_i d(x, K_i)$, where $\{K_i\}$ is a finite family of compact convex sets in $\mathbb{R}^n$.
9. The purpose of this problem is to prove Jung's inequality

$$D(S) \geq \sqrt{2(n+1)/n}\, R(S)$$

using Helly's theorem. (The same problem is treated in Section 12.2 by semi-infinite programming techniques.)
Let $S \subset \mathbb{R}^n$ be a compact set. Define the diameter of S as $D(S) := \max\{\|x - y\| : x, y \in S\}$, and the inradius $R(S) := \min_{x\in\mathbb{R}^n} \max_{y\in S} \|x - y\|$, the radius of the smallest ball containing S.

(a) Prove the claim that if Jung's inequality holds for sets with cardinality at most $n+1$, then it holds for all compact sets.
Hint: For $x \in S$, define the closed ball $\overline{B}(x) := \overline{B}_{\sqrt{n/(2(n+1))}D(S)}(x)$. Show that the hypothesis of the claim implies that $\cap_{x\in\omega}\overline{B}(x) \neq \emptyset$ for all $\omega = \{x_1, \ldots, x_{n+1}\} \subseteq S$.
(b) Let $S = \{y_1, \ldots, y_{n+1}\} \subset \mathbb{R}^n$. Show that there exists a point x^* minimizing the function $\varphi(x) := \max_i \|x - y_i\|$.
(c) Show that the point x^* in (b) lies in $\text{co}(S)$, so that $\sum_{i=1}^{n+1} \lambda_i(x^* - y_i) = 0$ for some nonnegative λ_i, not all zero.
Hint: If not, then separate x^* and $\text{co}(S)$ by a hyperplane H; then, moving x^* orthogonally slightly toward H, we can decrease the distance between x^* and each y_i. Why?
(d) Complete the problem using the technique in the proof of Theorem 12.5 starting with the equations $\sum_{i,j} \lambda_i\lambda_j\|y_i - y_j\|^2 = \cdots = 2R(S)^2$.

10. State and prove versions of Bárány's and Tverberg's theorems for convex cones.
11. It is customary to prove Hölder's inequality and then deduce Minkowski's inequality from it. In this problem, we reverse the order of the proofs. Let $p > 1$ and $q > 1$ be conjugate exponents, that is,

$$\frac{1}{p} + \frac{1}{q} = 1, \quad p > 1, q > 1.$$

Hölder's inequality states

$$\langle x, y\rangle \leq \|x\|_p \cdot \|y\|_q$$

for all nonnegative, nonzero $x, y \in \mathbb{R}^n$.
(a) Assuming Minkowski's inequality, prove that Lemma 13.15 implies

$$\|y\|_q \geq \|z\|_q^{1-q}\Big\langle (z_1^{q-1}, \ldots, z_n^{q-1}), (y_1, \ldots, y_n)\Big\rangle,$$

for all $z, y \in \mathbb{R}^n_+ \backslash \{0\}$, with equality holding if and only if z and y are proportional.
(b) Define $x = (z_1^{q-1}, \ldots, z_n^{q-1})$, and show that $\|x\|_p^p = \|z\|_q^q$. Use these and (a) to conclude that Hölder's inequality holds for $x, y \in \mathbb{R}^n_+$, with equality holding if and only if the vectors $(x_1^p, \ldots, x_n^p)$ and $(y_1^q, \ldots, y_n^q)$ are proportional.
(c) Show that Hölder's inequality holds for vectors in $\mathbb{R}^n$, with equality holding if and only if the vectors $(|x_1|^p, \ldots, |x_n|^p)$ and $(|y_1|^q, \ldots, |y_n|^q)$ are proportional.

12. Let $p > 1$, and define the homogeneous function of degree one

$$f_p(x) = (x_1^p - |x_2|^p - \cdots - |x_n|^p)^{1/p}$$

on the set

$$K_p = \left\{ x = (x_1, \tilde{x}) = (x_1, x_2, \ldots, x_n) : x_1 \geq \|\tilde{x}\|_p \right\}.$$

The goal of this problem is to show that f_p is a concave function on K_p.

(a) Show that K_p is a convex cone. (When $p = 2$ and $n = 4$, this is the forward *light cone*, or the *Lorentz cone*, in physics.)

(b) The *hypograph* of f is the set

$$\mathrm{hypo}(f_p) := \{(x, \mu) : x \in K, f_p(x) \geq \mu\}.$$

Note that $f_p(x) \geq 0$, and use it to show that f_p is a concave function if and only if $\mathrm{hypo}(f) \cap \{(x, \mu) : \mu \geq 0\}$ is convex.

(c) Show that

$$\mathrm{hypo}(f_p) \cap \{(x, \mu) : \mu \geq 0\} = \{(x_1, \tilde{x}, \mu) : \mu \geq 0, x_1 \geq \|(\mu, \tilde{x})\|_p\}.$$

Show that this set is convex; conclude that f_p is a *superlinear function*,

$$f_p(x + y) \geq f_p(x) + f_p(y) \quad \text{for all} \quad x, y \in K_p.$$

This is called *Bellman's inequality*.

(d) Show that the above results and Lemma 13.15 imply *Popoviciu's inequality*

$$x_1 y_1 - x_2 y_2 - \cdots - x_n y_n \geq f_p(x) \cdot f_q(y) \quad \forall x \in K_p,\ y \in K_q,$$

where $p, q > 1$ are conjugate exponents, that is, $p^{-1} + q^{-1} = 1$.

13. **(Hilbert metric)** Let K be a compact convex body in $\mathbb{R}^n$. Let us first define an asymmetric distance function $D(a, b)$ on $C = \mathrm{int}\, K$,

$$D(a, b) = \ln \frac{|bf|}{|af|};$$

see Figure 13.1. Note that $D(a, b) \to \infty$ if $a \to \partial K$. Show that

$$D(a, c) \leq D(a, b) + D(b, c),$$

by verifying the following assertions:

(a) Define $a = (1 - \alpha) f + \alpha b$; then $D(a, b) = -\ln \alpha$. Similarly, define $b = (1 - \beta) g + \beta c$, $a = (1 - \gamma) k + \gamma c$, and $k = (1 - \delta) f + \delta g$. Show that

$$\gamma = \alpha\beta, \quad k = \frac{1 - \alpha}{1 - \alpha\beta} f + \frac{\alpha(1 - \beta)}{1 - \alpha\beta} g.$$

(b) Show that

$$D(a, c) \leq \ln \frac{|ck|}{|ak|} = -\ln(\alpha\beta) = D(a, b) + D(b, c).$$

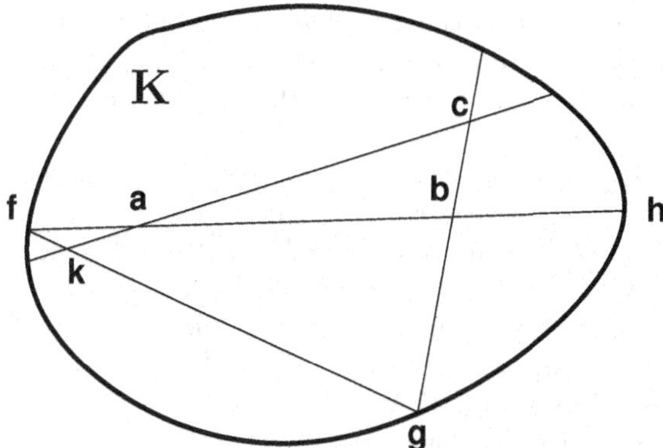

Fig. 13.1. Hilbert metric on a convex body.

Consequently, $D(a,b)$ satisfies the triangular inequality, and it is clear that $D(a,b)=0$ if and only $a=b$. However, $D(a,b)\neq D(b,a)$, so D is not a metric in the usual sense. It is easy to fix this defect by defining a new distance

$$d(a,b) := D(a,b) + D(b,a) = \ln\left(\frac{|bf|\cdot|ah|}{|af|\cdot|bh|}\right).$$

(c) Show that $d(a,b)$ is a metric in the usual sense.

It is called the *Hilbert metric* on C. Note that $d(a,b)\to\infty$ when a or b approaches the boundary of C.

14. This problem constructs an interesting isomorphism between $\mathbb{R}^n$ and an open convex polyhedral cone, and another one between $\mathbb{R}^n$ and the interior of a solid convex polytope.

Consider the functions $f,g:\mathbb{R}^n\to\mathbb{R}$,

$$f(x)=\sum_{i=1}^{k}\alpha_i e^{\langle u_i,x\rangle} \quad\text{and}\quad g(x)=\ln f(x),$$

where $\{\alpha_i\}_{i=1}^{k}$ are positive numbers. Assume that the convex cone

$$K=\left\{\sum_{i=1}^{k}\mu_i u_i : \mu_i>0,\ i=1\ldots,k\right\}$$

is a proper ($K\neq\mathbb{R}^n$) and open polyhedral cone in $\mathbb{R}^n$.

Define the maps

$$F(x):=\nabla f:\mathbb{R}^n\to\mathbb{R}^n \quad\text{and}\quad G(x):=\nabla g:\mathbb{R}^n\to\mathbb{R}^n.$$

(a) Show that the function f is strictly convex.

Hint: Use the Taylor series for $f(x+td)$ to compute $d^T Hf(x)d$, and prove that $d^T Hf(x)d = d^T DF(x)d = 0$ implies $d=0$.

(b) Use (a) to prove that the mapping $F(x)$ is one-to-one. Use the inverse function theorem to prove that the range $F(\mathbb{R}^n)$ is an open set.

(c) If u_1 is an extreme direction of the polyhedral cone $\overline{K}$, then show that $u_1 \in \overline{F(\mathbb{R}^n)}$.
Hint: Argue that the ray $\{tu_1 : t \geq 0\}$ is an exposed face of $\overline{K}$, so that there exists a vector v such that $\langle u_1, v\rangle = 0$, but $\langle u_i, v\rangle < 0$ for $j = 2, \dots, k$. See what happens to $F(w + tv)$ as $t \to \infty$, and show that one can choose $w \in \mathbb{R}^n$ such that $F(w + tv) \to u_1$ as $t \to \infty$.

(d) Prove that $F(\mathbb{R}^n)$ is almost convex in the sense that the sets $\overline{F(\mathbb{R}^n)}$ and $\operatorname{int} F(\mathbb{R}^n)$ are both convex.
Hint: Use Proposition 11 in Section 6.7 of Aubin and Ekeland [10]; even better, prove the assertion directly.

(e) So far, it has been shown that $F(\mathbb{R}^n)$ is an open convex set contained in K whose closure is $\overline{K}$. Prove that $F(\mathbb{R}^n) = K$. Argue that F is an analytic isomorphism between $\mathbb{R}^n$ and K.

(f) Prove that the function $g(x)$ is convex.
Hint: Either compute the Hessian $Hg(x)$ using the Taylor series for $g(x + td)$, or use Exercise 19 on p. 111.

(g) Show that

$$G(x) = \nabla g(x) = \sum_{i=1}^{k} \frac{\alpha_i}{f(x)} e^{\langle u_i, x\rangle} u_i$$

maps $\mathbb{R}^n$ one-to-one onto the interior of the convex polytope $P = \operatorname{co}(u_1, \dots, u_k)$.
Hint: Let $\tilde{K}$ be the open polyhedral cone generated by the vectors $\{(u_i, 1)\}_{i=1}^k$ in $\mathbb{R}^{n+1}$ and use the results above to show that

$$F(x, x_{n+1}) := e^{x_{n+1}} \Big(\sum_{i=1}^{k} \alpha_i e^{\langle u_i, x\rangle} u_i, \sum_{i=1}^{k} \alpha_i e^{\langle u_i, x\rangle} \Big)$$

maps $\mathbb{R}^{n+1}$ one-to-one onto $\tilde{K}$. To finish the proof, note that P can be considered as the intersection of $\tilde{K}$ with the plane $y_{n+1} = 1$.

14

Three Basic Optimization Algorithms

This chapter has a different focus from that of the rest of the book. While all the other chapters deal with the *theory* of optimization, this chapter deals with *numerical algorithms* for finding solutions of optimization problems. Although, for space reasons, we cannot go into this vast, central aspect of optimization in great depth, it is important to give an introduction to it, and to show how numerical methods and the theory of optimization fit together.

In this chapter, we discuss three central algorithms: the *steepest-descent method, Newton's method*, and the *conjugate-gradient method.* The first two are chosen because many algorithms in optimization are based on them. The last algorithm is a powerful method for solving the very specific problem of minimizing convex quadratic functions in $\mathbb{R}^n$ (or equivalently for solving linear equations $Ax = b$ where A is a symmetric positive definite matrix). It is chosen because of its intrinsic importance, and to show that, sometimes, a simple modification of a slow algorithm (in this case the steepest-descent method) can lead to a method with much better theoretical and practical properties.

The steepest-descent method is indeed a very slow algorithm and is not used much in practice. Its merit lies in the fact that it is simple and reliable. Once we choose any starting point and an appropriate line-search routine, the method will decrease the value of the function we are minimizing and eventually will find a local minimizer. The success of an optimization algorithm often hinges on the quality of its line-search algorithm, and the line-search routines we discuss here are also useful in other algorithms.

We discuss the convergence rate of the steepest-descent method for minimizing a very simple function, a strongly convex quadratic function. The estimate we give, due to Kantorovich, hints that the steepest-descent method may be very slow; a later result of Akaike [2] confirms that the steepest-descent method often performs at the rate given by Kantorovich's estimate.

We also examine the steepest-descent method for minimizing a class of convex functions, and show that it is possible to extend the convergence rate results for quadratic functions to this class of problems. Moreover, it is even pos-

O. Güler, *Foundations of Optimization*, Graduate Texts in Mathematics 258,
DOI 10.1007/978-0-387-68407-9_14,

sible to obtain conjugate-gradient-like methods for the same class of problems with improved convergence rates; see for example [206, 204, 207, 117, 208].

Newton's method (sometimes called the Newton–Raphson method) is actually a method for the more general problem of finding solutions to a system of nonlinear equations. It is a *local* method, meaning that it is guaranteed to work in a neighborhood (perhaps a very small neighborhood) of a solution; it could even be chaotic far away from a solution. It has the merit is that it is a very *fast* method; in technical terms, it converges *quadratically* in a neighborhood of the solution. The celebrated theory of Kantorovich can be used to estimate the size of the quadratic convergence region. We give an account of his theory in this chapter. There have been two important recent advances in Newton methods that we do not treat here. One is Smale's theory [33] for estimating the quadratic convergence region using data (function values and derivatives) at a single point, and the other is the theory of *self-concordant* functions due to Nesterov and Nemirovski [209], which is the cornerstone of the theory of *interior-point methods.* Self-concordant functions are a very special class of convex functions for which Newton's method converges globally, at a predetermined linear rate far away from a solution and at a quadratic rate locally. The interested reader can find a wealth of information in the book [209] and a concise introduction in [226].

The conjugate-gradient method of Hestenes and Stiefel [132] was originally a method for minimizing a strongly convex quadratic function. In this context, it is a vast improvement over the steepest-descent method discussed above. Although it has been extended to more general classes of functions as discussed above, its properties are most remarkable and best understood in its original setting, and we restrict our treatment to this case. Nowadays, the conjugate-gradient method is a very important numerical method. It is used in practice as an iterative numerical method for solving a linear system of equations $Ax = b$, where A in an $n \times n$ symmetric, positive definite matrix, with A sparse and n large.

We provide a fairly complete treatment of the conjugate-gradient method here. We derive the algorithm, show its remarkable properties, and give a complete derivation of a convergence-rate estimate for it. To a beginner, the conjugate-gradient method may seem intricate and hard to understand. In Section 14.10, we give an independent treatment of the method that we believe is fairly straightforward, certainly shorter.

14.1 Gradient-Descent Methods

Let U be a nonempty open subset of $\mathbb{R}^n$, and $f : U \to \mathbb{R}$ a function with continuous partial derivatives $\partial f/\partial x_i$, $i = 1, \ldots, n$; we recall that such a function is Fréchet differentiable.

Gradient-descent methods try numerically to find a local minimizer of f iteratively, using only the function value and gradient information; they generate a sequence of points $\{x_k\}_0^\infty \subseteq U$ such that if x_k is not already a local minimizer of f, then

$$x_{k+1} = F(x_k, \nabla f(x_k))$$

for a suitable function F such that

$$f(x_{k+1}) < f(x_k).$$

Thus, the function values decrease at each step of the method.

14.1.1 Descent Directions

Definition 14.1. *Let $x \in U$ such that $\nabla f(x) \neq 0$, so that x is not a critical point of f. A* descent direction *for f is a nonzero vector $d \in \mathbb{R}^n$ such that there exists $\bar{t} > 0$ with the property*

$$f(x + td) < f(x) \quad \text{for all } t, \; 0 < t < \bar{t}. \tag{14.1}$$

Thus, f strictly decreases along the half-line $\mathbb{R}_{++}d := \{x + td : t > 0\}$ for sufficiently small step sizes $t > 0$.

If f is differentiable, it is easy to characterize descent directions by calculus.

Lemma 14.2. *Let $x \in U$ be a noncritical point of f, and $d \in \mathbb{R}^n$ a nonzero vector.*

If $\langle \nabla f(x), d\rangle < 0$ (d makes an obtuse angle with the gradient $\nabla f(x)$), then d is a descent direction of f at x.

Conversely, if d is a descent direction of f at x, then $\langle \nabla f(x), d\rangle \leq 0$.

Proof. Since f is Gâteaux differentiable,

$$f(x + td) = f(x) + t\langle \nabla f(x), d\rangle + o(t). \tag{14.2}$$

Thus, if d satisfies $\langle \nabla f(x), d\rangle < 0$, then $f(x + td) < f(x)$ for all sufficiently small $t > 0$; conversely, if $f(x+td) < f(x)$ for all sufficiently small $t > 0$, then letting $t \downarrow 0$ in (14.2) gives $\langle \nabla f(x), d\rangle \leq 0$. □

Remark 14.3. If $x(t)$ is a curve on the level set $f^{-1}(\alpha) = \{x : f(x) = \alpha\}$, then differentiating the equation $f(x(t)) = \alpha$ gives

$$\langle \nabla f(x(t)), x'(t)\rangle = 0.$$

Since $x'(t)$ is tangent to the level set $f^{-1}(\alpha)$, we see that the gradient $\nabla f(x(t))$ is a normal vector to the tangent plane of $f^{-1}(\alpha)$ at the point x.

The Steepest-Descent Direction

In (14.2), we have

$$f(x + td) \approx f(x) + t\langle \nabla f(x), d\rangle \quad \text{for small} \quad t > 0.$$

Thus, it seems that we can make the *most* decrease in the function value of f for a fixed small $t > 0$ if we minimize the quantity $\langle \nabla f(x), d\rangle$ over all directions $d \in \mathbb{R}^n$ with $\|d\| = 1$.

Since

$$\langle \nabla f(x), d\rangle = \|\nabla f(x)\| \cdot \|d\| \cos\theta = \|\nabla f(x)\| \cos\theta,$$

we would choose $\cos(\theta) = -1$, that is,

$$d = -\frac{\nabla f(x)}{\|\nabla f(x)\|};$$

this is the reason why the (unnormalized) direction $d = -\nabla f(x)$ is called the *steepest-descent direction* of f at the point x.

14.1.2 Step-Size (Step-Length) Selection Rules

Once a descent direction d_k is somehow chosen at the point x_k, the next iterate is given by

$$x_{k+1} := x_k + t_k d_k, \quad t_k > 0,$$

where the *step size* t_k is a suitably chosen quantity. One must exercise great caution in choosing t_k; improperly chosen t_k may in fact lead to methods that do *not* converge to a local minimizer.

We list here some step-length rules that have been proposed over the years:

1. (***Exact minimization rule***) Choose t_k such that

   $$f(x_k + t_k d_k) \le f(x_k + t d_k) \ \text{ for all } \ t \ge 0,$$

   that is, $x_k + t_k d_k$ is the global minimizer of t on the half-line $\{x_k + td_k : t \ge 0\}$. This rule may be very expensive or even impossible, except for some special classes of functions such as the quadratic function we will encounter in Section 14.2.
2. (***Limited minimization rule***) Choose t_k such that

   $$f(x_k + t_k d_k) \le f(x_k + t d_k) \ \text{ for all } \ 0 \le t \le s,$$

   where $s > 0$ is some predetermined quantity.
3. (***Constant step-length rule***)

   $$x_{k+1} = x_k + \alpha d_k,$$

   where $\alpha > 0$ is a fixed, predetermined constant.

This rule works for some functions having special properties. For example, if $\nabla f(x)$ is a Lipschitz continuous function, that is, there exists a constant $L > 0$ such that

$$\|\nabla f(y) - \nabla f(x)\| \le L\|x - y\|,$$

then one can choose $\alpha_k = \|\nabla f(x_k)\|/L$, that is, $x_{k+1} = x_k - \nabla f(x_k)/L$; see Theorem 14.13 on page 373.

4. (***Armijo's rule***) Fix $s > 0$, $0 < \beta < 1$, and $0 < \sigma < 1$, and test the inequality

$$f(x_k) - f(x_k + \beta^i s d_k) \ge -\sigma(\beta^i s)\langle \nabla f(x_k), d_k\rangle, \quad i = 0, 1, 2, \ldots, \tag{14.3}$$

iteratively, starting with $i = 0$.
The idea here is that the decrease $f(x_k) - f(x_{k+1})$ in the function value should be *sufficiently* large. We first test the inequality (14.3) with $i = 0$, that is, with step size s,

$$f(x_k) - f(x_k + s d_k) \ge -\sigma s\langle \nabla f(x_k), d_k\rangle.$$

If the above inequality is satisfied, we set $x_{k+1} = x_k + s d_k$; otherwise, the decrease $f(x_k) - f(x_{k+1})$ in the function value is deemed not large enough, and we cut back the step size to βs: we set $x_{k+1} := x_k + \beta s d_k$ and test the inequality (14.3) with $i = 1$,

$$f(x_k) - f(x_k + \beta s d_k) \ge -\sigma\langle \nabla f(x_k), \beta s d_k\rangle,$$

etc. This method is very practical. However, we need to prove the implicit claim that the iterations (14.3) will eventually terminate. Suppose the claim is false. If i in (14.3) is large, then $t := \beta^i s > 0$ is small, and (14.3) gives

$$f(x_k + t d_k) > f(x_k) + \sigma t\langle \nabla f(x_k), d_k\rangle.$$

Comparing this with

$$f(x_k + t d_k) = f(x_k) + t\langle \nabla f(x_k), d_k\rangle + o(t),$$

we obtain

$$(\sigma - 1)\langle \nabla f(x_k), d_k\rangle + \frac{o(t)}{t} < 0;$$

letting $t \searrow 0$ gives

$$(\sigma - 1)\langle \nabla f(x_k), d_k\rangle \le 0,$$

a contradiction, because $\sigma - 1 < 0$ and $\langle \nabla f(x_k), d_k\rangle < 0$. The claim is proved.

5. (***Goldstein's rule***) Fix $c \in (0, 1/2)$. Choose t_k such that

$$\begin{aligned} f(x_k) + (1 - c)t_k\langle \nabla f(x_k), d_k\rangle &\le f(x_k + t d_k) \\ &\le f(x_k) + c t_k\langle \nabla f(x_k), d_k\rangle. \end{aligned}$$

As in Armijo's rule, the second inequality here requires that the function value of f decrease sufficiently in going from the point x_k to the new point $x_k + t_k d_k$. The first inequality can be thought of as requiring that the step size t_k not be too small.

6. (***Wolfe's rule***) Fix two constants $0 < c_1 < c_2 < 1$. Choose t_k such that

$$\begin{aligned} f(x_k + td_k) &\leq f(x_k) + c_1 t_k \langle \nabla f(x_k), d_k \rangle, \\ \langle \nabla f(x_k + td_k), d_k \rangle &\geq c_2 \langle \nabla f(x_k), d_k \rangle. \end{aligned} \tag{14.4}$$

The first inequality requires that the function value of f decrease sufficiently, and the second one requires that the step size t_k be not too small; see Figure 14.1.
This is a popular step-size selection rule in quasi-Newton methods.

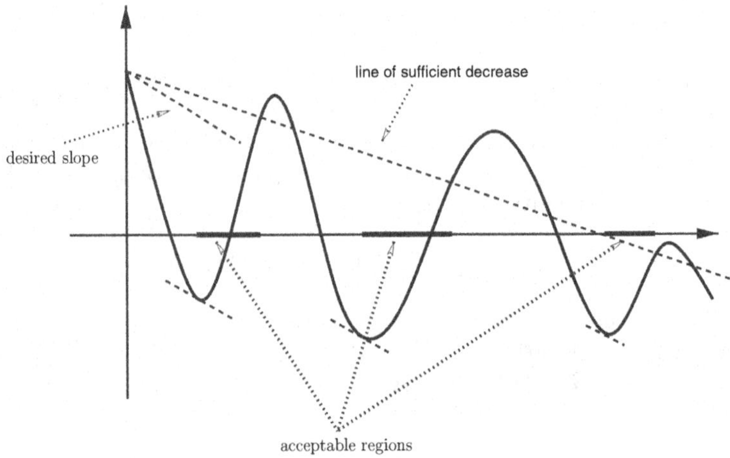

Fig. 14.1. Wolfe's step-length selection rule.

14.1.3 Convergence of Descent Methods

In this section we prove the convergence of descent methods under several step-size rules, including the popular Armijo's and Wolfe's rules.

Let the descent direction d_k satisfy the condition that the angle θ_k between the vectors $-\nabla f(x_k)$ and d_k is positive and bounded away from zero, that is, there exists $\epsilon > 0$ such that

$$\cos \theta_k := \frac{\langle -\nabla f(x_k), d_k \rangle}{\|\nabla f(x_k)\|} \in (\epsilon, 1], \quad \|d_k\| = 1. \tag{14.5}$$

Theorem 14.4. *Let $\{x_k\}_0^\infty$ be the sequence of vectors generated by a descent method*

$$x_{k+1} = x_k + \alpha_k d_k,$$

where d_k satisfies (14.5), *and α_k is chosen by the Armijo rule with parameters s, β, σ. If $\lim_{i\to\infty} x_{k_i} = x^*$ is a limit point of the iteration sequence $\{x_k\}_0^\infty$, then x^* is a critical point of f, that is, $\nabla f(x^*) = 0$.*

Proof. By taking a further subsequence if necessary, we may assume that d_{k_i} converges to some direction vector d^*, $\|d^*\| = 1$. Since the ith step is successful in (14.3), we have

$$f(x_{k_i}) - f(x_{k_i} + \alpha_{k_i} d_{k_i}) \geq -\sigma\alpha_{k_i}\langle \nabla f(x_{k_i}, d_{k_i}\rangle \geq \epsilon\sigma\alpha_{k_i}\|\nabla f(x_{k_i}\|, \quad (14.6)$$

and if there is a previous unsuccessful step, we also have

$$f(x_{k_i}) - f\Big(x_{k_i} + \frac{\alpha_{k_i}}{\beta} d_{k_i}\Big) < -\sigma\frac{\alpha_{k_i}}{\beta}\langle \nabla f(x_{k_i}), d_{k_i}\rangle. \quad (14.7)$$

Suppose that $\nabla f(x^*) \neq 0$. Since $x_{k_i} \to x^*$ and $f(x_k)$ is a decreasing sequence, we have $f(x_{k_i}) \searrow f(x^*)$ and the left-hand side of (14.6) converges to zero as $i \to \infty$. The other two terms of the same inequality converge to zero as well, and by virtue of our assumption (14.5), we obtain

$$\langle \nabla f(x^*), d^*\rangle < 0 \quad \text{and} \quad \alpha_{k_i} = \|x_{k_i+1} - x_{k_i}\| \to 0, \quad (14.8)$$

and the second fact implies that step-length selection requires backtracking at x_{k_i} when i is large.

We can use (14.7) to arrive at a contradiction: by the mean value theorem,

$$-\frac{\alpha_{k_i}}{\beta}\langle \nabla f(z_{k_i}), d_{k_i}\rangle < -\sigma\frac{\alpha_{k_i}}{\beta}\langle \nabla f(x_{k_i}), d_{k_i}\rangle$$

for some $z_{k_i} \in (x_{k_i}, x_{k_i} + \alpha_{k_i}\beta^{-1}d_{k_i})$. We have $\lim_{i\to\infty} z_{k_i} = x^*$; simplifying the above inequality, and taking limits as $i \to \infty$, we obtain $\langle \nabla f(x^*), d^*\rangle \geq 0$. This contradicts (14.8) and proves that $\nabla f(x^*) = 0$. □

Corollary 14.5. *Suppose that the search directions satisfy* (14.5). *If the step size is chosen by the exact minimization rule or limited minimization rule, then any limit x^* of the sequence $\{x_k\}_0^\infty$ is a critical point of f, that is, $\nabla f(x^*) = 0$.*

Proof. Suppose that $\lim_{i\to\infty} x_{k_i} = x^*$, but $\nabla f(x^*) \neq 0$. Let $\hat{x}_{k_i} = x_{k_i} + \hat{\alpha}_{k_i} d_{k_i}$ be the point that would be chosen using an Armijo-type rule. Then

$$f(x_{k_i}) - f(x_{k_i+1}) \geq f(x_{k_i}) - f(\hat{x}_{k_i+1}) \geq -\sigma\hat{\alpha}_{k_i}\langle \nabla f(x_{k_i}), d_{k_i}\rangle.$$

Since $f(x_k) \searrow f(x^*)$, we have $f(x_{k_i}) - f(x_{k_{i+1}}) \searrow 0$, and as in the proof of the previous theorem, we obtain $\lim_{i\to\infty} \hat{\alpha}_{k_i} = 0$. Also, we have, from (14.7)

$$f(x_{k_i}) - f(x_{k_i} + \frac{\hat{\alpha}_{k_i}}{\beta} d_{k_i}) < -\sigma \frac{\hat{\alpha}_{k_i}}{\beta} \langle \nabla f(x_{k_i}), d_{k_i} \rangle.$$

As in the previous theorem, we arrive at a contradiction to the descent condition $\langle \nabla f(x^*), d^* \rangle < 0$. □

Finally, we consider the convergence of descent methods under Wolfe's step-size selection rule. We start with

Lemma 14.6. (***Zoutendijk***) *Let f be a function bounded from below on the sublevel set $M := \{x : f(x) \leq f(x_0)\}$, with a Lipschitz continuous gradient there, that is, for some $L > 0$,*

$$\|\nabla f(y) - \nabla f(x)\| \leq L\|y - x\| \quad \textit{for all } x, y \in M.$$

Let $\{x_k\}_0^\infty$ be the sequence of points generated by a descent algorithm using Wolfe's step-selection rule (14.4)*. Then*

$$\sum_{k=0}^{\infty} \cos^2 \theta_k \|\nabla f(x_k)\|^2 < \infty,$$

where θ_k is defined in (14.5)*.*

Proof. Recalling $\|d_k\| = 1$, we have

$$(c_2 - 1)\big\langle \nabla f(x_k), d_k \big\rangle \leq \big\langle \nabla f(x_{k+1}) - \nabla f(x_k), d_k \big\rangle \leq t_k L,$$

where the inequalities follow from the second condition in Wolfe's rule and the Lipschitz condition on f, respectively. This gives a lower bound for the step size,

$$t_k \geq \frac{1 - c_2}{L} \langle -\nabla f(x_k), d_k \rangle,$$

and substituting it into the first condition in Wolfe's rule (14.4) gives

$$\begin{aligned} f(x_k) - f(x_{k+1}) &\geq \frac{c_1(1 - c_2)}{L} |\langle \nabla f(x_k), d_k \rangle|^2 \\ &= \frac{c_1(1 - c_2)}{L} \cos^2 \theta_k \|\nabla f(x_k)\|^2. \end{aligned}$$

The lemma follows, since

$$f(x_0) - f(x_K) = \sum_{k=0}^{K-1} [f(x_k) - f(x_{k+1})] \geq \frac{c_1(1 - c_2)}{L} \sum_{k=0}^{K-1} \cos^2 \theta_k \|\nabla f(x_k)\|^2$$

and $f(x_0) - f(x_K) \leq f(x_0) - \inf_M f < \infty$. □

We immediately have the following corollary.

Corollary 14.7. *Let f be a function satisfying the conditions of Lemma 14.6. If the search directions $\{d_k\}$ satisfy the condition* (14.5) *and $x^* \in M$ is a limit point of the sequence $\{x_k\}$, then $\nabla f(x^*) = 0$.*

14.2 Convergence Rate of the Steepest-Descent Method on Convex Quadratic Functions

The results above on the gradient-descent methods above have been very modest; we have showed only that if the minimizing sequence $\{x_i\}_0^\infty$ has a limit point x^*, then x^* is a critical point. Any result on the *rate of convergence* to x^* has been absent. In fact, it is impossible to give such results for *general* problems. The situation is radically different for convex problems. Since the introduction of the famous ellipsoid method of Nemirovski and Yudin [205], we know that there are methods to solve convex programming problems in polynomial time.

In this section, we analyze the convergence rate of the steepest-descent algorithm on the minimization of a strongly convex quadratic function.

Consider the quadratic function

$$q(x) = \frac{1}{2}\langle Qx, x\rangle - \langle b, x\rangle + a, \tag{14.9}$$

where Q is an $n \times n$ symmetric positive definite matrix, $b \in \mathbb{R}^n$, and $a \in \mathbb{R}$. We write

$$r(x) := \nabla q(x) = Qx - b$$

throughout this subsection.

Of course, the global minimizer x^* of q over $\mathbb{R}^n$ satisfies $r(x^*) = Qx^* - b = 0$, that is, $x^* = Q^{-1}b$. The steepest-descent method for minimizing q does not need any line searches; if x_k is at hand, then x_{k+1} may be computed exactly by minimizing q on the line $\ell = \{x_k - \alpha r_k : \alpha \in \mathbb{R}\}$. It is easy to compute x_{k+1}: letting

$$h(\alpha) := q(x_k - \alpha r_k) = q(x_k) - \alpha\|r_k\|^2 + \frac{\alpha^2}{2}\langle Qr_k, r_k\rangle,$$

we see that the exact minimizer α_k of h is given by

$$0 = h'(\alpha_k) = -\|r_k\|^2 + \alpha_k\langle Qr_k, r_k\rangle;$$

thus

$$x_{k+1} = x_k - \alpha_k r_k, \quad \text{where} \quad \alpha_k = \frac{\|r_k\|^2}{\langle Qr_k, r_k\rangle}.$$

We need the following technical result for estimating the convergence rate of the steepest-descent algorithm for minimizing q.

Lemma 14.8. ***(Kantorovich's inequality)*** *If Q is a symmetric positive definite $n \times n$ matrix with eigenvalues $\{\lambda_i\}_1^n$ in the interval $[m, M]$, then*

$$\frac{\langle Qx, x\rangle \cdot \langle Q^{-1}x, x\rangle}{\|x\|^4} \le \frac{(m+M)^2}{4mM}.$$

Proof. Since the above inequality remains unchanged if we replace each Q with τQ, where τ is any positive constant, we assume that $mM = 1$. Let $Q = U^T \Lambda U$ be the spectral decomposition of Q, where $\Lambda = \mathrm{diag}(\lambda_1, \dots, \lambda_n)$. Setting $s = Ux$, where $\|x\| = \|s\| = 1$, and defining $t_i = s_i^2$, $i = 1, \dots, n$, the lemma reduces to proving the inequality

$$\Big(\sum_{i=1}^{n} t_i \lambda_i\Big) \cdot \Big(\sum_{i=1}^{n} \frac{t_i}{\lambda_i}\Big) \le \frac{\left(m + \frac{1}{m}\right)^2}{4} \quad \text{for all} \quad t_i \ge 0, \ \sum_{i=1}^{n} t_i = 1,$$

where $\lambda_i \in [m, 1/m]$, $i = 1, \dots, n$.

This inequality follows from the claim that

$$\Big(\sum_{i=1}^{n} t_i \lambda_i\Big) \cdot \Big(\sum_{i=1}^{n} \frac{t_i}{\lambda_i}\Big) \le \frac{\left(\sum_{i=1}^{n} t_i (\lambda_i + \lambda_i^{-1})\right)^2}{4} \le \frac{\left(m + \frac{1}{m}\right)^2}{4}.$$

The first inequality follows because of the inequality $ab \le (a+b)^2/4$; the second inequality follows since $\lambda_i + \lambda_i^{-1} \le m + m^{-1}$, due to the fact that the maximum of the function $\lambda + \lambda^{-1}$ over the interval $[m, 1/m]$ is attained at the endpoints m and $1/m$. □

A different proof of Kantorovich's inequality is given in Section 10.4 using nonlinear programming techniques.

Theorem 14.9. *In the steepest-descent method for minimizing a strongly convex quadratic function $q(x)$ in (14.9), the optimality gap $E(x) = q(x) - \min_{\mathbb{R}^n} q$ decreases at a geometric rate,*

$$E(x_{k+1}) \le \left(\frac{\kappa - 1}{\kappa + 1}\right)^2 E(x_k),$$

where $\kappa = \lambda_{\max}/\lambda_{\min}$ is the condition number of the matrix Q.

Proof. Since $x_{k+1} - x_k = -\alpha_k r_k$, we have

$$E(x_{k+1}) = E(x_k) - \alpha_k \|r_k\|^2 + \frac{\alpha_k^2}{2} \langle Q r_k, r_k \rangle = E(x_k) - \frac{1}{2} \frac{\|r_k\|^4}{\langle Q r_k, r_k \rangle},$$

where the first equality follows from Taylor's formula and the second equality follows because $\alpha_k = \|r_k\|^2 / \langle Q r_k, r_k \rangle$. We also have

$$\begin{aligned} E(x_k) &= E(x^*) + \langle \nabla E(x^*), x_k - x^* \rangle + \frac{1}{2} \langle Q(x_k - x^*), x_k - x^* \rangle \\ &= \frac{1}{2} \langle Q^{-1} r_k, r_k \rangle, \end{aligned}$$

where the first equation is Taylor's formula, the second one follows because $E(x^*) = 0$, $\nabla E(x^*) = 0$, and $r_k = Q(x_k - x^*)$. We deduce that

$$\begin{aligned}\frac{E(x_{k+1})}{E(x_k)} &= 1 - \frac{\|r_k\|^4}{\langle Qr_k, r_k\rangle \cdot \langle Q^{-1}r_k, r_k\rangle} \le 1 - \frac{4mM}{(M+m)^2} \\ &= \left(\frac{M-m}{M+m}\right)^2 = \left(\frac{\kappa-1}{\kappa+1}\right)^2,\end{aligned}$$

where the inequality follows from Lemma 14.8. □

Corollary 14.10. *In the steepest-descent method for minimizing a strongly convex quadratic function q on $\mathbb{R}^n$, the optimality gap $E(x) = q(x) - \min_{\mathbb{R}^n} q$ is halved in every $O(\kappa)$ iterations, where $\kappa = \lambda_{\max}/\lambda_{\min}$ is the condition number of the matrix Q.*

Proof. Let m be the smallest integer satisfying the condition $(1-2/(\kappa+1))^m \le 1/2$. We have $\frac{E(x_m)}{E(x_0)} \le \frac{1}{2}$, and if κ is large, then

$$-\ln 2 \approx m \ln\Big(1 - \frac{2}{\kappa+1}\Big) \approx \frac{-2m}{\kappa+1} \approx \frac{-2m}{\kappa},$$

where we have used

$$\ln(1-t) = \int \frac{dt}{1-t} = \int (1 + t + t^2 + \cdots)\, dt = t + \frac{t^2}{2} + \cdots \approx t$$

for small $|t|$. This proves that $m = O(\kappa)$. □

The convergence rates given in Theorem 14.9 and Corollary 14.10 suggest that the steepest-descent method converges very slowly. For example, if Q has a condition number $\kappa = 10^6$, then our estimates indicate that the steepest-descent method would require on the order of a million iterations to reduce the initial optimality gap $E(x_0)$ by half. Numerical computations support this pessimistic view. Moreover, it was shown by Akaike [2] that the estimate given in Theorem 14.9 is actually realized most of the time. Consequently, the steepest-descent method is a very *slow* algorithm, even for minimizing a convex quadratic function.

Convergence rate estimates for $E(x)$ can also be given for minimizing a convex quadratic function that is not strongly convex, that is, the matrix Q is positive semidefinite but not positive definite. We do not derive this estimate here, since a similar estimate is given in Section 14.3 for a much larger class of problems.

We will see in Section 14.7 that the conjugate-gradient method, which is a simple modification of the steepest-descent method, is a vast improvement upon the steepest-descent method for minimizing convex quadratic functions.

14.3 Convergence Rate of the Steepest-Descent Method on Convex Functions

Let $f : \mathbb{R}^n \to \mathbb{R}$ be a convex function with a Lipschitz continuous derivative, that is, there exists a constant $L > 0$ such that

$$\|\nabla f(x) - \nabla f(y)\| \leq L\|y - x\| \quad \text{for all} \quad x, y \in \mathbb{R}^n.$$

In this section, we consider a version of the steepest-descent method for minimizing f, and provide a convergence-rate estimate in terms of the optimality gap $E(x) = f(x) - \min_{\mathbb{R}^n} f$.

We start with some technical results.

Lemma 14.11. *If f is a convex function with a Lipschitz continuous derivative satisfying*

$$\|\nabla f(x) - \nabla f(y)\| \leq L\|x - y\| \quad \textit{for all} \quad x, y \in \mathbb{R}^n,$$

then

$$f(y) \leq f(x) + \langle \nabla f(x), y - x \rangle + \frac{L}{2}\|y - x\|^2.$$

Proof. Define the function $g(t) := f(x + t(y - x))$, and note that

$$g(1) - g(0) = \int_0^1 g'(t)dt = g'(0) + \int_0^1 (g'(t) - g'(0))dt.$$

Since $g(1) = f(y)$, $g(0) = f(x)$, and $g'(t) = \langle \nabla f(x + t(y - x)), y - x \rangle$, we obtain

$$\begin{aligned} f(y) - f(x) - \langle \nabla f(x), y - x \rangle &= \int_0^1 \langle \nabla f(x + t(y - x)) - \nabla f(x), y - x \rangle dt \\ &\leq \|y - x\| \int_0^1 \|\nabla f(x + t(y - x)) - \nabla f(x)\|\, dt \\ &\leq L\|y - x\|^2 \int_0^1 t\, dt = \frac{L}{2}\|y - x\|^2. \end{aligned}$$

□

Lemma 14.12. *Let $f : C \to \mathbb{R}$ be a differentiable convex function on the convex set C. The following conditions are equivalent:*

$$\|\nabla f(y) - \nabla f(x)\| \leq L\|y - x\| \quad \textit{for all} \quad x, y \in C. \tag{14.10}$$

$$\begin{aligned} \frac{1}{2L}\|\nabla f(y) - \nabla f(x)\|^2 &\leq f(y) - f(x) - \langle \nabla f(x), y - x \rangle \\ &\leq \frac{L}{2}\|y - x\|^2 \quad \textit{for all} \quad x, y \in C. \end{aligned} \tag{14.11}$$

$$\begin{aligned} \frac{1}{L}\|\nabla f(y) - \nabla f(x)\|^2 &\leq \langle \nabla f(y) - \nabla f(x), y - x \rangle \\ &\leq L\|y - x\|^2 \quad \textit{for all} \quad x, y \in C. \end{aligned} \tag{14.12}$$

Proof. (14.10) ⇒ (14.11): Notice that Lemma 14.11 gives the second inequality in (14.11); to prove the first inequality, define the function

$$g(y) := f(y) - \langle \nabla f(x), y \rangle,$$

with the gradient $\nabla g(y) = \nabla f(y) - \nabla f(x)$. Since $\nabla g(x) = 0$, x is a global minimizer of g, and we have

$$g(x) \le g\Big(y - \frac{\nabla g(y)}{L}\Big) \le g(y) - \frac{\|\nabla g(y)\|^2}{2L},$$

where the second inequality follows from Lemma 14.11. The first inequality in (14.11) is obtained by substituting the values of $g(x)$, $g(y)$, and $\nabla g(y)$, and rearranging the terms of the resulting inequality.

(14.11) ⇒ (14.12): Obtain an inequality by switching x and y in (14.11), and add it term by term to (14.11), and simplify.

(14.12) ⇒ (14.10): This is immediate. □

The first inequality in (14.11) seems to be due to Nesterov [208].

We are now ready to state our convergence-rate result.

Theorem 14.13. *Let $f : \mathbb{R}^n \to \mathbb{R}$ be a convex differentiable function with Lipschitz continuous derivative, that is,*

$$\|\nabla f(y) - \nabla f(x)\| \le L\|y - x\| \quad \text{for all} \quad x, y \in \mathbb{R}^n,$$

where $L > 0$. Suppose that f has a minimizer on $\mathbb{R}^n$, and consider the following gradient-descent method for minimizing f:

$$\begin{aligned} &x_0 \text{ given,} \\ &x_{k+1} = x_k - \frac{1}{L}\nabla f(x_k), \; k \ge 0. \end{aligned} \tag{14.13}$$

Then the minimizing sequence $\{x_k\}_0^\infty$ converges to a minimizer x^* of f, $\|x_k - x^*\|$ is a strictly decreasing sequence converging to zero, and we have the global convergence rate estimates

$$f(x_k) - f(x^*) \le \frac{5L}{2\sum_{i=0}^{k-1} \|x_i - x^*\|^{-2}}, \tag{14.14}$$

$$f(x_k) - f(x^*) = o(1/k). \tag{14.15}$$

Proof. Lemma 14.11 gives the estimate

$$\begin{aligned} f(x_{i+1}) - f(x_i) &\le \frac{-1}{L}\|\nabla f(x_i)\|^2 + \frac{L}{2}\|L^{-1}\nabla f(x_i)\|^2 \\ &= \frac{-1}{2L}\|\nabla f(x_i)\|^2 = \frac{-L}{2}\|x_i - x_{i+1}\|^2. \end{aligned} \tag{14.16}$$

Since f is convex, we also have

$$f(x^*) \geq f(x_i) + \langle \nabla f(x_i), x^* - x_i \rangle.$$

For brevity, let us define $w_i := f(x_i) - f^*$; the above inequality gives

$$w_i \leq \|\nabla f(x_i)\| \cdot \|x_i - x^*\| = L\|x_i - x_{i+1}\| \cdot \|x_i - x^*\|,$$

and it follows from this inequality and (14.16) that

$$\begin{aligned} f(x_i) - f(x_{i+1}) = w_i - w_{i+1} &\geq \frac{L}{2}\|x_i - x_{i+1}\|^2 \\ &\geq \frac{L}{2}\left(\frac{w_i}{L\|x_i - x^*\|}\right)^2 = \frac{w_i^2}{2L\|x_i - x^*\|^2} \\ &\geq \frac{w_{i+1}^2}{2L\|x_i - x^*\|^2}, \end{aligned}$$

which we rewrite in the form

$$w_i^{-1} \leq w_{i+1}^{-1}\left(1 + \frac{w_{i+1}}{2L\|x_i - x^*\|^2}\right)^{-1}. \tag{14.17}$$

Note that

$$w_{i+1} \leq w_i \leq \frac{L}{2}\|x_i - x^*\|^2,$$

where the second inequality follows from Lemma 14.11; thus, the fraction in (14.17) is at most 1/4. Using calculus, it is easily verified that $(1+t)^{-1} \leq 1 - (4t)/5$ when $0 \leq t \leq 1/4$; plugging this in (14.17) gives

$$w_i^{-1} \leq w_{i+1}^{-1} - \frac{2}{5L\|x_i - x^*\|^2}.$$

Summing this inequality from $i = 0$ to $i = k-1$, we get

$$0 \leq w_0^{-1} \leq w_k^{-1} - \frac{2}{5L}\sum_{i=0}^{k-1}\|x_i - x^*\|^{-2}.$$

Recalling $w_k = f(x_k) - f(x^*)$, we arrive at (14.14).

Next, we show that the sequence $\|x_k - x^*\|$ is strictly decreasing, that is, $\|x_{k+1} - x^*\| < \|x_k - x^*\|$: this follows from

$$\begin{aligned} \|x_{k+1} - x^*\|^2 &= \|x_k - x^* - \nabla f(x_k)/L\|^2 \\ &= \|x_k - x^*\|^2 + \frac{\|\nabla f(x_k)\|^2}{L^2} - \frac{2}{L}\langle x_k - x^*, \nabla f(x_k) - \nabla f(x^*)\rangle \\ &\leq \|x_k - x^*\|^2 + \frac{\|\nabla f(x_k)\|^2}{L^2} - 2\frac{\|\nabla f(x_k)\|^2}{L^2} \\ &< \|x_k - x^*\|^2, \end{aligned}$$

where the first inequality follows from Lemma 14.12.

Consequently, the sequence $\{x_k\}_1^\infty$ is bounded, and there exists a subsequence x_{k_i} converging to a point $\tilde{x}$. Taking the limit in (14.16) as $k_i \to \infty$ gives $0 \le -\|\nabla f(\tilde{x})\|^2/L$, or $\nabla f(\tilde{x}) = 0$, meaning that $\tilde{x}$ is a minimizer of f. Since $\|x_k - \tilde{x}\|$ is decreasing as shown above, we see that the whole sequence $\{x_k\}$ converges to $\tilde{x}$.

It remains to prove (14.15). Let us take x^* to be the point $\tilde{x}$, so that $\|x_k - x^*\| \to 0$. Define $a_k := \|x_k - x^*\|^{-2}$, and note that a_k is an increasing sequence diverging to infinity. Let us rewrite (14.14) in the form

$$k(f(x_k) - f^*) \le \frac{5L}{2}\Big(\frac{1}{k}\sum_{i=0}^{k-1} a_i\Big)^{-1}. \tag{14.18}$$

We claim that $\overline{a}_k := (\sum_{i=0}^{k-1} a_i)/k \to \infty$. Let $M > 0$ be arbitrary and suppose that $a_i \ge M$ for all $i \ge N$. Then

$$\overline{a}_{2N} = \frac{1}{2N}\sum_{i=0}^{2N-1} a_i \ge \frac{1}{2N}\sum_{i=N}^{2N-1} a_i \ge \frac{M}{2}.$$

It is easily verified that $\{\overline{a}_k\}$ is an increasing sequence, and hence the claim is proved. Consequently, the right-hand side of (14.18) goes to zero as $k \to \infty$, proving (14.15). □

Remark 14.14. One may wonder why the convergence rate above is given in terms of the optimal value gap $f(x_k) - f(x^*)$, and not in terms of the distance of the iterates to the optimal solution set. It turns out that under the assumptions of the theorem, it is not possible to give convergence-rate estimates in terms of $\|x_k - x^*\|$ unless some strong convexity conditions are present in the function f.

It is possible to design algorithms for the same class of problems with an improved complexity estimate $f(x_k) - f(x^*) = O(1/k^2)$; see [205, 220, 206, 207, 208].

It is even possible to give the same kinds of complexity estimates, either $f(x_k) - f(x^*) = o(1/k)$ or $f(x_k) - f(x^*) = O(1/k^2)$, for minimization of *nondifferentiable* convex functions, by regularizing f through the *proximal mapping*

$$f_\lambda(x) := \min_z\Big\{f(z) + \frac{1}{2\lambda}\|x - z\|^2\Big\},$$

which gives a differentiable, convex function f_λ whose derivative is Lipschitz continuous with constant $1/\lambda$. The resulting algorithms are the *proximal-point methods* [116, 117], or more general proximal-point methods [57]. Using convex duality, these algorithms make it possible to give global convergence rates for *augmented Lagrangian methods* for general convex programming.

14.4 Gradient Projection Method

This algorithm is an extension of the gradient-descent method to the minimization of a function over a closed convex set $C \subseteq \mathbb{R}^n$.

It has the following description:

Step 0: Choose $x_0 \in C$, $s > 0$, $\beta > 0$, and $0 < \sigma < 1$.
Step k: Given x_k, compute

$$\overline{x}_k = (x_k - s\nabla f(x_k))^+ := \Pi_C(x_k - s\nabla f(x_k)).$$

Perform an Armijo-type line search by recursively testing the inequality

$$f(x_k) - f\big(x_k + \beta^m(\overline{x}_k - x_k)\big) \geq -\sigma\beta^m\langle\nabla f(x_k), \overline{x}_k - x_k\rangle, \qquad (14.19)$$

for $m = 0, 1, \ldots$, until it is satisfied, say at $m_k := m$.
Set

$$x_{k+1} = x_k + \beta^{m_k}(\overline{x}_k - x_k).$$

We need the following technical result.

Lemma 14.15. *Let $C \subseteq \mathbb{R}^n$ be a closed convex set, and $f : C \to \mathbb{R}$ a differentiable function. Let $s > 0$. A point $x^* \in C$ satisfies the variational inequality*

$$\langle\nabla f(x^*), x - x^*\rangle \geq 0 \quad \textit{for all} \quad x \in C \qquad (14.20)$$

if and only if

$$(x^* - s\nabla f(x^*))^+ := \Pi_C(x^* - s\nabla f(x^*)) = x^*. \qquad (14.21)$$

Proof. By Theorem 6.1, $\Pi_C(x^* - s\nabla f(x^*)) = x^*$ if and only if

$$\Big\langle (x^* - s\nabla f(x^*)) - x^*, x - x^* \Big\rangle \leq 0 \quad \text{for all} \quad x \in C,$$

which is clearly equivalent to (14.20). □

Theorem 14.16. *Let $C \subseteq \mathbb{R}^n$ be a closed convex set, and $f : C \to \mathbb{R}$ a differentiable function. Let $\{x_k\}_0^\infty$ be a sequence generated by the gradient projection method with Armijo's step-size selection rule described above.*

If x^* is a limit point of the sequence $\{x_k\}_0^\infty$, then x^* is a critical point, that is, it satisfies the necessary condition for a local minimizer of f on C given by the variational inequality (14.20).

Proof. The inequality (14.19) gives

$$f(x_k) - f(x_{k+1}) = f(x_k) - f(x_k + \alpha_k d_k) \geq -\sigma\alpha_k\langle\nabla f(x_k), d_k\rangle, \qquad (14.22)$$

where

$$d_k := \overline{x}_k - x_k = (x_k - s\nabla f(x_k))^+ - x_k \quad \text{and} \quad \alpha_k = \beta^{m_k}.$$

First, we show that d_k is a descent direction (assuming that x_k is not a local minimizer of f on C, in which case we would stop the algorithm at x_k). Theorem 6.1 implies

$$\left\langle x_k - s\nabla f(x_k) - (x_k - s\nabla f(x_k))^+,\ x_k - (x_k - s\nabla f(x_k))^+ \right\rangle \leq 0,$$

and recalling the definition of d_k, this gives $\langle d_k + s\nabla f(x_k), d_k\rangle \leq 0$; therefore

$$0 < \|d_k\|^2 \leq -s\langle \nabla f(x_k), d_k\rangle, \tag{14.23}$$

where the first inequality follows from Lemma 14.15 because x_k is not a local minimizer of f on C. This proves that d_k is a strict descent direction at x_k.

Let $\{x_{k_i}\}$ be a subsequence of $\{x_k\}$ converging to x^*. Since d_k is a strict descent direction, the right-hand side of (14.22) is positive. Thus, $f(x_{k_i}) > f(x_{k_i+1}) \geq f(x_{k_{i+1}})$, and since $\lim_{i\to\infty} f(x_{k_i}) = f(x^*)$, the left-hand side of (14.22) converges to zero. This implies that

$$\lim_{i\to\infty} \alpha_{k_i}\langle \nabla f(x_{k_i}), d_{k_i}\rangle = 0. \tag{14.24}$$

We have $x_{k_i} \to x^*$, $d_{k_i} \to d^* := x^* - (x^* - s\nabla f(x^*))^+$, and by virtue of (14.23),

$$0 \leq \|d^*\|^2 \leq -s\langle \nabla f(x^*), d^*\rangle. \tag{14.25}$$

We claim that $\langle \nabla f(x^*), d^*\rangle = 0$. If α_{k_i} does not converge to zero, then the claim follows from (14.24); if $\alpha_{k_i} \to 0$, then the mean value theorem implies

$$f(x_{k_i} + \alpha_{k_i} d_{k_i}) - f(x_{k_i}) = \langle \nabla f(x_{k_i} + t_{k_i} d_{k_i}), d_{k_i}\rangle,$$

and letting $i \to \infty$, both sides of the equation converge to zero, and we have again proved the claim.

Finally, (14.25) implies that $d^* = 0$, and Lemma 14.15 implies that x^* satisfies (14.20). □

14.5 Newton's Method

Newton's method (or the Newton–Raphson method) is a method for finding roots of systems of equations. In its more general form, it could be applied, for example, to finding solutions to a nonlinear equation

$$f(x) = y_0,$$

where $f : U \to Y$ is a differentiable map from an open set U in a Banach space X into another Banach space Y, and $y_0 \in Y$ is given, under conditions that will be spelled out in Section 14.6.

In this section, we consider Newton's method in finite dimensions, where the map $f\colon U \to \mathbb{R}^n$ is defined on an open set $U \subseteq \mathbb{R}^n$.

By replacing $f(x)$ with $f(x)-y_0$, we will assume from now on that $y_0 = 0$, that is, we are solving an equation

$$f(x) = 0.$$

Newton's method is an iterative method having the form

$$x_{k+1} = x_k - Df(x_k)^{-1}f(x_k), \quad k \geq 0, \tag{14.26}$$

starting from an initial solution $x_0 \in U$. Thus, $Df(x_k)$ must be invertible at each step k.

For functions of a single variable ($n = 1$), Newton's method is normally written in the form

$$x_{k+1} = x_k - \frac{f(x_k)}{f'(x_k)}. \tag{14.27}$$

The geometric interpretation of Newton's method stems from Taylor's formula. At an approximate solution x_k, we replace the graph of the function f with its tangent line

$$\ell_k(x) = f(x_k) + f'(x_k)(x - x_k),$$

and determine the next approximate solution x_{k+1} by solving the linear equation $\ell_k(x) = 0$; that is, we expand $f(x)$ in a Taylor series around x_k to the linear term, and set it to zero:

$$f(x) \approx f(x_k) + f'(x_k)(x - x_k) = 0.$$

If we define

$$N_f(x) = x - \frac{f(x)}{f'(x)},$$

then Newton's method for f is the discrete dynamical system obtained by iterating $N_f(x)$, that is,

$$x_{k+1} = N_f(x_k).$$

Note that if $f'(x) \neq 0$, $f(x) = 0$ is equivalent to $N_f(x) = x$, so that a root of f is a fixed point of N_f, and vice versa.

Assuming that f'' exists, we calculate

$$N_f'(x) = 1 - \frac{f'(x)^2 - f(x)f''(x)}{f'(x)^2} = \frac{f(x)f''(x)}{f'(x)^2}.$$

If $f(x^*) = 0$, we have $N_f(x^*) = x^*$ and $N_f'(x^*) = 0$; Taylor's formula gives

$$\begin{aligned} N_f(x) &= N_f(x^*) + N_f'(x^*)(x - x^*) + \frac{1}{2}N_f''(z)(x - x^*)^2 \\ &= x^* + \frac{1}{2}N_f(z)(x - x^*)^2 \end{aligned}$$

for some $z \in (x^*, x)$. Putting $x = x_k$ in the above equation gives

$$x_{k+1} = N_f(x_k) = x^* + \frac{1}{2} N_f''(z_k)(x_k - x^*)^2,$$

that is,

$$x_{k+1} - x^* = \frac{1}{2} N_f''(z_k)(x_k - x^*)^2.$$

Assuming that x_k is close to x^*, and N_f'' is continuous, we have $\|\frac{1}{2}N_f''(z_k)\| \leq M$ for some $M > 0$, so that

$$\|x_{k+1} - x^*\| \leq M(x_k - x^*)^2; \tag{14.28}$$

this shows that Newton's method is *quadratically convergent* in a neighborhood of the root x^*, under the assumption that $f'(x^*) \neq 0$.

The inequality (14.28) shows why Newton's method is extremely fast: if $|x_k - x^*| \approx 10^{-d}$, that is, x_k and x^* have d common digits after the decimal point, then $|x_{k+1} - x^*| \approx 10^{-2d}$, meaning that x_{k+1} and x^* have $2d$ common digits. Thus, Newton's method approximately *doubles* the number of accurate digits at every step.

We give an illustration of this: consider approximating the numerical value of $\sqrt{2}$. This can be done by applying Newton's method to the equation $f(x) = 0$, where $f(x) = x^2 - 2$. In this case, Newton's method has the form

$$x_{k+1} = x_k - \frac{f(x_k)}{f'(x_k)} = x_k - \frac{x_k^2 - 2}{2x_k} = \frac{x_k}{2} + \frac{1}{x_k}.$$

Starting from the initial point $x_0 = 1$, we generate the following table:

k	x_k	$x_k - \sqrt{2}$
0	1	−0.414213562
1	1.5	0.085786438
2	1.41666	0.002453105
3	1.414215686	0.2124×10^{-5}
4	1.414213562	0

Note that in this case $f'(x^*) = 2\sqrt{2} \neq 0$.

Remark 14.17. The quadratic convergence of Newton's method is not guaranteed when $f'(x^*) = 0$. For example, the root of the equation $f(x) = x^2$ is $x^* = 0$. In this case Newton's method generates the sequence

$$x_{k+1} = x_k - \frac{x_k^2}{2x_k} = \frac{x_k}{2},$$

which is only a *linear* rate of convergence. If $x_0 = 1$, then $x_{20}^* \approx 0.38 \times 10^{-5}$, much slower than a quadratic rate of convergence.

We now consider the multidimensional version of Newton's method. Let $f(x) = (f_1(x), ..., f_n(x)$ with the coordinate functions $f_i(x)$, $i = 1, \ldots, n$. Again, we develop $f(x)$ in a Taylor series around $x_k \in \mathbb{R}$, cut off at the linear term, and determine the solution of the approximate, linear, map. This can be done coordinatewise:

$$f_i(x_{k+1}) \approx f_i(x_k) + \langle \nabla f_i(x_k), x_{k+1} - x_k \rangle = 0, \quad i = 1, \ldots, n,$$

or in vector notation,

$$\begin{pmatrix} f_1(x_k) \\ \vdots \\ f_n(x_k) \end{pmatrix} + \begin{bmatrix} \nabla f_1(x_k)^T \\ \vdots \\ \nabla f_n(x_k)^T \end{bmatrix} (x_{k+1} - x_k) = 0;$$

the matrix is the Jacobian $Df(x_k)$, so that this gives

$$f(x_k) + Df(x_k)(x_{k+1} - x_k) = 0. \tag{14.29}$$

If $Df(x_k)$ is invertible, we multiply the above equation on the left by $Df(x_k)^{-1}$ to obtain Newton's method (14.26) for solving $f(x) = 0$.

We can expect fast convergence of Newton's method here, too.

Theorem 14.18. *Let $f : U \to \mathbb{R}^n$ be a continuously differentiable function on an open set $B \subseteq \mathbb{R}^n$. If $x^* \in U$ is a root of f, that is, $f(x^*) = 0$, such that $Df(x^*)$ is invertible, then*

(a) There exists an open ball $B := B_\delta(x^*)$ around x^* such that if $x_0 \in B$, then Newton's method initiated at x_0 is well defined, and the iterates $\{x_k\}_0^\infty$ stay in B and converge to x^* superlinearly, that is,

$$\lim_{k\to\infty} \frac{\|x_{k+1} - x^*\|}{\|x_k - x^*\|} = 0. \tag{14.30}$$

(b) If $Df(x)$ is Lipschitz continuous in B, that is, there exists $L > 0$ such that

$$\|Df(y) - Df(x)\| \le L\|y - x\| \quad \textit{for all} \quad x, y \in B, \tag{14.31}$$

then Newton's method converges quadratically, that is, there exists $K > 0$ such that

$$\|x_{k+1} - x^*\| \le K\|x_k - x^*\|^2. \tag{14.32}$$

Let us first prove a useful technical result.

Lemma 14.19. *If f and x^* are as in the above theorem, then*

$$f(x) = \int_0^1 Df(x^* + t(x - x^*))(x - x^*)\, dt.$$

Proof. For the coordinate function f_i, define $\alpha(t) = f_i(x^* + t(x - x^*))$. We have $\alpha'(t) = \langle \nabla f_i(x^* + t(x - x^*)), x - x^* \rangle$ by the chain rule. Since $\alpha(1) - \alpha(0) = \int_0^1 \alpha'(t) dt$ by the fundamental theorem of calculus, and $\alpha(1) = f_i(x)$, $\alpha(0) = f_i(x^*) = 0$, we have

$$f_i(x) = \int_0^1 \langle \nabla f_i(x^* + t(x - x^*)), x - x^* \rangle \, dt, \quad i = 1, \ldots, n;$$

the lemma follows when we write these equations in matrix notation. □

We now prove Theorem 14.18.

Proof. We calculate

$$\begin{aligned}
&x_{k+1} - x^* \\
&= x_k - x^* - Df(x_k)^{-1} f(x_k) \\
&= Df(x_k)^{-1} \Big(Df(x_k)(x_k - x^*) - f(x_k) \Big) \\
&= Df(x_k)^{-1} \Big(Df(x_k)(x_k - x^*) - \int_0^1 Df(x^* + t(x_k - x^*))(x_k - x^*) \, dt \Big) \\
&= Df(x_k)^{-1} \int_0^1 [Df(x_k) - Df(x^* + t(x_k - x^*))] \cdot (x_k - x^*) \, dt.
\end{aligned}$$

This implies

$$\begin{aligned}
\|x_{k+1} - x^*\| \le \|Df(x_k)^{-1}\| \cdot \|x_k - x^*\| \\
\times \int_0^1 \|Df(x_k) - Df(x^* + t(x_k - x^*))\| \, dt.
\end{aligned}$$

If $\|Df(x)^{-1}\| \le M$ on B, and δ defining $B = B_\delta(x^*)$ is small enough that $\|Df(x_k) - Df(x^* + t(x_k - x^*))\| < 1/2M$, then $\|x_{k+1} - x^*\| < \|x_k - x^*\|/2$ and all Newton's iterates $\{x_k\}_0^\infty$ stay in B and converge to x^* geometrically. Since

$$\frac{\|x_{k+1} - x^*\|}{\|x_k - x^*\|} \le M \int_0^1 \|Df(x_k) - Df(x^* + t(x_k - x^*))\| \, dt, \tag{14.33}$$

and the integrand above will be as small as desired as $k \to \infty$, we see that (14.30) holds.

If Df is Lipschitz continuous, then (14.31) and (14.33) imply

$$\begin{aligned}
\frac{\|x_{k+1} - x^*\|}{\|x_k - x^*\|} &\le M \int_0^1 L \|x_k - (x^* + t(x_k - x^*))\| \, dt \\
&= ML \|x_k - x^*\| \int_0^1 (1 - t) \, dt \\
&= \frac{ML}{2} \|x_k - x^*\|;
\end{aligned}$$

setting $K = ML/2$, we obtain

$$\|x_{k+1} - x^*\| \le K\|x_k - x^*\|^2.$$

This proves the theorem. ☐

Here is a numerical example illustrating the quadratic convergence of the method for a function of two variables. Let us try to compute a solution to the system of equations

$$\begin{aligned} 3x^2y + y^2 &= 1, \\ x^4 + xy^3 &= 1. \end{aligned}$$

We apply Newton's method to the function

$$f(x, y) = (3x^2y + y^2 - 1, x^4 + xy^3 - 1):$$

k	(x_k, y_k)	$(x_k - x^*, y_k - x^*)$
0	$(.98, .32)$	$(-.01277, .0135)$
1	$(.99309, .3060802)$	$(.00031, -.00036)$
2	$(.9927801, .3064402)$	$(.18 \times 10^{-6}, -.21 \times 10^{-6})$
3	$(.9927799948, .306440465)$	$(10^{-10}, 10^{-10})$
4	$(.9927799949, .306440466)$	$(0, 0)$

Let us point out the use of Newton's method in optimization. Let $f : U \to \mathbb{R}$ be a function with continuous second partial derivatives $\partial^2 f/\partial x_i \partial x_j$. A local minimizer (or a maximizer) $x^* \in U$ is a critical point, that is, $\nabla f(x^*) = 0$, which means that x^* is a root of the map $F : U \to \mathbb{R}^n$,

$$F(x) = \nabla f(x).$$

Newton's method applied to $F(x) = 0$ is given by

$$x_{k+1} = x_k - DF(x_k)^{-1} F(x_k).$$

Since $F(x_k) = \nabla f(x_k)$ and

$$DF(x) = \begin{bmatrix} \nabla F_1(x)^T \\ \vdots \\ \nabla F_n(x)^T \end{bmatrix} = \begin{bmatrix} \frac{\partial}{\partial x_1}\left(\frac{\partial f(x)}{\partial x_1}\right) & \cdots & \frac{\partial}{\partial x_n}\left(\frac{\partial f(x)}{\partial x_1}\right) \\ \vdots & \vdots & \vdots \\ \frac{\partial}{\partial x_1}\left(\frac{\partial f(x)}{\partial x_n}\right) & \cdots & \frac{\partial}{\partial x_n}\left(\frac{\partial f(x)}{\partial x_n}\right) \end{bmatrix}$$

$$= \begin{bmatrix} \frac{\partial^2 f(x)}{\partial x_1^2} & \cdots & \frac{\partial^2 f(x)}{\partial x_n \partial x_1} \\ \vdots & \vdots & \vdots \\ \frac{\partial^2 f(x)}{\partial x_1 \partial x_n} & \cdots & \frac{\partial^2 f(x)}{\partial x_n^2} \end{bmatrix} = Hf(x),$$

Newton's method for finding a critical point of f becomes

$$x_{k+1} = x_k - HF(x_k)^{-1}\nabla f(x_k).$$

This Newton's method is also connected with Taylor's series:

$$\begin{aligned} f(x) &\approx f(x_k) + \nabla f(x_k)(x - x_k) + \frac{1}{2}(x - x_k)^T Hf(x_k)(x - x_k) \\ &=: q(x), \end{aligned}$$

and the quadratic function $q(x)$ has the gradient

$$\nabla q(x) = \nabla f(x_k) + Hf(x_k)(x - x_k),$$

and the critical point $\overline{x}$ satisfies $\nabla f(x_k) + Hf(x_k)(\overline{x} - x_k) = 0$, that is,

$$Hf(x_k)^{-1}\nabla f(x_k) + \overline{x} - x_k = 0.$$

Therefore,

$$\overline{x} = x_k - Hf(x_k)^{-1}\nabla f(x_k) = x_{k+1},$$

that is, x_{k+1} is the critical point of the second-order Taylor approximation

$$f(x_k) + \nabla f(x_k)(x - x_k) + \frac{1}{2}(x - x_k)Hf(x_k)(x - x_k).$$

14.6 Convergence Theory of Kantorovich

In Theorem 14.18, we need to know that a root x^* exists, and we do not know the size of the convergence region of quadratic convergence around x^*. There exist more sophisticated results on the convergence of Newton's method that remove these drawbacks. A famous class of such results was initiated by Kantorovich [154, 155]. Another approach, with a different set of assumptions, was advanced by Smale; see [33]. Moreover, there exist very sophisticated versions of Newton's method that are known as the *"hard" implicit function theorems* due to Nash and Moser that have important applications to partial differential equations, differential geometry (embedding of Riemannian manifolds into $\mathbb{R}^n$), and dynamical systems (KAM theory, the stability of the solar system); see [239, 126, 71].

Another important application of Newton's method occurs in interior-point methods, where Newton's method with a step-size selection rule is applied to minimize a very special class of convex functions called *self-concordant functions* on a convex set, in such a manner that enables us to obtain polynomial-time algorithms for a large class of convex programming problems. The two important properties of the self-concordant functions responsible for these desirable outcomes are the inequalities

$$\begin{aligned} D^3F(x)[h,h,h] &\leq 2D^2F(x)[h,h]^{3/2}, \\ DF(x)[h]^2 &\leq \theta D^2F(x)[h,h], \end{aligned}$$

among the first three directional derivatives of F at the point x along the direction h, and where $\theta > 0$ is a characteristic property of f called the self-concordance parameter, which is responsible for the speed of the algorithm. We do not go into more detail regarding these important methods; the interested reader will find a wealth of information in the books [209, 226]. However, one remark should be made: the distinguishing feature of self-concordant functions that sets them apart from all other Newton's methods we are aware of, is the idea that the norm used to measure distances should come from the function itself (this requires that F be strongly convex), and should change from point to point. The self-concordant functions use the Euclidean norm that comes from the Hessian D^2F, by defining the local norm at x by the formula

$$\|u\|_x^2 := \langle D^2F(x)u, u\rangle;$$

the directional derivative inequalities are best viewed in terms of these local norms.

The theorem below and its assumptions belong to the circle of ideas initiated by Kantorovich. The proof below is an adaptation of the one given in [71]; however, we have formulated the hypotheses of the theorem in an affine invariant manner, following [75]. Affine invariance simply means that since Newton's method is affine invariant, that is, Newton's method applied to $F(x)$ and $G(x) = AF(x)$, where A is an invertible linear map, generates in both cases the same solution sequence $\{x_n\}$, the hypotheses (14.34) of the theorem should remain the same if we replace $F(x)$ with the map $G(x)$.

The natural setting of the following theorem below is in Banach spaces. A reader who is interested only in finite dimensions may simply assume throughout that X and Y are finite-dimensional Euclidean spaces, say $X = Y = \mathbb{R}^n$. For brevity, we will denote the derivative DF by F' in this section.

Theorem 14.20. *Let $F : D \to Y$ be a Fréchet differentiable map with continuous derivative from an open, convex set D in a Banach space X to a Banach space Y. Let $x_0 \in D$ be a point such that $F'(x_0)$ be invertible, and assume that*

$$\begin{aligned} \|F'(x_0)^{-1}F(x_0)\| &\le \alpha, \\ \|F'(x_0)^{-1}(F'(y) - F'(x))\| &\le \kappa\|y - x\| \quad \text{for all } x, y \in D, \\ \alpha\kappa &< \frac{1}{2}, \\ \overline{B}_{2\alpha}(x_0) &\subset D. \end{aligned} \tag{14.34}$$

Then F has a unique root x^* in $\overline{B}_{2\alpha}(x_0)$, and the Newton iterates $\{x_n\}_0^\infty$ converge quadratically *to x^* such that*

$$\|x_n - x^*\| \le \frac{\alpha}{2^{n-1}} q^{2^n - 1}, \quad \text{where} \quad q = 2\alpha\kappa < 1. \tag{14.35}$$

Proof. Consider the map

$$G(x) := F'(x_0)^{-1}F(x).$$

Since Newton's method is affine invariant, applying it to the map $G(x)$ generates the same sequence $\{x_n\}$. The hypotheses of the theorem can be rewritten in the form

$$\begin{aligned} &(a) \quad \|G(x_0)\| \le \alpha, \\ &(b) \quad \|G'(x) - G'(y)\| \le \kappa\|x - y\| \quad \text{for all } x, y \in D, \\ &(c) \quad \alpha\kappa < \frac{1}{2} \quad \text{and} \quad \overline{B}_{2\alpha}(x_0) \subset D. \end{aligned}$$

Define the quantities

$$\begin{aligned} \Delta x_n &:= x_{n+1} - x_n, \\ \alpha_n &:= \|x_{n+1} - x_n\| = \|\Delta x_n\|, \\ \beta_n &:= \|G'(x_n)^{-1}\|, \\ \gamma_n &:= \alpha_n \beta_n \kappa. \end{aligned}$$

The idea of the proof is to estimate α_n, then β_n, and then use these two estimates to obtain a recurrence for γ_n. Since $G(x_{n-1}) + G'(x_{n-1})\Delta x_{n-1} = 0$, from Newton's formula we have

$$\begin{aligned} \alpha_n &= \|G'(x_n)^{-1}G(x_n)\| \le \beta_n\|G(x_n)\| \\ &= \beta_n\|G(x_n) - G(x_{n-1}) - G'(x_{n-1})\Delta x_{n-1}\| \\ &= \beta_n\left\|\int_0^1 [G'(x_{n-1} + t\Delta x_{n-1}) - G'(x_{n-1})]\Delta x_{n-1}\, dt\right\| \\ &\le \beta_n \int_0^1 \|[G'(x_{n-1} + t\Delta x_{n-1}) - G'(x_{n-1})]\Delta x_{n-1}\|\, dt \\ &\le \beta_n\kappa\|\Delta x_{n-1}\|^2 \int_0^1 t\, dt = \frac{\alpha_{n-1}^2\beta_n\kappa}{2}. \end{aligned} \tag{14.36}$$

Since

$$\begin{aligned} G'(x_n) &= G'(x_{n-1}) + (G'(x_n) - G'(x_{n-1})) \\ &= G'(x_{n-1})\left[I + G'(x_{n-1})^{-1}(G'(x_n) - G'(x_{n-1}))\right], \end{aligned}$$

we also have

$$\begin{aligned} \beta_n &= \|G'(x_n)^{-1}\| \\ &\le \beta_{n-1}\left\|\left[I + G'(x_{n-1})^{-1}(G'(x_n) - G'(x_{n-1}))\right]^{-1}\right\| \\ &\le \frac{\beta_{n-1}}{1 - \|G'(x_{n-1})^{-1}\| \cdot \|G'(x_n) - G'(x_{n-1})\|} \\ &\le \frac{\beta_{n-1}}{1 - \beta_{n-1}\kappa\alpha_{n-1}} = \frac{\beta_{n-1}}{1 - \gamma_{n-1}}, \end{aligned} \tag{14.37}$$

where in the second inequality we used Neumann's formula $(I + A)(I - A + A^2 - A^3 + \cdots) = I$ for an operator satisfying $\|A\| < 1$, and the resulting inequality

$$\|(I + A)^{-1}\| = \left\|\sum_{i=0}^{\infty}(-1)^i A^i\right\| \le \sum_{i=0}^{\infty} \|A\|^i = \frac{1}{1 - \|A\|}.$$

The inequalities (14.36) and (14.37) imply

$$\alpha_n \le \frac{\kappa\alpha_{n-1}^2\beta_{n-1}}{2(1-\gamma_{n-1})} = \frac{\gamma_{n-1}}{2(1-\gamma_{n-1})}\alpha_{n-1}$$

and

$$\gamma_n = \kappa\alpha_n\beta_n \le \kappa\frac{\gamma_{n-1}}{2(1-\gamma_{n-1})}\cdot\alpha_{n-1}\cdot\frac{\beta_{n-1}}{1-\gamma_{n-1}} = \frac{\gamma_{n-1}^2}{2(1-\gamma_{n-1})^2}.$$

In summary, we have proved that

$$\begin{aligned} \alpha_n &\le \frac{\gamma_{n-1}}{1-\gamma_{n-1}}\cdot\frac{\alpha_{n-1}}{2}, \\ \gamma_n &\le \frac{\gamma_{n-1}^2}{2(1-\gamma_{n-1})^2}. \end{aligned} \tag{14.38}$$

Note that $\beta_0 = 1$ and thus $\gamma_0 = \alpha\kappa < \frac{1}{2}$. Using induction, it follows from (14.38) that

$$\gamma_n < \frac{1}{2} \quad \text{and} \quad \alpha_n \le \frac{\alpha_{n-1}}{2} \quad \text{for all} \quad n \ge 1;$$

this proves that $\alpha_n \le \alpha/2^n$, and

$$\begin{aligned} \|x_n - x_0\| &\le \|x_n - x_{n-1}\| + \cdots + \|x_1 - x_0\| \\ &\le \|x_1 - x_0\|(1 + 2^{-1} + 2^{-2} + \cdots) \\ &= 2\alpha, \end{aligned}$$

which shows that the sequence $\{x_n\}_0^\infty$ is well defined. Similarly,

$$\begin{aligned} \|x_n - x_m\| &\le \|x_n - x_{n-1}\| + \cdots + \|x_{m+1} - x_m\| \\ &\le \|x_{m+1} - x_m\|(2^{-m} + 2^{-m-1} + \cdots) \\ &\le \frac{\alpha}{2^{m-1}}, \end{aligned}$$

so that $\{x_n\}_0^\infty$ is a Cauchy sequence, with limit $x^* \in \overline{B}_{2\alpha}(x_0)$. Since

$$\|G(x_n)\| \le \|G'(x_n)\| \cdot \|G'(x_n)^{-1}G(x_n)\| = \alpha_n\|G'(x_n)\| \to 0,$$

because $\{G'(x_n)\}_0^\infty$ is bounded, we see that $G(x^*) = 0$.

Using the technique in the proof of Theorem 14.18, we can prove that

$$\|x_{n+1} - x^*\| \le c\|x_n - x^*\|^2 \quad \text{with} \quad c = \kappa \max\{\beta_n\}/2;$$

therefore $\{x_n\}_0^\infty$ is quadratically convergent to x^*; here $c < \infty$ because $\beta_n \to \|G'(x^*)^{-1}\| < \infty$.

If $\tilde{x} \in \overline{B}_{2\alpha}(x_0)$ is another root of G, then $G(\tilde{x}) = G(x^*) = 0$, $G'(x_0) = I$, and

$$\begin{aligned} \|\tilde{x} - x^*\| &= \|G(\tilde{x}) - G(x^*) - G'(x_0)(\tilde{x} - x^*)\| \\ &\le \int_0^1 \|[G'(x^* + t(\tilde{x} - x^*)) - G'(x_0)](\tilde{x} - x^*)\| \, dt \\ &\le \kappa\|\tilde{x} - x^*\| \int_0^1 \|x^* + t(\tilde{x} - x^*) - x_0\| \, dt \\ &\le 2\alpha\kappa\|\tilde{x} - x^*\|; \end{aligned}$$

since $2\alpha\kappa < 1$, we have $\tilde{x} = x^*$, proving the uniqueness of the root of F in $B_{2\alpha}(x_0)$.

It remains to prove the estimate (14.35). Define

$$\delta_n := \frac{\gamma_n}{1 - \gamma_n}.$$

Then, (14.38) and the fact that $\gamma_n < 1/2$ for all n imply $\delta_n \le \delta_{n-1}^2$, which in turn implies by induction that $\delta_n \le (\delta_0)^{2^n}$ for all $n \ge 0$; therefore (14.38) gives

$$\alpha_n \le \frac{1}{2}\alpha_{n-1}\delta_0^{2^{n-1}} \le \cdots \le 2^{-n}\delta_0^{2^n - 1}\alpha_0 \le 2^{-n}q^{2^n - 1}\alpha$$

and

$$\begin{aligned} \|x_n - x^*\| &\le \|x_n - x_{n+1}\| + \|x_{n+1} - x_{n+2}\| + \cdots = \sum_{i=n}^{\infty} \alpha_i \\ &\le 2 \cdot 2^{-n} q^{2^n - 1} \alpha = \frac{\alpha}{2^{n-1}} q^{2^n - 1}. \end{aligned}$$

The theorem is proved. □

14.7 Conjugate-Gradient Method

Throughout this section, we consider solving the system of linear equations

$$Qx = b, \tag{14.39}$$

where Q is an $n \times n$ symmetric positive definite matrix. Hestenes and Stiefel [132] invented the *conjugate-gradient method (CGM)* in order to solve this system, especially when Q is a sparse large-scale matrix. Such large-scale, sparse linear systems come up, for example, in the numerical solution of partial differential equations. The conjugate-gradient method is an important iterative method, with many desirable properties that will be highlighted below.

Note that the equation (14.39) is equivalent to minimizing the strongly convex quadratic function

$$q(x) = \frac{1}{2}\langle Qx, x\rangle - \langle b, x\rangle + a, \tag{14.40}$$

where $a \in \mathbb{R}$, because $\nabla q(x) = Qx - b$; thus the equation $Qx = b$ has the variational characterization $\nabla q(x) = 0$. We set

$$r(x) := \nabla q(x)$$

throughout this section.

The conjugate-gradient method has been extended to nonquadratic minimization problems; see for example [190, 213] for further information.

14.7.1 Q-Inner Product and Q-Norm

Definition 14.21. *Let Q be a symmetric, positive definite $n \times n$ matrix. The* Q-inner product *on $\mathbb{R}^n$ is given by*

$$\langle x, y\rangle_Q := \langle Qx, y\rangle;$$

then $(\mathbb{R}^n, \langle \cdot, \cdot_Q\rangle)$ is a finite-dimensional Euclidean space. The corresponding norm is called the Q-norm,

$$\|x\|_Q := \sqrt{\langle Qx, x\rangle}.$$

This inner product can also be viewed as follows: let $Q = U\Lambda U^T$ be the spectral decomposition of Q. (Recall that $U = [u_1, \ldots, u_n]$, $\Lambda = \operatorname{diag}(\lambda_1, \ldots, \lambda_n)$, where u_i, $\|u_i\| = 1$, is an eigenvector of Q with the corresponding eigenvalue λ_i.) The matrix $R = U \operatorname{diag}(\sqrt{\lambda_1}, \ldots, \sqrt{\lambda_n})U^T$ is the unique symmetric matrix satisfying $R^2 = Q$; it is called the *square root* of Q, and denoted by $Q^{1/2}$. Thus,

$$\langle x, y\rangle_Q = \langle Q^{1/2}x, Q^{1/2}y\rangle \quad \text{and} \quad \|x\|_Q = \|Q^{1/2}x\|,$$

that is, $\|x\|_Q$ is simply the ordinary Euclidean norm of the vector $Q^{1/2}x$ and $\langle x, y\rangle_Q$ is the ordinary inner product of the vectors $Q^{1/2}x$ and $Q^{1/2}y$.

Definition 14.22. *A set of directions $\{d_i\}_{i=1}^k \subset \mathbb{R}^n$ is called* Q-conjugate *or* Q-orthogonal *if they are orthogonal in the Q-inner product, that is,*

$$\|d_i\|_Q \neq 0, \ \langle d_i, d_j\rangle_Q = 0 \quad \textit{for all} \quad i \neq j, \ \ i, j = 1, \ldots, k. \tag{14.41}$$

Lemma 14.23. *A set $\{d_i\}_{i=1}^k$ of Q-conjugate directions is linearly independent.*

Proof. If $\alpha_1 d_1 + \cdots + \alpha_k d_k = 0$, then

$$0 = \Big\langle \sum_{i=1}^{k} \alpha_i d_i, Qd_j \Big\rangle = \sum_{i=1}^{k} \alpha_i \langle d_i, Qd_j\rangle = \alpha_j \langle d_j, Qd_j\rangle;$$

since $\langle d_j, Qd_j\rangle \neq 0$, it follows that $\alpha_j = 0$ for all j, $j = 1, \ldots, k$. □

An important feature of conjugate directions $\{d_i\}_1^k$ is that if we have them at hand, then it is *trivial* to minimize the quadratic function $q(x)$ on an affine subspace $A = x_0 + \mathrm{span}(\{d_i\}_{i=1}^k)$: if $x = x_0 + \sum_{i=1}^k \gamma_i d_i \in A$, then by Taylor's formula,

$$\begin{aligned} q(x) &= q(x_0) + \sum_{i=1}^{k} \gamma_i \langle r_0, d_i\rangle + \frac{1}{2} \sum_{i,j=1}^{k} \gamma_i \gamma_j \langle Qd_i, d_j\rangle \\ &= q(x_0) + \sum_{i=1}^{k} \Big[\gamma_i \langle r_0, d_i\rangle + \frac{1}{2} \sum_{i=1}^{k} \gamma_i^2 \langle Qd_i, d_i\rangle \Big], \end{aligned}$$

where the second equation follows since the vectors $\{d_i\}_1^k$ are Q-conjugate. Thus, the minimizer x^* of q on A can be obtained by minimizing *separately* over each γ_i; we have

$$x^* = x_0 + \sum_{i=1}^{k} \gamma_i d_i, \qquad \text{where } \gamma_i = -\frac{\langle r_0, d_i\rangle}{\langle Qd_i, d_i\rangle}. \tag{14.42}$$

Gram–Schmidt Process

Suppose we are given linearly independent vectors $\{r_i\}_0^k$. The Gram–Schmidt orthogonalization process generates a set of Q-conjugate direction $\{d_i\}_0^k$ out of $\{r_i\}_0^k$ such that $\mathrm{span}(\{r_i\}_0^k) = \mathrm{span}(\{d_i\}_0^k)$.

The process starts by setting $d_0 = r_0$; if $\{d_i\}_0^{i-1}$ is at hand, then we define

$$d_i = r_i + \sum_{j=0}^{i-1} \beta_j d_j,$$

where the coefficients β_j are determined as follows: since we require $\langle d_i, d_l\rangle_Q = 0$ for $l < i$, we must have

$$\begin{aligned} 0 = \langle d_i, d_l\rangle_Q &= \Big\langle r_i + \sum_{j=1}^{i-1} \beta_j d_j, d_l \Big\rangle_Q \\ &= \langle r_i, d_l\rangle_Q + \sum_{j=1}^{i-1} \beta_j \langle d_j, d_l\rangle_Q = \langle r_i, d_l\rangle_Q + \beta_l \langle d_l, d_l\rangle_Q, \end{aligned}$$

where the last equation follows because $\langle d_j, d_l \rangle_Q = 0$ for $j \neq l$; therefore $\beta_l = -\langle r_i, d_l \rangle_Q / \langle d_l, d_l \rangle_Q$, and

$$d_i = r_i - \sum_{j=1}^{i-1} \frac{\langle r_i, d_j \rangle_Q}{\langle d_j, d_j \rangle_Q} d_j = r_i - \sum_{j=1}^{i-1} \frac{\langle Qr_i, d_j \rangle}{\langle Qd_j, d_j \rangle} d_j.$$

14.7.2 Conjugate-Direction Methods

It is useful to discuss a more general class of algorithms known as *conjugate-direction methods* before studying the conjugate-gradient method in detail.

Let $\{d_i\}_{i=0}^{n-1}$ be set of Q-conjugate search directions, either preselected, or selected recursively, somehow, as the algorithm proceeds. A conjugate-direction method simply minimizes $q(x)$ in (14.40) by successively minimizing it using $\{d_k\}$ as search directions.

The precise description of the method follows.

Conjugate-Direction Method

Let x_0 be a point and $\{d_i\}_{i=0}^{n-1}$ a set of Q-conjugate directions in $\mathbb{R}^n$. For $k = 0, 1 \ldots, n-1$, set

$$x_{k+1} = x_k - \frac{\langle r_k, d_k \rangle}{\langle Qd_k, d_k \rangle} d_k, \qquad \textit{where} \qquad r_k = \nabla q(x_k). \tag{14.43}$$

Here x_{k+1} is the exact minimizer of the function q on the line $\ell = \{x_k + \alpha d_k : \alpha \in \mathbb{R}\}$, because defining

$$h(\alpha) := q(x_k + \alpha d_k) = q(x_k) + \alpha \langle r_k, d_k \rangle + \frac{\alpha^2}{2} \langle Qd_k, d_k \rangle,$$

we see that the exact minimizer α_k of h is given by the equation $h'(\alpha_k) = \langle r_k, d_k \rangle + \alpha_k \langle Qd_k, d_k \rangle = 0$.

We now come to one of the key results in the theory of the conjugate-gradient method.

Theorem 14.24. *Let $\{d_i\}_{i=0}^{n-1}$ be a set of Q-conjugate directions in $\mathbb{R}^n$, and $\{x_i\}_{i=0}^{n-1}$ the points generated by the conjugate-direction method using these directions.*

Then x_k is the global minimizer of q on the k-dimensional affine subspace

$$M_k := x_0 + \operatorname{span}(\{d_i\}_0^{k-1});$$

consequently, the solution to the equation $Qx = b$ can be found in at most n steps of the conjugate-direction method.

Moreover, r_k is orthogonal to the directions $\{d_i\}_0^{k-1}$, that is, $\langle r_k, d_i \rangle = 0$ for $i = 0, \ldots, k-1$.

Proof. Let $\overline{x} = x_0 + \gamma_0 d_0 + \cdots + \gamma_{k-1} d_{k-1}$ be the minimizer of q on M_k. We will show that $\gamma_i = \alpha_i$, $i = 0, \ldots, k-1$, which will imply the first statement of the theorem.

By (14.42), we have

$$\gamma_i = -\frac{\langle r_0, d_i \rangle}{\langle Qd_i, d_i \rangle}, \qquad i = 0, \ldots, k-1.$$

Now, if $x \in M_i$ such that $x = x_0 + \delta_0 d_0 + \cdots + \delta_{i-1} d_{i-1}$, then

$$\begin{aligned}\nabla q(x) &= Qx_0 + \delta_0 Qd_0 + \cdots + \delta_{i-1} Qd_{i-1} - b \\ &= r_0 + \delta_0 Qd_0 + \cdots + \delta_{i-1} Qd_{i-1},\end{aligned}$$

and

$$\langle \nabla q(x), d_i \rangle = \langle r_0, d_i \rangle + \sum_{j=0}^{i-1} \delta_j \langle Qd_j, d_i \rangle = \langle r_0, d_i \rangle;$$

this shows that

$$\langle \nabla q(x), d_i \rangle = \langle r_0, d_i \rangle \quad \text{for all} \quad x \in M_i,$$

and setting $x = x_i \in M_i$ gives $\langle r_i, d_i \rangle = \langle r_0, d_i \rangle$. It follows from (14.43) that $\gamma_i = \alpha_i$, $i = 0, 1, \ldots, k-1$.

To prove the second statement, note that x_k is minimizes the function

$$h(\delta_0, \ldots, \delta_{k-1}) := q(x_0 + \delta_0 d_0 + \cdots + \delta_{k-1} d_{k-1})$$

on $\mathbb{R}^k$. This implies

$$0 = \frac{\partial h(\gamma)}{\partial \delta_i} = \langle \nabla q(x_k), d_i \rangle = \langle r_k, d_i \rangle, \quad i = 0, \ldots, k-1.$$

This completes the proof of the theorem. □

14.7.3 The Conjugate-Gradient Method

The conjugate-gradient method (CGM) of Hestenes and Stiefel is a conjugate-direction method that successively manufactures the Q-conjugate directions $\{d_i\}_0^k$ out of the negative gradient directions $\{-r_i\}_0^k = \{-\nabla q(x_i)\}_0^k$ encountered during the running of the algorithm, using the Gram–Schmidt process described above. Note that this is possible, since the Gram–Schmidt process never looks ahead; only the knowledge of $\{-r_j\}_0^k$ is needed to generate d_k.

Consequently, we have the following, *preliminary* version of the conjugate-gradient method:

Choose a point $x_0 \in \mathbb{R}^n$; *define* $d_0 = -\nabla q(x_0) := -r_0$.
while $r_k \neq 0$,

$$d_k = -r_k + \sum_{j=0}^{k-1} \frac{\langle Qr_k, d_j \rangle}{\langle Qd_j, d_j \rangle} d_j, \tag{14.44}$$

$$\alpha_k = \frac{\|r_k\|^2}{\langle Qd_k, d_k \rangle}, \tag{14.45}$$

$$x_{k+1} = x_k + \alpha_k d_k. \tag{14.46}$$

end (while)

We note that the formula for α_k in (14.45) follows, because Theorem 14.24 implies

$$\langle r_k, d_k \rangle = \langle r_k, -r_k + \beta_0 d_0 + \cdots + \beta_{k-1} d_{k-1} \rangle = -\|r_k\|^2.$$

As it stands, the above algorithm is not practical, because the computation of d_k by the Gram–Schmidt process requires too much computational effort. Happily, (14.44) can be simplified considerably; this is the *magical* feature of the CGM.

Theorem 14.25. *Let the gradient vectors* $\{r_i\}_{i=0}^k$ *be all nonzero. Then* $\{r_i\}_{i=0}^k$ *are mutually orthogonal, that is,*

$$\langle r_i, r_j \rangle = 0 \quad \textit{for all} \quad i \neq j. \tag{14.47}$$

Moreover,

$$d_k = -r_k + \beta_k d_{k-1}, \qquad \textit{where} \qquad \beta_k = \frac{\|r_k\|^2}{\|r_{k-1}\|^2}.$$

Thus, the search direction d_k at the kth step of CGM is a linear combination of the current gradient vector $\nabla q(x_k)$ and the last search direction d_{k-1}.

Proof. Note that the vectors $\{d_i\}_0^{k-1}$ are generated from the vectors $\{r_i\}_0^{k-1}$ by the Gram–Schmidt process, so that

$$\text{span}\{d_0, \ldots, d_{k-1}\} = \text{span}\{r_0, \ldots, r_{k-1}\}.$$

Since $\{d_i\}_{i=0}^{k-1}$ are Q-conjugate, Theorem 14.24 implies that $r_k \perp \{d_i\}_{i=0}^{k-1}$; thus,

$$r_k \perp \{r_0, r_1, \ldots, r_{k-1}\},$$

proving the first statement of the theorem.

We can now simplify equation (14.44). We must have $\alpha_j \neq 0$ in the equation $x_{j+1} = x_j + \alpha_j d_j$, because otherwise $r_{k+1} = r_k$, and this contradicts (14.47); then

$$Qd_j = Q\Big(\frac{x_{j+1} - x_j}{\alpha_j}\Big) = \frac{(Qx_{j+1} - b) - (Qx_j - b)}{\alpha_j} = \frac{r_{j+1} - r_j}{\alpha_j}. \tag{14.48}$$

Therefore, we have in (14.44)

$$\langle Qr_k, d_j\rangle = \langle r_k, Qd_j\rangle = \frac{1}{\alpha_j}\langle r_k, r_{j+1} - r_j\rangle, \quad j = 0, \ldots, k-1.$$

Consequently,

$$\langle Qr_k, d_j\rangle = 0 \quad \text{for all} \quad j < k-1,$$
$$\langle Qr_k, d_{k-1}\rangle = \frac{\|r_k\|^2}{\alpha_{k-1}}.$$

Substituting these in (14.44) gives

$$d_k = -r_k + \frac{\|r_k\|^2}{\alpha_{k-1}\langle Qd_{k-1}, d_{k-1}\rangle} d_{k-1} = -r_k + \frac{\|r_k\|^2}{\|r_{k-1}\|^2} d_{k-1},$$

where the second equality follows from (14.45). □

Here is the final description of the conjugate-gradient method.

The Conjugate-Gradient Method

Choose a point $x_0 \in \mathbb{R}^n$*; define* $d_0 = -\nabla q(x_0) =: -r_0$.
while $r_k \neq 0$,

$$\begin{aligned}
\alpha_k &= \frac{\|r_k\|^2}{\langle Qd_k, d_k\rangle}, \\
x_{k+1} &= x_k + \alpha_k d_k, \\
r_{k+1} &= r_k + \alpha_k Q d_k, \\
\beta_{k+1} &= \frac{\|r_{k+1}\|^2}{\|r_k\|^2}, \\
d_{k+1} &= -r_{k+1} + \beta_{k+1} d_k,
\end{aligned} \tag{14.49}$$

end (while)

For easy reference, we state some of the important results we proved above.

Corollary 14.26. *Suppose the CGM generates the points* $\{x_i\}_0^{k+1}$*, where the last point* x_{k+1} *is the optimal solution. The vectors* $\{r_i\}_0^k$ $(r_i = \nabla q(x_i))$ *are orthogonal, the directions* $\{d_i\}_0^k$ *are Q-conjugate, and*

$$\operatorname{span}\{r_0, \ldots, r_i\} = \operatorname{span}\{d_0, \ldots, d_i\}, \quad i = 0, \ldots, k.$$

14.8 Convergence Rate of the Conjugate-Gradient Method

In theory, the conjugate-gradient method (14.49) terminates in at most n steps by Theorem 14.24. This is rarely true in practice, however, because of numerical roundoff errors. Since the 1970s, the prevailing view is that the conjugate-gradient method is an *iterative* method for solving linear equations $Qx = b$, where the matrix Q is large-scale, sparse, symmetric, and positive definite (or positive semidefinite). Viewed from this perspective, it becomes important to estimate the convergence rate of the algorithm based on the properties of the matrix Q, not merely its dimension n, and to obtain convergence-rate results that are independent of n. It even becomes relevant to investigate the behavior the conjugate-gradient method for solving a linear equation $Qx = b$ where $Q : H \to H$ is a self-adjoint positive semidefinite operator on an infinite-dimensional Hilbert space H. Theorem 14.24 will be a key tool in this section.

A consequence of Corollary 14.26, which in turn is a consequence of Theorem 14.24 and Theorem 14.25, is the following result.

Corollary 14.27. *Let $\{x_i\}_{i=0}^k$, $\{r_i\}_{i=0}^k$, and $\{d_i\}_{i=0}^k$ be the iterates, gradients, and conjugate direction vectors, respectively, generated by a conjugate-gradient method.*

If all the gradient vectors r_i are nonzero, then

$$\operatorname{span}\{d_0, d_1, \ldots, d_k\} = \operatorname{span}\{r_0, r_1, \ldots, r_k\} = \operatorname{span}\{r_0, Qr_0, \ldots, Q^k r_0\}.$$

Proof. The first equality was already proved in Corollary 14.26; to prove the second, we use induction on k. The equality is trivially true for $k = 0$; assuming that it is true for $k - 1$, let us prove its truth for k. Let $\{r_i\}_{i=0}^k$ all be nonzero vectors. It suffices to show that $\operatorname{span}\{r_i\}_0^k \subseteq \operatorname{span}\{Q^i r_0\}_0^k$, because $\operatorname{span}\{r_i\}_0^k$ has dimension $k + 1$ by Corollary 14.26, while the subspace $\operatorname{span}\{Q^i r_0\}_0^k$ has dimension at most $k + 1$. By the induction hypothesis, we have $\operatorname{span}\{r_i\}_0^{k-1} = \operatorname{span}\{Q^i r_0\}_0^{k-1}$, so it suffices to prove that $r_k \in \operatorname{span}\{Q^i r_0\}_0^k$; this last assertion is true, because $r_k = r_{k-1} + \alpha_{k-1} Q d_{k-1}$ from the description of CGM, and both r_{k-1} and d_{k-1} lie in $\operatorname{span}\{Q^i r_0\}_0^{k-1}$ by the induction hypothesis. □

Definition 14.28. *Let A be an $n \times n$ matrix and $d \in \mathbb{R}^n$ a vector. The* Krlov subspace *of A of order k in the direction d is the linear subspace*

$$\mathcal{K}(A, d, k) := \operatorname{span}\{d, Ad, \ldots, A^{k-1}d\}.$$

Let

$$E(x) := q(x) - q(x^*)$$

be the objective value gap $q(x) - q(x^*)$ at the point x, where x^* is the minimizer of q on $\mathbb{R}^n$. By Taylor's formula,

$$E(x) = q(x) - q(x^*) = \frac{1}{2}\langle Q(x - x^*), x - x^* \rangle = \frac{1}{2}\|x - x^*\|_Q^2. \qquad (14.50)$$

We now come to perhaps the most central result concerning the conjugate-gradient method, sometimes called the *expanding subspace theorem.* We shall use it in Section 14.10 to give a short and independent derivation of the conjugate-gradient method itself.

Theorem 14.29. (*Expanding subspace theorem*) *The point x_k generated by the conjugate-gradient method has the variational characterization*

$$E(x_k) = \frac{1}{2}\min\{\|p(Q)(x_0 - x^*)\|_Q^2 : p \in \mathcal{P}_k,\ p(0) = 1\},$$

where $\mathcal{P}_k$ is the vector space of polynomials of degree at most k.

Proof. It follows from Theorem 14.24 and Corollary 14.27 that x_k is the minimizer q on the affine subspace $M_{k-1} = x_0 + \mathcal{K}(Q, r_0, k-1)$. If $x \in M_{k-1}$, write

$$x = x_0 + \gamma_0 r_0 + \gamma_1 Q r_0 + \cdots + \gamma_{k-1} Q^{k-1} r_0 = x_0 + p(Q) r_0,$$

where p is the polynomial

$$p(t) := \gamma_0 + \gamma_1 t + \cdots + \gamma_{k-1} t^{k-1}.$$

Since $r_0 = Qx_0 - b = Qx_0 - Qx^* = Q(x_0 - x^*)$, this gives

$$x - x^* = (x_0 - x^*) + p(Q) r_0 = [I + Qp(Q)](x_0 - x^*); \qquad (14.51)$$

thus $x - x^* = \bar{p}(Q)(x_0 - x^*)$, where $\bar{p}(t) = 1 + tp(t) \in \mathcal{P}_k$ and $\bar{p}(0) = 1$. □

Theorem 14.29 is a powerful result on the convergence rate of the conjugate-gradient method. However, it has the drawback that the gap $E(x_{k+1})$ is estimated by the displacement vector $x_0 - x^*$; it may be preferable to estimate the gap in terms of a known similar quantity instead, say $E(x_0)$. This is done below.

Theorem 14.30. *Let x_{k+1} be the $(k+1)$th point generated by the conjugate-gradient method. If $\{\lambda_i\}_{i=1}^n$ are the eigenvalues of the matrix Q, then*

$$\frac{E(x_{k+1})}{E(x_0)} \le \min_{p \in \mathcal{P}_k} \max_i \left(1 + \lambda_i p(\lambda_i)\right)^2.$$

Proof. Let $Q = U \Lambda U^T$ be the spectral decomposition of Q, where $\Lambda = \operatorname{diag}(\lambda_1, \ldots, \lambda_n)$, and write $x_0 - x^* = U\delta$.

If $p \in \mathcal{P}_k$ is such that $1 + tp(t)$ is the polynomial realizing the minimum in Theorem 14.29, then $p(Q) = Up(\Lambda)U^T$, and we have

$$\begin{aligned}
&\Big\langle Q(I+p(Q)Q)(x_0-x^*),\,(I+p(Q)Q)(x_0-x^*)\Big\rangle \\
&=\Big\langle \Lambda(I+p(\Lambda)\Lambda)\delta,\,(I+p(\Lambda)\Lambda)\delta\Big\rangle \\
&=\sum_{i=1}^{n}(1+\lambda_i p(\lambda_i))^2\lambda_i\delta_i^2 \\
&\le\Big(\max_i(1+\lambda_i p(\lambda_i))^2\Big)\sum_{i=1}^{n}\lambda_i\delta_i^2.
\end{aligned}$$

Theorem 14.29 implies that

$$\begin{aligned}
E(x_{k+1}) &\le \Big(\max_i(1+\lambda_i p(\lambda_i))^2\Big)\cdot\Big(\frac{1}{2}\sum_{i=1}^{n}\lambda_i\delta_i^2\Big) \\
&= \Big(\max_i(1+\lambda_i p(\lambda_i))^2\Big)E(x_0),
\end{aligned}$$

where the equality follows from

$$2E(x_0)=\|x_0-x^*\|_Q^2=\|U\delta\|_Q^2=\langle\Lambda\delta,\delta\rangle=\sum_{i=1}^{n}\lambda_i\delta_i^2.$$

□

Corollary 14.31. *Let x_k be the kth point generated by the conjugate-gradient method. Let $\kappa=\lambda_{\max}/\lambda_{\min}$ be the condition number of Q, where $\lambda_{\max}$ and $\lambda_{\min}$ are the largest and smallest eigenvalues of Q, respectively.*

Then

$$\frac{E(x_k)}{E(x_0)}\le\frac{1}{T_k\left(\frac{\kappa+1}{\kappa-1}\right)}\le 2\left(\frac{\sqrt{\kappa}-1}{\sqrt{\kappa}+1}\right)^k, \tag{14.52}$$

where T_k is Chebyshev's polynomial given by

$$T_k(x)=\cos(k\cos^{-1}x).$$

Proof. Write $m=\lambda_{\min}$ and $M=\lambda_{\max}$. Theorem 14.30 implies

$$\frac{E(x_k)}{E(x_0)}\le\min_p\max_i p(\lambda_i)^2\le\min_p\max_{x\in[m,M]}p(x)^2,$$

where the minimization is over polynomials $p\in\mathcal{P}_k$ satisfying $p(0)=1$.

The minimax problem $\min_p\max_{x\in[m,M]}p(x)^2$ above is a classical problem in approximation theory whose solution is the polynomial given by

$$p(x)=\frac{T_k\left(1+2\frac{x-M}{M-m}\right)}{T_k\left(1+2\frac{-M}{M-m}\right)}=\frac{T_k\left(1+2\frac{x-M}{M-m}\right)}{T_k\left(-\frac{\kappa+1}{\kappa-1}\right)}$$

with

$$\max_{x \in [m,M]} |p(x)| = \frac{1}{\left|T_k\left(-\frac{\kappa+1}{\kappa-1}\right)\right|} = \frac{1}{T_k\left(\frac{\kappa+1}{\kappa-1}\right)};$$

see Theorem 12.15 on page 327. The last equality above follows from the fact that T_k is an even or odd function, according to whether k is even or odd. (It is easy to prove this using $T_0(x) = 1$, $T_1(x) = x$, and the recursive formulas $T_{k+1}(x) = 2xT_k(x) - T_{k-1}(x)$.) This proves the first inequality in (14.52).

Toward proving the second inequality in (14.52), we express $T_k(x)$ in a different form. Defining $\theta = \cos^{-1} x$, we have

$$\begin{aligned} e^{i\theta} &= \cos\theta + i\sin\theta = \cos\theta + i\sqrt{1-\cos^2\theta} = \cos\theta + \sqrt{\cos^2\theta - 1} \\ &= x + \sqrt{x^2-1} \end{aligned}$$

and

$$\begin{aligned} T_k(x) = \cos(k\theta) &= \frac{e^{ik\theta} + e^{-ik\theta}}{2} = \frac{\left(e^{i\theta}\right)^k + \left(e^{-i\theta}\right)^{-k}}{2} \\ &= \frac{\left(x+\sqrt{x^2-1}\right)^k + \left(x+\sqrt{x^2-1}\right)^{-k}}{2}. \end{aligned}$$

Thus, we have

$$\begin{aligned} T_k\left(\frac{\kappa+1}{\kappa-1}\right) &\geq \frac{1}{2}\left(\frac{\kappa+1}{\kappa-1} + \sqrt{\left(\frac{\kappa+1}{\kappa-1}\right)^2 - 1}\right)^k = \frac{1}{2}\left(\frac{\kappa+1+2\sqrt{\kappa}}{\kappa-1}\right)^k \\ &= \frac{1}{2}\cdot\left(\frac{\sqrt{\kappa}+1}{\sqrt{\kappa}-1}\right)^k. \end{aligned}$$

The corollary is proved. □

In general, it is impossible to improve upon the convergence-rate estimates given in Corollary 14.31; see for example [153], pp. 373–374. Another "optimality" property of the conjugate-gradient method is given in [205], pp. 261–263, where it is shown that the conjugate-gradient method is essentially an *optimal* algorithm in the black-box, oracle model of computational complexity for optimization problems. Nevertheless, as one would expect from Theorem 14.30, it is possible to obtain better convergence rates for the conjugate-gradient method if one knows more about the distribution of the eigenvalues of Q. The interested reader is referred to the extensive literature on the subject for more information.

Corollary 14.32. *The optimality gap* $E(x) = q(x) - q(x^*)$ *is halved in every* $O(\sqrt{\kappa})$ *iterations of the conjugate-gradient method.*

Proof. See the proof of Corollary 14.10. □

Thus, if $\kappa = 10^6$, say, then the conjugate-gradient method is a thousand fold improvement over the steepest-descent method.

14.9 The Preconditioned Conjugate-Gradient Method

We have seen that the convergence rate of the conjugate-gradient method for solving the equation $Qx = b$ is proportional to the square root of the condition number of the matrix Q. For ill-conditioned matrices, this rate may still be slow. In order to accelerate the convergence rate, we may instead solve an equivalent linear system

$$\bar{Q}\bar{x} = \bar{b}, \tag{14.53}$$

where

$$\bar{Q} = C^{-1}QC^{-T}, \quad \bar{x} = C^T x, \quad \bar{b} = C^{-1}b, \tag{14.54}$$

where C is an invertible $n \times n$ matrix and $C^{-T} = (C^{-1})^T = (C^T)^{-1}$, such that the matrix $\bar{Q}$ has smaller, hopefully much smaller, condition number than Q.

It will be seen below (see (14.56)) that the matrix C itself is not needed in the preconditioned CGM; we need only to solve linear equations of the form $Mz = r$, where

$$M = CC^T$$

for z. The matrix C should be chosen so that the equation $Mz = r$ is easy to solve, and of course $\bar{Q}$, equivalently $M^{-1}Q$, has reasonably small condition number. There exists an extensive literature on how to choose C, but we cannot pursue this topic in any depth here; see for example [14, 234].

The quadratic function corresponding to $(\bar{Q}, \bar{b})$ is given by

$$\bar{q}(\bar{x}) := \frac{1}{2}\langle \bar{Q}\bar{x}, \bar{x}\rangle - \langle \bar{b}, \bar{x}\rangle.$$

Note that we have

$$\begin{aligned}
\bar{q}(\bar{x}) &= \frac{1}{2}\langle C^{-1}QC^{-T}(C^T x), C^T x\rangle - \langle C^{-1}b, C^T x\rangle \\
&= \frac{1}{2}\langle Qx, x\rangle - \langle b, x\rangle \\
&= q(x)
\end{aligned}$$

and

$$\bar{r}(\bar{x}) = \bar{Q}\bar{x} - \bar{b} = C^{-1}QC^{-T}(C^T x) - C^{-1}b = C^{-1}r(x). \tag{14.55}$$

The conjugate-gradient method applied to (14.53) becomes

Choose a point $\bar{x}_0 \in \mathbb{R}^n$. *Define* $\bar{d}_0 = -\bar{r}_0 = -\nabla\bar{q}(\bar{x}_0)$.
while $\bar{r}_k \neq 0$,

$$\begin{aligned}
\alpha_k &= \frac{\|\bar{r}_k\|^2}{\langle \bar{Q}\bar{d}_k, \bar{d}_k\rangle}, \\
\bar{x}_{k+1} &= \bar{x}_k + \alpha_k \bar{d}_k, \\
\bar{r}_{k+1} &= \bar{r}_k + \alpha_k \bar{Q}\bar{d}_k, \\
\beta_{k+1} &= \frac{\|\bar{r}_{k+1}\|^2}{\|\bar{r}_k\|^2}, \\
\bar{d}_{k+1} &= -\bar{r}_{k+1} + \beta_{k+1}\bar{d}_k.
\end{aligned}$$

end (while)

We can describe the above algorithm in the original variables. For this purpose, we first define

$$x_k = C^{-T}\bar{x}_k, \quad r_k = C\bar{r}_k,$$

in accordance with the equations (14.54) and (14.55). We also define

$$d_k = C^{-T}\bar{d}_k,$$

so that the equation $\bar{x}_{k+1} = \bar{x}_k + \alpha_k \bar{d}_k$ transforms into the equation $x_{k+1} = x_k + \alpha_k d_k$, that is, d_k is the direction of movement from x_k to x_{k+1}. Finally, we need an auxiliary variable z_k, which we define as $z_k := (CC^T)^{-1} r_k$, that is, z_k is the solution of the linear system

$$Mz_k = r_k, \quad \text{where } M = CC^T.$$

By making these substitutions in the preconditioned conjugate-gradient method above, we arrive at the following description.

The Preconditioned Conjugate-Gradient Method

Choose a point $x_0 \in \mathbb{R}^n$. *Define* $d_0 = -\nabla q(x_0) := -r_0$, *and solve* $Mz_0 = r_0$ *for* z_0.
while $r_k \neq 0$,

$$\begin{aligned}
\alpha_k &= \frac{\langle z_k, r_k\rangle}{\langle Qd_k, d_k\rangle}, \\
x_{k+1} &= x_k + \alpha_k d_k, \\
r_{k+1} &= r_k + \alpha_k Q d_k, \\
\textit{solve } Mz_{k+1} &= r_{k+1} \textit{ for } z_{k+1}, \\
\beta_{k+1} &= \frac{\langle z_{k+1}, r_{k+1}\rangle}{\langle z_k, r_k\rangle}, \\
d_{k+1} &= -z_{k+1} + \beta_{k+1} d_k.
\end{aligned} \tag{14.56}$$

end (while)

14.10 The Conjugate-Gradient Method and Orthogonal Polynomials

We now give an independent derivation of the conjugate-gradient method, inspired by the expanding subspace theorem, Theorem 14.29. This approach to the conjugate-gradient method has considerable advantages: the derivation is short, clear, self-contained, and provides a link between the conjugate-gradient method and *orthogonal polynomials*.

Choose a point $x_0 \in \mathbb{R}^n$, and consider the algorithm for minimizing $q(x)$ such that it generates the sequence $\{x_k\}$, where x_k is the global minimizer $q(x)$ on the affine subspace

$$M_k := x_0 + \mathcal{K}(Q, r_0, k) = x_0 + \operatorname{span}\{r_0, Qr_0, \ldots, Q^{k-1}r_0\}, \quad k \geq 1,$$

where $r_0 = \nabla q(x_0) = Qx_0 - b$.

A priori, this is a conceptual algorithm, because it is not clear how to minimize $q(x)$ on M_k *effectively*. (Of course, we know by Theorem 14.29 that x_k can be computed recursively, and the resulting algorithm is the conjugate-gradient method, but we do not need or use this knowledge here.)

Denote by

$$e_k := x_k - x^*$$

the displacement vector at step k, where x^* is the minimizer of q on $\mathbb{R}^n$; then

$$r_0 = Qx_0 - b = Q(x_0 - x^*) = Qe_0.$$

Now, any point $x \in M_k$ has a representation

$$\begin{aligned} x &= x_0 + c_1 r_0 + c_2 Q r_0 + \cdots + c_k Q^{k-1} r_0 \\ &= x^* + e_0 + c_1 Q e_0 + \cdots + c_k Q^k e_0 \\ &= x^* + p(Q) e_0, \end{aligned}$$

where

$$p(t) = 1 + c_1 t + \cdots + c_k t^k \in \mathcal{P}_k, \quad p(0) = 1.$$

We denote by p_k the polynomial corresponding to x_k, that is, $x_k = x^* + p_k(Q)e_0$. We conclude that

$$e_k = p_k(Q)e_0 \text{ and } r_k = p_k(Q)r_0, \tag{14.57}$$

where the second equation follows since $r_k = Qe_k = Qp_k(Q)e_0 = p_k(Q)Qe_0 = p_k(Q)r_0$.

If $r(t) \in \mathcal{P}_{k-1}$, then the polynomials p_k and $p_k(t) + \epsilon t r(t)$ are both in $\mathcal{P}_k$ and have value 1 at $t = 0$. Since

$$E(x) = \frac{1}{2}\|x - x^*\|_Q^2 = \frac{1}{2}\|p(Q)(e_0)\|_Q^2,$$

where the first equation follows from (14.50) and $E(x)$ is minimized at p_k, it follows that

$$
\begin{aligned}
0 &\leq \|[p_k(Q) + \epsilon r(Q)Q]e_0\|_Q^2 - \|p_k(Q)e_0\|_Q^2 \\
&= 2\epsilon\langle p_k(Q)e_0, r(Q)Qe_0\rangle_Q + \epsilon^2 \|r(Q)Qe_0\|_Q^2
\end{aligned}
$$

for all $\epsilon \in \mathbb{R}$, which is possible if and only if

$$\langle p_k(Q)e_0, r(Q)Qe_0\rangle_Q = \langle p_k(Q)r_0, r(Q)r_0\rangle = 0$$

for all polynomials $r(t) \in \mathcal{P}_{k-1}$.

Now we define an inner product on polynomials as follows:

$$\langle p, q\rangle_P := \langle p(Q)r_0, q(Q)r_0\rangle. \tag{14.58}$$

Then

$$\langle p_k, r\rangle_P = 0 \text{ for all } r \in \mathcal{P}_{k-1},$$

that is, the polynomials $\{p_k\}$ are *orthogonal* in the inner product $\langle\cdot,\cdot\rangle_P$.

Note that (14.57) and (14.58) give

$$\langle p_i, p_j\rangle_P = \langle r_i, r_j\rangle.$$

The vectors $\{r_k\}_0^{l-1}$ thus form an orthogonal set of vectors (hence a basis) in $\mathcal{K}(Q, r_0, l)$ as long as they are nonzero. Similarly, the polynomials $\{p_k\}_0^{l-1}$ are also linearly independent, and the polynomial $p_k(t)$ thus has degree k. Therefore, our algorithm stops precisely at step m, where m is the dimension of the Krylov space $\mathcal{K}(Q, r_0) := \text{span}\{Q^i r_0 : i \geq 1\}$, that is, the first time we have $M_k = M_{k+1}$.

Write

$$tp_k(t) = \sum_{i=0}^{k+1} c_{k,i}p_i(t);$$

we have

$$\langle tp_k, p_i\rangle_P = \langle p_k, tp_i\rangle_P = 0$$

for $i = 0, \ldots, k-2$, so that only the last three coefficients $c_{k,k-1}$, $c_{k,k}$, and $c_{k,k+1}$ may be nonzero. In fact, $c_{k,k+1} \neq 0$, since $tp_k(t)$ has degree $k+1$. (We remark that this three-term recurrence for $\{p_k\}$ is a characteristic property of orthogonal polynomials; see [250].)

Let us write $p_{k+1}(t) = (\gamma_k t + \xi_k)p_k(t) + \delta_k p_{k-1}(t)$. Since all three polynomials take the value 1 at $t = 0$, we have $\xi_k = 1 - \delta_k$, and

$$p_{k+1}(t) = [\gamma_k t + (1 - \delta_k)]p_k(t) + \delta_k p_{k-1}(t). \tag{14.59}$$

It follows from (14.57) and (14.59) that

$$\begin{aligned}
x_{k+1} - x^* = e_{k+1} &= p_{k+1}(Q)e_0 \\
&= \gamma_k Q p_k(Q)e_0 + (1-\delta_k)p_k(Q)e_0 + \delta_k p_{k-1}(Q)e_0 \\
&= \gamma_k r_k + (1-\delta_k)e_k + \delta_k e_{k-1} \\
&= \gamma_k r_k + (1-\delta_k)(x_k - x^*) + \delta_k(x_{k-1} - x^*).
\end{aligned}$$

Therefore,

$$\begin{aligned}
x_{k+1} &= x_k + \gamma_k r_k - \delta_k(x_k - x_{k-1}), \\
r_{k+1} &= r_k + \gamma_k Q r_k - \delta_k(r_k - r_{k-1}).
\end{aligned} \tag{14.60}$$

The coefficients γ_k and δ_k can be computed as follows: since $\langle r_{k+1}, r_k\rangle = 0$ and $\langle r_{k+1}, r_{k-1}\rangle = 0$, we have, using (14.60), the equations

$$\gamma_k\langle Qr_k, r_k\rangle + (1-\delta_k)\|r_k\|^2 = 0, \quad \gamma_k\langle Qr_k, r_{k-1}\rangle + \delta_k\|r_{k-1}\|^2 = 0,$$

which have the solution

$$\begin{aligned}
\beta_k &= \frac{\|r_k\|^2}{\|r_{k-1}\|^2}, \\
\gamma_k &= \frac{-\|r_k\|^2}{\langle Qr_k, r_k + \beta_k r_{k-1}\rangle}, \\
\delta_k &= \frac{\beta_k\langle Qr_k, r_{k-1}\rangle}{\langle Qr_k, r_k + \beta_k r_{k-1}\rangle}.
\end{aligned} \tag{14.61}$$

The equations (14.60) *and* (14.61) *give an alternative description of the conjugate-gradient method.*

In the remainder of this section, we show the exact correspondence between the data of the algorithm here and the one given in (14.49). (This is not strictly necessary but rounds out our treatment of the conjugate-gradient method.)

First, note that the sequences $\{x_k\}_0^{m-1}$ and $\{r_k\}_0^{m-1}$ are the same, since x_k minimizes $q(x)$ on M_k in both algorithms and $r_k = Qx_k - b$. Next, we claim that $\gamma_k = -\alpha_k$, where α_k is defined in (14.49). Using the last equation in (14.49), we have

$$\begin{aligned}
\langle Qr_k, r_k\rangle &= \langle Q(d_k - \beta_k d_{k-1}), d_k - \beta_k d_{k-1}\rangle \\
&= \langle Qd_k, d_k\rangle + \beta_k^2\langle Qd_{k-1}, d_{k-1}\rangle,
\end{aligned}$$

where the second equation follows by the Q-conjugacy of the vectors $\{d_i\}$ in (14.49); similarly,

$$\begin{aligned}
\langle Qr_k, r_{k-1}\rangle &= \langle Q(d_k - \beta_k d_{k-1}), d_{k-1} - \beta_{k-1}d_{k-2}\rangle \\
&= -\beta_k\langle Qd_{k-1}, d_{k-1}\rangle.
\end{aligned}$$

Adding, we get

$$\langle Qr_k, r_k + \beta_k r_{k-1}\rangle = \langle Qd_k, d_k\rangle.$$

It follows from (14.61) that $\gamma_k = -\|r_k\|^2/\langle Qd_k, d_k\rangle = -\alpha_k$, proving our claim. Similarly,

$$\delta_k = \frac{-\beta_k^2\langle Qd_{k-1}, d_{k-1}\rangle}{\langle Qd_k, d_k\rangle} = \frac{-\alpha_k\beta_k}{\alpha_{k-1}}. \tag{14.62}$$

We *define* d_k as in (14.49) by the formula

$$x_{k+1} = x_k + \alpha_k d_k \text{ (or equivalently, } r_{k+1} = r_k + \alpha_k Q d_k).$$

Then we have

$$d_k = \frac{x_{k+1} - x_k}{\alpha_k} = \frac{\gamma_k}{\alpha_k} r_k - \frac{\delta_k}{\alpha_k}(\alpha_{k-1}d_{k-1}) = -r_k + \beta_k d_{k-1},$$

where the second equality follows from (14.60). This proves the last equation in (14.49) (with $k+1$ replaced by k), and completes the equivalence of the formulas (14.60) and (14.61) with (14.49).

Remark 14.33. The attentive reader will note that all the results of this section can be worked out without orthogonal polynomials ever being mentioned. However, the connection between the conjugate-gradient method and orthogonal polynomials is important and should be kept in mind. We remark that this connection was noted already in [132], the first paper on the subject, and is treated in more detail later in [246]. The classic book [250] is still a good reference on orthogonal polynomials. The inspiration to write this section comes from the book [186], but we have removed the restrictions there on the matrix Q.

14.11 Exercises

1. Consider the steepest-descent method to compute a minimizer of the function $f(x, y) = x^4 + y^2 - 8y$ starting at the point $(x^0, y^0) = (0, 1)$. Determine the next iterate (x_1, y_1) using several of the step-size selection rules.
2. Newton's method is used to compute a solution to the system $3x^2y + y^2 = 1$, $x^4 + xy^3 = 1$ starting at the point $(x_0, y_0) = (1, 1)$. Determine the next iterate (x_1, y_1).
3. Consider the function $f(x, y) = 2x^2 + y^2 - 2xy + 2x^3 + x^4$.
   (a) Determine all the critical points of f.
   (b) Suppose we initiate the Newton's method for minimizing f at a point (x_0, y_0). Write down the equation describing this Newton's method.
   (c) Suppose that the Newton's method in (b) is initiated at the point $(-1, 0)$. Compute numerically the next point (x_1, y_1).
   (d) Near which critical points can we expect a quadratic convergence rate from Newton's method?
4. Consider using Newton's method for finding roots of the function

$$f(x, y) = (x^2 - y - 1, x + y^2 - 1).$$

(a) If the initial point is $(1, 1)$, compute the next iterate.
(b) Note that the points $(1, 0)$ and $(0, -1)$ are roots of f. If the initial point (x_0, y_0) is very close to any of these roots, will Newton's method be fast in the sense that it will converge quadratically?
(c) Find all other roots of f, if they exist.

5. Here is a problem in which we have quadratic convergence even if we modify Newton's method, such that

$$x_{k+1} = x_k - g(x_k)f(x_k),$$

where g is a suitable function.

Suppose we would like to invert a nonsingular matrix $A \in \mathbb{R}^{n\times n}$. We can view this problem as solving the equation

$$F(X) = 0, \quad \text{where} \quad F(X) = I - AX, \quad F : \mathbb{R}^{n\times n} \to \mathbb{R}^{n\times n}.$$

Since F is linear, we have $DF(X) = -A$. Obviously, if we apply Newton's method, we obtain

$$X_{k+1} = X_k - DF(X_k)^{-1}F(X_k) = X_k + A^{-1}(I - AX_k) = A^{-1},$$

that is, Newton's method converges in one step. The drawback is, of course, that calculating $DF(X_k)^{-1} = -A^{-1}$ is computationally costly. Suppose we replace $DF(X_k)^{-1}$ by $G(X_K) := -X_k$. (Since $DF(X)^{-1} = -A^{-1}$ and $X_k \approx A^{-1}$, this is reasonable.)

(a) Show that if X_0 is close enough to A^{-1}, then the method

$$X_{k+1} = X_k + X_kF(X_k) = X_k + X_k(I - AX_k) = X_k(2I - AX_k)$$

is quadratically convergent, in the sense that

$$\|F(X_{k+1})\| \le \|F(X_k)\|^2.$$

How would you quantify the "closeness" criterion for X_0 so that the method actually converges, that is, what conditions do you need on X_0 so that the above "Newton's method" converges and converges quadratically?
Hint: Find an equation connecting $F(X_k)$ and $F(X_{k+1})$.

(b) How can we choose X_{k+1} so that the resulting method is cubically convergent in the sense that

$$\|F(X_{k+1})\| \le \|F(X_k)\|^3?$$

(c) How can we choose X_{k+1} so that the method is nth-order convergent?

6. Prove the following facts about the conjugate-gradient method using its description in (14.49):

$$(a) \qquad d_k = -\sum_{i=0}^{k} \frac{\|r_k\|^2}{\|r_i\|^2} r_i.$$

$$(b) \qquad \|d_k\|^2 = \|r_k\|^2 + \beta_k^2 \|d_{k-1}\|^2 = \sum_{i=0}^{k} \frac{\|r_k\|^4}{\|r_i\|^2}.$$

$$(c) \qquad \langle d_k, d_l \rangle = \sum_{i=0}^{k} \frac{\|r_k\|^2 \cdot \|r_l\|^2}{\|r_i\|^2} = \frac{\|r_l\|^2}{\|r_k\|^2} \cdot \|d_k\|^2 \quad \text{for} \quad k \le l.$$

Thus, $\|d_k\|$ increases with k, and the angle between two different directions d_k and d_l is acute.

7. Prove the following facts about the conjugate-gradient method:

$$(a) \qquad \alpha_k = \frac{|r_k|^2}{\langle Qd_k, d_k \rangle} = \frac{-\langle r_k, d_k \rangle}{\langle Qd_k, d_k \rangle} = \frac{-\langle r_0, d_k \rangle}{\langle Qd_k, d_k \rangle},$$

$$(b) \qquad \beta_{k+1} = \frac{|r_{k+1}|^2}{|r_k|^2} = \frac{\langle r_{k+1}, Qd_k \rangle}{\langle Qd_k, d_k \rangle} = \frac{-\langle r_{k+1}, Qr_k \rangle}{\langle Qd_k, d_k \rangle}.$$

8. Prove the following three-term recurrence formulas for $\{d_k\}$ and $\{r_k\}$ in the conjugate-gradient method:

$$(a) \qquad \begin{aligned} d_1 &= (1 + \beta_1) d_0 - \alpha_0 Q d_0, \\ d_{k+1} &= (1 + \beta_{k+1}) d_k - \alpha_k Q d_k - \beta_k d_{k-1}, \quad k > 0, \end{aligned}$$

$$(b) \qquad \begin{aligned} r_1 &= r_0 - \alpha_0 Q r_0, \\ r_{k+1} &= \left(1 + \frac{\alpha_k \beta_k}{\alpha_{k-1}}\right) r_k - \alpha_k Q r_k - \frac{\alpha_k \beta_k}{\alpha_{k-1}} r_{k-1}, \quad k > 0. \end{aligned}$$

9. Prove that in the conjugate-gradient method, the iterates x_k get closer to the optimal solution x^* at each step, that is, $\|x_k - x^*\|$ is a strictly decreasing sequence.
Hint: Prove that $\langle x^* - x_k, x_k - x_{k-1} \rangle > 0$ by expressing the vectors $x^* - x_k$ and $x_k - x_{k-1}$ as linear combinations of $\{d_i\}$ and using the property $\langle d_i, d_j \rangle > 0$, $i \neq j$, indicated in Exercise 6 on page 404.

A

Finite Systems of Linear Inequalities in Vector Spaces

In this appendix, we characterize the consistency of finitely many linear inequalities in an arbitrary vector space E over $\mathbb{R}$ by proving the central Motzkin's transposition theorems in this setting. This is done using elementary combinatorial techniques that go back at least to the work of Carver [55] in the 1920s. A significant merit of this approach, besides its elementary character, is the fact that the completeness of $\mathbb{R}$ is not used, so that the proofs can be extended to vector spaces over other fields, say the field $\mathbb{Q}$ of rational numbers.

We first need a definition and a technical lemma.

Definition A.1. *Let E be a real vector space. A system of inequalities and equalities*

$$\ell_i(x) < \alpha_i, \ i \in I, \quad \ell_j(x) \leq \alpha_j, \ j \in J, \quad \ell_k(x) = \alpha_k, \ k \in K,$$

where $\ell_i, \ell_j, \ell_k : E \to \mathbb{R}$ are linear functionals and the index sets I, J, K are finite, is called irreducibly inconsistent *if it has no solution, but dropping any one of the constraints leads to a system that has a solution.*

Lemma A.2. *Let E be a real vector space, and let $\{\ell_i\}_1^k$ be linear functionals on E. The set*

$$L := \{y \in \mathbb{R}^k : y_i = \ell_i(x), \ i = 1, \ldots, k\}$$

is a linear subspace of $\mathbb{R}^k$, which is a proper subspace of $\mathbb{R}^k$ if and only if $\{\ell_i\}_1^k$ is linearly dependent.

Proof. The fact that L is a linear subspace of $\mathbb{R}^k$ is an immediate consequence of the linearity of $\{\ell_i\}_1^k$. We have $L \neq \mathbb{R}^k$ if and only if there exists a nonzero $\lambda \in \mathbb{R}^k$ such that λ is orthogonal to L, that is, $\sum_{i=1}^k \lambda_i \ell_i(x) = 0$ for all $x \in E$, which is equivalent to the linear dependence of $\{\ell_i\}_1^k$. □

We are ready to state and prove the homogeneous version of Motzkin's transposition theorem.

Theorem A.3. (***Motzkin's transposition theorem, homogeneous version***) *Let E be a real vector space, and let $\{\ell_i\}_{i\in I}$, $\{\ell_j\}_{i\in J}$, and $\{\ell_i\}_{k\in K}$ be linear functionals on E, where I, J, K are finite sets.*

Then the linear system

$$\ell_i(x) < 0,\ i \in I, \quad \ell_j(x) \leq 0,\ j \in J, \quad \ell_k(x) = 0,\ k \in K, \tag{A.1}$$

is inconsistent if and only if there exist multipliers $\lambda := (\lambda_i : i \in I)$, $\mu := (\mu_j : j \in J)$, and $\delta := (\delta_k : k \in K)$ such that

$$\sum_{i\in I} \lambda_i \ell_i + \sum_{j\in J} \mu_j \ell_j + \sum_{k\in K} \delta_k \ell_k = 0, \quad (\lambda, \mu) \geq 0, \ \lambda \neq 0. \tag{A.2}$$

Proof. If $x \in E$ satisfies (A.1) and the multipliers (λ, μ, δ) satisfy (A.2), then

$$\begin{aligned} 0 &= \Big(\sum_1^l \lambda_i \ell_i + \sum_1^m \mu_j \ell_j + \sum_1^p \delta_k \ell_k\Big)(x) \\ &= \sum_{i\in I} \lambda_i \ell_i(x) + \sum_{j\in J} \mu_j \ell_j(x) + \sum_{k\in K} \delta_k \ell_k(x) \\ &\leq \sum_{i\in I} \lambda_i \ell_i(x) < 0, \end{aligned}$$

where the last inequality follows since at least one λ_i is positive.

To complete the proof of the theorem, it remains to prove the claim that if (A.1) is inconsistent, then there exist multipliers (λ, μ, δ) satisfying (A.2). Let us make some observations that will simplify the proof of the claim.

First of all, we may assume that $K = \emptyset$. Suppose that we have succeeded in proving our claim for this case. If $K \neq \emptyset$, then each equality $\ell_k(x) = 0$, $k \in K$, can be written as two inequalities $\ell_k(x) \leq 0$ and $-\ell_k(x) \leq 0$, thus reducing (A.1) to a system with no equality constraints, and obtain multipliers (λ, μ), including the multipliers $\mu_{k_1} \geq 0$ and $\mu_{k_2} \geq 0$ corresponding to the inequalities $\ell_k(x) \leq 0$ and $-\ell_k(x) \leq 0$, respectively. If we let $\delta_k := \mu_{k_1} - \mu_{k_2}$ be the multiplier for the equality $\ell_k(x) = 0$, then it is clear that the multipliers $(\lambda, \{\mu_j\}_{j\in J}, \delta)$ satisfy the required properties, proving the claim in the case $K \neq \emptyset$.

Secondly, we may assume that the system (A.1) is *irreducibly* inconsistent, if necessary by getting rid of constraints one at a time while still preserving the inconsistency of the system. If we succeed in proving the claim in this case, then it is clear that setting the multipliers to zero for the omitted inequalities proves the claim for the original case.

Thirdly, we may assume that $|I| = 1$, that is, there exists exactly one strict inequality in (A.1). Suppose that $|I| > 1$; pick $p \in I$ and consider the system obtained by relaxing all strict inequalities except $\ell_p(x) < 0$ to non-strict inequalities,

$$\ell_p(x) < 0,\ \ell_i(x) \leq 0,\ i \in I \setminus \{p\}, \quad \ell_j(x) \leq 0,\ j \in J. \tag{A.3}$$

If this system is consistent, then there is a point $a \in E$ satisfying it, and since (A.1) (with $K = \emptyset$) is irreducibly inconsistent, there exists a point $b \in E$ satisfying the inequalities

$$\ell_p(b) \geq 0,\ \ell_i(b) < 0,\ i \in I \setminus \{p\}, \quad \ell_j(b) \leq 0,\ j \in J.$$

But then a point $c \in (a, b)$ sufficiently near a satisfies all the inequalities in (A.1), a contradiction. This proves that (A.3) is inconsistent, in fact irreducibly inconsistent, since (A.1) is.

Thus far, we have succeeded in reducing our claim to proving that if the system

$$\ell(x) < 0, \quad \ell_j(x) \leq 0,\ j \in J, \tag{A.4}$$

is irreducibly inconsistent, then there exist multipliers $\lambda > 0$, $\mu_j \geq 0$, $j \in J$, such that

$$\lambda\ell + \sum_{j \in J} \mu_j \ell_j = 0; \tag{A.5}$$

this is a special case of Farkas's lemma. Note that if $\ell = 0$, then $J = \emptyset$, and the claim is obvious. We consider the remaining case in the rest of the proof, where $\ell \neq 0$, $J \neq \emptyset$, and $\ell_j \neq 0$ for every $j \in J$.

For any two distinct indices $r, k \in J$, there exist, by virtue of irreducible inconsistency of (A.4), $a, b \in E$ such that

$$\begin{aligned} &\ell(a) < 0, \quad \ell_r(a) > 0,\ \ell_k(a) \leq 0, \quad \ell_j(a) \leq 0,\ j \in J \setminus \{r, k\}, \\ &\ell(b) < 0, \quad \ell_r(b) \leq 0,\ \ell_k(b) > 0, \quad \ell_j(a) \leq 0,\ j \in J \setminus \{r, k\}. \end{aligned}$$

If we move on the line segment $[a, b]$ from a to b, we obtain a point c satisfying the equality $\ell_k(c) = 0$; the point c satisfies the inequalities

$$\ell(c) < 0, \quad \ell_r(c) > 0,\ \ell_k(c) = 0, \quad \ell_j(x) \leq 0,\ j \in J \setminus \{r, k\},$$

where the second inequality follows because (A.4) is inconsistent.

Next, consider two consistent systems like the one above,

$$\begin{aligned} &\ell(x) < 0, \quad \ell_r(x) > 0,\ \ell_{k_1}(x) = 0,\ \ell_{k_2}(x) \leq 0, \quad \ell_j(x) \leq 0,\ j \in J \setminus \{r, k_1, k_2\}, \\ &\ell(x) < 0, \quad \ell_r(x) \leq 0,\ \ell_{k_1}(x) = 0,\ \ell_{k_2}(x) > 0, \quad \ell_j(x) \leq 0,\ j \in J \setminus \{r, k_1, k_2\}, \end{aligned}$$

and apply the same idea to obtain a consistent system

$$\ell(x) < 0,\ \ell_r(x) > 0, \quad \ell_{k_1}(x) = 0,\ \ell_{k_2}(x) = 0, \quad \ell_j(x) \leq 0,\ j \in J \setminus \{r, k_1, k_2\}.$$

In this manner, we can replace, one at a time, all but one of the nonstrict inequalities with equality, and arrive at the consistent system, for any $k \in J$,

$$\ell(x) < 0, \quad \ell_k(x) > 0, \quad \ell_j(x) = 0,\ j \in J \setminus \{k\}. \tag{A.6}$$

Since (A.4) is inconsistent, Lemma A.2 implies that the linear functionals $\{\ell\} \cup \{\ell_j\}_{j \in J}$ are linearly dependent; thus there exist scalars $\{\lambda, \mu_j, j \in J\}$, not all zero, such that

$$\lambda \ell + \sum_{j \in J} \mu_j \ell_j = 0.$$

Clearly, at least two multipliers are nonzero, and we may assume without loss of generality that $\mu_r > 0$ for some $r \in J$. If $a \in E$ satisfies (A.6) for $k = r$, then

$$\lambda \ell(a) = -\mu_r \ell_r(a) < 0,$$

and since $\ell(a) < 0$, we have $\lambda > 0$.

Finally, if $s \in J$ is arbitrary, and $b \in E$ satisfies (A.6) for $k = s$, then

$$\mu_s \ell_s(b) = -\lambda \ell(b) > 0,$$

and since $\ell_s(b) > 0$, we have $\mu_s > 0$. The theorem is proved. □

Note that two other proofs of Theorem A.3 in the case that E is a finite-dimensional vector space over $\mathbb{R}$ have been given in Theorem 3.15 on page 72 and Theorem 7.17 on page 183. All three proofs are independent of one another. The proof above is the most general one, and has the added virtue that it needs no prerequisites.

Now that the central result Theorem A.3 is established, the following four results follow as easy corollaries.

Corollary A.4. (***Gordan's lemma***) *Let E be a real vector space, and let $\{\ell_i\}_{i \in I}$, $\{\ell_j\}_{j \in J}$ be a finite set of linear functionals on E. Then*

$$\{d : \ell_i(d) < 0,\ i \in I,\ \ell_j(d) = 0,\ j \in J\} = \emptyset$$

if and only if there exist a nonnegative, nonzero vector $\lambda \in \mathbb{R}^{|I|}$ and a vector $\mu \in \mathbb{R}^{|J|}$ such that

$$\sum_{i \in I} \lambda_i \ell_i + \sum_{j \in J} \mu_j \ell_j = 0.$$

Corollary A.5. (***Farkas's lemma, homogeneous version***) *Let E be a real vector space, and let $\{\ell_i\}_{i=1}^k$, $\{\ell\}$ be linear functionals on E. The following statements are equivalent:*

(*a*) *any $x \in E$ satisfying $\ell_i(x) \le 0,\ i = 1, \ldots, k,$ also satisfies $\ell(x) \le 0$,*

(*b*) *there exists $\lambda \ge 0$ such that $\ell = \sum_{i=1}^k \lambda_i \ell_i$.*

Proof. Note that (a) is equivalent to the inconsistency of the system

$$-\ell(x) < 0,\ \ell_i(x) \le 0,\ i = 1, \ldots, k.$$

□

Theorem A.6. (***Motzkin's transposition theorem, affine version***) *Let E be a vector space over $\mathbb{R}$. Let $\{\ell_i\}_{i\in I}$, $\{\ell_j\}_{j\in J}$, and $\{\ell_k\}_{k\in K}$ be linear functionals on E, and let $\{\alpha_i\}_{i\in I}$, $\{\alpha_j\}_{j\in J}$, and $\{\alpha_k\}_{k\in K}$ be real scalars, where I, J, K are finite sets.*

The linear system

$$\ell_i(x) < \alpha_i,\ i \in I, \quad \ell_j(x) \le \alpha_j,\ j \in J, \quad \ell_k(x) = \alpha_k,\ k \in K, \tag{A.7}$$

is inconsistent if and only if there exist multipliers $\lambda_0 \in \mathbb{R}$, $\lambda \in \mathbb{R}^{|I|}$, $\mu \in \mathbb{R}^{|J|}$, $\delta \in \mathbb{R}^{|K|}$, satisfying

$$\begin{aligned}
&\sum_{i\in I} \lambda_i \ell_i + \sum_{j\in J} \mu_j \ell_j + \sum_{k\in K} \delta_k \ell_k = 0,\\
&\sum_{i\in I} \lambda_i \alpha_i + \sum_{j\in J} \mu_j \alpha_j + \sum_{k\in K} \delta_k \alpha_k + \lambda_0 = 0, \qquad \text{(A.8)}\\
&(\lambda_0, \lambda, \mu) \ge 0, \quad (\lambda_0, \lambda) \ne 0.
\end{aligned}$$

Proof. The inconsistency of the system (A.7) is equivalent to that of the homogenized system $t > 0$, $\ell_i(x) < t\alpha_i$, $i \in I$, $\ell_j(x) \le t\alpha_j$, $j \in J$, and $\ell_k(x) = t\alpha_k$, $k \in K$, that is, of the homogeneous linear system in the vector space $E \times \mathbb{R}$,

$$\tilde{\ell}_0(x,t) < 0,\ \tilde{\ell}_i(x,t) < 0,\ \tilde{\ell}_j(x,t) \le 0,\ \tilde{\ell}_k(x,t) = 0,\ (i,j,k) \in I \times J \times K, \tag{A.9}$$

where $\tilde{\ell}_0(x,t) = -t$, $\tilde{\ell}_i(x,t) = \ell_i(x) - t\alpha_i$, $i \in I \cup J \cup K$. This follows from the fact that if x solves (A.7), then $(x,1)$ solves (A.9), and conversely, if (x,t) solves (A.9), then x/t solves (A.7). Theorem A.3 implies that there exist nonnegative multipliers $\lambda_0 \in \mathbb{R}$, $\lambda \in \mathbb{R}^{|I|}$, $\mu \in \mathbb{R}^{|J|}$, and a multiplier $\delta \in \mathbb{R}^{|K|}$ such that $(\lambda_0, \lambda) \ne 0$ and

$$\lambda_0 \tilde{\ell}_0 + \sum_{i\in I} \lambda_i \tilde{\ell}_i + \sum_{j\in J} \mu_j \tilde{\ell}_j + \sum_{k\in K} \delta_k \tilde{\ell}_k = 0.$$

Setting $(x,t) = (x,0)$ gives the first equality in (A.8), while setting $(x,t) = (0,1)$ gives the second one. □

Theorem A.7. (***Farkas's lemma, affine version***) *Let E be a real vector space. Let $\{\ell_i\}_{i=1}^k$, ℓ be linear functionals on E, and let $\{\alpha_i\}_{i=1}^k$, γ be real scalars. Suppose that the linear inequality system*

$$\ell_i(x) \le \alpha_i, \quad i = 1, \dots, k,$$

is consistent.

Then the following statements are equivalent:

(a) *any $x \in E$ satisfying $\ell_i(x) \le \alpha_i$, $1 \le i \le k$, also satisfies $\ell(x) \le \gamma$,*

(b) *$\exists (\lambda_1, \dots, \lambda_k) \ge 0$ such that $\sum_{i=1}^k \lambda_i \ell_i = \ell$, $\sum_{i=1}^k \lambda_i \alpha_i \le \gamma$.*

Proof. Note that (a) is equivalent to the inconsistency of the system

$$-\ell(x) < -\gamma, \ \ell_i(x) \leq \alpha_i, \ i = 1, \ldots, k.$$

By Theorem A.6, there exist nonnegative multipliers $\{\lambda_i\}_{i=-1}^{k}$ such that $(\lambda_{-1}, \lambda_0) \neq 0$ and

$$-\lambda_{-1}\ell + \sum_{i=1}^{k} \lambda_i \ell_i = 0, \quad -\lambda_{-1}\gamma + \sum_{i=1}^{k} \lambda_i \alpha_i + \lambda_0 = 0.$$

Since the system $\ell_i(x) \leq \alpha_i$, $i = 1, \ldots, k$, is consistent, we must have $\lambda_{-1} > 0$ by Theorem A.6, and we may set $\lambda_{-1} = 1$. The theorem follows. □

B

Descartes's Rule of Sign

In this appendix, we give a simple proof of Descartes's rule of sign following Vinberg [260].

Recall Definition 2.22:

Definition B.1. *Let $a_0, a_1, \ldots, a_n$ be a sequence of real numbers. If all the numbers in the sequence are nonzero, the total number of variations of sign in the sequence, denoted by $V(a_0, a_1, \ldots, a_n)$, is the number of times consecutive numbers a_{k-1} and a_k differ in sign, that is,*

$$V(a_0, a_1, \ldots, a_n) := |\{k : a_{k-1}a_k < 0, k = 1, \ldots, n\}|.$$

If the sequence $a_0, a_1, \ldots, a_n$ contains zeros, then $V(a_0, a_1, \ldots, a_n)$ is defined to be the variations of the reduced sequence by ignoring all zero elements in the sequence. Also, we define $V(a_0) = 0$ for any $a_0 \in \mathbb{R}$.

Let $p(x) = a_0 + a_1x + a_2x^2 + \cdots + a_nx^n$ be a polynomial of degree n with real coefficients. We write

$$V_p := V(a_0, a_1, \ldots, a_n).$$

We recall Theorem 2.23:

Theorem B.2. (***Descartes's rule of sign***) *Let $p(x) = a_0 + a_1x + a_2x^2 + \cdots + a_nx^n$ be a nonzero polynomial of degree n with real coefficients. Then the number of positive roots $N_p(0, \infty)$ of p is given by*

$$N_p(0, \infty) = V_p - 2\kappa,$$

where κ is a nonnegative integer.

Moreover, if the roots of p are all real, then $\kappa = 0$, that is, $N_p(0, \infty) = V_p$.

The following simple technical lemmas will be used in its proof.

Lemma B.3. *Let $p(x) = a_0 + a_1x + \cdots + a_nx^n$ be a polynomial with real coefficients such that $a_0 \neq 0$ and $a_n > 0$. Then V_p and $N_p(0,\infty)$ differ by an even integer.*

Proof. We have $p(0) = a_0$ and $p(x) > 0$ for large enough x. As we move along the real line to the right, $p(x)$ changes sign when we pass a simple root. When we pass a root of multiplicity k, the sign of $p(x)$ changes $(-1)^k$ times. This means that $N_p(0,\infty)$ is even if $a_0 > 0$ and odd if $a_0 < 0$. A little thought shows that the same thing is true about V_p. □

Lemma B.4. $N_p(0,\infty) \leq N_{p'}(0,\infty) + 1$ *and* $V_{p'} \leq V_p$.

Proof. The second inequality is clear; to prove the first, note that by Rolle's theorem, there exists a root of p' strictly between two distinct roots of p, and if x is a root of p with multiplicity $k > 1$, then x is a root of p' with multiplicity $k-1$. □

The number of negative roots of the polynomial p is equal to the number of positive roots of the polynomial

$$\overline{p}(x) := p(-x).$$

Lemma B.5. $V_p + V_{\overline{p}} \leq n$.

Proof. A change of sign occurs in the sequence $a_0, \ldots, a_{k-1}, a_k, \ldots, a_n$ of p at the kth position if and only if no change of sign occurs in the coefficients of $\overline{p}$ at the kth position. If all a_i are nonzero, we have $V_p + V_{\overline{p}} = n$; in the case that some $a_i = 0$, we have $V_p + V_{\overline{p}} \leq n$. □

We are ready to give the proof of Theorem B.2.

Proof. We first prove the inequality $N_p(0,\infty) \leq V_p$ by induction on the degree of the polynomial p. If $\deg p = 0$, then $N_p(0,\infty) = V_p = 0$. If $\deg p = n > 0$, then $\deg p' = n-1$, and we have

$$N_p(0,\infty) \leq N_{p'}(0,\infty) + 1 \leq V_{p'} + 1 \leq V_p + 1,$$

where the first and last inequalities follow from Lemma B.4, and the middle one from the induction hypothesis. Lemma B.3 implies that $N_p(0,\infty) = V_p + 1$ is impossible. This establishes the first statement of the theorem.

If all roots of p are real, we can assume that 0 is not a root of p. Then,

$$n = N_p(0,\infty) + N_{\overline{p}}(0,\infty) \leq V_p + V_{\overline{p}} \leq n;$$

thus $N_p(0,\infty) = V_p$ and $N_{\overline{p}}(0,\infty) = V_{\overline{p}}$. □

Let p be a polynomial of degree n with real coefficients. For $a < b$, let $N_p(a,b]$ be the number of roots of p in the interval $(a,b]$, and if $c \in \mathbb{R}$, let $V_p(c) = V(p(c), p'(c), \ldots, p^{(n)}(c))$. It is easy to verify that $V_p(c)$ is the number of sign variations in the coefficients of the polynomial $x \mapsto p(x+c)$.

The following two results follow easily from Theorem B.2.

Theorem B.6. (***Budan–Fourier***) *If p is a polynomial of degree n with real coefficients, then*

$$N_p(a,b] = V_p(a) - V_p(b) - 2\kappa,$$

where $\kappa \geq 0$ is a nonnegative integer.

Theorem B.7. (***Loewy–Curtiss***) *Let p be a polynomial of degree n with real coefficients. Then*

$$N_p(a,b] = V_p(a) - V_p(b)$$

for every real interval $(a,b]$ if and only if all roots of p are real.

Chapter 10 of the book [225] is a good resource for the root-counting results on polynomials.

B.1 Exercises

1. Prove Theorem B.6.
2. Prove Theorem B.7.

C

Classical Proofs of the Open Mapping and Graves's Theorems

Theorem C.1. (***Open mapping theorem***) *Let X and Y be Banach spaces and let $A : X \to Y$ be a continuous linear mapping onto Y.*

Then there exists $\tau > 0$ such that

$$\tau B_Y \subseteq A(B_X), \tag{C.1}$$

where $B_X = \{x \in X : \|x\| < 1\}$ and $B_Y = \{y \in Y : \|y\| < 1\}$ are the open unit balls in X and Y, respectively.

Consequently, A is an open mapping, that is, if $O \subseteq X$ is open set, then $A(O)$ is open set in Y.

Proof. Since A is a linear map, it suffices to prove (C.1). Since A is an onto mapping, we have

$$Y = A(X) = A(\cup_{n=1}^{\infty} nB_X) = \cup_{n=1}^{\infty} A(nB_X) = \cup_{n=1}^{\infty} nA(B_X).$$

It follows from the Baire category theorem that at least one set $\overline{nA(B_X)}$ contains an open set, or equivalently, $\overline{A(B_X)}$ contains an open set, say $O_1 = y + \tau B_Y \subseteq \overline{A(B_X)}$. Since $B_X = -B_X$, we have $O_2 = -y - \tau B_Y = -y + \tau B_Y \subseteq \overline{A(B_X)}$. If $z \in Y$ such that $\|z\| < \tau$, then there exist $\{u_k\}_1^{\infty}$ and $\{v_k\}_1^{\infty}$ in X such that $y + z = \lim Au_k$ and $-y + z = \lim Av_k$, and thus $z = \lim A(u_k + v_k)/2 \in \overline{A(B_X)}$. This proves that

$$\tau B_Y \subseteq \overline{A(B_X)}. \tag{C.2}$$

It remains to show that $A(B_X)$ contains some open ball θB_Y. We claim that this is true for any $0 < \theta < \tau$. Pick an arbitrary $y \in Y$, $\|y\| < \theta$. We have $y \in \overline{A(rB_X)}$, where $r = \theta/\tau$, so that there exists $x_1 \in rB_X$ such that $\|y - Ax_1\| < \alpha\theta$, where $0 < \alpha < 1$ is chosen such that $\theta < \tau(1 - \alpha)$. Next, $y - Ax_1 \in \overline{A((\alpha\theta/\tau)rB_X)} = \overline{A(\alpha^2 rB_X)}$, so there exists $x_2 \in \alpha^2 rB_X$ such that $\|y - Ax_1 - Ax_2\| < \alpha^2\theta$. Continuing in this manner, we obtain a sequence $\{x_n\}_1^{\infty}$ in X such that

$$\|x_n\| < \alpha^n r, \text{ and } \|y - A(x_1 + \cdots + x_n)\| < \alpha^n \theta, \quad n \geq 1.$$

Then the point $x := \sum_1^\infty x_n$ satisfies $Ax = y$ and

$$\|x\| \leq r + \alpha r + \alpha^2 r + \cdots = \frac{r}{1-\alpha} = \frac{\theta}{\tau(1-\alpha)} < 1.$$

This proves our claim. □

We remark that the inclusion (C.1) also holds if the open balls B_X and B_Y are replaced by the closed units balls $\overline{B}_X = \{x \in X : \|x\| \leq 1\}$ and $\overline{B}_Y = \{y \in Y : \|y\| \leq 1\}$: if $0 < \tau' < \tau$, then (C.1) gives

$$\tau' \overline{B}_Y \subseteq \tau B_Y \subseteq A(B_X) \subseteq A(\overline{B}_X).$$

The following theorem, proved first by Graves [110] (see also Dontchev [79]), is an important generalization of the open mapping theorem.

Theorem C.2. (***Graves's theorem***) *Let X and Y be Banach spaces, $r > 0$, and let $f : r\overline{B}_X \to Y$ be a mapping such that $f(0) = 0$. Let $A : X \to Y$ be a continuous linear mapping onto Y satisfying*

$$\tau \overline{B}_Y \subseteq A(\overline{B}_X). \tag{C.3}$$

Let $f - A$ be Lipschitz continuous on $r\overline{B}_X$ with a constant δ, $0 \leq \delta < \tau$, that is,

$$\|f(x_1) - f(x_2) - A(x_1 - x_2)\| \leq \delta \|x_1 - x_2\| \ \ \textit{for all} \ \ x_1, x_2 \in r\overline{B}_X. \tag{C.4}$$

Then

$$(\tau - \delta) r \overline{B}_Y \subseteq f(r\overline{B}_X),$$

that is, the equation $y = f(x)$ has a solution $\|x\| \leq r$ whenever $\|y\| \leq (\tau - \delta) r$.

Proof. Define $c := \tau - \delta$ and let $y \in Y$ be any point satisfying $\|y\| \leq cr$. We will show that there exists $x \in r\overline{B}_X$ such that $f(x) = y$. Toward that goal, we recursively generate a sequence $\{x_n\}_0^\infty$ in $r\overline{B}_X$ converging to x.

We start with $x_0 = 0$, and using (C.3), pick a point x_1 satisfying

$$Ax_1 = y, \quad \text{and} \quad \|x_1\| \leq \frac{\|y\|}{\tau} \leq \frac{cr}{\tau} < r.$$

Assuming that $\{x_j\}_0^k$ has been generated, we generate x_{k+1} from the equation

$$A(x_{k+1} - x_k) = (A - f)(x_k) - (A - f)(x_{k-1}), \quad k \geq 1. \tag{C.5}$$

Here, the right-hand-side vector above has norm at most $\delta \|x_k - x_{k-1}\|$ by (C.4), so that by virtue of (C.3), we can choose x_{k+1} such that $\tau \|x_{k+1} - x_k\| \leq \delta \|x_k - x_{k-1}\|$. Thus,

$$\|x_{k+1} - x_k\| \leq \frac{\delta}{\tau}\|x_k - x_{k-1}\|$$

and

$$\|x_n - x_{n-1}\| \leq \left(\frac{\delta}{\tau}\right)^{n-1} \|x_1\|, \quad n \geq 1.$$

Since $\delta/\tau < 1$, $\{x_n\}$ is a Cauchy sequence, and hence converges to a point $x \in X$. In addition, we have

$$\|x_n\| \leq \sum_{k=1}^{n} \|x_k - x_{k-1}\| \leq \sum_{k=1}^{n} \left(\frac{\delta}{\tau}\right)^{k-1} \|x_1\| \leq \frac{\|x_1\|}{1 - \frac{\delta}{\tau}} \leq \frac{cr}{\tau - \delta} = r.$$

Thus, x_{k+1} satisfies the required property. We also have $\|x\| \leq r$.

Summing the equation (C.5) from $k = 1$ to $k = n - 1$, and using the facts $Ax_1 = y$ and $(A-f)(x_0) = (A-f)(0) = 0$, we obtain $Ax_n - y = (A-f)(x_{n-1})$. Since $x_n \to x$, and A, f are continuous, we have $f(x) = y$. $\square$

References

1. S. Agmon, *The relaxation method for linear inequalities*, Canadian J. Math. **6** (1954), 382–392.
2. H. Akaike, *On a successive transformation of probability distribution and its application to the analysis of the optimum gradient method*, Ann. Inst. Statist. Math. Tokyo **11** (1959), 1–16.
3. F. Albrecht and H. G. Diamond, *A converse of Taylor's theorem*, Indiana Univ. Math. J. **21** (1971/72), 347–350.
4. V. M. Alekseev, V. M. Tikhomirov, and S. V. Fomin, *Optimal control*, Contemporary Soviet Mathematics, Consultants Bureau, New York, 1987, translated from the Russian by V. M. Volosov.
5. A. D. Alexandrov, *Convex polyhedra*, Springer Monographs in Mathematics, Springer-Verlag, Berlin, 2005, translated from the 1950 Russian edition by N. S. Dairbekov, S. S. Kutateladze, and A. B. Sossinsky, with comments and bibliography by V. A. Zalgaller and appendices by L. A. Shor and Yu. A. Volkov.
6. D. D. Ang and V. T. Tuan, *An elementary proof of the Morse-Palais lemma for Banach spaces*, Proc. Amer. Math. Soc. **39** (1973), 642–644.
7. L. Armijo, *Minimization of functions having Lipschitz continuous first partial derivatives*, Pacific J. Math. **16** (1966), 1–3.
8. K. J. Arrow, L. Hurwicz, and H. Uzawa, *Studies in linear and nonlinear programming*, with contributions by H. B. Chenery, S. M. Johnson, S. Karlin, T. Marschak, R. M. Solow. Stanford Mathematical Studies in the Social Sciences, vol. II, Stanford University Press, Palo Alto, Calif., 1958.
9. S. Artstein-Avidan and V. Milman, *The concept of duality in convex analysis, and the characterization of the Legendre transform*, Ann. of Math. (2) **169** (2009), no. 2, 661–674.
10. J.-P. Aubin and I. Ekeland, *Applied nonlinear analysis*, Pure and Applied Mathematics, John Wiley & Sons Inc., New York, 1984, A Wiley-Interscience Publication.
11. V. I. Averbuch and O. G. Smolyanov, *The theory of differentiation in linear topological spaces*, Russian Math. Surveys **22** (1967), no. 6, 201–258.
12. ———, *The various definitions of the derivative in linear topological spaces*, Russian Math. Surveys **23** (1968), no. 4, 67–113.

13. M. Avriel, *Nonlinear programming*, Prentice-Hall Inc., Englewood Cliffs, N.J., 1976, Analysis and methods, Prentice-Hall Series in Automatic Computation.
14. O. Axelsson, *Iterative solution methods*, Cambridge University Press, Cambridge, 1994.
15. D. Azé and J.-N. Corvellec, *On the sensitivity analysis of Hoffman constants for systems of linear inequalities*, SIAM J. Optim. **12** (2002), no. 4, 913–927 (electronic).
16. ______, *Variational methods in classical open mapping theorems*, J. Convex Anal. **13** (2006), no. 3-4, 477–488.
17. J. Bair and R. Fourneau, *Étude géométrique des espaces vectoriels*, Springer-Verlag, Berlin, 1975.
18. ______, *Étude géometrique des espaces vectoriels. II*, Lecture Notes in Mathematics, vol. 802, Springer-Verlag, Berlin, 1980.
19. K. Ball, *Ellipsoids of maximal volume in convex bodies*, Geom. Dedicata **41** (1992), no. 2, 241–250.
20. S. Banach, *Sur les fonctionelles linéaires II*, Studia Math. **1** (1929), 223–229.
21. ______, *Über homogene polynome in* (l^2), Studia Math. **7** (1937), 36–44.
22. I. Bárány, *A generalization of Carathéodory's theorem*, Discrete Math. **40** (1982), no. 2-3, 141–152.
23. I. Bárány and S. Onn, *Colourful linear programming and its relatives*, Math. Oper. Res. **22** (1997), no. 3, 550–567.
24. V. Barbu and Th. Precupanu, *Convexity and optimization in Banach spaces*, second ed., Mathematics and Its Applications (East European Series), vol. 10, D. Reidel Publishing Co., Dordrecht, 1986.
25. A. Barvinok, *A course in convexity*, Graduate Studies in Mathematics, vol. 54, American Mathematical Society, Providence, RI, 2002.
26. H. H. Bauschke, O. Güler, A. S. Lewis, and H. S. Sendov, *Hyperbolic polynomials and convex analysis*, Canad. J. Math. **53** (2001), no. 3, 470–488.
27. M. S. Bazaraa, H. D. Sherali, and C. M. Shetty, *Nonlinear programming – theory and applications*, second ed., John Wiley & Sons Inc., New York, 1993.
28. A. Ben-Tal and J. Zowe, *A unified theory of first and second order conditions for extremum problems in topological vector spaces*, Math. Programming Stud. **19** (1982), 39–76.
29. Y. Benyamini and J. Lindenstrauss, *Geometric nonlinear functional analysis. Vol. 1*, American Mathematical Society Colloquium Publications, vol. 48, American Mathematical Society, Providence, RI, 2000.
30. C. Berge, *Topological spaces*, Dover Publications Inc., Mineola, NY, 1997.
31. P. Bernhard and A. Rapaport, *On a theorem of Danskin with an application to a theorem of von Neumann-Sion*, Nonlinear Anal. **24** (1995), no. 8, 1163–1181.
32. D. P. Bertsekas, *Nonlinear programming*, second ed., Athena Scientific, Belmont, Massachusetts, 1999.
33. L. Blum, F. Cucker, M. Shub, and S. Smale, *Complexity and real computation*, Springer-Verlag, New York, 1998, with a foreword by R. M. Karp.
34. J. Bochnak, M. Coste, and M.-F. Roy, *Real algebraic geometry*, Ergebnisse der Mathematik und ihrer Grenzgebiete (3) [Results in Mathematics and Related Areas (3)], vol. 36, Springer-Verlag, Berlin, 1998, translated from the 1987 French original, Revised by the authors.
35. J. Bochnak and J. Siciak, *Polynomials and multilinear mappings in topological vector spaces*, Studia Math. **39** (1971), 59–76.

36. V. G. Boltjanskiĭ, *The method of "tents" in the theory of extremal problems*, Uspehi Mat. Nauk **30** (1975), no. 3(183), 3–55, translated in Russian Math. Surveys 30 1975, 1–54.
37. V. G. Boltyanskiĭ, *Optimal control of discrete systems*, John Wiley & Sons, New York-Toronto, Ont., 1978.
38. E. Bonan, *Comparaison d'un corps convexe avec ses deux ellipsoïdes optimaux*, C. R. Acad. Sci. Paris Sér. I Math. **315** (1992), no. 5, 557–560.
39. J. F. Bonnans and A. Shapiro, *Perturbation analysis of optimization problems*, Springer Series in Operations Research, Springer-Verlag, New York, 2000.
40. T. Bonnesen and W. Fenchel, *Theory of convex bodies*, BCS Associates, Moscow, ID, 1987, translated from the German and edited by L. Boron, C. Christenson, and B. Smith.
41. J. M. Borwein and A. S. Lewis, *Convex analysis and nonlinear optimization*, CMS Books in Mathematics/Ouvrages de Mathématiques de la SMC, 3, Springer-Verlag, New York, 2000, Theory and examples.
42. J. M. Borwein and D. Preiss, *A smooth variational principle with applications to subdifferentiability and to differentiability of convex functions*, Trans. Amer. Math. Soc. **303** (1987), no. 2, 517–527.
43. J. M. Borwein and Q. J. Zhu, *Techniques of variational analysis*, CMS Books in Mathematics/Ouvrages de Mathématiques de la SMC, 20, Springer-Verlag, New York, 2005.
44. N. Bourbaki, *Topological vector spaces. Chapters 1–5*, Elements of Mathematics (Berlin), Springer-Verlag, Berlin, 1987.
45. ———, *General topology. Chapters 1–4*, Elements of Mathematics (Berlin), Springer-Verlag, Berlin, 1998, translated from the French, Reprint of the 1989 English translation.
46. ———, *General topology. Chapters 5–10*, Elements of Mathematics (Berlin), Springer-Verlag, Berlin, 1998, translated from the French, Reprint of the 1989 English translation.
47. H. J. Bremermann, *Complex convexity*, Trans. Amer. Math. Soc. **82** (1956), 17–51.
48. H. Brézis, *Opérateurs maximaux monotones et semi-groupes de contractions dans les espaces de Hilbert*, North-Holland Publishing Co., Amsterdam, 1973, North-Holland Mathematics Studies, No. 5. Notas de Matemática (50).
49. H. Brézis and L. Nirenberg, *Remarks on finding critical points*, Comm. Pure Appl. Math. **44** (1991), no. 8-9, 939–963.
50. A. Brøndsted, *An introduction to convex polytopes*, Graduate Texts in Mathematics, vol. 90, Springer-Verlag, New York, 1983.
51. M. Buchner, J. Marsden, and S. Schecter, *Applications of the blowing-up construction and algebraic geometry to bifurcation problems*, J. Differential Equations **48** (1983), no. 3, 404–433.
52. G. Buskes, *The Hahn-Banach theorem surveyed*, Dissertationes Math. (Rozprawy Mat.) **327** (1993), 49.
53. C. Carathéodory, *Über den Variabilitätsbereich der Fourier'schen Konstanten von positiven harmonischen Funktionen*, Rend. Circ. Mat. Palermo **32** (1911), 193–217.
54. ———, *Calculus of variations and partial differential equations of first order*, Chelsea, New York, 1989.
55. W. B. Carver, *Systems of linear inequalities*, Ann. of Math. **23** (1922), no. 2, 212–220.

56. J. W. S. Cassels, *An elementary proof of some inequalities*, J. London Math. Soc. **23** (1948), 285–290.
57. G. Chen and M. Teboulle, *Convergence analysis of a proximal-like minimization algorithm using Bregman functions*, SIAM J. Optim. **3** (1993), no. 3, 538–543.
58. V. Chvátal, *Linear programming*, A Series of Books in the Mathematical Sciences, W. H. Freeman and Company, New York, 1983.
59. F. H. Clarke, *A new approach to Lagrange multipliers*, Math. Oper. Res. **1** (1976), no. 2, 165–174.
60. ———, *Optimization and nonsmooth analysis*, Canadian Mathematical Society Series of Monographs and Advanced Texts, John Wiley & Sons Inc., New York, 1983, A Wiley-Interscience Publication.
61. F. H. Clarke, Yu. S. Ledyaev, R. J. Stern, and P. R. Woleński, *Nonsmooth analysis and control theory*, Graduate Texts in Mathematics, vol. 178, Springer-Verlag, New York, 1998.
62. A. R. Conn, N. I. M. Gould, and P. L. Toint, *Trust-region methods*, MPS/SIAM Series on Optimization, Society for Industrial and Applied Mathematics (SIAM), Philadelphia, PA, 2000.
63. W. A. Coppel, *Foundations of convex geometry*, Cambridge University Press, Cambridge, 1998.
64. R. Courant and D. Hilbert, *Methods of mathematical physics. Vol. I*, Wiley Classics Library, John Wiley & Sons Inc., New York, 1989.
65. J. M. Danskin, *The theory of max-min, with applications*, SIAM J. Appl. Math. **14** (1966), 641–664.
66. ———, *The theory of max-min and its application to weapons allocation problems*, Econometrics and Operations Research, Vol. V, Springer-Verlag New York, Inc., New York, 1967.
67. G. B. Dantzig, *Linear programming and extensions*, Princeton University Press, Princeton, N.J., 1963.
68. ———, *Reminiscences about the origins of linear programming*, Oper. Res. Lett. **1** (1981/82), no. 2, 43–48.
69. L. Danzer, B. Grünbaum, and V. Klee, *Helly's theorem and its relatives*, Proc. Sympos. Pure Math., Vol. VII, Amer. Math. Soc., Providence, R.I., 1963, pp. 101–180.
70. E. De Giorgi, A. Marino, and M. Tosques, *Problems of evolution in metric spaces and maximal decreasing curve*, Atti Accad. Naz. Lincei Rend. Cl. Sci. Fis. Mat. Natur. (8) **68** (1980), no. 3, 180–187.
71. K. Deimling, *Nonlinear functional analysis*, Springer-Verlag, Berlin, 1985.
72. V. F. Dem′yanov and V. N. Malozemov, *Introduction to minimax*, Dover Publications Inc., New York, 1990, translated from the Russian by D. Louvish, Reprint of the 1974 edition.
73. J. E. Dennis, Jr. and J. J. Moré, *Quasi-Newton methods, motivation and theory*, SIAM Rev. **19** (1977), no. 1, 46–89.
74. J. E. Dennis, Jr. and R. B. Schnabel, *Numerical methods for unconstrained optimization and nonlinear equations*, Classics in Applied Mathematics, vol. 16, Society for Industrial and Applied Mathematics (SIAM), Philadelphia, PA, 1996, Corrected reprint of the 1983 original.
75. P. Deuflhard, *Newton methods for nonlinear problems*, Springer Series in Computational Mathematics, vol. 35, Springer-Verlag, Berlin, 2004, Affine invariance and adaptive algorithms.

76. R. Deumlich, K.-H. Elster, and R. Nehse, *Recent results on separation of convex sets*, Math. Operationsforsch. Statist. Ser. Optim. **9** (1978), no. 2, 273–296.
77. J. Dieudonné, *Foundations of modern analysis*, Pure and Applied Mathematics, Vol. X, Academic Press, New York, 1960.
78. A. V. Dmitruk, A. A. Miljutin, and N. P. Osmolovskiĭ, *Ljusternik's theorem and the theory of the extremum*, Uspekhi Mat. Nauk **35** (1980), no. 6(216), 11–46, 215, translated in Russian Math. Surveys 35, 1980 11–51.
79. A. L. Dontchev, *The Graves theorem revisited*, J. Convex Anal. **3** (1996), no. 1, 45–53.
80. A. Y. Dubovitskii and A. A. Milyutin, *Extremum problems in the presence of restrictions*, USSR Comp. Math. and Math. Phys. **5** (1965), 1–80.
81. ———, *Second variations in extremal problems with constraints*, Soviet Math. Dokl. **6** (1965), 12–16.
82. ———, *A translation of Euler's equations*, Ž. Vyčisl. Mat. i Mat. Fiz. **9** (1969), 1263–1284.
83. J. Eckhoff, *Helly, Radon, and Carathéodory type theorems*, Handbook of convex geometry, Vol. A, B, North-Holland, Amsterdam, 1993, pp. 389–448.
84. C. H. Edwards, Jr., *Advanced calculus of several variables*, Dover Publications Inc., New York, 1994, corrected reprint of the 1973 original.
85. M. Eidelheit, *Zur Theorie der konvexen Mengen in linearen normierten Räumen*, Studia Math. **6** (1936), 104–111.
86. I. Ekeland, *On the variational principle*, J. Math. Anal. Appl. **47** (1974), 324–353.
87. ———, *Nonconvex minimization problems*, Bull. Amer. Math. Soc. (N.S.) **1** (1979), no. 3, 443–474.
88. ———, *The ϵ-variational principle revisited*, Methods of nonconvex analysis (Varenna, 1989), Lecture Notes in Math., vol. 1446, Springer, Berlin, 1990, with notes by S. Terracini, pp. 1–15.
89. I. Ekeland and R. Témam, *Convex analysis and variational problems*, Classics in Applied Mathematics, vol. 28, Society for Industrial and Applied Mathematics (SIAM), Philadelphia, PA, 1999, translated from the French.
90. H. Everett, III, *Generalized Lagrange multiplier method for solving problems of optimum allocation of resources*, Operations Res. **11** (1963), 399–417.
91. K. Fan, I. Glicksberg, and A. J. Hoffman, *Systems of inequalities involving convex functions*, Proc. Amer. Math. Soc. **8** (1957), 617–622.
92. C. Fefferman, *An easy proof of the fundamental theorem of algebra*, Amer. Math. Monthly **74** (1967), 854–855.
93. W. Fenchel, *On conjugate convex functions*, Canadian J. Math. **1** (1949), 73–77.
94. ———, *Convex cones, sets, and functions*, 1951, Mimeographed Notes, Princeton University, Princeton, New Jersey.
95. ———, *A remark on convex sets and polarity*, Comm. Sém. Math. Univ. Lund [Medd. Lunds Univ. Mat. Sem.] **1952** (1952), no. Tome Supplementaire, 82–89.
96. A. V. Fiacco and G. P. McCormick, *Nonlinear programming*, second ed., Classics in Applied Mathematics, vol. 4, Society for Industrial and Applied Mathematics (SIAM), Philadelphia, PA, 1990, Sequential unconstrained minimization techniques.
97. R. Fletcher, *Practical methods of optimization*, second ed., A Wiley-Interscience Publication, John Wiley & Sons Ltd., Chichester, 1987.

98. M. Frank and P. Wolfe, *An algorithm for quadratic programming*, Naval Res. Logistics Quart. **3** (1956), 95–110.
99. J. Frehse, *An existence theorem for a class of non-coercive optimization and variational problems*, Math. Z. **159** (1978), no. 1, 51–63.
100. W. Fulton, *Eigenvalues, invariant factors, highest weights, and Schubert calculus*, Bull. Amer. Math. Soc. (N.S.) **37** (2000), no. 3, 209–249 (electronic).
101. L. Gårding, *Linear hyperbolic partial differential equations with constant coefficients*, Acta Math. **85** (1951), 1–62.
102. ———, *An inequality for hyperbolic polynomials*, J. Math. Mech. **8** (1959), 957–965.
103. I. V. Girsanov, *Lectures on mathematical theory of extremum problems*, Springer-Verlag, Berlin, 1972, edited by B. T. Poljak, translated from the Russian by D. Louvish, Lecture Notes in Economics and Mathematical Systems, Vol. 67.
104. A. M. Gleason, *The definition of a quadratic form*, Amer. Math. Monthly **73** (1966), 1049–1056.
105. A. J. Goldman and A. W. Tucker, *Theory of linear programming*, Linear inequalities and related systems, Annals of Mathematics Studies, no. 38, Princeton University Press, Princeton, N.J., 1956, pp. 53–97.
106. A. A. Goldstein, *Constructive real analysis*, Harper & Row Publishers, New York, 1967.
107. E. G. Gol′šteĭn, *Theory of convex programming*, American Mathematical Society, Providence, R.I., 1972, translated from the Russian by K. Makowski, Translations of Mathematical Monographs, Vol. 36.
108. G. H. Golub and C. F. Van Loan, *Matrix computations*, third ed., Johns Hopkins Studies in the Mathematical Sciences, Johns Hopkins University Press, Baltimore, MD, 1996.
109. A. Granas and J. Dugundji, *Fixed point theory*, Springer Monographs in Mathematics, Springer-Verlag, New York, 2003.
110. L. M. Graves, *Some mapping theorems*, Duke Math. J. **17** (1950), 111–114.
111. P. Gritzmann and V. Klee, *Separation by hyperplanes in finite-dimensional vector spaces over Archimedean ordered fields*, J. Convex Anal. **5** (1998), no. 2, 279–301.
112. M. Grötschel, L. Lovász, and A. Schrijver, *Geometric algorithms and combinatorial optimization*, second ed., Algorithms and Combinatorics, vol. 2, Springer-Verlag, Berlin, 1993.
113. P. Gruber, *Zur Charakterisierung konvexer Körper. Über einen Satz von Rogers und Shephard. II*, Math. Ann. **184** (1970), 79–105.
114. P. M. Gruber, *Convex and discrete geometry*, Grundlehren der Mathematischen Wissenschaften [Fundamental Principles of Mathematical Sciences], vol. 336, Springer, Berlin, 2007.
115. B. Grünbaum, *Convex polytopes*, second ed., Graduate Texts in Mathematics, vol. 221, Springer-Verlag, New York, 2003, prepared and with a preface by V. Kaibel, V. Klee, and G. M. Ziegler.
116. O. Güler, *On the convergence of the proximal point algorithm for convex minimization*, SIAM J. Control Optim. **29** (1991), no. 2, 403–419.
117. ———, *New proximal point algorithms for convex minimization*, SIAM J. Optim. **2** (1992), no. 4, 649–664.
118. ———, *Barrier functions in interior point methods*, Math. Oper. Res. **21** (1996), no. 4, 860–885.

119. ———, *Hyperbolic polynomials and interior point methods for convex programming*, Math. Oper. Res. **22** (1997), no. 2, 350–377.
120. O. Güler, F. Gürtuna, and O. Shevchenko, *Duality in quasi-Newton methods and new variational characterizations of the DFP and BFGS updates*, Optim. Methods Softw. **24** (2009), no. 1, 45–62.
121. O. Güler, A. J. Hoffman, and U. G. Rothblum, *Approximations to solutions to systems of linear inequalities*, SIAM J. Matrix Anal. Appl. **16** (1995), 688–696.
122. O. Güler and Y. Ye, *Convergence behavior of interior-point algorithms*, Math. Programming **60** (1993), no. 2, Ser. A, 215–228.
123. H. Hahn, *Über lineare Gleichungssysteme in linearen Räumen*, J. Reine Angew. Math. **157** (1927), 214–229.
124. H. Halkin, *A satisfactory treatment of equality and operator constraints in the Dubovitskii-Milyutin optimization formalism*, J. Optimization Theory Appl. **6** (1970), 138–149.
125. P. R. Halmos, *Naive set theory*, Springer-Verlag, New York, 1974.
126. R. S. Hamilton, *The inverse function theorem of Nash and Moser*, Bull. Amer. Math. Soc. (N.S.) **7** (1982), no. 1, 65–222.
127. G. H. Hardy, J. E. Littlewood, and G. Pólya, *Inequalities*, Cambridge Mathematical Library, Cambridge University Press, Cambridge, 1988, reprint of the 1952 edition.
128. R. A. Hauser and O. Güler, *Self-scaled barrier functions on symmetric cones and their classification*, Found. Comput. Math. **2** (2002), no. 2, 121–143.
129. E. Helly, *Über linearer Funktionaloperationen*, Akad. Wiss. Wien **121** (1912), 265–297.
130. ———, *Über Systeme linearer Gleichungen mit unendlich vielen Unbekannten*, Monatsh. für Math. und Phys. **31** (1921), 60–91.
131. M. R. Hestenes, *Optimization theory*, Wiley-Interscience [John Wiley & Sons], New York, 1975, The finite dimensional case, Pure and Applied Mathematics.
132. M. R. Hestenes and E. Stiefel, *Methods of conjugate gradients for solving linear systems*, J. Research Nat. Bur. Standards **49** (1952), 409–436 (1953).
133. J.-B. Hiriart-Urruty, *A short proof of the variational principle for approximate solutions of a minimization problem*, Amer. Math. Monthly **90** (1983), no. 3, 206–207.
134. J.-B. Hiriart-Urruty and C. Lemaréchal, *Fundamentals of convex analysis*, Grundlehren Text Editions, Springer-Verlag, Berlin, 2001.
135. W. M. Hirsch and A. J. Hoffman, *Extreme varieties, concave functions, and the fixed charge problem*, Comm. Pure Appl. Math. **14** (1961), 355–369.
136. A. J. Hoffman, *On approximate solutions of systems of linear inequalities*, J. Research Nat. Bur. Standards **49** (1952), 263–265.
137. R. B. Holmes, *A course on optimization and best approximation*, Springer-Verlag, Berlin, 1972, Lecture Notes in Mathematics, Vol. 257.
138. ———, *Geometric functional analysis and its applications*, Springer-Verlag, New York, 1975, Graduate Texts in Mathematics, No. 24.
139. L. Hörmander, *On a theorem of Grace*, Math. Scand. **2** (1954), 55–64.
140. ———, *The analysis of linear partial differential operators. I*, Classics in Mathematics, Springer-Verlag, Berlin, 2003, Distribution theory and Fourier analysis, reprint of the second (1990) edition.
141. ———, *Notions of convexity*, Modern Birkhäuser Classics, Birkhäuser Boston Inc., Boston, MA, 2007, reprint of the 1994 edition.

142. A. Horn, *Eigenvalues of sums of Hermitian matrices*, Pacific J. Math. **12** (1962), 225–241.
143. R. A. Horn and C. R. Johnson, *Matrix analysis*, Cambridge University Press, Cambridge, 1990, corrected reprint of the 1985 original.
144. A. D. Ioffe, *Regular points of Lipschitz functions*, Trans. Amer. Math. Soc. **251** (1979), 61–69.
145. ______, *Metric regularity and subdifferential calculus*, Uspekhi Mat. Nauk **55** (2000), no. 3(333), 103–162.
146. A. D. Ioffe and V. M. Tihomirov, *Theory of extremal problems*, Studies in Mathematics and Its Applications, vol. 6, North-Holland Publishing Co., Amsterdam, 1979, translated from the Russian by K. Makowski.
147. J.-Y. Jaffray and J.-Ch. Pomerol, *A direct proof of the Kuhn-Tucker necessary optimality theorem for convex and affine inequalities*, SIAM Rev. **31** (1989), no. 4, 671–674.
148. F. John, *Extremum problems with inequalities as subsidiary conditions*, Studies and Essays Presented to R. Courant on his 60th Birthday, January 8, 1948, Interscience Publishers, Inc., New York, N. Y., 1948, pp. 187–204.
149. P. Jordan and J. von Neumann, *On inner products in linear, metric spaces*, Ann. of Math. (2) **36** (1935), no. 3, 719–723.
150. F. Juhnke, *Embedded maximal ellipsoids and semi-infinite optimization*, Beiträge Algebra Geom. **35** (1994), no. 2, 163–171.
151. ______, *Polarity of embedded and circumscribed ellipsoids*, Beiträge Algebra Geom. **36** (1995), no. 1, 17–24.
152. S. Kakutani, *Ein Beweis des Satzes von M. Edelheit über konvexe Mengen*, Proceedings of the Imperial Academy of Tokyo **13** (1937), 93–94.
153. Shmuel Kaniel, *Estimates for some computational techniques in linear algebra*, Math. Comp. **20** (1966), 369–378.
154. L. V. Kantorovich, *Functional analysis and applied mathematics*, NBS Rep. 1509, U. S. Department of Commerce National Bureau of Standards, Los Angeles, Calif., 1952, translated by C. D. Benster.
155. L. V. Kantorovich and G. P. Akilov, *Functional analysis*, second ed., Pergamon Press, Oxford, 1982, translated from the Russian by H. L. Silcock.
156. R. N. Karasev, *Topological methods in combinatorial geometry*, Uspekhi Mat. Nauk **63** (2008), no. 6(384), 39–90.
157. S. Karlin and L. S. Shapley, *Some applications of a theorem on convex functions*, Ann. of Math. (2) **52** (1950), 148–153.
158. S. Karlin and W. J. Studden, *Tchebycheff systems: With applications in analysis and statistics*, Pure and Applied Mathematics, Vol. XV, Interscience Publishers John Wiley & Sons, New York-London-Sydney, 1966.
159. N. Karmarkar, *A new polynomial-time algorithm for linear programming*, Combinatorica **4** (1984), no. 4, 373–395.
160. W. Karush, *Minima of functions of several variables with inequalities as side conditions*, Master's thesis, Department of Mathematics, The University of Chicago, 1939.
161. J. L. Kelley, *General topology*, Springer-Verlag, New York, 1975, Reprint of the 1955 edition [Van Nostrand, Toronto, Ont.], Graduate Texts in Mathematics, No. 27.
162. O. D. Kellogg, *On bounded polynomials in several variables*, Math. Zeit. **27** (1927), 55–64.

163. L. G. Khachiyan, *A polynomial algorithm in linear programming*, Dokl. Akad. Nauk SSSR **244** (1979), no. 5, 1093–1096.
164. L. G. Khachiyan and M. J. Todd, *On the complexity of approximating the maximal inscribed ellipsoid for a polytope*, Math. Programming **61** (1993), no. 2, Ser. A, 137–159.
165. M. D. Kirszbraun, *Über die zusammenziehenden und Lipschitzschen Transformationen*, Fund. Math. **22** (1934), 77–108.
166. Daniel A. Klain and Gian-Carlo Rota, *Introduction to geometric probability*, Lezioni Lincee. [Lincei Lectures], Cambridge University Press, Cambridge, 1997.
167. V. L. Klee, Jr., *Convex sets in linear spaces*, Duke Math. J. **18** (1951), 443–466.
168. ———, *Convex sets in linear spaces. II*, Duke Math. J. **18** (1951), 875–883.
169. ———, *Convex sets in linear spaces. III*, Duke Math. J. **20** (1953), 105–111.
170. ———, *Separation properties of convex cones*, Proc. Amer. Math. Soc. **6** (1955), 313–318.
171. ———, *Strict separation of convex sets*, Proc. Amer. Math. Soc. **7** (1956), 735–737.
172. ———, *Extremal structure of convex sets*, Arch. Math. **8** (1957), 234–240.
173. ———, *Extremal structure of convex sets. II*, Math. Z. **69** (1958), 90–104.
174. ———, *Some characterizations of convex polyhedra*, Acta Math. **102** (1959), 79–107.
175. ———, *Maximal separation theorems for convex sets*, Trans. Amer. Math. Soc. **134** (1968), 133–147.
176. ———, *Separation and support properties of convex sets – a survey*, Control theory and the calculus of variations (A. V. Balakrishnan, ed.), Academic Press, 1969, pp. 235–303.
177. G. Köthe, *Topological vector spaces. I*, translated from the German by D. J. H. Garling. Die Grundlehren der mathematischen Wissenschaften, Band 159, Springer-Verlag New York Inc., New York, 1969.
178. M. G. Krein and D. P. Milman, *On extreme points of regularly convex sets*, Studia Math. **9** (1940), 133–138.
179. M. G. Kreĭn and A. A. Nudel′man, *The Markov moment problem and extremal problems*, American Mathematical Society, Providence, R.I., 1977, Ideas and problems of P. L. Čebyšev and A. A. Markov and their further development, translated from the Russian by D. Louvish, Translations of Mathematical Monographs, Vol. 50.
180. H. W. Kuhn, *Solvability and consistency for linear equations and inequalities*, Amer. Math. Monthly **63** (1956), 217–232.
181. ———, *Nonlinear programming: a historical view*, Nonlinear programming (Proc. Sympos., New York, 1975), Amer. Math. Soc., Providence, R. I., 1976, pp. 1–26. SIAM–AMS Proc., Vol. IX.
182. H. W. Kuhn and A. W. Tucker, *Nonlinear programming*, Proceedings of the Second Berkeley Symposium on Mathematical Statistics and Probability, 1950 (Berkeley and Los Angeles), University of California Press, 1951, pp. 481–492.
183. S. Lang, *Differential manifolds*, Addison-Wesley Publishing Co., Inc., Reading, Mass.-London-Don Mills, Ont., 1972.
184. ———, *Algebra*, third ed., Graduate Texts in Mathematics, vol. 211, Springer-Verlag, New York, 2002.
185. D. Laugwitz, *Differential and Riemannian geometry*, translated by F. Steinhardt, Academic Press, New York, 1965.

186. P. D. Lax, *Linear algebra*, Pure and Applied Mathematics (New York), John Wiley & Sons Inc., New York, 1997, A Wiley-Interscience Publication.
187. V. L. Levin, *The application of E. Helly's theorem in convex programming, problems of best approximation, and related questions*, Mat. Sb. (N.S.) **79 (121)** (1969), 250–263.
188. E. S. Levitin, A. A. Miljutin, and N. P. Osmolovskiĭ, *Higher order conditions for local minima in problems with constraints*, Uspekhi Mat. Nauk **33** (1978), no. 6(204), 85–148, 272.
189. Y. Li and S. Shi, *A generalization of Ekeland's ϵ-variational principle and its Borwein-Preiss smooth variant*, J. Math. Anal. Appl. **246** (2000), no. 1, 308–319.
190. D. G. Luenberger, *Linear and nonlinear programming*, second ed., Addison–Wesley, Reading, Massachusetts, 1984.
191. L. A. Lyusternik, *On conditional extrema of functionals*, Mathematicheskii Sbornik **41** (1934), 390–401.
192. O. L. Mangasarian, *Nonlinear programming*, Classics in Applied Mathematics, vol. 10, Society for Industrial and Applied Mathematics (SIAM), Philadelphia, PA, 1994, Corrected reprint of the 1969 original.
193. O. L. Mangasarian and S. Fromovitz, *The Fritz John necessary optimality conditions in the presence of equality and inequality constraints*, J. Math. Anal. Appl. **17** (1967), 37–47.
194. S. Mazur and W. Orlicz, *Grundlegende Eigenschaften der polynomischen Operationen*, Studia Math. **7** (1935), 36–44.
195. E. J. McShane, *The Lagrange multiplier rule*, Amer. Math. Monthly **80** (1973), 922–925.
196. D. P. Mil′man, *Facial characterization of convex sets; extremal elements*, Transactions of Moscow Mathematical Society **22** (1970), 69–139.
197. J. Milnor, *Morse theory*, Based on lecture notes by M. Spivak and R. Wells. Annals of Mathematics Studies, No. 51, Princeton University Press, Princeton, N.J., 1963.
198. H. Minkowski, *Gesammelte Abhandlungen*, AMS Chelsea Publishing, American Mathematical Society, Providence, R.I., 1967.
199. B. S. Mordukhovich, *Variational analysis and generalized differentiation. I*, Grundlehren der Mathematischen Wissenschaften [Fundamental Principles of Mathematical Sciences], vol. 330, Springer-Verlag, Berlin, 2006.
200. ———, *Variational analysis and generalized differentiation. II*, Grundlehren der Mathematischen Wissenschaften [Fundamental Principles of Mathematical Sciences], vol. 331, Springer-Verlag, Berlin, 2006.
201. J. J. Moré and D. C. Sorensen, *Computing a trust region step*, SIAM J. Sci. Statist. Comput. **4** (1983), no. 3, 553–572.
202. M. Morse, *Relations between the critical points of a real function of n independent variables*, Trans. Amer. Math. Soc. **27** (1925), no. 3, 345–396.
203. E. Nelson, *Topics in dynamics. I: Flows*, Mathematical Notes, Princeton University Press, Princeton, N.J., 1969.
204. A. S. Nemirovskiĭ and Y. E. Nesterov, *Optimal methods for smooth convex minimization*, Zh. Vychisl. Mat. i Mat. Fiz. **25** (1985), no. 3, 356–369, 477.
205. A. S. Nemirovsky and D. B. Yudin, *Problem complexity and method efficiency in optimization*, A Wiley-Interscience Publication, John Wiley & Sons Inc., New York, 1983, translated from the Russian and with a preface by E. R. Dawson, Wiley-Interscience Series in Discrete Mathematics.

206. Y. E. Nesterov, *A method for solving the convex programming problem with convergence rate $O(1/k^2)$*, Dokl. Akad. Nauk SSSR **269** (1983), no. 3, 543–547.
207. ———, *An approach to constructing optimal methods for minimization of smooth convex functions*, Èkonom. i Mat. Metody **24** (1988), no. 3, 509–517.
208. ———, *Introductory lectures on convex optimization*, Applied Optimization, vol. 87, Kluwer Academic Publishers, Boston, MA, 2004, A basic course.
209. Y. E. Nesterov and A. Nemirovskii, *Interior-point polynomial algorithms in convex programming*, SIAM Studies in Applied Mathematics, vol. 13, Society for Industrial and Applied Mathematics (SIAM), Philadelphia, PA, 1994.
210. Y. E. Nesterov and M. J. Todd, *Self-scaled barriers and interior-point methods for convex programming*, Math. Oper. Res. **22** (1997), no. 1, 1–42.
211. J. von Neumann, *Zur Theorie der Gesellschaftsspiele*, Math. Ann. **100** (1928), 295–320.
212. J. von Neumann and O. Morgenstern, *Theory of games and economic behavior*, Princeton University Press, Princeton, NJ, 2004, Reprint of the 1980 edition [Princeton Univ. Press, Princeton, NJ].
213. J. Nocedal and S. J. Wright, *Numerical optimization*, second ed., Springer Series in Operations Research and Financial Engineering, Springer, New York, 2006.
214. W. Oettli, *Optimality conditions for programming problems involving multivalued mappings*, Modern applied mathematics (Bonn, 1979), North-Holland, Amsterdam, 1982, pp. 195–226.
215. J. M. Ortega and W. C. Rheinboldt, *Iterative solution of nonlinear equations in several variables*, Academic Press, New York, 1970.
216. A. M. Ostrowski, *Solution of equations in Euclidean and Banach spaces*, Academic Press [A Subsidiary of Harcourt Brace Jovanovich, Publishers], New York-London, 1973, Third edition of *Solution of equations and systems of equations*, Pure and Applied Mathematics, Vol. 9.
217. A. L. Peressini, F. E. Sullivan, and J. J. Uhl, Jr., *The mathematics of nonlinear programming*, Undergraduate Texts in Mathematics, Springer-Verlag, New York, 1988.
218. G. Peters and J.H. Wilkinson, *The least squares problem and pseudoinverses*, Comput. J. **13** (1970), 309–316.
219. B. J. Pettis, *Separation theorems for convex sets*, Math. Mag. **29** (1956), 233–247.
220. B. T. Polyak, *Introduction to optimization*, Translations Series in Mathematics and Engineering, Optimization Software Inc. Publications Division, New York, 1987, translated from the Russian, with a foreword by D. P. Bertsekas.
221. J. Ponstein and W. K. Klein Haneveld, *On a general saddle-point condition in normed spaces*, Math. Programming **9** (1975), no. 1, 118–122.
222. L. S. Pontryagin, V. G. Boltyanskii, R. V. Gamkrelidze, and E. F. Mishchenko, *The mathematical theory of optimal processes*, translated from the Russian by K. N. Trirogoff; edited by L. W. Neustadt, Interscience Publishers John Wiley & Sons, Inc. New York-London, 1962.
223. B. N. Pshenichnyi, *Necessary conditions for an extremum*, translated from the Russian by K. Makowski. Translation edited by L. W. Neustadt. Pure and Applied Mathematics, vol. 4, Marcel Dekker Inc., New York, 1971.
224. H. Rademacher and I. J. Schoenberg, *Helly's theorems on convex domains and Tchebycheff's approximation problem*, Canadian J. Math. **2** (1950), 245–256.

225. Q. I. Rahman and G. Schmeisser, *Analytic theory of polynomials*, London Mathematical Society Monographs. New Series, vol. 26, The Clarendon Press Oxford University Press, Oxford, 2002.
226. J. Renegar, *A mathematical view of interior-point methods in convex optimization*, MPS/SIAM Series on Optimization, Society for Industrial and Applied Mathematics (SIAM), Philadelphia, PA, 2001.
227. F. Riesz and B. Sz.-Nagy, *Functional analysis*, Dover Books on Advanced Mathematics, Dover Publications Inc., New York, 1990, translated from the second French edition by L. F. Boron, Reprint of the 1955 original.
228. R. T. Rockafellar, *Convex analysis*, Princeton Mathematical Series, No. 28, Princeton University Press, Princeton, N.J., 1970.
229. ______, *Lagrange multipliers and optimality*, SIAM Rev. **35** (1993), no. 2, 183–238.
230. R. T. Rockafellar and R. J.-B. Wets, *Variational analysis*, Grundlehren der Mathematischen Wissenschaften [Fundamental Principles of Mathematical Sciences], vol. 317, Springer-Verlag, Berlin, 1998.
231. P. C. Rosenbloom, *Quelques classes de problèmes extrémaux*, Bull. Soc. Math. France **79** (1951), 1–58; 80, 183–215 (1952).
232. H. L. Royden, *Real analysis*, third ed., Macmillan Publishing Company, New York, 1988.
233. W. Rudin, *Functional analysis*, International Series in Pure and Applied Mathematics, McGraw-Hill Inc., New York, 1991.
234. Y. Saad, *Iterative methods for sparse linear systems*, second ed., Society for Industrial and Applied Mathematics, Philadelphia, PA, 2003.
235. K. S. Sarkaria, *Tverberg's theorem via number fields*, Israel J. Math. **79** (1992), no. 2-3, 317–320.
236. R. Schneider, *Convex bodies: the Brunn-Minkowski theory*, Encyclopedia of Mathematics and Its Applications, vol. 44, Cambridge University Press, Cambridge, 1993.
237. I. J. Schoenberg, *On a theorem of Kirszbraun and Valentine*, Amer. Math. Monthly **60** (1953), 620–622.
238. A. Schrijver, *Theory of linear and integer programming*, Wiley-Interscience Series in Discrete Mathematics, John Wiley & Sons Ltd., Chichester, 1986, A Wiley-Interscience Publication.
239. J. T. Schwartz, *Nonlinear functional analysis*, Gordon and Breach Science Publishers, New York, 1969, Notes by H. Fattorini, R. Nirenberg and H. Porta, with an additional chapter by H. Karcher, notes on Mathematics and Its Applications.
240. I. R. Shafarevich, *Basic algebraic geometry. 1*, Springer-Verlag, Berlin, 1994.
241. N. Z. Shor, *Minimization methods for nondifferentiable functions*, Springer Series in Computational Mathematics, vol. 3, Springer-Verlag, Berlin, 1985, translated from the Russian by K. C. Kiwiel and A. Ruszczyński.
242. M. Sion, *On general minimax theorems*, Pacific J. Math. **8** (1958), 171–176.
243. M. Slater, *Lagrange multipliers revisited: a contribution to nonlinear programming*, Discussion paper math. 403, Cowles Commission, Yale University, 1950.
244. D. C. Sorensen, *Newton's method with a model trust region modification*, SIAM J. Numer. Anal. **19** (1982), no. 2, 409–426.
245. M. Spivak, *Calculus on manifolds. A modern approach to classical theorems of advanced calculus*, W. A. Benjamin, Inc., New York-Amsterdam, 1965.

246. E. L. Stiefel, *Kernel polynomials in linear algebra and their numerical applications*, Nat. Bur. Standards Appl. Math. Ser. **1958** (1958), no. 49, 1–22.
247. J. Stoer and C. Witzgall, *Convexity and optimization in finite dimensions. I*, Die Grundlehren der mathematischen Wissenschaften, Band 163, Springer-Verlag, New York, 1970.
248. M. H. Stone, *Convexity*, Mimeographed lecture notes, The University of Chicago, 1946.
249. F. Sullivan, *A characterization of complete metric spaces*, Proc. Amer. Math. Soc. **83** (1981), no. 2, 345–346.
250. G. Szegő, *Orthogonal polynomials*, fourth ed., American Mathematical Society, Providence, R.I., 1975, American Mathematical Society, Colloquium Publications, Vol. XXIII.
251. S. P. Tarasov, L. G. Khachiyan, and I. I. Èrlikh, *The method of inscribed ellipsoids*, Dokl. Akad. Nauk SSSR **298** (1988), no. 5, 1081–1085.
252. M. Teboulle, *A simple duality proof for quadratically constrained entropy functionals and extension to convex constraints*, SIAM J. Appl. Math. **49** (1989), no. 6, 1845–1850.
253. F. Terkelsen, *The fundamental theorem of algebra*, Amer. Math. Monthly **83** (1976), no. 8, 647.
254. T. V. Tuan and D. D. Ang, *A representation theorem for differentiable functions*, Proc. Amer. Math. Soc. **75** (1979), no. 2, 343–350.
255. A. W. Tucker, *Dual systems of homogeneous linear relations*, Linear inequalities and related systems, Annals of Mathematics Studies, no. 38, University Press, Princeton, N. J., 1956, pp. 3–18.
256. Hoàng Tuy, *Convex inequalities and the Hahn-Banach theorem*, Dissertationes Math. (Rozprawy Mat.) **97** (1972), 35.
257. H. Tverberg, *A generalization of Radon's theorem*, J. London Math. Soc. **41** (1966), 123–128.
258. J. G. van der Corput and G. Schaake, *Ungleichungen für polynome und trigonometrische polynome*, Compositio Math. **2** (1935), 321–361.
259. J. van Tiel, *Convex analysis*, John Wiley & Sons Inc., New York, 1984, An introductory text.
260. E. B. Vinberg, *A course in algebra*, Graduate Studies in Mathematics, vol. 56, American Mathematical Society, Providence, RI, 2003, translated from the 2001 Russian original by A. Retakh.
261. O. V. Viskov, *The noncommutative approach to classical problems of analysis*, Trudy Mat. Inst. Steklov. **177** (1986), 21–32, 207, Proc. Steklov Inst. Math. **1988**, no. 4, 21–32, Probabilistic problems of discrete mathematics.
262. M. Vlach, *A separation theorem for finite families*, Comment. Math. Univ. Carolinae **12** (1971), 655–660.
263. W. C. Waterhouse, *Do symmetric problems have symmetric solutions?*, Amer. Math. Monthly **90** (1983), no. 6, 378–387.
264. ———, *The absolute-value estimate for symmetric multilinear forms*, Linear Algebra Appl. **128** (1990), 97–105.
265. J. D. Weston, *A characterization of metric completeness*, Proc. Amer. Math. Soc. **64** (1977), 186–188.
266. H. Weyl, *Elementare Theorie der konvexen Polyeder*, Comment. Math. Helv. **7** (1935), 290–306.

267. ———, *The elementary theory of convex polyhedra*, Contributions to the Theory of Games, Annals of Mathematics Studies, no. 24, Princeton University Press, Princeton, N.J., 1950, pp. 3–18.
268. J. Wolfe, *A proof of Taylor's formula*, Amer. Math. Month. **60** (1953), 415.
269. G. Xue and Y. Ye, *An efficient algorithm for minimizing a sum of Euclidean norms with applications*, SIAM J. Optim. **7** (1997), no. 4, 1017–1036.
270. Y. Ye, *Interior point algorithms*, Wiley-Interscience Series in Discrete Mathematics and Optimization, John Wiley & Sons Inc., New York, 1997, Theory and analysis, A Wiley-Interscience Publication.
271. D. B. Yudin and A. S. Nemirovskiĭ, *Estimation of the informational complexity of mathematical programming problems*, Èkonom. i Mat. Metody **12** (1976), no. 1, 128–142.
272. ———, *Informational complexity and effective methods for the solution of convex extremal problems*, Èkonom. i Mat. Metody **12** (1976), no. 2, 357–369.
273. W. I. Zangwill and B. Mond, *Nonlinear programming: a unified approach*, Prentice-Hall Inc., Englewood Cliffs, N.J., 1969, Prentice-Hall International Series in Management.
274. G. M. Ziegler, *Lectures on polytopes*, Graduate Texts in Mathematics, vol. 152, Springer-Verlag, New York, 1995.

Index